Sustainable Soil Management

Sustainable Soil Management

Edited by Lester Bane

SYRAWOOD
PUBLISHING HOUSE
New York

Published by Syrawood Publishing House,
750 Third Avenue, 9th Floor,
New York, NY 10017, USA
www.syrawoodpublishinghouse.com

Sustainable Soil Management
Edited by Lester Bane

International Standard Book Number: 978-1-68286-385-5 (Hardback)

Cataloging-in-publication Data

Sustainable soil management / edited by Lester Bane.
p. cm.
Includes bibliographical references and index.
ISBN 978-1-68286-385-5
1. Soil management. 2. Soil management--Environmental aspects. 3. Soil science. 4. Soils--Analysis. I. Bane, Lester.
S596 .S87 2017
631.4--dc23

Printed in China.

TABLE OF CONTENTS

PREFACE

Soil sciences and soil studies involve analyzing soil profiles that facilitate the growth of various ecosystems. Soil management is the study of various soil processes and techniques to sustain soil quality. Overgrowth and bad arbological practices also affect soil quality. This book provides new insight into various horticultural and agricultural techniques that contribute to sustainable soil management. This book is a valuable compilation of topics, ranging from the basic to the most complex advancement in the field of soil management. It will provide in-depth knowledge about theory and practice of this discipline. It also presents researches that have transformed this field and added its advancements. Any reader interested in soil conservation and agricultural soil sciences will find this book extremely helpful.

This book has been the outcome of endless efforts put in by authors and researchers on various issues and topics within the field. The book is a comprehensive collection of significant researches that are addressed in a variety of chapters. It will surely enhance the knowledge of the field among readers across the globe.

It gives us an immense pleasure to thank our researchers and authors for their efforts to submit their piece of writing before the deadlines. Finally in the end, I would like to thank my family and colleagues who have been a great source of inspiration and support.

Editor

1

Effect of Byproducts of Flue Gas Desulfurization on the Soluble Salts Composition and Chemical Properties of Sodic Soils

Jinman Wang[1,2]*, Zhongke Bai[1,2], Peiling Yang[3]

1 College of Land Science and Technology, China University of Geosciences, Beijing, People's Republic of China, **2** Key Laboratory of Land Consolidation and Rehabilitation, Ministry of Land and Resources, Beijing, People's Republic of China, **3** College of Hydraulic and Civil Engineering, China Agricultural University, Beijing, People's Republic of China

Abstract

The byproducts of flue gas desulfurization (BFGD) are a useful external source of Ca^{2+} for the reclamation of sodic soils because they are comparatively cheap, generally available and have high gypsum content. The ion solution composition of sodic soils also plays an important role in the reclamation process. The effect of BFGD on the soluble salts composition and chemical properties of sodic soils were studied in a soil column experiment. The experiment consisted of four treatments using two different sodic soils (sodic soil I and sodic soil II) and two BFGD rates. After the application of BFGD and leaching, the soil soluble salts were transformed from sodic salts containing Na_2CO_3 and $NaHCO_3$ to neutral salts containing NaCl and Na_2SO_4. The sodium adsorption ratio (SAR), pH and electrical conductivity (EC) decreased at all soil depths, and more significantly in the top soil depth. At a depth of 0–40 cm in both sodic soil I and sodic soil II, the SAR, EC and pH were less than 13, 4 dS m^{-1} and 8.5, respectively. The changes in the chemical properties of the sodic soils reflected the changes in the ion composition of soluble salts. Leaching played a key role in the reclamation process and the reclamation effect was positively associated with the amount of leaching. The soil salts did not accumulate in the top soil layer, but there was a slight increase in the middle and bottom soil depths. The results demonstrate that the reclamation of sodic soils using BFGD is promising.

Editor: Stephen J. Johnson, University of Kansas, United States of America

Funding: This research was supported by the Fundamental Research Funds for the Central Universities of China (2652012072), and the National Natural Science Foundation of China (50749032). The funders had no role in study design, data collection and analysis, decision to publish, or preparation of the manuscript.

Competing Interests: The authors have declared that no competing interests exist.

* E-mail: wangjinman2002@163.com

Introduction

Wet flue gas desulfurization (FGD) is the dominant technology used in the control of SO_2 emissions from coal-fired power plants. The major byproduct of the process is $CaSO_4$, or a mixture of $CaSO_3$ and $CaSO_4$ [1]. With the rapid development of the energy and power industries in China, the installed capacity of power plants with FGD devices, and therefore the amount of byproducts of flue gas desulfurization (BFGD), is expected to increase rapidly. By the end of 2010, the installed capacity of power plants in China with FGD devices was about 57.8 GW, and the annual production of BFGD was about 7.1 million tons. According to the National Development Program of China, the installed capacity of power plants with FGD devices will be 530 GW by 2020, with an annual production of 90 million tons of BFGD. As BFGD contain large amounts of moisture and ash, they can only be used as building gypsum after purification and dehydration, hence they are more expensive compared with natural gypsum produced in China. If the BFGD were to be directly disposed of rather than being treated and used, a vast area of land would be required. Such an approach would be a waste of valuable land resources and represent a potential threat of secondary pollution to the environment [2–4].

Significantly, there are large areas of sodic soils in China. Excessive exchangeable sodium (ES) has gradually been adsorbed in soil colloid, causing deterioration in the soil's physical properties and inducing the formation of sodic soils [5]. Sodic soils contain an excess of exchangeable Na^+ in soil colloids and have soluble carbonates in the form of $NaHCO_3$ and Na_2CO_3. The pH value, SAR and exchangeable sodium percentage (ESP) are greater than 8.5, 13 and 15, respectively, and the electrical conductivity of the saturated paste (EC_{sat}) is less than 4.0 dS m^{-1} [5–6]. Amendments for sodic soils are materials that directly or indirectly supply divalent cations (usually Ca^{2+}) to replace ES [3,7]. The dissolution-exchange reaction in soil colloids is represented as [5].

$$2NaX + CaSO_4 = CaX_2 + Na^+ + SO_4{}^{2-}, \qquad (1)$$

where X is the soil exchange phase. The exchange of Ca^{2+} for Na^+ in soil colloid results in the flocculation of soil particles, which restores the porous structure and high permeability of the soil. Because the main component of BFGD is $CaSO_4$, substantial evidence exists that these BFGD can, with proper use, be valuable for the reclamation of sodic soils [1,3,8–12].

The degree of ES absorbed by soil colloid depends on the soil soluble salts composition. ES is most easily absorbed into soil colloid when the soil solution contains large amounts of Na_2CO_3 and $NaHCO_3$. In soil containing neutral salts (such as NaCl and

Table 1. Physical and chemical properties of the tested soils.

Soil type	Soil depth (cm)	Bulk density ($g \cdot cm^{-3}$)	pH	EC ($dS \cdot m^{-1}$)	ES ($cmol \cdot kg^{-1}$)	CEC ($cmol \cdot kg^{-1}$)	ESP (%)
Sodic soil I	0–5	1.41	8.95	11.45	1.65	17.14	9.63
	5–20	1.46	9.32	5.18	3.72	17.86	20.83
	20–40	1.41	9.29	4.31	2.44	16.95	14.38
	40–60	1.47	8.96	2.52	2.55	16.84	15.12
	60–100	1.44	8.96	2.31	2.19	16.71	13.12
Sodic soil II	0–5	1.58	9.07	13.09	9.46	21.23	44.55
	5–20	1.52	8.91	4.52	6.07	21.63	28.06
	20–40	1.46	9.23	1.89	5.60	20.15	29.74
	40–60	1.45	9.63	1.26	5.40	20.34	26.53
	60–100	1.46	9.44	0.84	4.91	19.86	24.71

Na_2SO_4), Na^+ can only be absorbed by soil colloid and cause soil sodication if $Na^+/(Ca^{2+}+Mg^{2+}) \geq 4$ in the cationic soil solution [13]. Therefore, the ion solution composition of sodic soils plays an important role in the reclamation process. Although several studies have examined the changes in the physicochemical properties of sodic soils and crop yield after applying BFGD, less attention has been paid to the ion solution composition [3,6,10,12–13,14–16].

The objective of this study was to determine whether the application of BFGD causes changes in the soluble salts composition, SAR, EC and pH value of sodic soils at different depths; to analyze the relationship between the soil's chemical properties and soluble salts composition; and to elucidate the mechanism of sodic soil reclamation by BFGD.

Materials and Methods

Ethics Statements

This study was conducted with the permission of the Changsheng Experimental Station of the Baoyannur League Institute of Water Resources. No endangered or protected species were present in the field studies.

The Physical and Chemical Properties of the Soils Tested and the BFGD

The soils tested were sampled from the Changsheng Experimental Station of the Baoyannur League Institute of Water Resources in northwestern China (N 40°20′, E 108°31′). Through on-site investigation, two different sodic soils (sodic soil I and sodic soil II) were sampled using spade. The soil sampling depth was divided into five layers (0–5 cm, 5–20 cm, 20–40 cm, 40–60 cm and 60–100 cm). The soils in different layers were air-dried, crushed, mixed and passed through a 2-mm sieve before the soil column experiments. The physical and chemical properties of the tested soils are listed in Table 1. The sodication degree of sodic soil II was higher than that of sodic soil I.

The BFGD used in the test were from Huaneng Power International, Inc. The $CaSO_4$ content in the by-product was 89.8% and the moisture content was almost 10.1% (Table 2). The concentrations of pollution elements were far below the tolerance limits regulated by the control standards for pollutants in fly ash for agricultural use (GB8173-87) and the control standards for pollutants in sludge for agricultural use (GB4284-84) [17–18].

The BFGD Application Rate

The principle underlying the reclamation of sodic soils by BFGD is that the calcium contained in the BFGD replaces the sodium in soil colloids. The application rate of the BFGD was determined using a modified method based on gypsum application rates [11,15]. It was calculated according to the relative content of calcium in the BFGD and gypsum as follows:

$$R_{BFGD} = 1.11R_G = 0.095H\gamma(CEC)(E_{Nai} - E_{Naf}), \quad (2)$$

where R_{BFGD} is the BFGD application rate in t hm^{-2}; R_G is the gypsum application rate in t hm^{-2}; 1.11 is the modified coefficient, determined by comparing the content of $CaSO_4$ in gypsum and BFGD; H is the soil depth to be reclaimed in cm, 20 cm in this study; γ is the soil bulk density in g cm^{-3}, and the values for sodic soil I and sodic soil II at a depth of 0–20 cm are shown in Table 1;

Table 2. Component and properties of the BFGD.

pH	Density ($g \cdot cm^{-3}$)	Free water (%)	$CaSO_4 \cdot 2H_2O$ (%)		$CaSO_4 \cdot 1/2H_2O$ (%)		$CaCO_3$ (%)
5.90	1.02	10.10	89.8		0.20		5.55
Cd ($mg \cdot kg^{-1}$)	Cr ($mg \cdot kg^{-1}$)	As ($mg \cdot kg^{-1}$)	Se ($mg \cdot kg^{-1}$)	Ni ($mg \cdot kg^{-1}$)	Cu ($mg \cdot kg^{-1}$)	Hg ($mg \cdot kg^{-1}$)	Pb ($mg \cdot kg^{-1}$)
0.01	83.4	5.04	4.24	21.35	83.26	0.32	99.38
5[a]	250[a]	75[a]	15[a]	200[a]	250 [a]	5.0[b]	250[a]

Note: [a]Control Standards for Pollutants in Fly Ash for Agricultural Use (GB8173-87);
[b]Control Standards for Pollutants in Sludge for Agricultural Use (GB4284-84).

Table 3. Experimental design.

Experimental treatment	BFGD application rate (g·kg^{-1})	BFGD application rate (t·hm^{-2}) (g·kg^{-1})application rate (g·kg^{-1})	Sodic soils types
T1	3.00	1.65	Sodic soil I
T2	3.60	1.98	Sodic soil I
T3	7.00	3.86	Sodic soil II
T4	8.40	4.63	Sodic soil II

CEC is the cation exchange capacity in cmol kg^{-1}, and the capacities of sodic soil I and sodic soil II at 0–20 cm soil depth are shown in Table 1; $E_{Nai}$is the initial exchange sodium fraction, which is 1; and $E_{Naf}$is the desired final exchange sodium fraction, which was 0.1 in this study.

Experimental Design

The experiment consisted of four treatments (Table 3), with two sodic soils (sodic soil I and sodic soil II) and two BFGD rates. The BFGD application rates for sodic soil I and sodic soil II were divided into two levels. The amount of BFGD applied at the first level was determined according to the theoretical calculation of Formula 1. Taking into account the incomplete dissolution of the BFGD and deep percolation when it was applied, the amount of BFGD applied should be appropriately increased. The amount of the second application was determined by the theoretical calculation multiplied by a factor of 1.2. The BFGD application rates for different treatments are shown in Table 3 based on the initial properties of the sodic soil tested. The soil columns of four treatment were each leached four times.

The columns tested were contained in unplasticized polyvinyl chloride (UPVC) plastic pipes, with an outside diameter of 16 cm, inside diameter of 15.2 cm and height of 100 cm. The bottom of the soil column was closed with a UPVC board with a 1-cm drainage hole. A 5-cm layer of quartz sand was placed in the bottom of each column as the filter layer. The columns were filled with treated soils according to the soil depth, and the volumetric soil moisture content was 10%. The weight of dry soil to be filled at different depths was determined according to soil bulk density. The soils were evenly mixed in a sealed plastic container with water that is 10% by volume of the soils to be filled and maintained two days, and the soils with 10% of volumetric water content were obtained. Soil bulk density and soil weight in different column layers are shown in Table 4. The soil was filled in 5-cm depth at each time and tamped to a desired height in respective layers. The surface soil was loosened before the next soil layer was filled. During the filling process, BFGD was applied to and mixed evenly with the soil at 0–5 cm depth and 5–20 cm soil depth in the amount specified in the experimental design (Table 3). The soil column leaching tests were then carried out.

Because the soil water content in each soil column was only 10% (volumetric water content), the leaching water volume of each soil column was 4500 ml the first time (250 mm when translated into the field water volume; i.e. 2500 m^3·hm^{-2}), and was 1360 ml the second, third and fourth time (75 mm when translated into the field water volume; i.e. 750 m^3·hm^{-2}). The leaching interval was 12 d. The leaching soils were hierarchically sampled at soil depths of 0–5 cm, 5–10 cm, 10–20 cm, 20–40 cm, 40–60 cm, 60–80 cm and 80–100 cm. Each treatment soil column was sampled using an auger. The measured parameters were soluble salt composition (cations: CO_3^{2-}, HCO_3^-, Cl^- and SO_4^{2-}; anions: Ca^{2+}, Mg^{2+}, $K^+$and Na^+), pH, EC, ES and CEC.

Analytical Methods and Statistical Analyses

The samples were air-dried and passed through a 1-mm sieve. The EC, pH, soluble anions and soluble cations were measured using 1:5 water extracts. The soluble cations were measured using an atomic absorption spectrophotometer, soluble anions were determined by anion chromatography, and exchangeable cations were determined in 1 M ammonium acetate (pH = 7) extract. Following this extraction and washing with 96% alcohol, the cation exchange capacity was determined by the removal of ammonium ions by distillation [19]. Na and K were determined by flame emission spectroscopy in the extract, and Ca and Mg were determined by atomic absorption spectrophotometer. Soil pH was determined using the glass electrode method.

The SAR of 1:5 soil to water extracts (SAR1:5) was calculated as [5]

Table 4. The bulk density and weight of soils filled in each soil depth.

Sodic soil I				Sodic soil II			
Soil depth (cm)	Bulk density (g·cm^{-3})	Soil Weight (kg)	Total weight (kg)	Soil depth (cm)	Bulk density (g·cm^{-3})	Soil Weight (kg)	Total weight (kg)
0–5	1.41	1.28	26.13	0–5	1.58	1.43	26.72
5–20	1.46	3.97		5–20	1.52	4.12	
20–40	1.41	5.13		20–40	1.46	5.28	
40–60	1.47	5.32		40–60	1.45	5.28	
60–100	1.44	10.43		60–100	1.46	10.61	

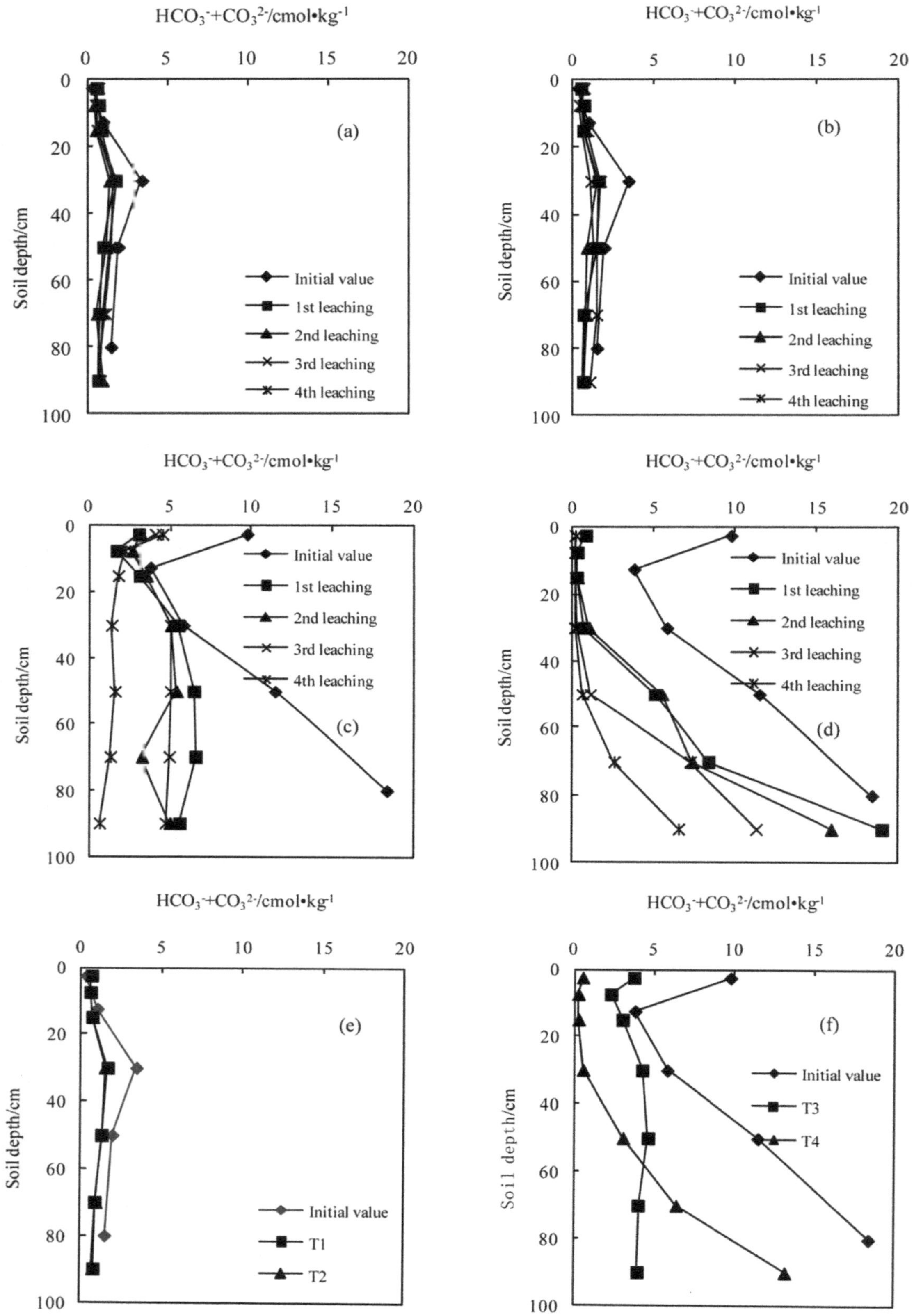

Figure 1. Variation in the concentration of carbonate and bicarbonate with successive leaching times (a.T1; b.T2; c.T3; d.T4); e. The average values in sodic soil I after leaching four times; f. The average value in sodic soil II after leaching four times).

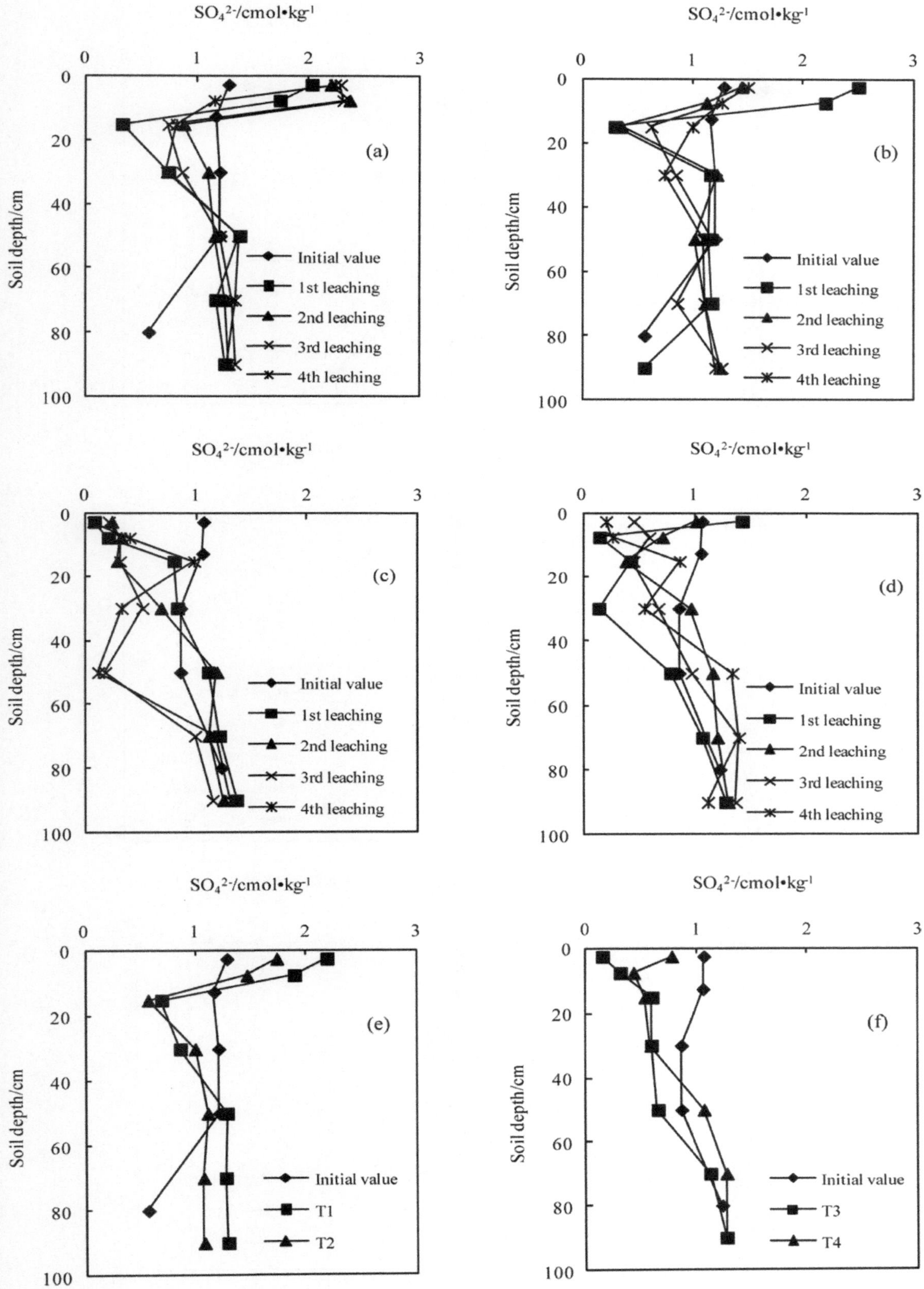

Figure 2. Variation in the concentration of sulfate with successive leaching times (a.T1; b.T2; c.T3; d.T4); e. The average values in sodic soil I after leaching four times; f. The average value in sodic soil II after leaching four times).

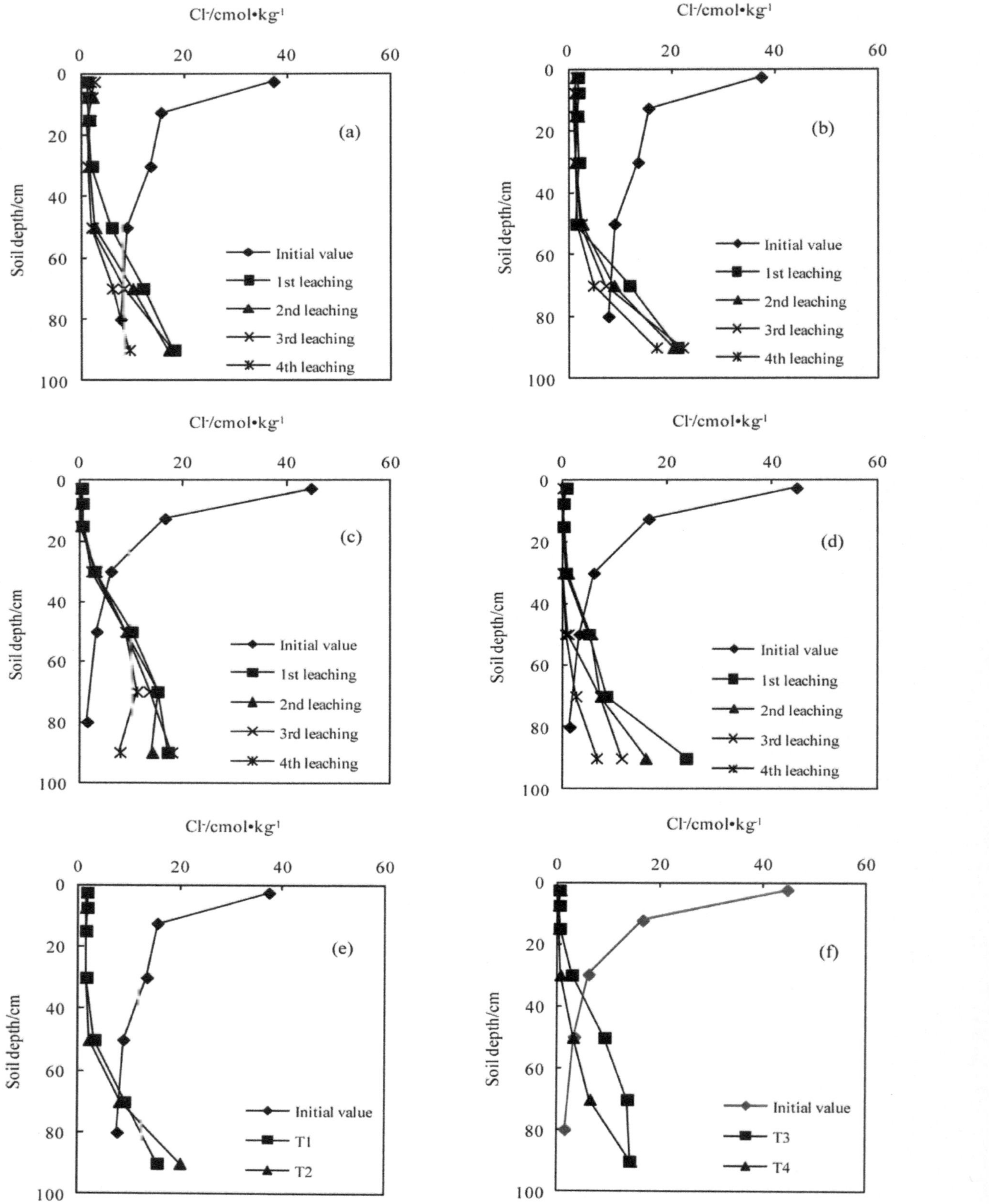

Figure 3. Variation in the concentration of chloride with successive leaching times (a.T1; b.T2; c.T3; d.T4); e. The average values in sodic soil I after leaching four times; f. The average value in sodic soil II after leaching four times).

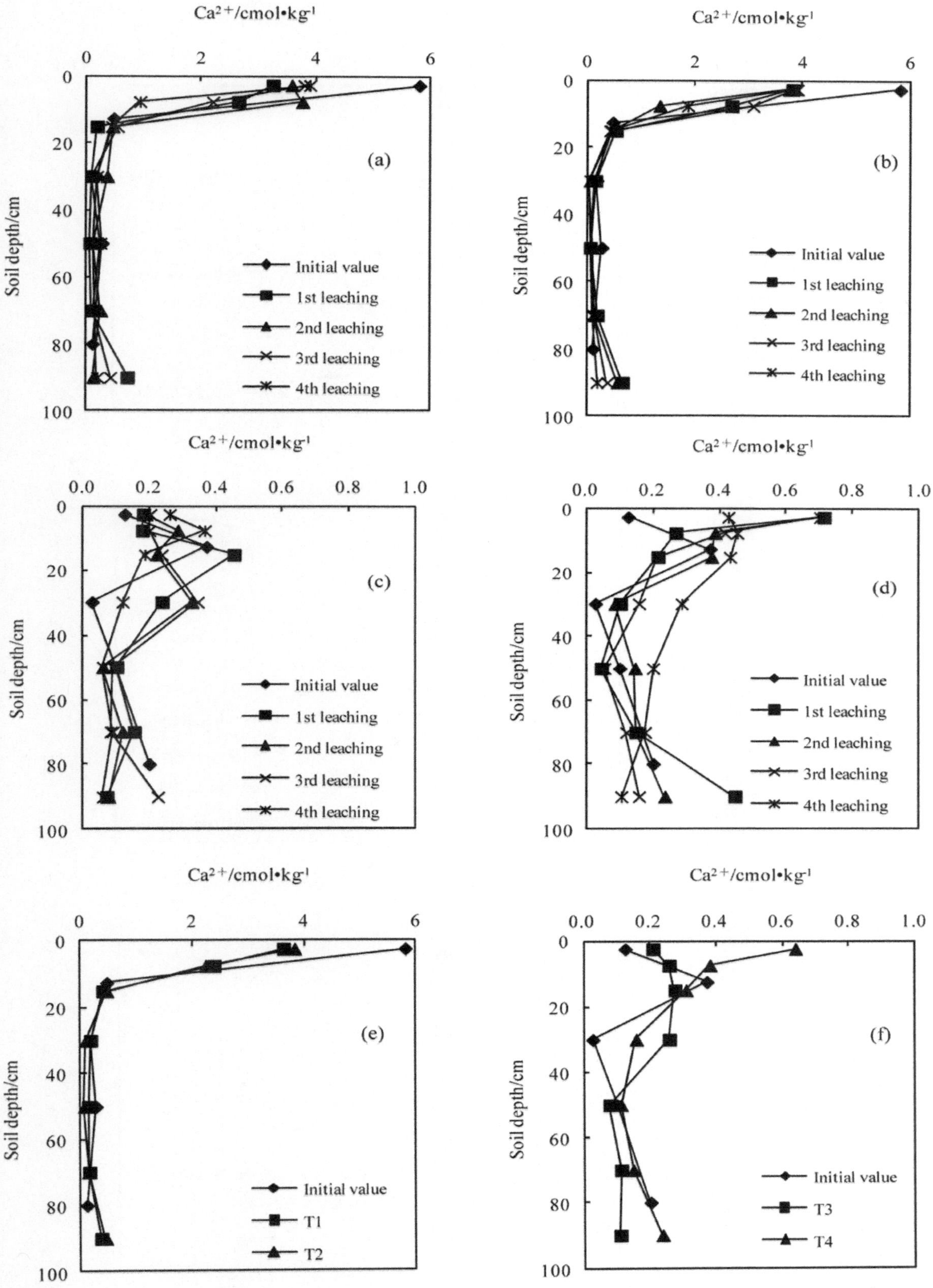

Figure 4. Variation in the concentration of calcium with successive leaching times (a.T1; b.T2; c.T3; d.T4); e. The average values in sodic soil I after leaching four times; f. The average value in sodic soil II after leaching four times).

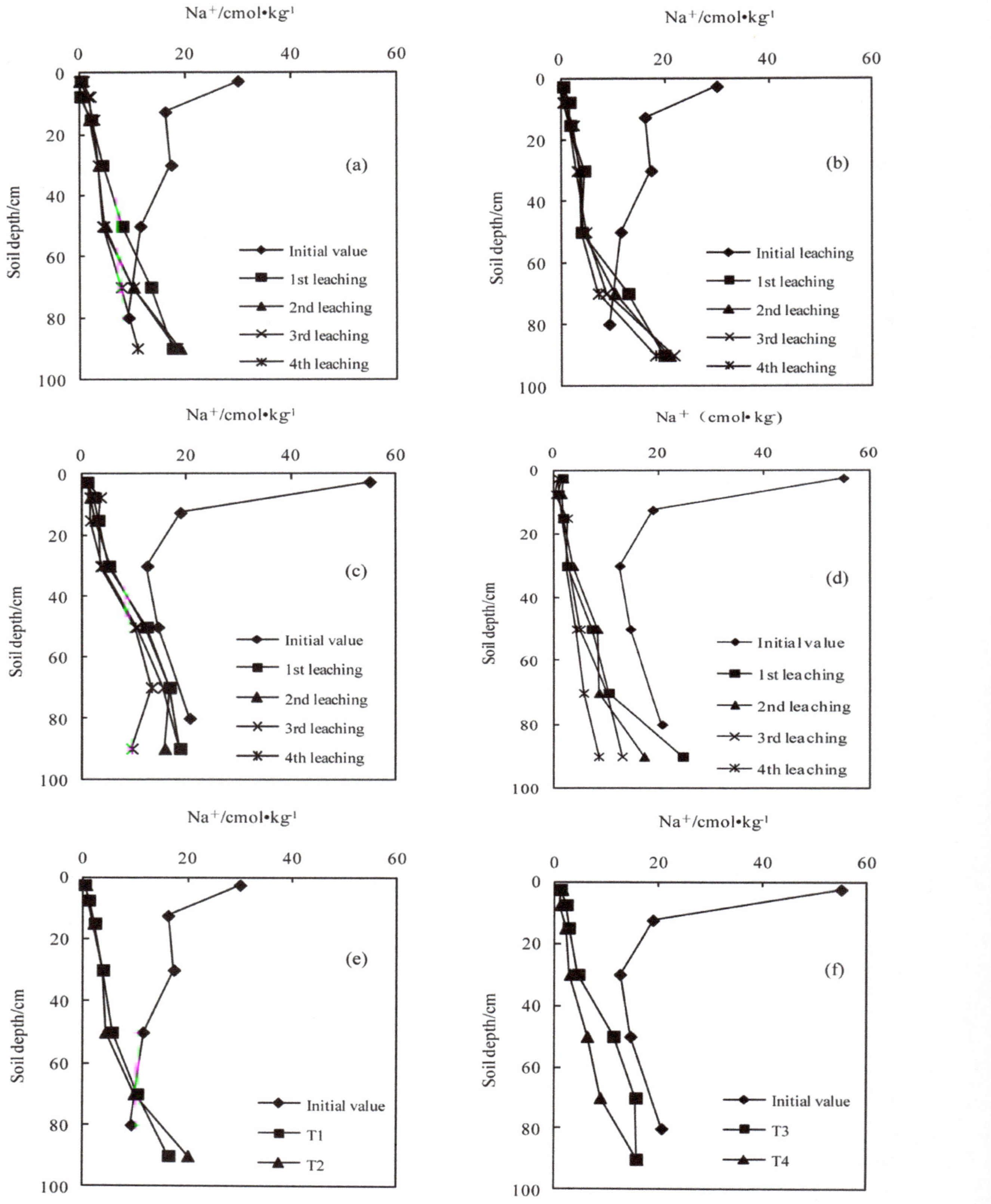

Figure 5. Variation in the concentration of sodium with successive leaching times (a.T1; b.T2; c.T3; d.T4); e. The average values in sodic soil I after leaching four times; f. The average value in sodic soil II after leaching four times).

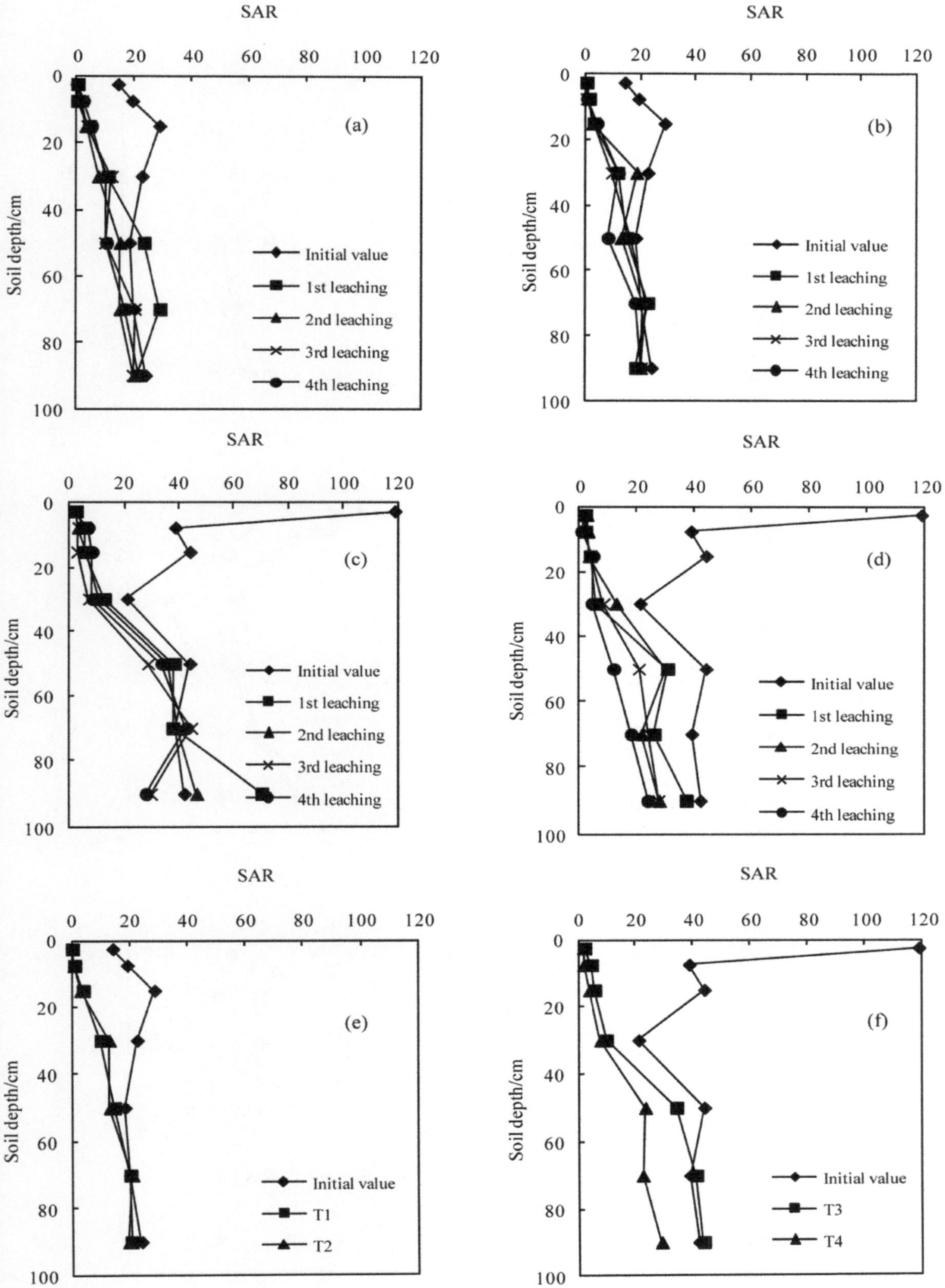

Figure 6. Variation in SAR with successive leaching times (a.T1; b.T2; c.T3; d.T4); e. The average values in sodic soil I after leaching four times; f. The average value in sodic soil II after leaching four times).

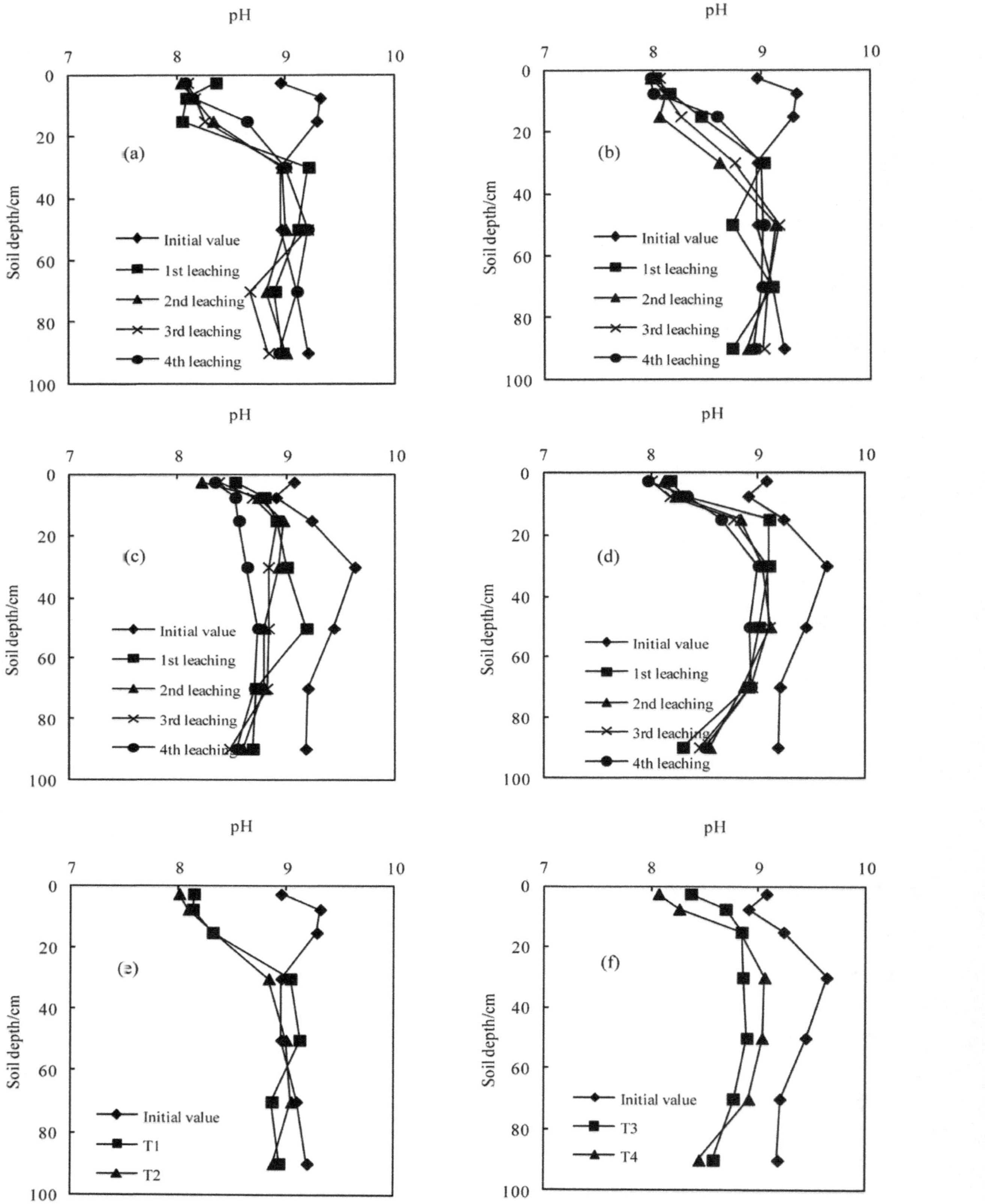

Figure 7. Variation in pH value with successive leaching times (a.T1; b.T2; c.T3; d.T4); e. The average values in sodic soil I after leaching four times; f. The average value in sodic soil II after leaching four times).

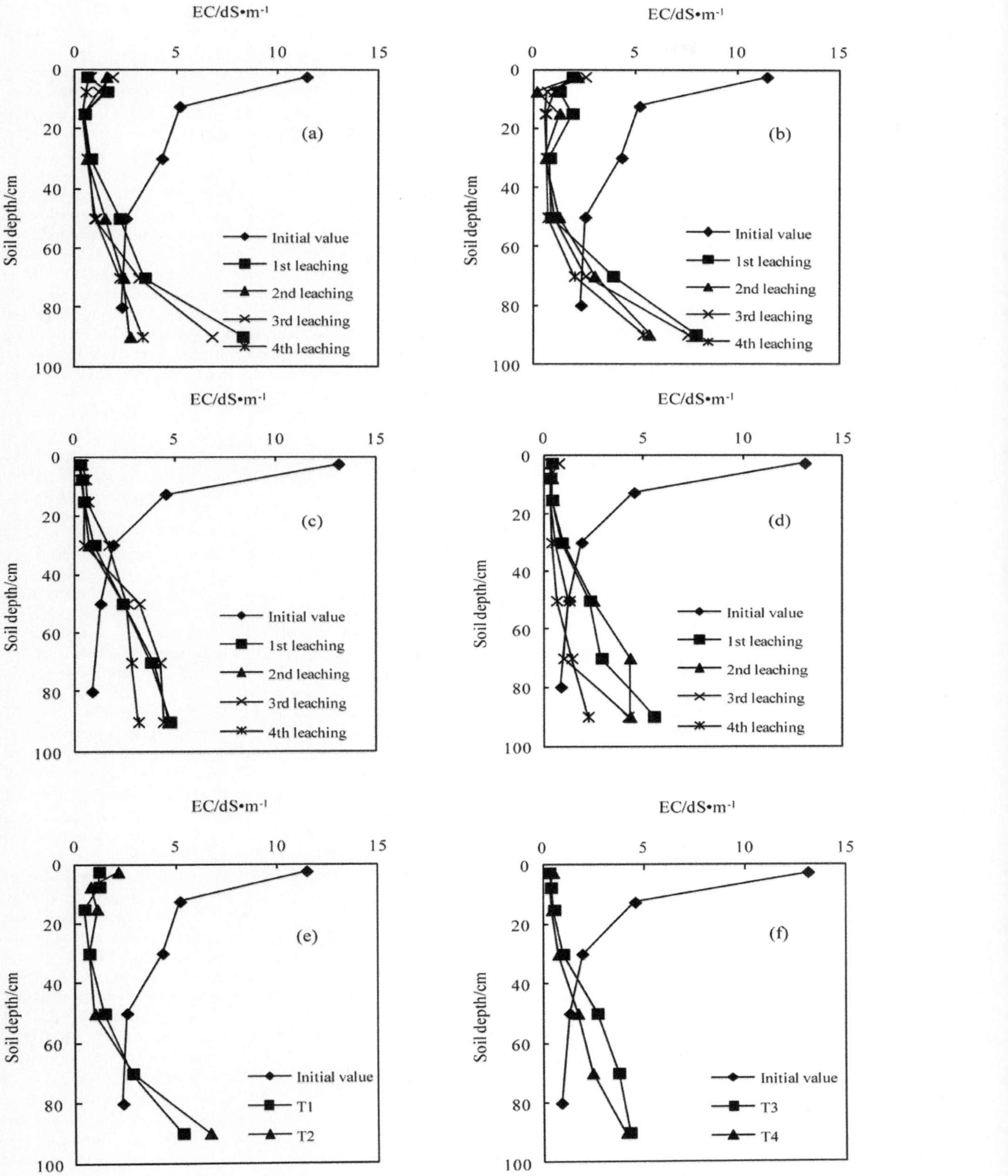

Figure 8. Variation in EC value with successive leaching times (a.T1; b.T2; c.T3; d.T4); e. The average values in sodic soil I after leaching four times; f. The average value in sodic soil II after leaching four times).

$$SAR_{1:5} = Na^+ / \sqrt{\frac{Ca^{2+} + Mg^{2+}}{2}}. \qquad (3)$$

The physical and chemical items were measured twice, and the chemical analysis was replicated three times. The standard errors of the means of the three samples from each treatment were calculated. Descriptive statistics (mean, range, standard deviation (SD) and coefficient of variation (CV)) on soil parameters were analyzed with SAS9.1 (Statistical Analysis System; SAS Institute Inc.) and they were considered to be significantly different at $P<0.05$.

Results

Variations in the Soluble Ion Composition

After applying the BFGD to the sodic soils, the main component of BFGD, $CaSO_4$, was constantly dissolved under the leaching action of the water flow, producing a rapid chemical reaction between the $CaSO_4$ and the soluble salts in the soil solution. As the $CaSO_4$ exchanged with the sodium in the soil colloid, it changed the soluble salts composition of the soil solution. The variation in the soluble salts composition (anions: HCO_3^-+CO_3^{2-}, SO_4^{2-} and Cl^-; cations: Ca^{2+} and Na^+)of the soil solution are shown in Figs. 1–5.

The concentrations of HCO_3^- and CO_3^{2-}. The concentrations of HCO_3^- and CO_3^{2-} in sodic soil II were higher than those in sodic soil I at the initial stage. After application of the BFGD, the concentrations of HCO_3^- and CO_3^{2-} in the two sodic soils decreased with each leaching, and the concentrations in the surface soils (0–30 cm) were lower than those in the deep soils (30–100 cm).

The concentrations of SO_4^{2-}. During the initial leaching stage, the concentration of SO_4^{2-} at a depth of 0–20 cm in sodic soil I increased significantly, but there was no significant increase in sodic soil I. With each successive leaching, the concentration of SO_4^{2-} gradually reduced, and that in sodic soil II at a depth of 0–20 cm also began to decrease. The concentrations of SO_4^{2-} with a high BFGD application rate were higher than those with a low application rate at all soil depths in the two sodic soils.

The concentrations of Cl^-. The concentration of Cl^- gradually reduced from the top to the bottom soils during the initial stage. After application of the BFGD and leaching, the Cl^- concentrations in the top soils of the two sodic soils decreased. The Cl^- accumulated in the bottom soil layer and gradually increased from the top to the bottom. The Cl^- concentrations with a high BFGD application rate were lower than those with a low application rate at all soil depths in the two sodic soils.

The concentrations of Ca^{2+}. After application of the BFGD, the concentration of Ca^{2+} slightly increased in the top soil layer, but there was no obvious change in the other layers. The concentration of Ca^{2+} decreased significantly in the top soil layer with each subsequent leaching. The Ca^{2+} concentrations with a high BFGD application rate were higher than those with a low BFGD application rate at all soil depths in the two sodic soils.

The concentrations of Na^+. The concentration of Na^+ gradually reduced from the top to the bottom soils during the initial stage. After application of the BFGD and leaching, the concentration of Na^+in each soil layer decreased, with a more obvious decrease in the top soil layer, with a gradual increase from the top to the bottom. The Na^+ concentrations with a high BFGD application rate were lower than those with a low BFGD application rate at all soil depths in the two sodic soils.

Variations in the Properties of the Sodic Soils

Soil SAR. Under soil dispersion conditions, high SAR levels in the soil solution will cause severe soil clay dispersion and decrease hydraulic conductivity [20]. The variations in SAR following the application of BFGD are shown in Fig. 6. Leaching greatly reduced the SAR in the top soils, and to a much lesser extent in the bottom soils. The SAR was significantly reduced below the initial value at depths of 0–50 cm, 0–60 cm, 0–40 cm and 0–50 cm in T1, T2, T3 and T4 respectively, and the SAR was below 13 at depths of 0–50 cm, 0–50 cm, 0–30 cm and 0–30 cm respectively. For those soil depths at which the SAR was below the initial value, sodic soil I with a low ESP was greater than sodic soil II with a slightly higher ESP, and the treatments with high BFGD application rates were greater than those with low BFGD application rates. The extent of the SAR decrease was positively correlated with the amount of leaching. Under the same soil conditions, the reclamation effect improved with each leaching and the SAR showed an increasing trend from the top soil to the bottom soil. Under different treatments, the SAR of T2 was lower than that of T1 in sodic soil I and the SAR of T4 was lower than that of T3 in sodic soil II. The soil SAR was generally below the initial value in T4 at all depths, but only at a depth of 0–50 cm in T3. This indicates that the $Na^+$was constantly being swapped and moved to the bottom layer of soil, thus increasing the relative proportion of Na^+ in the bottom layer. Therefore, applying the BFGD and leaching reduced the Na^+ in the soil solution by hindering the diffusion of the soil particles. Hence, the physical structure of the soil was improved and the sodic soil was reclaimed.

Soil EC. The variations in soil EC after the application of BFGD are shown in Fig. 7. The soil EC gradually increased from the top to the bottom soils during the initial stage. The EC in the surface soil significantly reduced under different treatments with the application of the BFGD and leaching, but increased in the bottom soil. The EC significantly decreased at depths of 0–60 cm, 0–60 cm, 0–40 cm and 0–40 cm in T1, T2, T3 and T4 respectively, and was below 4 dS m^{-1} at depths of 0–70 cm, 0–70 cm, 0–50 cm and 0–50 cm respectively. Therefore, application of the BFGD resulted in consistent improvement in the physical and chemical properties of sodic soils, and leaching did not cause an accumulation of salt in the surface soil. However, leaching should be strengthened to ensure the rapid discharge of salt accumulated in the bottom soil, prevent soil salinization and consolidate the reclamation effect on sodic soils.

Soil pH. The variations in soil pH after the application of BFGD are shown in Fig. 8. The soil surface pH significantly reduced under different treatments with the application of the BFGD and leaching. The soil pH significantly decreased at depths of 0–40 cm, 0–40 cm, 0–20 cm and 0–20 cm in T1, T2, T3 and T4 respectively, and was below 8.5 at depths of 0–20 cm, 0–20 cm, 0–10 cm and 0–10 cm respectively. With different BFGD application rates, the pH value of T2 was less than that of T1 in sodic soil I at depths of 0–60 cm, but the value for T2 was higher than that for T1 at depths below 60 cm. The pH value of T4 was less than that of T3 in sodic soil II at depths of 0–20 cm, whereas the pH value of T4 was more than that of T3 in sodic soil II at depths of 20–100 cm. In the process of improving sodic soils, the Ca^{2+} concentration in the soil solution forms carbonate and precipitate due to excessive HCO_3^- concentration in the irrigation water, thereby increasing the relative content of Na^+in the soil solution [5,16,21]. The soil solution had a pH below 8.5, illustrating that the carbonate and bicarbonate had disappeared and the soluble salt had gradually made the transition from sodic salt to neutral salt, thus the sodic soils had been reclaimed.

Discussion

The Mechanism of Change in the Ionic Composition of Soluble Salts

The application of BFGD in sodic soils caused the main component of the BFGD, $CaSO_4$, to dissolve due to the water flow infiltration. The $CaSO_4$ comprised the dissolved part and the residual part, and the dissolved part reacted rapidly with the sodic salts (Na_2CO_3 and $NaHCO_3$); see Equations 4 and 5 [5,22–23].

$$Na_2CO_3 + CaSO_4 \rightarrow Na_2SO_4 + CaCO_3 \quad (4)$$

$$2NaHCO_3 + CaSO_4 \rightarrow Na_2SO_4 + Ca(HCO_3)_2 \quad (5)$$

The concentrations of HCO_3^- and CO_3^{2-} rapidly decreased with these reactions, thus also reducing the soil pH value. After the disappearance of the sodic salt, there was an exchange reaction between the BFGD and the exchangeable Na^+ in the soil colloids (see Equation 6), which generated the soluble salt sodium sulfate [3]. As the sodium sulfate leached into the deep soil layer, the sodic level of the soil continuously decreased.

$$[\text{Soil colloid}]_{Na}^{Na} + CaSO_4 \rightarrow [\text{Soil colloid}]\ Ca + Na_2SO_4 \quad (6)$$

The application of the BFGD should increase the concentration of SO_4^{2-} and Ca^{2+}due to the dissolution of $CaSO_4$. However, because the solubility of $CaSO_4$ is relatively low, only 2 $g \cdot l^{-1}$ (25°C), the dissolved Ca^{2+} reacted with the sodic salts in the soil solution and the exchangeable Na^+ in the soil colloids, thus consuming the Ca^{2+} in the soil solution. The chemical reaction between SO_4^{2-} and Na^+ generated soluble sodium sulfate, which was then leached into the deeper soil levels. Therefore, rather than increasing the concentrations of $Ca^{2+}$and SO_4^{2-}, the concentrations were actually lower than the initial values in some soil layers. This effect is predominantly due to the increased permeability of the sodic soil once the salt has been leached from it and the concentration of Ca^{2+} and SO_4^{2-} has decreased [3,6,16,24–26].

In sodic soil areas there is also a certain amount of salinization, in which the soils contain some salt and the NaCl content is often very high. NaCl has a salting effect on the BFGD, which improves the solubility of $CaSO_4$, increases the infiltration characteristics of sodic soils and accelerates their reclamation [5,26–30]. With the reclamation of sodic soils, NaCl was constantly leached from the soil, thus reducing the NaCl content in the surface soil, and accumulating it in the deeper soil layers. There was a decrease in the Cl^- content at depths of 0–60 cm and 0–40 cm for sodic soil I and sodic soil II respectively, and an increase in the Cl^- content at depths of 60–100 cm and 40–100 cm.

As a result, the soil soluble salts made the transition from sodic salts containing Na_2CO_3 and $NaHCO_3$ to neutral salts containing NaCl and Na_2SO_4. Leaching of the soluble salts from the soil resulted in continuous improvement in the sodic soils.

The Relationship between Soil Properties and Soluble Salts Composition

The BFGD also immediately increased the EC of the soil solution and increased the ionic strength, which leads to proton generation and a reduction in the soil pH value [31]. The higher BFGD application rate reduced the soil SAR significantly more efficiently than the lower BFGD application rate because the BFGD increased Ca^{2+} in the soil solution, which promoted the displacement of the adsorbed Na^+ and subsequent leaching. A decrease in SAR_e was also found in the control treatment. This was attributed to the "valence dilution" mechanism reported by Reeve and Bower [32]. In a soil-water system, there is equilibrium between the monovalent and divalent cations at the soil exchange site and in the solution. The equilibrium is altered by adding water to the system. This dilution of the soil solution favors the adsorption of divalent cations such as Ca^{2+} at the cost of monovalent cations such as Na^+. Furthermore, if the soil solution is diluted by a factor of ΔC while maintaining the same ionic ratios, the initial SAR decreases by the factor $(\Delta C)^{1/2}$, i.e.,$SAR_{final} = (\Delta)^{0.5} \times SAR_{initial}$. Because of the dilution, the $(\Delta C)^{1/2}$ is less than 1. Thus, the SAR_{final} is less than the $SAR_{initial}$.

The application of BFGD was also found to be effective in lowering the soil pH. The significant decrease in soil pH may be attributed to leaching of the exchangeable Na^+, which was replaced by Ca^{2+} from the BFGD, and the decreased concentrations of HCO_3^- and CO_3^{2-}. This pH decrease was partly associated with the degree of soil reclamation and especially with the reduction in soil sodicity [5,33]. The pH of sodic soils is a function of the activity of CO_3^{2-}+HCO_3^-, and partially due to CO_2 and ionic strength [34]. The presence of soluble Na+ increases the pH of calcareous soils by increasing the activity of carbonate and bicarbonate. Sodium carbonate (Na_2CO_3) and sodium bicarbonate ($NaHCO_3$) are relatively soluble, and their removal by leaching can substantially reduce pH. Thus, the decrease in pH was in part associated with the leaching of soluble salts [35–36].

Conclusions

The following conclusions can be drawn from our findings.

(1) The application of BFGD and leaching caused the soil soluble salts to make the transition from sodic salts containing Na_2CO_3 and $NaHCO_3$ to neutral salts containing NaCl and Na_2SO_4. The poor physical properties of sodic soils meant that long leaching times were needed to obtain good reclamation effects.

(2) The soil SAR, pH and EC decreased at all soil depths, although it was more significant in the top soil. At a depth of 0–40 cm, the soil SAR, EC and pH were reduced to less than 13, 4 dS m^{-1} and 8.5 respectively, in both sodic soils I and II.

(3) Leaching played a key role in the reclamation process, and the reclamation effect was positively correlated with the amount of leaching. The soil salts did not accumulate in the top soil layer, and there was a slight increase in the soil salts in the middle and bottom layers. Stronger leaching is thus required to ensure the salt accumulated in the bottom soil is discharged rapidly, thus preventing soil salinization and consolidating the reclamation effect.

(4) The use of BFGD is inefficient, and the reclamation effect was improved by increasing the BFGD application rate by 20% above the theoretically calculated base rate.

(5) The results of this research demonstrate that the reclamation of sodic soils using BFGD is promising.

Author Contributions

Conceived and designed the experiments: JW PY. Performed the experiments: JW. Analyzed the data: JW ZB. Contributed reagents/materials/analysis tools: JW ZB. Wrote the paper: JW.

References

1. Wang SJ, Chen CH, Xu XC, L YJ (2008) Amelioration of alkali soil using flue gas desulfurization byproducts: Productivity and environmental quality. Environmental Pollution 151: 200–204.
2. Carlson CL, Adriano DC (1993) Environmental impacts of coal combustion residues. Journal of Environmental Quality 22: 227–247.
3. Chun S, Nishiyama M, Matsumoto S (2001) Sodic soils reclaimed with by-product from flue gas desulphurization: corn production and soil quality. Environmental Pollution 114: 453–459.
4. Stuczynski TL, McCarty GW, Wright RJ (1998) Impact of coal combustion product amendments on soil quality: Mobilization of soil organic nitrogen. Soil Science 163: 953–959.
5. USDA (1954) Diagnosis and improvement of saline and alkali soils. USDA, U.S. Govt. Printing Office, Washington, DC.
6. Wang JM, Bai ZK, Yang PL (2012) Sodic Soil Properties and Sunflower Growth as Affected by Byproducts of Flue Gas Desulfurization. PLoS ONE 7(12): e52437.
7. Frenke H, Gerstle Z, Alperovitch N (1989) Exchange-induced dissolution of gypsum and the reclamation of sodic soils. Soil Science 40: 599–611.
8. Clark RB, Ritchey KD, Baligar VC (2001) Benefits and constraints for use of FGD products on agricultural land. Fuel 80: 821–828.
9. Sloan JJ, Dowdy RH, Dolan MS, Rehm GW (1999) Plant and soil responses to field-applied flue gas desulphurization residue? Fuel 78: 169–174.
10. Wang JM, Yang PL, Ren SM, Xiang GM (2005) Variation of chemical indices of alkaline soil ameliorated with desulphurization byproducts. Acta Pedologica Sinica (in Chinese) 42(1): 98–105.
11. Wang JM, Yang PL (2004) The effect on physical and chemical properties of saline and sodic soils reclaimed with byproduct from flue gas desulphurization, In *Land and Water Management Decision Tools and Practices Vol.II*, Huang GH. and S.Pereria L (eds). China Agriculture Press, Beijing: 1015–1021.
12. Sakai Y, Matsumoto S, Sadakata M (2004) Alkali soil reclamation with flue gas desulphurization gypsum in China and assessment of metal content in corn grains. Soil & Sediment Contamination 13(1): 65–80.
13. Ilyas M, Qureshi RH, Qadir MA (1997) Chemical changes in saline-sodic soils after gypsum application and cropping. Soil Technologies 10: 247–260.
14. Chen L, Ramsier C, Bigham J, Slater B, Kost D, et al (2009) Oxidation of FGD-$CaSO_3$ and effect on soil chemical properties when applied to the soil surface. Fuel 88: 1167–1172.
15. Oster JD, Frenkel H (1980) The chemistry of the reclamation of sodic soils with gypsum and lime. Soil Science Society of America Journal 44: 41–45.
16. Chi CM, Zhao CW, Sun XJ, Wang ZC (2012) Reclamation of saline-sodic soil properties and improvement of rice (Oriza sativa L.) growth and yield using desulfurized gypsum in the west of Songnen Plain, northeast China. Geoderma 187–188: 24–30.
17. Ministry of Agriculture and Fisheries (1987) Control Standards for Pollutants in Fly Ash for Agricultural Use (GB8173–87). Beijing: China Standards Press (in Chinese).
18. Ministry of Agriculture and Fisheries (1984) Control Standards for Pollutants in Sludge for Agricultural Use (GB4284–84). Beijing: China Standards Press (in Chinese).
19. Rhodes DR (1982) Cation exchange capacity. In *Methods of soil analysis Part 2, Chemical methods*. Page AL, Miller RH and Keeney DR (eds). American Society of Agronomy:Madison, Wisconsin, 149–165.
20. Chorom M, Rengasamy P, Murry RS (1994) Clay dispersion as influenced by pH and net particle of sodic soils. Australian Journal of Soil Research 32: 1152–1163.
21. Truman CC, Nuti RC, Truman LR, Dean JD (2010) Feasibility of using FGD gypsum to conserve water and reduce erosion from an agricultural soil in Georgia. Catena 81: 234–239.
22. Carter MR, Pearen JR (1989) Amelioration of a saline-sodic soil with low application of calcium and nitrogen amendments. Arid Soil Research 3: 1–9.
23. Sakai Y, Aihara T, Yamamoto T, Sadakata M (2010) Evaluation of soil properties and additional application of desulfurization gypsum in an area of no amelioration after an initial application in sodic soil of China. Kagaku Kogaku Ronbunshu 36 (4): 275–280.
24. Gharaibeh MA, Eltaif NI, Shra'ah SH (2010) Reclamation of a calcareous saline-sodic soil using phosphoric acid and by-product gypsum. Soil Use and Management 26(2): 141–148.
25. Minashina NG (2005) Concentration and composition of salts in the solution of a gypsum-containing soil as dependent on the soil water content. Eurasian Soil Science 38(7): 718–726.
26. Ritvo G, Speed FM, Neill WH, Dixon JB, Lawrence AL, et al (1999) Regression analysis of soil chemical composition for two shrimp farms in Texas. Journal of the World Aquaculture Society 30(1): 26–35.
27. Reading LP, Baumgartl T, Bristow KL, Lockington DA (2012) Hydraulic conductivity increases in a sodic clay soil in response to gypsum applications: impacts of bulk density and cation exchange. Soil Science 177(3): 165–171.
28. Singh K, Singh B, Singh RR (2012) Changes in physical-chemical, microbial and enzymatic activities during restoration of degraded sodic lands: Ecological suitability of mixed forest over plantation. Catena 96: 57–67.
29. Rhoton FE, McChesney DS (2011) Erodibility of a Sodic Soil Amended With Flue Gas Desulfurization Gypsum. Soil Science 176(4): 190–195.
30. Muhammad D, Khattak RA (2011) Wheat yield and chemical composition as influenced by integrated use of gypsum, pressmud and FYM in saline-sodic soil. Journal of the Chemical Society of Pakistan 33(1): 82–89.
31. Simon M, Garcia I (1999) Physico-chemical properties of the soil-saturation extracts: estimation from electrical conductivity. Geoderma 90 (1–2): 99–109.
32. Reeve RC, Bower CA (1960) Use of high-salt waters as a flocculant and source of divalent cations for reclaiming sodic soils. Soil Science 90, 139–144.
33. Pandey VC, Singh K, Singh B, Singh RP (2011) New approaches to enhance eco- restoration efficiency of degraded sodic lands: critical research needs and future prospects. Restoration Ecology 29: 322–325.
34. Mashhady AS, Rowell DL (1978) Soil alkalinity. I. Equilibria and alkalinity development. Journal of Soil Science 29: 65–75.
35. Katerji N, Van Hoorn JW, Hamdy A, Mastrorilli M, Mou Karzel E (1997) Osmotic adjustment of sugar beets in response to soil salinity and its influence on stomata conductance, growth and yield. Agricultural Water Management 34: 57–69.
36. Shannon MC (1997) Adaptation of plants to salinity. In *Advances in Soil Science*. Lal R and Steward B A (eds). Springer-Verlag, 75–120.

2

A Novel Growing Device Inspired by Plant Root Soil Penetration Behaviors

Ali Sadeghi[1]*, Alice Tonazzini[1,2], Liyana Popova[1,2], Barbara Mazzolai[1]*

1 Center for Micro-BioRobotics, Istituto Italiano di Tecnologia, Pontedera, Italy, **2** The BioRobotics Institute, Scuola Superiore Sant'Anna, Pontedera, Italy

Abstract

Moving in an unstructured environment such as soil requires approaches that are constrained by the physics of this complex medium and can ensure energy efficiency and minimize friction while exploring and searching. Among living organisms, plants are the most efficient at soil exploration, and their roots show remarkable abilities that can be exploited in artificial systems. Energy efficiency and friction reduction are assured by a growth process wherein new cells are added at the root apex by mitosis while mature cells of the root remain stationary and in contact with the soil. We propose a new concept of root-like growing robots that is inspired by these plant root features. The device penetrates soil and develops its own structure using an additive layering technique: each layer of new material is deposited adjacent to the tip of the device. This deposition produces both a motive force at the tip and a hollow tubular structure that extends to the surface of the soil and is strongly anchored to the soil. The addition of material at the tip area facilitates soil penetration by omitting peripheral friction and thus decreasing the energy consumption down to 70% comparing with penetration by pushing into the soil from the base of the penetration system. The tubular structure provides a path for delivering materials and energy to the tip of the system and for collecting information for exploratory tasks.

Editor: Josh Bongard, University of Vermont, United States of America

Funding: This work was supported by the Future and Emerging Technologies (FET) programme within the Seventh Framework Programme for Research of the European Commission, under FET-Open grant number: 293431. The funders had no role in study design, data collection and analysis, decision to publish, or preparation of the manuscript.

Competing Interests: The authors have declared that no competing interests exist.

* E-mail: ali.sadeghi@iit.it (AS); barbara.mazzolai@iit.it (BM)

Introduction

Soil exploration mechanisms are among the most investigated systems because of their extremely wide range of applications. Soil is a source of vital elements (water, nutrients, and minerals) for all living systems, containing the major energy resources used by humanity and providing important elements that enable technological advancement. As a consequence of this exploitation, the effects of human activity on soil are extremely serious (soil features and functions can be recovered only after a long time and with high costs), and for this reason, long-term monitoring is needed, with particular attention to physical soil properties, the presence of contaminants (i.e., heavy metals and organic pollutants) and nutrients (i.e., phosphorous and nitrogen), and water conditions in shallow depths. The techniques used in soil monitoring commonly involve sensorized probes, which are pushed into the soil to certain depths. A multitude of developed mechanical penetration systems, most of which are based on electrical, pneumatic, and hydraulic actuators [1–7], have been developed, both for taking samples and for creating access for sensorized probes, which need to be introduced directly at a certain depth [8]. Among these systems, rotary drilling is the most widely used. Drilling devices are able to create straight and vertical boreholes in various substrates (e.g., sand, rock, and ice) [5–7]: the drill bit, which is typically actuated by an electric motor fixed on the body of the device, penetrates the medium perpendicularly to the drill axis, being forced against the bottom of the bore by the weight of the entire drill string. The required weight on the bit may be significant (depending on the type of soil), and in several operating conditions, the exploitable load is limited due to friction, buckling or a reduced vertical depth [9]. Moreover, drilling methods can produce local heating due to the cutting/shearing process and need lubricants or fluids to remove soil particles or dirty [10], these representing limiting factors when the aim is to find life or water signatures (e.g., in space missions) [2,11]. The most common alternative to drilling and sampling methods for soil testing in agricultural [12,13] and geotechnical practice [14,15] involves the use of sensorized steel probes (usually with a cone or a blunt tip) called 'penetrometers,' which are pushed from the top of the devices into the soil manually or by actuators on the soil surface. In agriculture, penetrometers are used at relatively shallow depths for characterizing soil strength properties (in particular, penetration resistance to root growth), which change under the influence of climate, plant growth, and soil management and which largely depend on bulk density, moisture content, and soil texture [16,17]. As previously reported [18], 40–80% of the measured soil resistance to a metal probe is frictional. To reduce this frictional component, which increases the penetrometer resistance to between 2 and 8 times that of the real root penetration resistance, penetrometers with a lubricated tip and with a rotating tip have been developed [19]. In geology and civil engineering, push rods are needed to achieve deep layers in field testing: the rods are inserted by a thrust mechanism counterbalanced by a reaction system (pushing equipment) [20]. The thrust capacity needed varies depending on the type of soil, and its maximum capacity is limited not only by the thrust mechanism performance but also by damage to the push rods, including buckling [14].

In this context, robotic devices could assist in identifying potentially useful or hazardous areas and in monitoring levels of essential elements or geological properties. However, robotic technology for soil exploration and monitoring has been scarcely developed compared with that which is available for exploration and monitoring above ground and underwater. This situation is partly a consequence of the physical constraints of underground operation, which make the task extremely challenging for an autonomous system [2]. The requirement for high dexterity in the exploration of underground unstructured environments has led to a growing interest in biologically inspired solutions [21,22]. Various penetration devices have been developed that mimic particular aspects of animals such as wood wasps, locusts, and clam [23–25]. These previous works were mainly focused on burrowing and drilling animals, which penetrate substrates by exploiting the movements produced by their muscular systems; but plants are the most efficient at soil exploration among living organisms, being able to grow and adapt inside the medium. They develop networks of growing and branching roots whose tips (apices) are highly sensitive. Plants efficiently explore the soil for minerals and water to fulfill their primary functions in a wide variety of environments. Each root apex is able to sense a variety of chemical, physical, and mechanical stimuli [26–28]. This information is integrated within genetically driven developmental rules and shared with the entire network of roots to improve plant fitness [29–31].

Unlike animals, for which growth is related to development, the distinctive feature of roots is the link between their growth and their movement [32]. Plant roots are able to find low-resistance pathways and exploit cracks in the soil, overcoming soil penetration resistance [33–35]. Their exploration capability arises from the root apex [36–38]; i.e., the plant root moves and penetrates soil by growing at the apical region [39–40]. The growth process in the tip enables the roots to adapt their morphology and organ development to the environmental conditions such as the soil texture and mechanical impedance because the cell division and morphology are directly influenced by the interaction with the surrounding environment [41–43]. In addition, the apex morphology has been demonstrated to have a significant role in soil penetration [44–46].

More specifically, root development is driven by two continuous processes in the apex: cell division and cell elongation, which occur in the meristematic and elongation regions (Fig. 1a, b), respectively. In this work, we term this growth process "elongation from the tip" (EFT) (Fig. 2, Video S1). Newly generated cells move from the meristematic region to the elongation region, where they expand axially because of the water absorbed by osmosis and the directional loosening of the cell wall [47,48]. This action allows the root to penetrate the soil with only a small part of its structure (the tip), while the remainder of the structure is stationary and in contact with the soil (the mature part, shown in Fig. 1a, b). This process provides the pressure required for the forward advancement of the root. The exerted pressure, up to 1 MPa, is dissipated in the expansion of a cavity in the soil and in the frictional resistance of the soil to the advancement of the root [49–51]. The penetration may be straight or curved, depending on whether the cell growth is symmetric or differential. Cap cells are continually produced in the meristematic region. These cells move to the root cap and then slough off from its outer surface while producing mucus [52]. In this way, the cap cells create an interface between the soil and the root apex (Video S1). The root cap protects the delicate cells in the meristematic region, and the mucus promotes root penetration by reducing friction [52–55]. Mature cells, which are situated behind the apex, are strongly anchored to the soil and thus allow the apex to move forward. This anchorage, also called root-soil adhesion [56], is achieved by root hairs, secondary roots, and the root architecture (Fig. 1b) [56–58]. Stronger root-soil adhesion may enable the root to penetrate harder soils [59].

The mechanical impedance of the soil directly influences the root growth rate and shapes the apex morphology and the overall root architecture [60,61]. Generally, when a root encounters a higher mechanical impedance, the elongation rate decreases and the apex diameter increases [12,62]. A larger root radius reduces the axial resistance of the soil immediately in front of the root and provides a stronger anchorage behind the penetrating apex [63,64].

A plethora of studies on the growth kinematics and the morphological features of plant roots have been performed to understand root penetration behaviors [49,57,65]. These studies mainly involved observations of living roots in various environmental conditions [66] and computational modeling and simulation of root movements [63,67,68]. A recent trend in this field is the development of new tools for studying root kinematics that follow the path taken over time by a recognizable part of the observed root system, which may lead to the discovery of new features of the mechanism of root growth [42,43,69,70,71]. Taking an engineering approach to the study of plant roots allows important principles to be extracted and used for designing innovative exploratory robots. In [72], the development of a self-anchoring system for soil penetration that was inspired by the sloughing cells in the apical root region described previously was reported. Tools for validating plant features were presented and the efficiency of this biological phenomenon was demonstrated quantitatively. In addition, the necessity of anchorage for successful substrate penetration was discussed. A quantitative evaluation of the role of EFT in decreasing the energy consumption in soil penetration was recently performed and described in [46]. That study demonstrated how penetration by EFT reduces dynamic friction and energy consumption in the root. Two sets of penetration tests were performed in granular substrates using a purposely built probe [46] to compare the performance of EFT with that of penetration by applying a downward force at the top of the probe (identified as "no elongation from the tip", NoEFT). These tests demonstrated that the penetration energy required for EFT was less than that for NoEFT in soil penetration tasks at various initial depths; the reduction increased from approximately 20% at an initial depth of 100 mm to 50% at an initial depth of 250 mm. The trend of these results shows that at greater depths, a significant reduction in energy consumption (more than 50%) can be obtained with the EFT approach. Therefore, EFT in a plant root represents an efficient solution for soil penetration, and the preliminary results reported in [46] were considered a starting point and a source of inspiration for designing and developing an innovative robotic soil exploration system. In this paper, we present a new generation of devices that are able to grow and construct their own body and were inspired by the growth and soil penetration behaviors in plant roots. The proposed root-like device grows through a monotonic process that continuously adds new material. This device can efficiently penetrate soil by taking advantage of root penetration behaviors, in particular EFT. However, because the design is based on plant roots principles, the system described represents a useful platform for the experimental validation of theories and hypotheses concerning actual plant root systems.

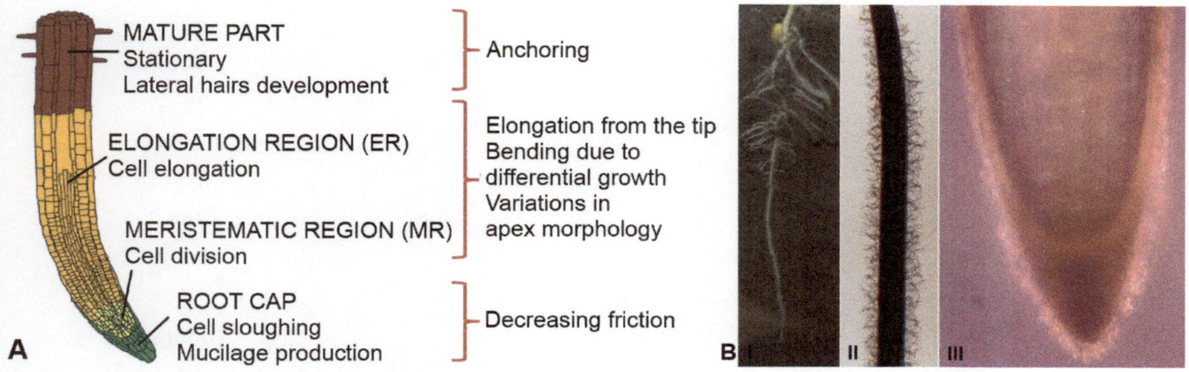

Figure 1. Plant root structures and functions. Figure (a) illustrates plant root structures and their functions. Four main regions can be identified, each with different roles. New cells are created by mitosis in the meristematic region (MR). These cells are then elongated by osmotic pressure and move to the elongation region (ER). Cell division and cell elongation provide the force to penetrate into the soil; asymmetries with respect to the root axis results in bending. The mature root consists of elongated cells that are stationary and provide a strong anchor to the soil, thus supporting tip penetration. Root cap cells exude mucilage and slough off, which decreases friction during penetration by lubricating the surrounding soil. Figure (b) shows images of an actual maize (*Zea mays*) root; the images were captured with a digital microscope (HIROX KH-7700). Specifically, image I is a two-week old maize root growing in soil, in which the primary root, the lateral roots and several seminal roots can be observed. Image II shows the mature region of the primary root with visible root hairs. Image III shows the apex of the primary root with a distinguishable root cap. Image IV shows the root cap of the primary root with sloughed cells visible around it.

Results

System Design

The decrease in energy observed in the EFT penetration trials (see the "Materials and Methods" section) was used as a guideline in the design of our exploratory robotic device. A root-growth approach at the apical region results in an energy-efficient solution that addresses the penetration issue. Based on these premises, we propose a bio-inspired design of a root-like robotic system that grows at the apical area by the addition of new material. Figure 3a shows the conceptual design of this system. The device is composed of a growing zone and a stationary mature zone. The mature zone consists of a hollow, tubular structure that allows the transfer of new material from a spool (external to the robotic root) and power to the growing zone.

The growing zone is a customized additive layering mechanism that generates a force for penetration into the soil and consists of: a) a rotating deposition head, b) a guiding nozzle on the deposition head, and c) a motor and a transmission mechanism (i.e., a gearset) (Fig. 3b). The material, which is in the form of a filament, can be easily transferred from the external spool to the nozzle at the periphery of the deposition head. The "growing" capability at the tip region is achieved with an additive manufacturing technique similar to Fusion Deposition Modeling (FDM) [73]. The layer-by-layer deposition creates the mature zone, which uses the tubular body as a support structure. The rotation of the deposition head is driven by a motor coupled to the transmission mechanism. The tubular body moves axially inside the mature zone in a passive manner, and any rotational movement between them is prevented by locking wires installed on the body. The axial slipping movement induced by the addition of filament material allows the new material to be automatically distributed on the surface of the deposition head, in the growing zone (Fig. 3a). To prevent any twisting between the filament and the motor power cables, the power is transferred by two annular electrodes (i.e., slip rings) that are fixed to the internal surface of the tubular body and are always

Figure 2. Elongation from the tip (EFT) process during three following steps of growth. The colors indicate the movement of cells from the MR to the mature root and sloughing; mature cells do not move with respect to the soil, whereas the apex is forced into the soil by the new cells originating in MR.

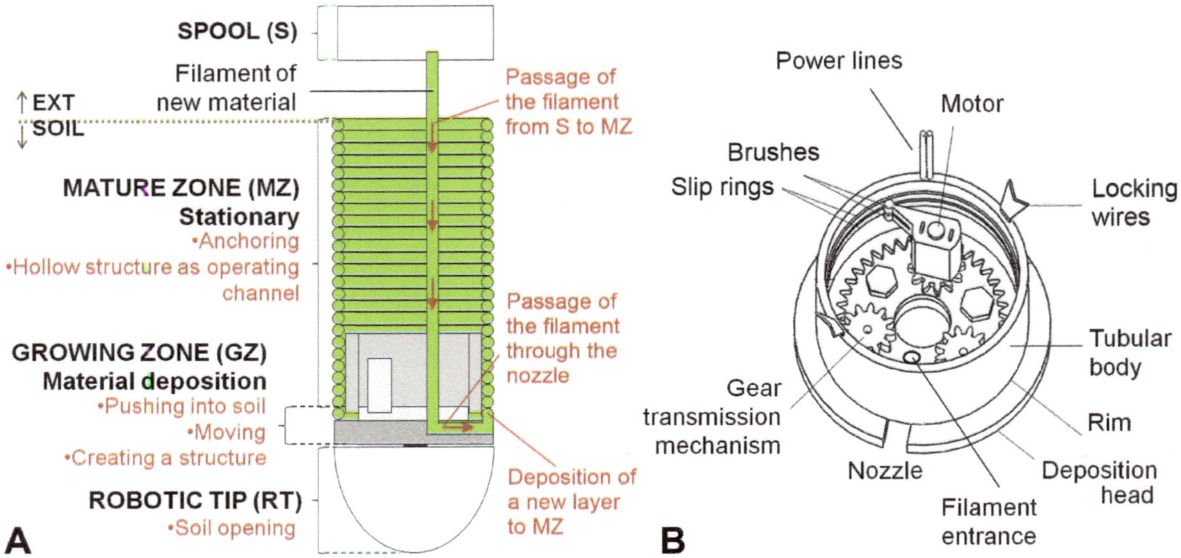

Figure 3. Robot Design. Figure (a) gives a schematic representation of the cross section of the root-like device that grows from the tip. The functional zones are the growing zone (GZ), the mature zone (MZ), the robotic tip (RT), and the spool (S) with the filament. The growing zone imitates the meristematic region in living plant roots; the GZ exerts an axial force on the robotic tip by adding new material. The mature zone is created by the layer-by-layer addition of material (the filament) and is stationary with respect to the soil. The hollow structure of the MZ allows the passage of the filament (new material) from the spool to the growing zone. Figure (b) shows the 3D design of the growing zone. The filament of new material passes through the filament entrance and the nozzle (located on the deposition head) to the external side of the tubular body between the rim and the previous layer of the mature zone. The rotation of the deposition head with respect to the tubular body, generated by the motor through the transmission mechanism, results in the deposition process that builds the mature zone. The motor receives power through brushes in contact with slip rings. During soil penetration, the growing zone slips inside the mature zone; any rotational movement between the two zones is suppressed by the locking wires.

in contact with the brushes connected to the rotating motor. The cables are connected to the electrodes through an axial hole in the tubular body (Fig. 3b).

The rotary motion of the deposition head is converted into a linear motion at the tip that provides the motive force for overcoming soil resistance (Fig. 4a). If we assume that the diameter of the filament remains constant during the growth process, one deposition cycle of the tip penetrates by a distance equal to the filament diameter. The penetration depth, P, resulting from the deposition of the filament (with length L and diameter d) around the tubular body (with external diameter D) is given as:

$$P = \frac{d}{\sqrt{(\pi(D+d))^2 + d^2}} L \tag{1}$$

Similar to a screw, the force transmitted from the anchored mature zone to the advancing tip is achieved by exploiting the helical deposition pathway of the filament material. If the deposited filament material from one complete turn is unwound, the helix angle α (the angle made by the helix of the deposited material with a plane perpendicular to the axis of the tubular body) can be related to the geometrical dimensions of the system:

$$\tan\alpha = \frac{d}{\pi(D+d)} \tag{2}$$

The two main sources of friction in the growing zone related to the passage and the deposition of the filament are the contact between the filament and the internal part of the nozzle and the contact between the filament and the rim of the deposition head, which can be considered a unique source of friction. The vertical resistance W is the result of soil resistance W_s and frictional losses W_f related to the slipping movement of the tubular body inside the mature zone. If we consider the equilibrium of the horizontal and vertical forces acting at a point on the newly deposited section of filament (Fig. 4b), then the torque M required by the deposition head to overcome the vertical resistance W caused by friction between the filament and the deposition head (with the coefficient of friction μ) is:

$$M = W \cdot \frac{D+d}{2} \cdot \frac{\mu + \tan\alpha}{1 - \mu\tan\alpha} = (W_s + W_f) \cdot \frac{D+d}{2} \cdot \frac{\mu + \tan\alpha}{1 - \mu\tan\alpha} \tag{3}$$

If a complete deposition cycle is considered and the vertical resistance is assumed to be locally constant (it increases for deeper layers), the efficiency of the deposition process is given by:

$$\eta_{deposition} = \frac{Work_{out}}{Work_{in}} = \frac{W \cdot d}{F \cdot \pi \cdot (D+d)} = \frac{\tan\alpha \cdot (1 - \mu\tan\alpha)}{\mu + \tan\alpha} \tag{4}$$

Thus, according to (2), the efficiency of the deposition mechanism depends on the diameter D of the tubular body, the diameter d of the material filament and the friction between the filament and the deposition head. The efficiency of the total actuation system (η_{tot_act}) is the result not only of the deposition process but also of the performances of the motor (η_{motor}) and of the transmission mechanism ($\eta_{transmission}$) as follows:

$$\eta_{tot\_act} = \eta_{motor} \times \eta_{transmission} \times \eta_{deposition} \quad (5)$$

The rotational movement of the deposition nozzle creates torsion in the filament that can cause unwanted nodes or curves. The torsion is countervailed by a rotary movement of the external spool on which the filament is wound.

The helical deposition of the filament in our system has similarities with a screw-based power transmission mechanism. Because of the large load carrying capacity, screws have traditionally been used as a mechanism to convey removed materials in tunneling and drilling [2]. Moreover, screws have been exploited as self-tapping and drill-free penetration systems. For example, in [7], the penetration into the soil occurs because of a strong interaction between the helical thread and the substrate: the movement of the whole body with respect to the environment determines an increase in the energy needed for penetration and this increase becomes larger for deeper probe penetration. For the proposed device, the similarity with a screw is exclusively related to the conversion of a rotary motion to a linear motion and the transmission of power. By contrast, from the tip-soil interaction viewpoint, our device penetrates by soil compaction around the tip, and no mass is moving along the axis of the system (i.e., soil removal along the borehole). In our case, the penetration process benefits from the absence of peripheral friction along the mature zone.

System Prototype

Based on the design proposed in the previous section (Fig. 3), a prototype of the root-like robotic system was developed and tested. The robotic root can be used to imitate the growth zone in actual plant roots and validate its role in root growth. The material used to grow the system is a polypropylene (PP) filament (nominal diameter $d = 2.5$ mm). The deposition head (external diameter $D_{hs} = 57$ mm) and the tubular body (external diameter $D = 50$ mm) are constructed of polytetrafluoroethylene (PTFE) to decrease the frictional effects during the filament deposition. A groove (width = 5 mm, length = 27 mm, depth = 4 mm) in the deposition head acts as a nozzle for guiding the filament onto the outer surface of the tubular body (Fig. 5, Video S2). Through the rim (thickness = 3.5 mm) created by the deposition head and the tubular body on the tip of the system, the deposited filament creates the penetration action. The rotation of the deposition head is driven by a 2.37 W motor with a 246:1 gear ratio (FAULHABER GMBH & CO.KG, Shonaich, Germany) coupled to a transmission mechanism containing a planetary gearset. The transmission mechanism consists of three spur gears that are mounted on a carrier disc and engage an internal gear (number of teeth for the internal gear = 37, number of teeth for the spur gears = 12). The gears and carrier were fabricated from Plexiglas using a laser cutting process (VersaLaser VLS3.50, Universal Laser Systems, Scottsdale, AZ). The internal gear is fixed to the bottom of the tubular body. The motor, which is mounted on the top of the carrier, turns one of the spur gears and determines the rotation of the carrier and the deposition head (Fig. 5b).

The locking wires are made from spring steel wire and mounted in the upper edge of the tubular body (Fig. 5b). The ends of these wires protrude into the mature zone and eliminate any relative rotational movements between the mature zone and the tubular body while allowing axial slipping.

Figure 6 shows the sequence of movements of the system as it grows.

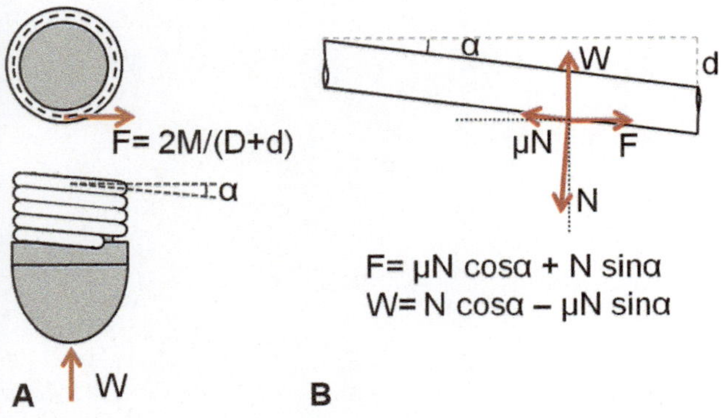

Figure 4. Force diagram for overcoming soil resistance. a) Schematic top view and side view of the growing mechanism. Red arrows represent forces acting on the system during the deposition process: F is the force applied by the motor and W is the vertical resistance during penetration, where M is the torque required for the deposition head (shown in Fig. 3b) to overcome W, d is the thickness of the filament and D is the external diameter of the tubular body b) Equilibrium of forces acting on the filament for one complete turn unwound, where α is the angle made by the helix of the deposited filament with respect to a plane perpendicular to the axis of the tubular body, N is the reaction force, and μ is the friction coefficient between filament and deposition head.

Validation

As reported in Materials and Methods, the prototype was tested in different depths and penetration conditions (EFT and NoEFT penetrations). Data are reported as means ± standard deviations (Fig. 7). For all the initial depths tested, the energy required for penetration was lower with EFT than with NoEFT (Fig. 7a). Because of the increasing soil pressure, the required energy increased with the initial depth. Notably, the increase in energy for the EFT penetrations had a gradual slope, from 54.7±5.5 J at a depth of 100 mm to 76.3±13.9 J at 200 mm, whereas for NoEFT, the results showed a rather constant and sharp slope, with energies between 98.8±7.8 J at 100 mm and 244.9±39.9 J at 200 mm. The reduction in energy consumption with EFT ranged from 45% at 100 mm depth to 70% at 200 mm depth. The standard deviations of the results, described by error bars in Fig. 7, increased with the initial depth.

Corresponding to greater depths, the penetration speed decreased in both types of penetration because of the higher mechanical impedance of the artificial soil. The EFT penetration speeds, averaged over the penetration time, were approximately 30–40% higher than the NoEFT speeds (Fig. 7b).

Discussion

The ability of roots to penetrate soil appears to be mainly attributable to growth at the apical region. In fact, if penetration of the root into the soil resulted from a force caused by growth of the root in the region proximal to the stem, the root would have to generate higher penetration forces because of the increased friction on the flanks. Furthermore, the deformation of the root structure due to the buckling and mechanical bending would compromise the root steering capabilities essential for the exploration. From a biological perspective, a hypothetical root that grows from the stem would be subject to a higher risk of tissue damage caused by both friction and buckling. Thus, root growth from the tip is a beneficial adaptation to the soil environment.

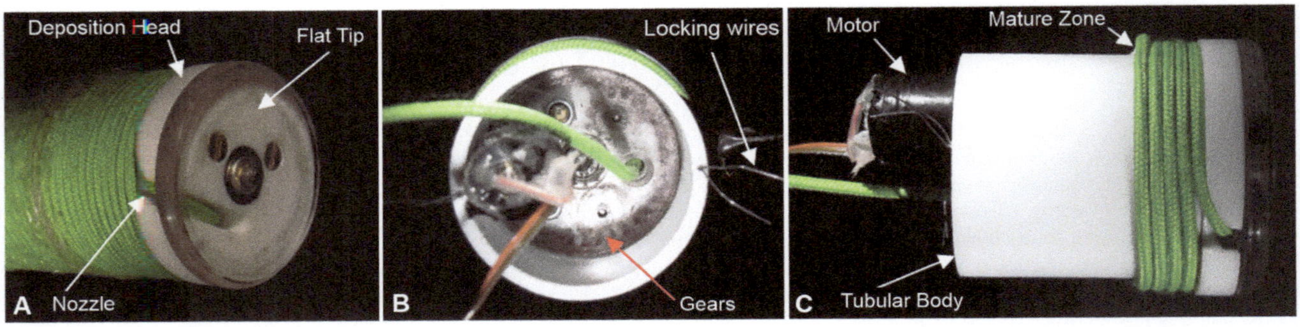

Figure 5. System Prototype. Figure (a) shows a picture of the root-like growing device. In the deposition head, a decoupled flat tip is assembled in order to eliminate any drilling effect in the penetration process. Figures (b) and (c) show the top and side views, respectively, of the device at the initiation of the mature zone. From the spool (not shown), the filament is drawn by the motor through the filament entrance into the nozzle. The locking wires are spring steel and prevent relative rotational movement between the tubular body and the mature zone.

Although growth is essential for the optimal interaction of the root with the soil, its contribution to soil penetration is difficult to quantify in living roots. To the best of our knowledge, studies concerning penetration capabilities have mainly addressed the rheological properties of the apex and its role in friction reduction [53,55], morphological changes [43,54,74], and root penetration pressure [49,66,75]. In the proposed biomimetic approach, the root-like growing device represents a translation (simplification) of biological concepts into an engineering system for soil penetration and exploration. This artificial growing system is able to create its own structure with a customized deposition technique. Similar to the probes commonly used for soil exploration, such as penetrometers [76,77], our device is able to create and follow a path into the soil without deviation or buckling: the self-developed structure anchors into the soil and provides a thrust force localized at the tip level. This device could provide a penetration resistance comparable to that experienced by growing roots. Moreover, the total thrust force can be used to surmount axial soil resistance without being dissipated by lateral friction: similar to living roots, the growing device succeeds in efficiently penetrating a granular medium. The experimental trials comparing EFT and NoEFT penetration demonstrated that the device penetrates soil faster and with less energy consumption using EFT penetration. We found lower energy consumption (from 45% to 70%) using EFT compared with NoEFT for initial depths ranging between 100 mm and 200 mm. This characteristic, obtained by the growing approach, is the key feature of our device, which is intended to perform an autonomous and nondestructive exploration of the shallow layers of soil.

In a previous study [45], we demonstrated the effect and contribution of the apex morphology in plant root soil penetration. The root-like device in the present study does not currently include a tip with an optimal shape for penetrating a granular medium, nor is it equipped with a sloughing mechanism such as that described in [72]. Therefore, the efficiency of this growing system can be increased by adding these two features. If supported by the growing mechanism presented here, a sloughing tip would burrow into soil with no peripheral friction because of the stationary interaction of the mature zone with the peripheral soil. Moreover, by depositing the filament asymmetrically around the outer surface

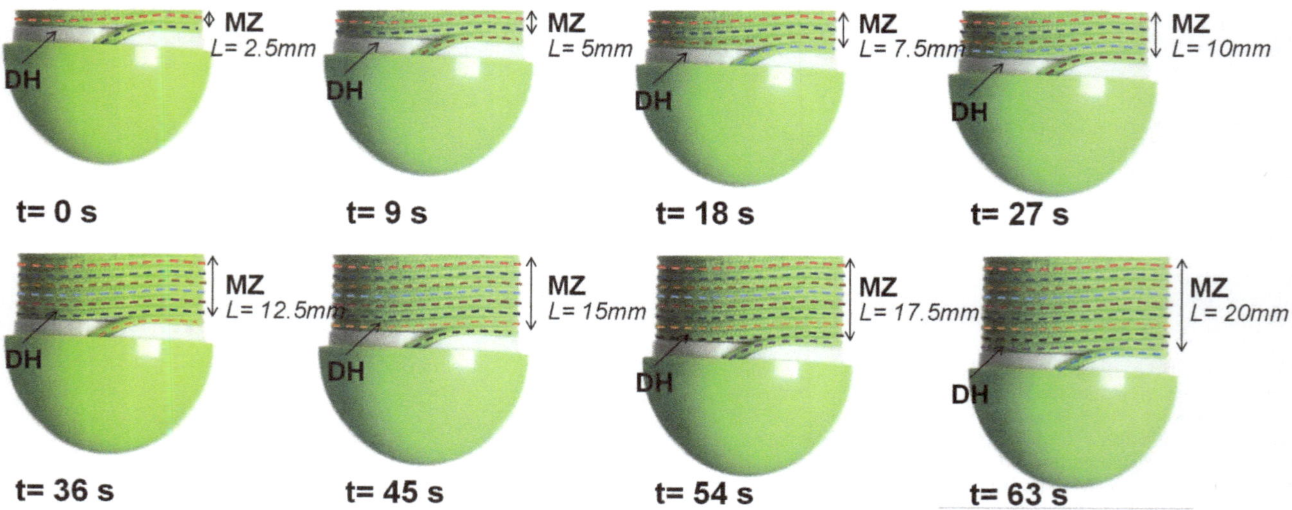

Figure 6. Deposition process and growth of the root-like device. The figure shows the sequence of the growth process. By rotating the deposition head (DH, in white), the filament passes through the nozzle and is deposited around the outer surface of the tubular body (shown in Figs. 3 and 5c). The penetration force is provided by the deposition of the filament at the growing zone (GZ), while the mature zone remains strongly anchored to the soil. The newly deposited filament is always located on the top of the DH, as shown by the colors.

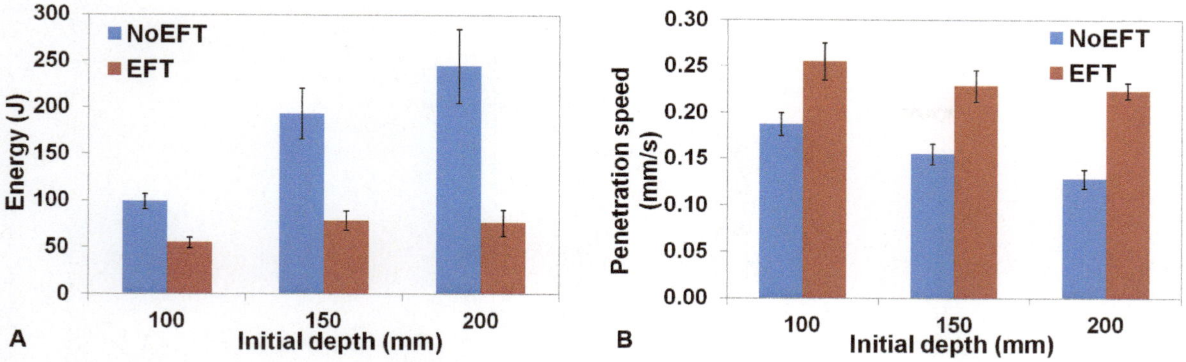

Figure 7. Energy consumption of the root-like device prototype in EFT and NoEFT penetration trials. Figure (a) shows the energy required for penetrating 30 mm of artificial soil (POM granules, diameter = 4 mm) in EFT (Red) and NoEFT trials (blue) at various initial depths (100, 150, 200 mm). For each initial depth, 5 tests were performed. Bars are the mean values over 5 repetition for each depth and condition (EFT and NoEFT). Error bars represent the standard error of the mean. Figure (b) shows the penetration speed achieved. The experimental conditions are described in the "Materials and Methods" section and shown in Fig. 8.

of the tubular body, bending can be achieved, which would enhance the exploration capabilities of the growing robot.

The proposed system could be used as an autonomous tunneling system, especially for granular substrates (e.g., deserts, regolith, or debris after natural disasters), or for space applications.

This work is based on a breakthrough concept of imitating the growth process in robots. Up to now, the concepts of reconfiguration have been developed in the self-reconfigurable robots, in which pre-existing/predefined modules are used to flexibly build and vary their structure [78–81]. These discrete modules can attach to each other by means of magnetic [82], mechanical [83], and passive chemical [84] bonding connectors. This powerful idea allows the system to adapt to its environment and is highly versatile for task orientation. Some unconventional materials and techniques have been used to progress beyond the lattice-type structure of conventional reconfigurable robots. In [85], the modules are linked and attached together with a second robot that injects foam to construct the desired structure. In [86], the concept of 'robotic body extension' was demonstrated by using an additive layering technique based on thermoplastic adhesives (TPAs). Unlike the conventional rapid prototyping processes, the structures fabricated in the present study can be reshaped and reconfigured by controlling a physical parameter, such as temperature. The flexibility provided by other additive manufacturing techniques, including shape deposition manufacturing (SDM) [87] and smart composite microstructure (SCM) [88], have also been extensively used to simultaneously fabricate and assemble the components of devices in an integrated manner. By contrast with the previous research, the proposed root-like device grows through a monotonic process that continuously adds new material. The technological approach used to grow the mature body is also responsible for the penetration function because the robot is pushing the soil while it is growing. Thus, the creation process has not only a functional meaning but also a structural one, and both of these features are implemented simultaneously on board the robotic system itself. The concept of a growing robot may spawn new approaches in assembly and robotics by incorporating the manufacturing technology inside the robot. Self-reconfigurable structures could use the growth process to construct themselves directly on the site of interest.

Being designed on the basis of biological functions, the root-like growing robot and its future developments can be applied as tools to validate hypotheses on plant growth that may be difficult to demonstrate *in vivo*. The full potential of biomimetic robotics remains to be explored and offers interesting and demanding research challenges.

Materials and Methods

Experimental Setup for Soil Penetration

The prototype with a flat tip was tested in a cohesionless granular medium consisting of polyoxymethylene (Ultraform N2320 003, BASF, Ludwigshafen, Germany) plastic beads (diameter = 4 mm). Repeatable results can be obtained with this substrate because of its low moisture absorption (0.2% under experimental conditions of $T = 23°C$ and 50% relative humidity), favorable mechanical properties and wear resistance properties [89]. The artificial soil was kept in a cylindrical container (height = 250 mm, diameter = 200 mm) at room temperature ($T = 21 \pm 1.5°C$) and had a bulk density of 870 g/l and a void ratio of 0.61.

Under these conditions, by using a 30° circular stainless steel cone (base area 323 mm^2 and 20.27 mm diameter) with a drive shaft, as defined by ASAE S313.3 [13], the artificial soil strength was characterized by measuring the resistance of the soil to penetration by a cone driven at a constant rate [90,91]. A resulting cone index (i.e., the force of insertion per unit cone base area, with the number of repetitions = 20 and the standard deviation <10%) of 0.017, 0.033, or 0.059 MPa at 100, 150, or 200 mm depths, respectively, was measured according to the standard ASAE EP542 procedure [92].

A dedicated setup was developed to perform the EFT and NoEFT penetration trials. This approach guaranteed that the effects of parameters such as manufacturing, materials, and components were equal in both experimental cases. In the EFT penetration tests, the mature zone (MZ) of the device was fixed to the holding structure (HS, Fig. 8a, Video S3), and the growing zone (GZ) was positioned in the artificial soil at a prescribed initial depth (h_0). In the NoEFT penetration tests (Fig. 8b, Video S3), the entire growing system (MZ+GZ) was fixed externally to the soil, and another probe (Fig. 8b) was added to the growing tip. This new probe and tip were geometrically equal to the growing system (MZ+GZ) used in the EFT penetration tests. The probe was inserted into the soil with the tip at the prescribed initial depth h_0 and was driven by the external growing system.

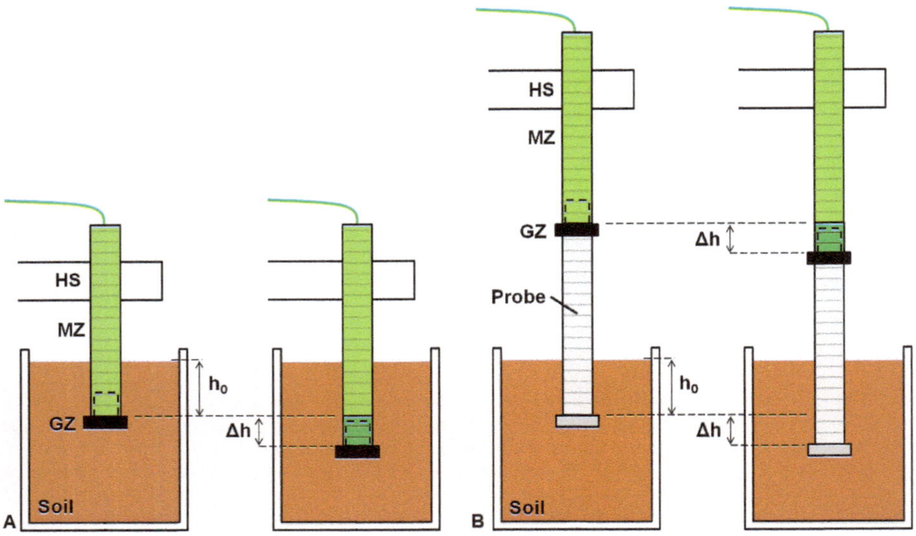

Figure 8. Schematic of the experimental setup for soil penetration with and without elongation from the tip (EFT and NoEFT, respectively). Figure (a) shows the setup for the evaluation of the energy required in the EFT tests, at the beginning (on the left) and at the end (on the right) of a soil penetration trial. In the initial position, the device is positioned in the soil. The growing zone (GZ) is placed at depth h_0 ($h_0 = 100$, 150, 200 mm), and the mature zone (MZ) is fixed to a holding structure (HS). The GZ adds new material at the tip, thus increasing the MZ length until it penetrates a total of Δh (30 mm). Figure (b) shows the setup for the evaluation of the energy required in the NoEFT tests, at the beginning (on the left) and at the end (on the right) of a soil penetration trial. In the initial position, a probe (F, light grey) with the same dimensions and geometric properties as the prototype is positioned at depth h_0 in the soil ($h_0 = 100$, 150, 200 mm). The prototype is attached to the HS and pushes the probe from the top. The GZ generated the vertical (penetration) force in both the EFT and NoEFT trials in order to have comparable results.

Energy Estimation in EFT and NoEFT Trials

The energy consumption of the penetration probe was calculated in both the EFT and NoEFT trials. In the former, the device penetrated axially by growing from the tip while the peripheral body of the device remained stationary relative to the soil. In the latter, a probe with same shape and dimensions as the device was pressed from the top; i.e., the entire body of the probe moved relative to the soil. For both the cases, the penetration tips were inserted into the soil at an initial depth h0 (100, 150 and 200 mm), and in each experiment the energy required to penetrate 30 mm soil was measured. To calculate the energy consumed, a precision power source/measurement unit (Agilent B2912A) with the capability to control and measure both voltage and current was used to measure the motor current and the time while the voltage was held constant (V = 4 V); the data were recorded on a PC. For each initial depth, 5 tests were performed; the mean values of energy consumed and standard deviations were calculated. The soil was redistributed after each penetration trial to restore the initial packing state of the grains (parameterized by bulk density and void ratio [93,94], which was locally changed because of the tip advancement and the consequent soil compaction during penetration.

Supporting Information

Video S1 Schematic animation of elongation from the tip (EFT) process.

Video S2 System Prototype. The sequences are speeded up 2X.

Video S3 Experimental Setup for Soil Penetration with and without elongation from the tip (NoEFT). The sequences are speeded up 8X.

Author Contributions

Conceived and designed the experiments: AS AT. Performed the experiments: AS AT. Analyzed the data: AS AT. Contributed reagents/materials/analysis tools: AS. Wrote the paper: AS AT LP BM. Design and development of robotic platform: AS.

References

1. Maurer WC (1980) Advanced drilling techniques. The Petroleum Publishing Company, Tulsa, Oklahoma.
2. Bar-Cohen Y, Zacny K (2009) Drilling in extreme environments. John Wiley & Sons.
3. Cooper GA (1994) Directional drilling. Sci Am 270: 56–61.
4. Ylikorpi T, Visentin G, Suomela J (2001) A robotic rover-based deep driller for Mars exploration. Proc of 35th Aerospace Mechanisms Symp.
5. Zacny K, Paulsen G, Chu P, Avanesyan A, Craft J, et al. (2012) Mars drill for the Mars sample return mission with a brushing and abrading bit, regolith and powder bit, core PreView Bit and a coring bit. Proc of IEEE Aerospace Conf: 1–8.
6. Kömle N, Kaufmann E, Kargl G, Gao Y, Rui X (2008) Development of thermal sensors and drilling systems for lunar and planetary regoliths. Adv Space Res 42: 363–368.
7. Nagaoka K, Kubota T, Otsuki M, Tanaka S (2008) Experimental study on autonomous burrowing screw robot for subsurface exploration on the Moon. IEEE RSJ 2008: 4104–4109.
8. Muñoz-Carpena R, Shukla S, Morgan K (2004) Field devices for monitoring soil water content. University of Florida Cooperative Extension Service, Institute of Food and Agricultural Sciences, EDIS.
9. Simon C, Goyallon D, Poloce JP, Sellami H, Gerbaud L, et al. (2007) Drilling without WOB, dream or reality? An effective field test by Total Angola. SPE/IADC Drilling Conf.

10. Zacny K, Quayle M, McFadden M, Neugebauer A, Huang K, et al. (2002) A novel method for cuttings removal from holes during percussive drilling on Mars. Revolutionary Aerospace Systems Concepts-Academic Linkage RASC-AL: 107–121.
11. Rykwalder PR, Blanford WJ (2005) Vadose zone monitoring techniques. Water Encyclopedia. John Wiley & Sons.
12. Bengough A, Mullins C (1990) The resistance experienced by roots growing in a pressurised cell. Plant and Soil 123: 73–82.
13. ASABE Standards, ASAE S313.3 (1999) Soil cone penetrometer. St. Joseph, MI, USA.
14. Lunne T, Berre T, Strandvik S (1997) Sample disturbance effects in soft low plastic Norwegian clay. Symp on Recent Developments in Soil and Pavement Mechanics: 81–102.
15. Fitzgerald M, Elsworth D (2010) Evolution of the pore-pressure field around a moving conical penetrometer of finite size. J Eng Mech 136(3): 263–272.
16. Ani AO, Mbajiorgu CC, Akubuo CO, Onwualu PA (2011) Proctor cone penetrometer for in-situ soil strength studies in Nigeria. Int J Agric & Biol Eng 4(3): 16–25.
17. Herrick JE, Jones TL (2002) A dynamic cone penetrometer for measuring soil penetration resistance. Am J Soil Sci Society 66(4): 1320–1324.
18. Whiteley GM, Utomo WH, Dexter AR (1981) A comparison of penetrometer pressures and the pressures exerted by roots. Plant and Soil 61(3): 351–364.
19. Bengough AG, Mullins CE, Wilson G, Wallace J (1991) The design, construction and use of a rotating-tip penetrometer. J Agr Eng Res 48: 223–227.
20. Tumay Kurup (2001) The design and operation of a novel miniature cone penetration test system. Electronic J of Geotechnical Eng 6.
21. Bar-Cohen Y (2006) Biomimetics - using nature to inspire human innovation. Bioinspir Biomim 1: 1–12.
22. Gouachea TP, Gaoa Y, Costec P, Gourinatb Y (2011) First experimental investigation of dual-reciprocating drilling in planetary regoliths: proposition of penetration mechanics. Planetary Space Sci 59: 1529–1541.
23. Gao Y, Ellery A, Sweeting M, Vincent J (2007) Bioinspired drill for planetary sampling: literature survey, conceptual design, and feasibility study. J Spacecraft and Rockets 44: 703–709.
24. Menon C, Vincent JFV, Lan N, Bilhaut L, Ellery A, et al. (2006) Bio-inspired micro-drills for future planetary exploration. Proc. of CANEUS Toulouse France.
25. Winter AG, Hosoi A (2011) Identification and evaluation of the Atlantic razor clam (Ensis directus) for biologically inspired subsea burrowing systems. Integrative and Comparative Biology 51: 151–157.
26. Gilroy S, Masson PH (2008) Plant tropisms. Wiley Online Library.
27. Arnaud C, Bonnot C, Desnos T, Nussaume L (2010) The root cap at the forefront. Comptes Rendus Biologies 333: 335–343.
28. Baluska F, Mancuso S, Volkmann D, Barlow PW (2010) Root apex transition zone: a signalling-response nexus in the root. Trends in Plant Science 15: 402–408.
29. Bengough AG, Bransby MF, Hans J, McKenna SJ, Roberts TJ, et al. (2006) Root responses to soil physical conditions; growth dynamics from field to cell. J Exp Bot 57: 437.
30. Hochholdinger F, Tuberosa R (2009) Genetic and genomic dissection of maize root development and architecture. Curr Opin Plant Biol 12: 172–177.
31. Zobel RW (1991) Root growth and development. The rhizosphere and plant growth. Springer: 61–71.
32. Darwin C (1880) The power of movement in plants. John Murray, London.
33. Inoue N, Arase T, Hagiwara M, Amano T, Hayashi T, et al. (1999) Ecological significance of root tip rotation for seedling establishment of Oryza sativa L. Ecol Res 14: 31–38.
34. Brown AH (1993) Circumnutations: from Darwin to space flights. Plant Physiol 101: 345–348.
35. Vollsnes A, Futsaether C, Bengough A (2010) Quantifying rhizosphere particle movement around mutant maize roots using time lapse imaging and particle image velocimetry. Europ J Soil Sci 61: 926–939.
36. Verbelen J-P, De Cnodder T, Le J, Vissenberg K, Baluška F (2006) Root apex of Arabidopsis thaliana consists of four distinct zones of growth activities: meristematic zone, transition zone, fast elongation zone, and growth terminating zone. Plant Sig Behav 1: 296–304.
37. Dexter A (1987) Mechanics of root growth. Plant Soil 98: 303–312.
38. Baluska F, Mancuso S, Volkmann D, Barlow P (2004) Root apices as plant command centres: the unique 'brain-like' status of the root apex transition zone. Biologia 59: 7–19.
39. Ishikawa H, Evans ML (1995) Specialized zones of development in roots. Plant Physiol 109: 725.
40. Barlow PW (2002) The root cap: cell dynamics, cell differentiation and cap function. J Plant Growth Reg 21: 261–286.
41. Barlow PW (1993) The response of roots and root systems to their environment–An interpretation derived from an analysis of the hierarchical organization of plant life. Env Exp Bot 33: 1–10.
42. Unger PW, Kaspar TC (1994) Soil compaction and root growth: a review. Agronomy J 86: 759–766.
43. Iijima M, Morita S, Barlow PW (2008) Structure and function of the root cap. Plant Production Sci 11: 17–27.
44. Iijima M, Higuchi T, Barlow PW (2004) Contribution of root cap mucilage and presence of an intact root cap in maize (Zea mays) to the reduction of soil mechanical impedance. Annals of Bot 94: 473–477.
45. Tonazzini A, Popova L, Mattioli F, Mazzolai B (2012) Analysis and characterization of a robotic probe inspired by the plant root apex. IEEE BioRob 2012: 1134–1139.
46. Tonazzini A, Sadeghi A, Popova L, Mazzolai B (2013) Plant root strategies for robotic soil penetration. Biomimetic and Biohybrid Systems: 447–449.
47. Schopfer P (2006) Biomechanics of plant growth. Am J of Bot 93: 1415–1425.
48. Croser C, Bengough AG, Pritchard J (2000) The effect of mechanical impedance on root growth in pea (Pisum sativum). II. Cell expansion and wall rheology during recovery. Physiol Plantarum 109: 150–159.
49. Clark L, Whalley W, Barraclough P (2003) How do roots penetrate strong soil? Plant and Soil 255: 93–104.
50. Greacen E, Oh J (1972) Physics of root growth. Nature 235: 24–25.
51. Barley K, Greacen E (1967) Mechanical resistance as a soil factor influencing the growth of roots and underground shoots. Advances in Agronomy 19: 1–43.
52. Hawes MC, Bengough G, Cassab G, Ponce G (2002) Root caps and rhizosphere. J Plant Growth Reg 21: 352–367.
53. Bengough A, McKenzie B (1997) Sloughing of root cap cells decreases the frictional resistance to maize (Zea mays L.) root growth. J Exp Bot 48: 885.
54. Iijima M, Higuchi T, Barlow PW, Bengough AG (2003) Root cap removal increases root penetration resistance in maize (Zea mays L.). J Exp Bot 54: 2105.
55. Bengough A, Kirby J (1999) Tribology of the root cap in maize (Zea mays) and peas (Pisum sativum). New phytologist 142: 421–425.
56. Goodman A, Ennos A (1999) The effects of soil bulk density on the morphology and anchorage mechanics of the root systems of sunflower and maize. Annals of Botany 83: 293–302.
57. Bengough AG, McKenzie B, Hallett P, Valentine T (2011) Root elongation, water stress, and mechanical impedance: a review of limiting stresses and beneficial root tip traits. J Exp Bot 62: 59–68.
58. Czarnes S, Hiller S, Dexter A, Hallett P, Bartoli F (1999) Root-soil adhesion in the maize rhizosphere: the rheological approach. Plant and Soil 211: 69–86.
59. Bingham I, Bengough A (2003) Morphological plasticity of wheat and barley roots in response to spatial variation in soil strength. Plant and Soil 250: 273–282.
60. Iijima M, Barlow PW, Bengough AG (2003) Root cap structure and cell production rates of maize (Zea mays) roots in compacted sand. New phytologist 160: 127–134.
61. Bengough A, Croser C, Pritchard J (1997) A biophysical analysis of root growth under mechanical stress. Plant Roots-From Cells to Systems: 107–116.
62. Materechera S, Alston A, Kirby J, Dexter A (1992) Influence of root diameter on the penetration of seminal roots into a compacted subsoil. Plant and Soil 144: 297–303.
63. Clark LJ, Ferraris S, Price AH, Whalley WR (2008) A gradual rather than abrupt increase in soil strength gives better root penetration of strong layers. Plant and Soil 307: 235–242.
64. Abdalla A, Hettiaratchi D, Reece A (1969) The mechanics of root growth in granular media. J Agr Eng Res 14: 236–248.
65. Chen G, Weil RR (2010) Penetration of cover crop roots through compacted soils. Plant and Soil 331: 31–43.
66. Bengough A, Mullins C (1990) Mechanical impedance to root growth: a review of experimental techniques and root growth responses. J Soil Sci 41: 341–358.
67. Bengough A (1997) Modelling rooting depth and soil strength in a drying soil profile. J Theoretical Bio 186: 327–338.
68. Kirby J, Bengough A (2002) Influence of soil strength on root growth: experiments and analysis using a critical-state model. Europ J Soil Sci 53: 119–127.
69. Lobet G, Pagès L, Draye X (2011) A novel image-analysis toolbox enabling quantitative analysis of root system architecture. Plant Physiology 157: 29–39.
70. French A, Ubeda-Tomás S, Holman TJ, Bennett MJ, Pridmore T (2009) High-throughput quantification of root growth using a novel image-analysis tool. Plant Physiol 150: 1784–1795.
71. French AP, Bennett M, Howells C, Patel D, Pridmore TP (2008) A probabilistic tracking approach to root measurement in images-particle filter tracking is used to measure roots, via a probabilistic graph. Biosignals 1: 108–115.
72. Sadeghi A, Tonazzini A, Popova L, Barbara M (2013) Innovative robotic mechanism for soil penetration inspired by plant roots. Proc in IEEE ICRA 2013.
73. Crump SS (1992) Apparatus and method for creating three-dimensional objects. US Patent 5,121,329.
74. Iijima M, Higuchi T, Barlow PW (2004) Contribution of root cap mucilage and presence of an intact root cap in maize (Zea mays) to the reduction of soil mechanical impedance. Ann Bot 94: 473.
75. Bengough A, Mullins C (1991) Penetrometer resistance, root penetration resistance and root elongation rate in two sandy loam soils. Plant and Soil 131: 59–66.
76. Minasny B (2012) Contrasting soil penetration resistance values acquired from dynamic and motor-operated penetrometers. Geoderma 177: 57–62.
77. Xu X, Lehane BM (2008) Pile and penetrometer end bearing resistance in two-layered soil profiles. Géotechnique 58(3): 187–197.
78. Murata S, Yoshida E, Tomita K, Kurokawa H, Kamimura A, et al. (2000) Hardware design of modular robotic system. Proc Int Conf on Intelligent Robots and Systems (IROS 2000) IEEE/RSJ 3: 2210–2217.
79. Fukuda T, Kawauchi Y (1990) Cellular robotic system (CEBOT) as one of the realization of self-organizing intelligent universal manipulator. IEEE: 662–667.

80. Lipson H, Pollack JB (2000) Automantic design and manufacture of robotic lifeforms. Nature 406: 6799.
81. Jones R, Haufe P, Sells E, Iravani P, Olliver V, et al. (2011) Reprap-the replicating rapid prototyping. Robotica 29(1): 177–191.
82. Suh JW, Homans SB, Yim M (2002) Telecubes: mechanical design of a module for self reconfigurable robotics. Proc IEEE Int Conf on Robotics and Automation ICRA'02: 4095–4101.
83. Jorgensen MW, Ostergaard EH, Lund HH (2004) Modular ATRON: Modules for a self-reconfigurable robot. Proc Int Conf on Intelligent Robots and Systems IEEE/RSJ IROS 2: 2068–2073.
84. Boncheva M, Bruzewicz DA, Whitesides GM (2003) Millimeter-scale self-assembly and its applications. Pure and Applied Chemistry 75(5): 621–630.
85. Campbell M (2011) Robot builds its own body from sprayable foam. New Scientist 212: 24.
86. Iida F, Wang L, Brodbeck L (2012) Self-reconfiguration robots based on thermoplastic adhesives. Proc Int Symp on Distributed Autonomous Robotic Systems (DARS 2012).
87. Bailey SA, Cham JG, Cutkosky MR, Full RJ (2000) Biomimetic robotic mechanisms via shape deposition manufacturing. Robotics Research-International Symposium 9: 403–410.
88. Wood RJ, Avadhanula S, Sahai R, Steltz E, Fearing RS (2008) Microrobot design using fiber reinforced composites. J Mech Design 130: 052304.
89. Sun LH, Yang ZG, Li XH (2008) Study on the friction and wear behavior of POM/Al2O3 nanocomposites. Wear 264: 693–700.
90. Lunne T, Robertson PK, Powell JJM (1997) Cone penetration testing. Geotechnical Practice.
91. Smith CW, Johnston MA, Lorentz S (1997) The effect of soil compaction and soil physical properties on the mechanical resistance of South African forestry soils. Geoderma 78: 93–111.
92. ASAE EP542 (1999) Procedures for using and reporting data obtained with the soil cone penetrometer. St. Joseph, MI.
93. Maladen RD, Ding Y, Umbanhowar PB, Kamor A, Goldman DI (2011) Mechanical models of sandfish locomotion reveal principles of high performance subsurface sand-swimming. J. R. Soc. Interface 8(62): 1332–1345.
94. Ding Y, Gravish N, Goldman DI (2011) Drag induced lift in granular media. Physical Review Letters 106: 028001.

3

Effects of Different Tillage and Straw Return on Soil Organic Carbon in a Rice-Wheat Rotation System

Liqun Zhu[1]*, Naijuan Hu[1], Minfang Yang[2], Xinhua Zhan[2], Zhengwen Zhang[1]

1 College of Agriculture, Nanjing Agricultural University, Nanjing, China, **2** College of Resources and Environmental Science, Nanjing Agricultural University, Nanjing, China

Abstract

Soil management practices, such as tillage method or straw return, could alter soil organic carbon (C) contents. However, the effects of tillage method or straw return on soil organic C (SOC) have showed inconsistent results in different soil/climate/cropping systems. The Yangtze River Delta of China is the main production region of rice and wheat, and rice-wheat rotation is the most important cropping system in this region. However, few studies in this region have been conducted to assess the effects of different tillage methods combined with straw return on soil labile C fractions in the rice-wheat rotation system. In this study, a field experiment was used to evaluate the effects of different tillage methods, straw return and their interaction on soil total organic C (TOC) and labile organic C fractions at three soil depths (0–7, 7–14 and 14–21 cm) for a rice-wheat rotation in Yangzhong of the Yangtze River Delta of China. Soil TOC, easily oxidizable C (EOC), dissolved organic C (DOC) and microbial biomass C (MBC) contents were measured in this study. Soil TOC and labile organic C fractions contents were significantly affected by straw returns, and were higher under straw return treatments than non-straw return at three depths. At 0–7 cm depth, soil MBC was significantly higher under plowing tillage than rotary tillage, but EOC was just opposite. Rotary tillage had significantly higher soil TOC than plowing tillage at 7–14 cm depth. However, at 14–21 cm depth, TOC, DOC and MBC were significantly higher under plowing tillage than rotary tillage except for EOC. Consequently, under short-term condition, rice and wheat straw both return in rice-wheat rotation system could increase SOC content and improve soil quality in the Yangtze River Delta.

Editor: Shuijin Hu, North Carolina State University, United States of America

Funding: This work was supported by Special Fund for Agro-scientific Research in the Public Interest (201103001). The funders had no role in study design, data collection and analysis, decision to publish, or preparation of the manuscript.

Competing Interests: The authors have declared that no competing interests exist.

* E-mail: zhulq@njau.edu.cn

Introduction

Soil organic carbon (C) has profound effects on soil physical, chemical and biological properties [1]. Maintenance of soil organic C (SOC) in cropland is important, not only for improvement of agricultural productivity but also for reduction in C emission [2]. However, short- and medium-term changes of SOC are difficult to detect because of its high temporal and spatial variability [3]. On the contrary, soil labile organic C fractions (i.e., microbial biomass C (MBC), dissolved organic C (DOC), and easily oxidizable C (EOC)) that turn over quickly can respond to soil disturbance more rapidly than total organic C (TOC) [1,3,4]. Therefore, these fractions have been suggested as early sensitive indicators of the effects of land use change on soil quality (e.g. [3,5,6]).

Agricultural practices such as tillage methods are conventionally used for loosening soils to grow crops. But long-term soil disturbance by tillage is believed to be one of the major factors reducing SOC in agriculture [7]. Frequent tillage may destroy soil organic matter (SOM) [8] and speed up the movement of SOM to deep soil layers [9]. As a consequence, agricultural practices that reduce soil degradation are essential to improve soil quality and agricultural sustainability. Crop residue plays an important role in SOC sequestration, increasing crop yield, improving soil organic matter, and reducing the greenhouse gas (e.g. [10–13]). As an important agricultural practice, straw return is often implemented with tillage in the production process. Although numerous studies have indicated that tillage methods combined with straw return had a significant effect on labile SOC fractions, the results varied under different soil/climate conditions. For example, both no-tillage and shallow tillage with residue cover had significantly higher SOC than conventional tillage without residue cover in Loess Plateau of China [14], while Wang et al. [15] reported that the difference between the treatments of plowing with straw return and no-tillage with straw return on TOC in central China was not significant. Rajan et al. [2] showed that in Chitwan Valley of Nepal, no-tillage with crop residue application at upper soil depth had distinctly higher SOC sequestration than conventional tillage with crop residue. The effects of tillage on soil labile organic C vary with regional climate [16], soil condition (e.g. [17–20]), residue management practice, and crop rotation (e.g. [21,22]). Therefore, the investigation on soil labile organic C for specific soil, climate, and cropping system is necessary to improve the soil quality.

The Yangtze River Delta of China is the main production region of rice and wheat, and rice-wheat rotation is the most important cropping system in this region [23]. The total sown area of rice and wheat in the Yangtze River Delta accounted for about 20.1% of that in China in 2011, and the total yield was 22.1% of the national yield for these two crops [24]. Many field experiments in this region (e.g. [25–27]) about the effects of tillage methods

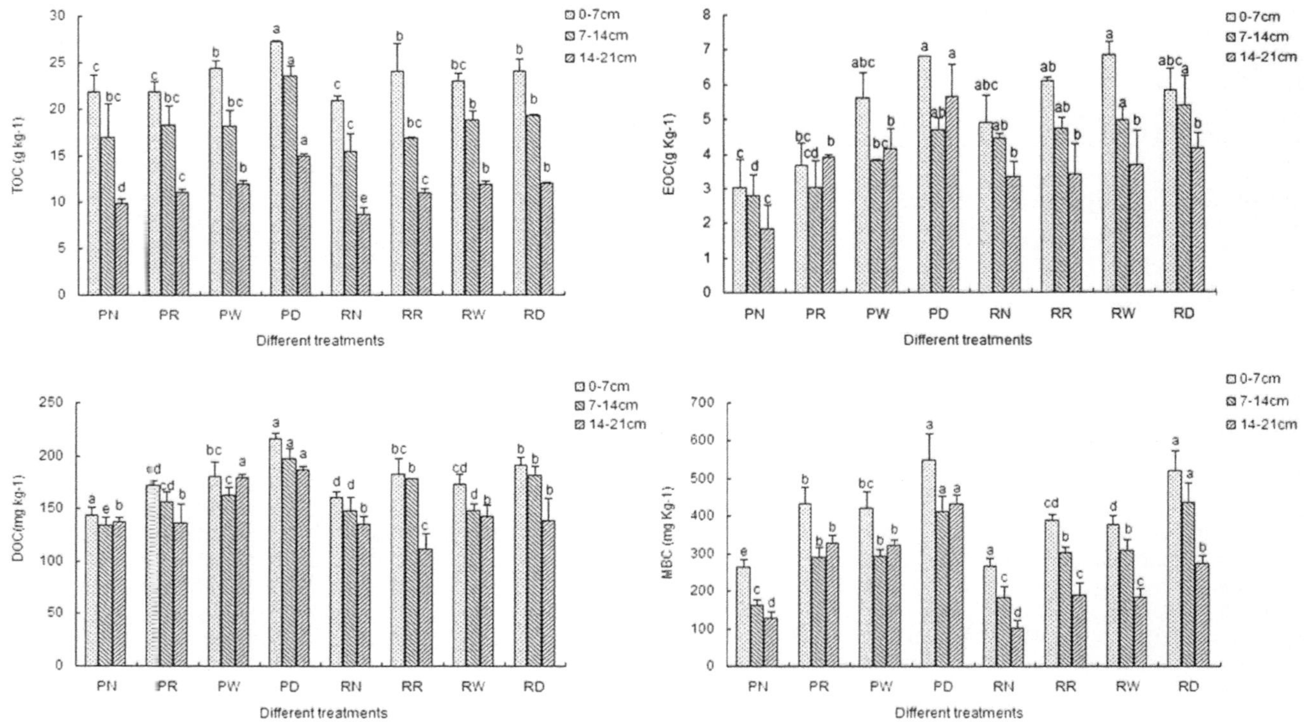

Figure 1. Effects of eight treatments on soil TOC, EOC, DOC and MBC contents at three depths.

combined with straw return on cropland ecosystem in rice-wheat rotation system have been studied during these years. However, most of them are focused on soil physical-chemical properties, soil nutrient and crop yield. To our knowledge, there is little information about the effects of different tillage and straw return on soil labile C fractions in rice-wheat rotation system. Thus, the objectives of this study were (1) to quantify the effects of tillage methods and straw return on soil TOC, MBC, DOC, EOC contents in the rice-wheat rotation system in the Yangtze River Delta, and (2) to explore an optimal management practice combination of tillage and straw return for improving the soil quality and increasing the local crop production.

Table 1. Linear correlations among soil TOC and labile organic C fractions at the 0–21 cm depth.

Index	TOC	EOC	DOC	MBC
TOC	1			
EOC	0.638**	1		
DOC	0.758**	0.684**	1	
MBC	0.741**	0.639**	0.908**	1

TOC: total organic carbon; MBC: microbial biomass carbon; DOC: dissolved organic carbon; EOC: easily oxidizable carbon.
* $P<0.05$.
** $P<0.01$.

Materials and Methods

Site Description

The experiment was conducted at Changwang Country, Youfang Town, Yangzhong City, Jiangsu Province, China (119°42′–119°58′E, 32°–2°19′N, 4–4.5 m above mean sea level) from November, 2009 to June, 2011. Access to the study site was obtained in the form of a rent contract, in which we had to confirm that our study did not involve endangered or protected species.

The experimental site had a subtropical monsoon climate with an average annual precipitation of 1000 mm, an average annual temperature of 15.1°C, and a mean annual sunshine hour of 2135 h. The soil of the experimental site was a loam and classified as an anthrosols. Rice-wheat double cropping system was the most important cropping system in the region. The main properties of soil (0–20 cm depth) sampled in November 2009 were as follows: soil organic matter 29.81 g kg^{-1}; alkali-hydrolyzale nitrogen 194.02 mg kg^{-1}; available phosphorus 13.60 mg kg^{-1}; available potassium 51.45 mg kg^{-1}; and pH 7.34.

The variety of wheat used in this study was Yangmai16 *(Triticum aestivum* L.) and rice was Nangeng47 (*Oryza sativa* L.).

Experimental Design and Field Managements

The experiment had a split-plot design with two tillage methods in the main plots and four straw return modes in subplots with three replications (6 m×5 m). Tillage methods included plowing tillage (P) and rotary tillage (R). Straw return modes were as follows: no straw return (N), only rice straw return (R), only wheat straw return (W), and rice and wheat straw both return (D). There were eight treatments in this study: (1) plowing tillage with no straw return (PN: rice with plowing tillage-wheat with plowing tillage); (2) plowing tillage with only rice straw return (PR: rice with

Table 2. Effects of different tillage factor and straw factor on soil TOC, EOC, DOC and MBC at 0–7 cm, 7–14 cm and 14–21 cm depth.

Soil dept(cm)	Factors	Treatments	TOC(g kg^{-1})	labile organic carbon fractions contents		
				EOC (g kg^{-1})	DOC(mg kg^{-1})	MBC(mg kg^{-1})
0–7	Tillage factor	P	23.87±2.67a	4.78±1.23a	178.36±33.74a	417.16±17.20a
		R	23.00±2.38a	5.92±1.48a	177.05±30.96a	389.22±36.56b
	Straw factor	N	21.40±1.87c	3.97±0.92b	152.25±10.76c	266.73±11.68c
		R	23.01±2.97bc	4.88±1.40ab	177.61±33.57b	411.67±27.32b
		W	23.65±1.52b	6.23±1.56a	176.92±30.87b	398.57±27.34b
		D	25.68±2.63a	6.32±1.68a	204.04±22.67a	535.79±20.31a
7–14	Tillage factor	P	19.30±1.23a	3.58±1.76b	163.05±37.30a	311.42±43.89a
		R	17.64±2.48b	4.89±1.52a	164.11±27.23a	306.50±36.09a
	Straw factor	N	16.28±3.20c	3.62±1.76c	141.49±18.49d	172.63±16.58c
		R	17.63±2.86bc	3.90±2.70bc	167.39±21.63b	315.19±20.00b
		W	18.50±1.75b	4.39±0.54ab	155.47±22.41c	313.93±14.31b
		D	21.47±2.02a	5.04±0.76a	189.98±17.71a	434.09±11.65a
14–21	Tillage factor	P	12.00±2.56a	3.89±1.42a	160.06±30.21a	280.68±58.67a
		R	10.90±2.61b	3.64±1.14a	132.00±25.31b	187.08±44.86b
	Straw factor	N	9.28±1.69d	2.58±1.18c	136.34±11.66b	114.99±18.05d
		R	10.99±1.96c	3.66±1.26b	123.67±29.32c	239.95±39.07b
		W	11.97±3.06b	3.91±1.39b	161.19±25.91a	238.92±35.06c
		D	13.56±1.75a	4.90±0.68a	162.93±26.49a	341.67±31.14a

Different letters in a line under a specific influence factor denote significant difference at the 5% level. Different capitals in a column at different soil depths and treatments present significant different at the 0.05 level. TOC: total organic carbon; MBC: microbial biomass carbon; DOC: dissolved organic carbon; EOC: easily oxidizable carbon.

plowing tillage - wheat with plowing tillage+rice straw return); (3) plowing tillage with only wheat straw return (PW: rice with plowing tillage+wheat straw return - wheat with plowing tillage); (4) plowing tillage with rice and wheat straw both return(PD: rice with plowing tillage+wheat straw return -wheat with plowing tillage+rice straw return); (5) rotary tillage with no straw return (RN: rice with rotary tillage - wheat with rotary tillage); (6) rotary tillage with only rice straw return (RR: rice with rotary tillage - wheat with rotary tillage+rice straw return); (7) rotary tillage with only wheat straw return (RW: rice with rotary tillage+wheat straw return - wheat with rotary tillage); (8) rotary tillage with rice and wheat straw both return (RD: rice with rotary tillage+wheat straw return - wheat with rotary tillage+rice straw return).

The experimental site was cultivated with a rice-wheat rotation prior to November 2009, where wheat was planted with plowing tillage from November to the following June, and rice was transplanted by plowing tillage from June to November. In this study, after wheat or rice was harvested, they were cultivated at a depth of 10–15 cm by rotary cultivation in rotary tillage plots while for the plowing tillage plots, cultivation was at a depth of 20–25 cm with a moldboard plough. Before the rice and wheat were sown, the plowing tillage plots were disked and moldboard plowed for weed control and bedding. This was followed by an application of fertilizer. For straw returned plots, the wheat and rice straw were cut into 8–10 cm after air-dried, and placed back on the surface of the soil in June or November of each year, with returned amount of 6000 kg·hm^{-2} for both wheat and rice straw.

In this study, wheat was sown on November 3, 2009 and November 24, 2010, respectively. The seed quantity was 150 kg hm^{-1} by machine. The base fertilizer applied before sowing was 135 kg·hm^{-2} pure N, 67.5 kg·hm^{-2} P_2O_5, and 67.5 kg·hm^{-2} K_2O, and topdressing was at the elongation stage with 135 kg hm^{-1} pure N. For all treatments, N was applied in the form of $CO(NH_2)_2$, and the fertilizer in wheat seasons was applied at the same rate. The wheat was harvested on June 3, 2010 and June 8, 2011, respectively. Rice was transplanted at about 3–4 seedlings per hole, 255,000 holes per hectare on June 15, 2010 and June 15, 2011, respectively. The rice was fertilized just before transplanting with a base fertilizer (120 kg hm^{-1} pure N; 60 kg hm^{-1} P_2O_5; 60 kg hm^{-1} K_2O), and at tillering stage and earing stage with a topdressing (180 kg hm^{-1} N (3:1 ratio)). The fertilizer applied in the two rice seasons were the same. The rice was harvested on November 12, 2010 and November 29, 2011, respectively. The pesticide management of both rice and wheat seasons was in accordance with the conventional, and all other management procedures were identical for the eight treatments.

Soil Sampling and Analytical Methods

Soil samples were collected by a geotome (5 cm diameter) on October 29, 2011 (just before the rice was harvested). Five random locations were chosen in each of the 24 observational plots and samples were taken from each location at three soil depths (0–7, 7–14, and 14–21 cm) separately. Soil samples from each depth were about 200 g, fully blended. The collected moist samples were ground and sieved through a 10 mesh screen. Sieved soil samples were divided into two sub-samples. One was air-dried and sieved again through 100 mesh screen for determining soil TOC and EOC. Another was immediately stored in 4°C refrigerators for determining DOC and MBC. During sieving, crop residues, root material and stones were removed.

Table 3. Affecting force analysis of different tillage and straw return and their interaction on soil TOC, EOC, DOC and MBC at 0–7 cm, 7–14 cm and 14–21 cm depth.

Soil depth(cm)	Difference source	Affecting force(%)			
		TOC	EOC	DOC	MBC
0–7	Block	10.35	8.53	6.55	0.26
	Tillage	0.34	28.06**	0.64	2.06**
	Straw return	19.69*	24.79**	38.58**	95.95**
	Straw return × tillage	10.71	4.34	4.86	0.93
	Error	58.92	34.28	49.37	0.79
7–14	Block	19.97**	5.36	11.53	0.29
	Tillage	0.02	13.03*	0.85	0.07
	Straw return	21.41**	13.16	39.21**	97.83**
	Straw return × tillage	26.30**	3.89	11.12*	0.96
	Error	32.32	64.56	37.28	0.86
14–21	Block	10.22	1.61	16.66**	0.27
	Tillage	1.05	3.21	11.77**	23.81**
	Straw return	32.35**	25.06**	35.65**	70.17**
	Straw return × tillage	14.72*	13.95*	9.60*	4.93**
	Error	41.66	56.17	26.33	0.82

The affecting force of tillage = tillage variables (square)/total variables (sum of total squares) ×100%; the affecting force of straw = straw variables (square)/ total variable (sum of total squares) ×100%; the affecting force of the interaction = interaction variable (square)/total variables ((total squares of sum) ×100%. TOC: total organic carbon; MBC: microbial biomass carbon; DOC: dissolved organic carbon; EOC: easily oxidizable carbon.
* $P<0.05$.
**$P<0.01$.

Total organic C (TOC) concentration was determined by oxidation with potassium dichromate and titration with ferrous ammonium sulphate [28].

Dissolved organic C (DOC) was extracted from 10 g of moist soil with 1:2.5 ratio of soil to water at 25.8°C [29]. After shaking for 1 h and centrifuging for 10 min at 4500 r min^{-1}, the supernatant was filtered with a 0.45 mm membrane filter. The filtrate was measured by oxidation with potassium dichromate and titration with ferrous ammonium sulphate.

Microbial biomass C (MBC) was analyzed by the fumigation extraction method [30]. Each sample was weighed into two equivalent portions, one was fumigated for 24 h with ethanol-free chloroform and the other was the unfumigated control. Both fumigated and unfumigated soils were shaken for 1 h with 0.5 M K_2SO_4 (2:5 soil: extraction ratio), centrifuged and filtered.

Easily oxidizable C (EOC) was measured as described by Blair et al. [3]. Finely ground air-dried soil samples were reacted with 333 mmol L^{-1} $KMnO_4$ by shaking at 60 r min^{-1} for 1 h. The suspension was then centrifuged at 2000 r min^{-1} for 5 min. The supernatant was diluted and measured spectrophotometrically at 565 nm. All soil samples were analyzed in triplicate.

Data Analysis

The SPSS 16.0 analytical software package was used for all statistical analyses. A 2-factor analysis of variance (ANOVA) was employed for difference test among eight treatments at $P<0.05$, with separation of means by least significant difference (LSD). Correlation analysis were performed to determine correlations among soil labile organic C fractions in the 0–21 cm soil depth, and the significant probability levels of the results were given at the $P<0.05$ (*) and $P<0.01$ (**), respectively. Moreover, the affecting force analysis of tillage factor, straw factor and their interaction influence on labile organic C fractions was calculated based on the method of Leng [31]: the affecting force of tillage = tillage variables (square)/total variables (sum of total squares)×100%; the affecting force of straw return = straw return variables (square)/total variable (sum of total squares)×100%; the affecting force of interaction = interaction variable (square)/total variables ((total squares of sum)×100%.

Results

Soil TOC, DOC, MBC and EOC Contents in Different Treatments

As shown in Fig. 1, the different treatments significantly affected the contents of soil TOC and labile organic C fractions, where PD generally had the highest contents of TOC, DOC, MBC and EOC at the three soil depths. Crop straw return treatments (PR, PW, PD, RR, RW, RD) had consistently higher amount of TOC and labile organic C fractions at the three soil depths than without crop straw return treatments (PN, RN). Moreover, PN had significantly lower TOC, DOC, MBC and EOC at 0–7 cm and 7–14 cm, and RN had the lowest TOC and MBC at 14–21 cm compared to other treatments (Fig. 1). Soil TOC and labile organic C fractions generally decreased with an increase in soil depth under all treatments. As expected, soil TOC and labile organic C fractions were significantly and positively correlated with each other (Table 1).

Effects of Different Tillage Methods on Soil TOC and Labile Organic C Fractions

Tillage had a significant effect on MBC at 0–7 cm soil depth, but seldom on soil TOC, DOC and EOC. Soil TOC, DOC and MBC contents were all higher under plowing tillage (P) than rotary tillage (R), while EOC was opposite at 0–7 cm soil depth (Table 2). At 7–14 cm, soil EOC under rotary tillage (R) was significantly higher than plowing tillage (P), but TOC had the contrary results, and there were no significant differences on DOC and MBC (Table 2). At 14–21 cm, soil TOC, DOC and MBC were significantly higher under plowing tillage (P) than rotary tillage (R), except EOC (Table 2).

Effects of Different Straw Return on Soil TOC and Labile Organic C Fractions

Straw return had significant effects on soil TOC and labile organic C at the three depths as shown in Table 2. In general, soil TOC and three labile organic C ranged in the following order: rice and wheat straw both return>only wheat or rice straw return>no straw return at three depths (Table 2). At 7–14 cm depth, only rice straw return in the wheat season had significantly higher DOC than only wheat straw return in the rice season (Table 2). However, at 14–21 cm depth, except for MBC, soil TOC, EOC, DOC under only rice straw return in the wheat season were lower than only wheat straw return in the rice season (Table 2). Moreover, there were significant differences in TOC and MBC among the four straw return at 14–21 cm depth (Table 2).

Affecting Force Analysis of Different Tillage, Straw Return and their Interaction on Soil TOC and Labile Organic C Fractions

Affecting force of different tillage, straw return and their interaction on soil TOC and labile organic C were different with increasing soil depth (Table 3). The affecting force of tillage increased with the increase of soil depth (Table 3). Tillage had significant affecting force on EOC and MBC at 0–7 cm depth, but seldom on TOC and DOC (Table 3). At 7–14 cm depth, the affecting force of tillage on EOC was lower than at 0–7 cm, and there was no significant affecting force on soil TOC and other labile organic C fractions (Table 3). At 14–21 cm depth, tillage had significant affecting force on DOC and MBC. However, there was no significant affecting force on TOC and EOC (Table 3).

Straw return had significant affecting force on soil TOC, DOC and MBC at the three depths, but there was no significant affecting force on EOC at 7–14 cm (Table 3). Among the four indictors, the affecting force of straw return on MBC was the greatest at the three depths, which reached 95.95%, 97.83% and 70.17%, respectively (Table 3).

The affecting force of the interaction generally increased with an increase in soil depth (Table 3). At 0–7 cm soil depth, the interaction had no significant affecting force on soil TOC and labile organic C (Table 3). Soil TOC and DOC were mainly dominated by the interaction at 7–14 cm depth, but there was no significant affecting force on MBC and EOC. At 14–21 cm depth, the interaction had significant affecting force on soil TOC and all labile organic C (Table 3).

Discussion

Suitable soil tillage practice can increase the SOC content, and improve SOC density of the plough layer [32]. The effect size of tillage methods on SOC dynamics depends on the tillage intensity [33]. Compared to conventional tillage (CT), no-tillage and reduced tillage could significantly improve the SOC content in cropland. Frequent tillage under CT easily exacerbate C-rich macroaggregates in soils broken down due to the increase of tillage intensity, then forming a large number of small aggregates with relatively low organic carbon content and free organic matter particles. Free organic matter particles have poor stability and are easy to degradation, thereby causing the loss of SOC [33,34]. In our study, at 0–7 cm soil depth, soil EOC under plowing tillage was lower than rotary tillage (Table 2). The reason could be attributed to the tillage method. Tillage increases the effect of drying–rewetting and freezing-thawing on soil, which increases macroaggregate susceptibility to disruption [21,35,36], and accelerates the labile organic C mineralization and SOM degradation, thus increasing the loss of EOC [14,37]. At 7–14 cm, rotary tillage had higher soil EOC and DOC than plowing tillage, but lower at 14–21 cm soil depth, indicating that tillage affected the vertical distribution of EOC and DOC (Table 2). The difference in soil condition after plowing tillage or rotary tillage affects the rate of straw decomposition, thereby resulting in a difference in the soil nutrient accumulation [38]. Similarly, Liu et al. [39] have found that SOM content under plowing and rotary tillage at deeper soil both were higher than that of the upper soil. The reason might be that rotary tillage and plowing tillage mixed crop straw into the deeper soil layer, making SOM well-distributed at different depths [40].

Carbon input can be increased by adopting straw return in cropland [14]. Fresh residues are C source for microbial activity and nucleation centers for aggregation when returned to cropland. The enhanced microbial activity induces the binding of residue and soil particles into macroaggregates [34,41], which could increase aggregates stability, fix the unstable C, thus improving the concentration of SOC [42] and increasing C sequestration [14]. In our research, straw return had significantly higher soil TOC and labile organic carbon fractions contents at the three soil depths than no crop straw return (Table 2). Soil TOC and labile organic C fractions in both rice and wheat straw return treatments were higher than only wheat or rice straw return (Table 2), indicating that straw return plays an very important role in increasing soil TOC and labile organic C fractions. Similar observations have been reported by other researchers [43–45]. At the three depths, soil TOC in the treatment of only wheat straw return in rice season was higher than only rice straw return in wheat season, moreover, the difference was significant ($p<0.05$) at 7–14 cm (Table 2). This was related to the relatively near-surface higher water content and favorable soil temperature during the rice growing season, resulting in relatively fast straw decomposition [46]. The decomposition of wheat straw provides enough energy and carbon source for soil microorganisms, thus increases the microorganisms' activities. Alternatively, after wheat straw return, the high temperature and humid conditions accelerates the reduction of the C/N ratio of the straw, allowing for sufficient decomposition. More nutrients are released and utilized by the crops, which therefore lightens the pressure of burning straw and improves the soil quality [47]. According to table 1, the study showed that MBC was affected by the straw return factor with an affecting force of 95.95% at 0–7 cm depth and 97.83% at 7–14 cm depth. The probable explanation maybe that crop residue might enter the labile C pool, provide substrate for the soil microorganisms, and contribute to the accumulation of labile C [48].

In our study, PD had the highest content of soil TOC at all the three soil depths (Fig. 1). The reason might be that plowing tillage made the soil and straw in the plow layer turned over quarterly, which increased the stability of the TOC content at each soil layer [49]. In addition, the rice and wheat straw were both returned under PD treatment from 2009, plowing tillage made much SOM enter into the soil and accumulate [45]. However, Tian et al. [50] found that rotary tillage with straw return had higher SOC than plowing tillage with straw return at 0–10 cm soil depth in wheat field. The diverse results might be due to the different regional climate, soil type, crop rotation and the length of study [22]. In this study, at upper soil layer, the interaction effect between tillage and straw return was not significant, but generally increased with an increase in soil depth (Table 3). Rajan et al [2] also found that single effect of residue application was not significant but its significance became apparent after its interaction with tillage system.

In our study, soil labile organic C fractions were significantly and positively correlated with TOC concentrations at 0–21 cm soil depth (Table 1). Such correlations suggested that TOC was a major determinant of soil labile organic C fractions. MBC, DOC and EOC were also significantly and positively correlated with each other in this study (Table 1). The results were consistent with Chen et al. [14], who reported similar correlations between soil TOC, labile organic C fractions (MBC, DOC, particulate organic C, EOC and hot-water extractable C), and macroaggregate C within 0–15 cm depth. Dou et al. [51] also observed the same results. MBC is the living part of SOM, which plays an important role in maintenance of soil fertility [52]. It serves as a sensitive indicator of change and future trends in organic matter level [53]. Dissolved organic C consists of organic compounds present in soil solution, acts as a substrate for microbial activity, and is the primary energy source for soil microorganisms [1]. Easily oxidizable C partly reflects enzymatic decomposition of labile SOC [54]. Therefore, it is not surprising to find the positive correlations among the labile C pools as they have a close association with each other.

Conclusions

In this study, after 2 years of a rice-wheat rotation, soil TOC and labile organic C fractions in PR, PW, PD, and RR, RW, RD were all higher than PN and RN. PD and RD had more significant effects on EOC DOC and MBC compared to other treatments at 0–21 cm depth. Soil TOC and labile organic C fractions were highly correlated with each other. Under short-term conditions, rice and wheat straw both return in rice-wheat rotation system can increase SOC content and improve soil quality in the Yangtze River Delta, which is a suitable agricultural practice in this region under rice-wheat cropping system.

Acknowledgments

We thank Mr. Kejun Gu (Jiangsu Province Academy of Agricultural Sciences, China) for supporting the field experiment, Dr. Changqing Chen (Nanjing Agricutural University, China) for his kind help in data analysis,and Mr. Michael Rickaille (University of the West Indies) for his critical reading.

Author Contributions

Conceived and designed the experiments: LQZ. Performed the experiments: MFY NJH ZWZ. Analyzed the data: MFY. Contributed reagents/materials/analysis tools: LQZ. Wrote the paper: LQZ NJH XHZ.

References

1. Haynes RJ (2005) Labile organic matter fractions as central components of the quality of agricultural soils: an overview. Adv. Agron 85: 221–268.
2. Rajan G, Keshav RA, Zueng-Sang C, Shree CS, Khem RD (2012) Soil organic carbon sequestration as affected by tillage, crop residue, and nitrogen application in rice–wheat rotation system. Paddy Water Environ 10: 95–102.
3. Blair GJ, Lefory RDB, Lise L (1995) Soil carbon fractions based on their degree of oxidation and the development of a carbon management index for agricultural system. Aust. J. Agric. Res 46: 1459–1466.
4. Ghani A, Dexter M, Perrott WK (2003) Hot-water extractable carbon in soils: a sensitive measurement for determining impacts of fertilization, grazing and cultivation. Soil Biol. Biochem 35: 1231–1243.
5. Rudrappa L, Purakayastha TJ, Singh D, Bhadraray S (2006) Long-term manuring and fertilization effects on soil organic carbon pools in a Typic Haplustept of semi-arid sub-tropical India. Soil Till. Res 88: 180–192.
6. Yang CM, Yang LZ, Zhu OY (2005) Organic carbon and its fractions in paddy soil as affected by different nutrient and water regimes. Geoderma 124: 133–142.
7. Baker JM, Ochsner TE, Venterea RT, Griffis TJ (2007) Tillage and soil carbon sequestration–what do we really know? Agric Ecosyst Environ 118, 1–5.
8. Hernanz JL, L'opez R, Navarrete L, S'anchez-Gir'on V (2002) Long-term effects of tillage systems and rotations on soil structural stability and organic carbon stratification in semiarid central Spain. Soil Till. Res 66 (2): 129–141.
9. Shan YH, Yang LZ, Yan TM, Wang JG (2005) Downward movement of phosphorus in paddy soil installed in large-scale monolith lysimeters. Agr. Ecosyst. Environ 111 (1–4): 270–278.
10. Zhang ZJ (1998) Effects of long-term wheat-straw returning on yield of crop and soil fertility. Chinese Journal of Soil Science 29(4): 154−155.
11. Sun X, Liu Q, Wang DJ, Zhang B (2007) Effect of long-term application of straw on soil fertility. Chinese Journal of Eco-Agriculture 16(3): 587−592.
12. West TO, Post WM (2002) Soil organic carbon sequestration rates by tillage and crop rotation. Soil Science Society of American Journal 66: 1930–1946.
13. Liu SP, Nie XT, Zhang HC, Dai QG, Huo ZY, et al. (2006) Effects of tillage and straw returning on soil fertility and grain yield in a wheat-rice double cropping system. Transactions of the CSAE 22(7): 48–51.
14. Chen HQ, Hou RX, Gong YS, Li HW, Fan MS, et al. (2009) Effects of 11 years of conservation tillage on soil organic matter fractions in wheat monoculture in Loess Plateau of China. Soil Till. Res 106: 85–94.
15. Wang DD, Zhou L, Huang SQ, Li CF, Cao CG (2013) Short-term Effects of Tillage Practices and Wheat-straw Returned to the Field on Topsoil Labile Organic Carbon Fractions and Yields in Central China. Journal of Agro-Environment Science 32(4): 735–740.
16. Miller AJ, Amundson R, Burke IC, Yonker C (2004) The effect of climate and cultivation on soil organic C and N. Biogeoche mistry 67: 57–72.
17. Diekow J, Mielniczuk J, Knicker H, Bayer C, Dick DP, et al. (2005) Soil C and N stocks as affected by cropping systems and nitrogen fertilisation in a southern Brazil Acrisol managed under no-tillage for 17 years. Soil and Tillage Research 81, 87–95.
18. Galantini JA, Senesi N, Brunetti G, Rosell R (2004) In fluence of texture on organic matter distribution and quality and nitrogen and sulphur status in semiarid Pampean grassland soils of Argentina. Geoderma 123, 143–152.
19. Ouédraogo E, Mando A, Stroosnijder L (2006) Effects of tillage, organic resources and nitrogen fertiliser on soil carbon dynamics and crop nitrogen uptake in semi-arid West Africa. Soil and Tillage Res 91: 57–67.
20. Yamashita T, Feiner H, Bettina J, Helfrich M, Ludwig B (2006) Organic matter in density fractions of water-stable aggregates in silty soils: effect of land use. Soil Biology and Biochemistry 38: 3222–3234.
21. Paustian K, Collins HP, Paul EA (1997) Management controls in soil carbon. In: Paul, E.A., Paustian, K.A., Elliott, E.T., Cole, C.V. (Eds). Soil Organic Matter in Temperate Ecosystems: Long Term Experiments in North America. RC, Boca Raton, FL 15–49.
22. Puget P, Lal R (2005) Soil organic carbon and nitrogen in a Mollisol in central Ohio as affected by tillage and land use. Soil Till. Res 80, 201–213.
23. Ding LL, Cheng H, Liu ZF, Ren WW (2013) Experimental warming on the rice-wheat rotation agro-ecosystem. Plant Science Journal 31(1): 49–56.
24. Editorial Board of China Agriculture Yearbook (2012) China Agriculture Yearbook 2009, Electronic Edition. China Agriculture Press, Beijing, China.
25. Hao JH, Ding YF, Wang QS, Liu ZH, Li GH, et al. (2010) Effect of wheat crop straw application on the quality of rice population and soil properties. Journal of Nanjing Agricultural University 33(3): 13–18.
26. Zhu LQ, Zhang DW, Bian XM (2011) Effects of continuous returning straws to field and shifting different tillage methods on changes of physical-chemical properties of soil and yield components of rice. Chinese Journal of Soil Science 42(1): 81–85.
27. Liu SP, Nie XT, Zhang HC, Dai QG, Huo ZY, et al. (2006) Effects of tillage and straw returning on soil fertility and grain yield in a wheat-rice double cropping system. Transactions of the CSAE 22(7): 48–51.
28. Lu RK (1999) Soil agricultural chemistry analysis. China's agricultural science and technology press, 106–110.
29. Jiang PK, Xu QF, Xu ZH, Cao ZH (2006) Seasonal changes in soil labile organic carbon pools within a Phyllostachys praecox stand under high rate fertilization and winter mulch in subtropical China. For. Ecol. Manage 236: 30–36.
30. Vance F, Brookes P, Jenkinson D (1987) Microbial biomass measurements in forest soils: the use of the chloroform fumigation-incubation method in strongly acid soils. Soil Biochem 19: 697–702.
31. Leng SC (1992) Biological Statistic and Field Experimental Design. Beijing: China Radio& Television Press (in chinese).
32. Duan HP, Niu YZ, Bian XM (2012) Effects of tillage mode and straw return on soil organic carbon and rice yield in direct seeding rice field. Bulletin of Soil and Water Conservation 32(3): 23–27.
33. Yang JC, Han XG, Huang JH, Pan QM (2003) The dynamics of soil organic matter in cropland responding to agricultural practices. Acta Ecologica Sinica 23(4): 787–796.
34. Six J, Elliott ET, Paustian K (1999) Aggregate and soil organic matter dynamics under conventional and no-tillage systems. Soil Sci. Soc. Am. J. 63, 1350–1358.
35. Beare MH, Hendrix PF, Coleman DC (1994) Water-stable aggregates and organic matter fractions in conventional-tillage and no-tillage soils. Soil Sci. Soc. Am. J. 58, 777–786.
36. Mikha MM, Rice CW (2004) Tillage and manure effects on soil and aggregate-associated carbon and nitrogen. Soil Sci. Soc, AM. J 68: 809–816.
37. Wang J, Zhang RZ, Li AZ (2008) Effect on soil active carbon and C cool management index of different tillages. Agricultural Research in the Arid Areas 26(6): 8–12.
38. Li XJ, Zhang ZG (1999) Influence on soil floods properties of mulching straws and soil-returning straw. Territory and Natural Resources Study 1: 43−45.
39. Liu DY (2009) Physiological and ecological mechanism of stable and high yield of broadcasted rice in paddy field with high standing-stubbles under no-tillage condition. PhD: Sichuan Agricultural University.
40. Gao YJ, Zhu PL, Huang DM, Wang ZM, Li SX (2000) Long-term impact of different soil management on organic matter and total nitrogen in rice-based cropping system. Soil and Environmental Sciences 9(1): 27−30.
41. Jastrow JD (1996) Soil aggregate formation and the accrual of particulate and mineral associated organic matter. Soil Biol. Biochem. 28, 656–676.
42. Govaerts B, Sayre KD, Lichter K, Dendooven L, Deckers J (2007) Influence of permanent raised bed planting and residue management on physical and chemical soil quality in rain fed maize/wheat systems. Plant Soil 291, 39–54.
43. Stockfisch N, Forstreuter T, Ehlers W (1999) Ploughing effects on soil organic matter after twenty years of conservation tillage in Lower Saxony, Germany. Soil Till. Res 52: 91−101.

44. Chen SH, Zhu ZL, Liu DH, Shu L, Wang CQ (2008) Influence of straw mulching with no-till on soil nutrients and carbon pool management index. Plant Nutrition and Fertilizer Science 14(4): 806−809.
45. Song MW, Li AZ, Cai LQ, Zhang RS (2008) Effects of different tillage methods on soil organic carbon pool. Journal of Agro-Environment Science 27(2): 622−626.
46. Zuo YP, Jia ZK (2004) Effect of soil moisture content o n straw decomposing and its dynamic changes. Joural of Northwest Sci-Tech University of Agri. and For. (Nat. Sci. Ed.) 32(5): 61–63.
47. Dai ZG (2009) Study on nutrient release characteristics of crop residue and effect of crop residue returning on crop yield and soil fertility. PhD: Huazhong Agricultural University (in Chinese).
48. Li CF, Yue LX, Kou ZK, Zhang ZS, Wang JP, et al. (2012) Short-term effects of conservation management practices on soil labile organic carbon fractions under a rape–rice rotation in central China. Soil Till, Res 119: 31–37.
49. Zhang P, Li H, Jia ZK, Wang W, Lu WT, et al. (2011) Effects of straw returning on soil organic carbon and carbon mineralization in Semi-arid areas of southern Ningxia, China. Journal of Agro-Environment Science 30(12): 2518−2525.
50. Tian SZ, Ning YT, Wang Y, Li HJ, Zhong WL, et al. (2010) Effects of different tillage methods and straw-returning on soil organic carbon content in a winter wheat field. Chinese Journal of Applied Ecology 21(2): 373–378.
51. Dou FG, Wright AL, Hons FM (2008) Sensitivity of labile soil organic carbon to tillage in wheat-based cropping systems. Soil Sci. Soc. Am. J. 72: 1445–1453.
52. Wu TY, Schoenau JJ, Li FM, Qian PY, Malhi SS, et al. (2004) Influence of cultivation and fertilization on total organic carbon and carbon fractions in soils from the Loess Plateau of China. Soil & Till, Res 77: 59–68.
53. Gregorich EG, Ellert BH, Gregorich EG, Carter MR, Monreal CM, et al. (1994) Towards a minimum data set to assess soil organic matter quality in agricultural soils. Can. J. Soil Sci 74: 367–385.
54. Loginow W, Wisniewski W, Gonet SS, Ciescinska B (1987) Fractionation of organic carbon based on susceptibility to oxidation. Pol. J. Soil Sci 20: 47–52.

4

Impacts of Edaphic Factors on Communities of Ammonia-Oxidizing Archaea, Ammonia-Oxidizing Bacteria and Nitrification in Tropical Soils

Vidya de Gannes[1], Gaius Eudoxie[1], William J. Hickey[2]*

1 Dept. Food Production, Faculty of Food and Agriculture, University of the West Indies, St. Augustine Campus, Republic of Trinidad and Tobago, **2** O.N. Allen Laboratory for Soil Microbiology, Dept. Soil Science, University of Wisconsin-Madison, Madison, Wisconsin, United States of America

Abstract

Nitrification is a key process in soil nitrogen (N) dynamics, but relatively little is known about it in tropical soils. In this study, we examined soils from Trinidad to determine the edaphic drivers affecting nitrification levels and community structure of ammonia-oxidizing bacteria (AOB) and ammonia-oxidizing archaea (AOA) in non-managed soils. The soils were naturally vegetated, ranged in texture from sands to clays and spanned pH 4 to 8. The AOA were detected by qPCR in all soils (*ca.* 10^5 to 10^6 copies archaeal *amoA* g^{-1} soil), but AOB levels were low and bacterial *amoA* was infrequently detected. AOA abundance showed a significant negative correlation ($p<0.001$) with levels of soil organic carbon, clay and ammonium, but was not correlated to pH. Structures of AOA and AOB communities, as determined by *amoA* terminal restriction fragment (TRF) analysis, differed significantly between soils ($p<0.001$). Variation in AOA TRF profiles was best explained by ammonium-N and either Kjeldahl N or total N ($p<0.001$) while variation in AOB TRF profiles was best explained by phosphorus, bulk density and iron ($p<0.01$). In clone libraries, phylotypes of archaeal *amoA* (predominantly *Nitrososphaera*) and bacterial *amoA* (predominanatly *Nitrosospira*) differed between soils, but variation was not correlated with pH. Nitrification potential was positively correlated with clay content and pH ($p<0.001$), but not to AOA or AOB abundance or community structure. Collectively, the study showed that AOA and AOB communities were affected by differing sets of edaphic factors, notably that soil N characteristics were significant for AOA, but not AOB, and that pH was not a major driver for either community. Thus, the effect of pH on nitrification appeared to mainly reflect impacts on AOA or AOB activity, rather than selection for AOA or AOB phylotypes differing in nitrifying capacity.

Editor: Christopher Quince, University of Glasgow, United Kingdom

Funding: These studies were supported by a grant (to Vidya de Gannes) from the Postgraduate Research Fund and the Department of Food Production, Faculty of Food and Agriculture University of the West Indies, St. Augustine Campus, Trinidad and Tobago, and an endowment from the O.N. Allen Professorship in Soil Microbiology (to William Hickey). The funders had no role in study design, data collection and analysis, decision to publish, or preparation of the manuscript.

Competing Interests: The authors have declared that no competing interests exist.

* E-mail: wjhickey@wisc.edu

Introduction

Nitrification is a key process in the nitrogen (N) cycle, as transformation of ammonium-N to nitrate-N can cause nitrate contamination of groundwater, and greenhouse gas production (*i.e.*, N_2O) directly and indirectly *via* denitrification. Thus, predicting the potential for N contamination of the environment necessitates a thorough understanding of the environmental factors affecting the prokaryotes mediating nitrification, a group that now includes ammonia-oxidizing archaea (AOA). The AOA are ubiquitous constituents of marine and terrestrial environments [1,2,3,4], and their discovery has changed the paradigm of aerobic nitrification mediated solely by ammonia-oxidizing bacteria (AOB). As such, for soils, there's now much interest in discerning the role AOA may play in nitrification relative to AOB, and predicting nitrification now requires an understanding of the environmental characteristics that may drive the structure and activity of both AOA and AOB communities (niche differentiation).

In soils, drivers of niche differentiation of AOA *vs.* AOB has centered largely on pH and, to a lesser extent, levels and forms of nitrogen. The former has been implicated to differentially shape AOA *vs.* AOB communities in strongly acidic soils, where AOA abundance increases, or remains unchanged, with decreasing pH, while that of AOB decreases [5,6,7,8]. However, in circum-neutral to alkaline soils, correlations of AOA or AOB abundance to pH have not been consistent [9,10]. Phylogenetic analyses have also implicated pH as a niche-shaping factor as some AOA phylotypes have been correlated with pH [11]. But, the strength of that correlation is uncertain in part because as the data set is skewed toward acidic environments, and it's often uncertain whether the correlations are attributed to pH alone and or other edaphic factors [5].

Potential differences between AOA and AOB in N relations have been evaluated by field and microcosm studies, in which large inputs of ammonium stimulated AOB, but not AOA [9]. In contrast, growth of AOA, but not AOB, was stimulated by addition of an organic N source [12]. These differences have been interpreted as reflecting a preference of AOA for low ammonium environments, which would be consistent with some physiological analyses of the few available AOA cultures [13,14]. But, field data establishing correlations of AOA to natural levels of ammonium or any other form of soil N are limited.

Soil is a highly complex microbial habitat and it's likely that AOA and AOB communities are affected by multiple edaphic factors [7,15,16]. Thus, studies have recently begun to focus on identification of multiple drivers of AOA and AOB community structure and function [17]. However, to date these studies have primarily focused on comparison of land use including soils with N inputs derived from management practices and/or varying in vegetation history [18,19,20,21,22]. Also, prior studies have spanned spatial scales sufficiently large as to include climatic variability as well [17,23].There has been relatively little focus specifically on effects of edaphic properties on AOA and AOB communities in soils with a common type of natural vegetation, and within a common climatic region. Furthermore, database surveys have revealed a strong biogeographical component to the distribution of AOA and AOB phylotypes [24,25]. But, the environmental characteristics affecting the biogeography of AOA and AOB are ill-defined, in part because the majority of studies on soil AOA and AOB communities have focused on soils from temperate regions, and relatively little on other biomes, such as the tropics. Studies of nitrifier communities in tropical soils could add much needed information on the biogeography of AOA and AOB. Moreover, a deeper understanding of nitrification in soils of the tropics is important as N cycling in these soils impacts food production for a large part of the world, and the dynamics of global climate change.

Thus, goals of the present study were to: 1) elucidate the structure of AOA and AOB communities in tropical soils from Trinidad, 2) identify edaphic drivers of variation in their composition and 3) determine the relation between nitrification levels, edaphic characteristics and AOA/AOB communities. The soils ranged from pH 4.8 to 8.2, encompassed a range of characteristics and were all sampled from non-managed locations under natural vegetation (primarily grasses) so that anthropogenic impacts were minimized. Soils were characterized for nitrification potential and a wide range of physico-chemical characteristics, while AOA and AOB communities were evaluated for abundance *via* quantitative PCR of bacterial *amoA* and archaeal *amoA*, and community structure assessed by *amoA* terminal restriction fragment analyses. Our hypotheses were: 1) soil pH would be the major factor affecting nitrifier communities, with acidic environments selecting for greater abundance of AOA and AOA phylotypes unique to low pH soils, and 2) nitrification levels would be correlated to pH as well as the abundance of AOA in acidic soils, and to abundance of AOB in alkaline soils.

Materials and Methods

Soil sampling, physical and chemical properties

Soils were collected from nine locations in Trinidad (Fig. 1, Table S1 in File S1). All soils were obtained from public land, for which no permissions were required for sampling, and no impacts on endangered species. At each location, three core samples (2.5 cm diameter × 10 cm length) were obtained and the top 0–10 cm were removed for analysis. Samples were composited, mixed, sieved (4.75 mm mesh size) and subsamples (20 g each) were taken for moisture content determination by weight loss after drying (105°C, 24 h) and for physical and chemical analyses (three subsamples per soil, 50 g each). Soil samples used for the nitrification potential assay (see below) were used immediately after field sampling. Soil pH was determined by the slurry method (1:5, w:v; sample:distilled deionized water) measured with an Eijkelkamp pH/mV/EC/Salinity/T/02m (Agrisearch Equipment ZG Giesbeek, the Netherlands). Total carbon (TC), organic carbon (OC) and inorganic carbon were determined by dry combustion using a LECO CNS-2000 analyzer. Total N (TN) was determined by combustion and by the Kjeldahl digestion method (TKN). Nitrate-N was measured in soil water extracts colorimetrically after reaction with phenoldisulphonic acid, and ammonium-N was determined by flow injection analysis. Major and minor trace elements were determined by inductively-coupled mass spectrometry. Cation exchange capacity (CEC) was determined by ammonium acetate extraction preceded by cation quantification *via* atomic absorption spectrometry.

Nitrification potential

Nitrification potential (NP) was determined using the shaken slurry procedure [26]. For each sample, 5 g soil was added to 100 mL assay solution (1 mM phosphate buffer adjusted to match pH of each soil, 800 µM ammonium) and then the flasks were loosely covered and incubated with mixing (150 rpm on a bench top orbital shaker) for 24 h, at 29°C, a temperature which approximated that of the soil. The experiment was conducted with 3 replicates per soil in a 9×3 randomized design. During incubation, a sterile pipet was used to remove 10 mL aliquots at 15 min, 2 h, 8 h and 24 h intervals. These samples were centrifuged (10,000× *g*, 10 min) and the supernatant were analyzed for nitrite and nitrate [27]. Nitrite and nitrate concentrations in the 15 min interval were used as baseline measurements for which nitrite and nitrate concentrations in all later time points were corrected.

DNA extraction, qPCR and *amoA* terminal restriction fragment analyses

The Power Soil DNA Isolation kit (MO BIO Laboratories Inc., West Carlsbad, CA) was used for DNA extraction and post-extraction clean-up was done with the Power Clean DNA Clean up kit (MO BIO Laboratories), with both procedures following the manufacturer's instructions. For each soil, triplicate 1 g soil sample extractions were done in batches of 0.25 g, giving 12 separate DNA extracts. These extracts were then pooled to give three samples (each the equivalent of one gram soil) and DNA concentrations were quantified fluorometrically by using a Quibit fluorometer (Invitrogen, Carlsbad, CA).

Each of the three replicate DNA extracts was analyzed by qPCR as described previously [1] using iQ SYBRgreen Supermix (BioRad Laboratories Inc., Hercules CA) and a MyiQ Real-Time qPCR detection system (Bio-Rad). Primer pairs used for qPCR of archaeal *amoA* were amoA_AF, amoA_AR [1], and bacterial *amoA* was amplified with primer pairs amoA_BF, amoA_BR [1]. The qPCR reactions for *amoA* quantification were done following the protocol of De Gannes et al. [1]. The qPCR analyses of archaeal 16S rRNA genes utilized primers 967F and 1060R [28] while qPCR of bacterial 16S rRNA genes was done with primers 338F and 518R [28]. Methods for qPCR of 16S rRNA genes followed those of Karlsson et al. [28]. All qPCR assays included a melting-curve protocol for analysis of primer specificity. Standard curves for qPCR were done with cloned segments of the gene of interest, with efficiencies of 91% to 100%, and r^2 values of 98% to 99.6%. Spiking experiments in which a dilution series of standard template was prepared in the sample matrix were done to verify the absence of PCR inhibitors.

Terminal restriction fragment (TRF) analyses utilized 6FAM-labelled forward primers for archael *amoA* and bacterial *amoA* and Phusion High Fidelity Master Mix with HF buffer (New England Biolabs, Ipswich, MA). Bacterial *amoA* was amplified using the primers and assay conditions of Horz et al. [29] while PCR of

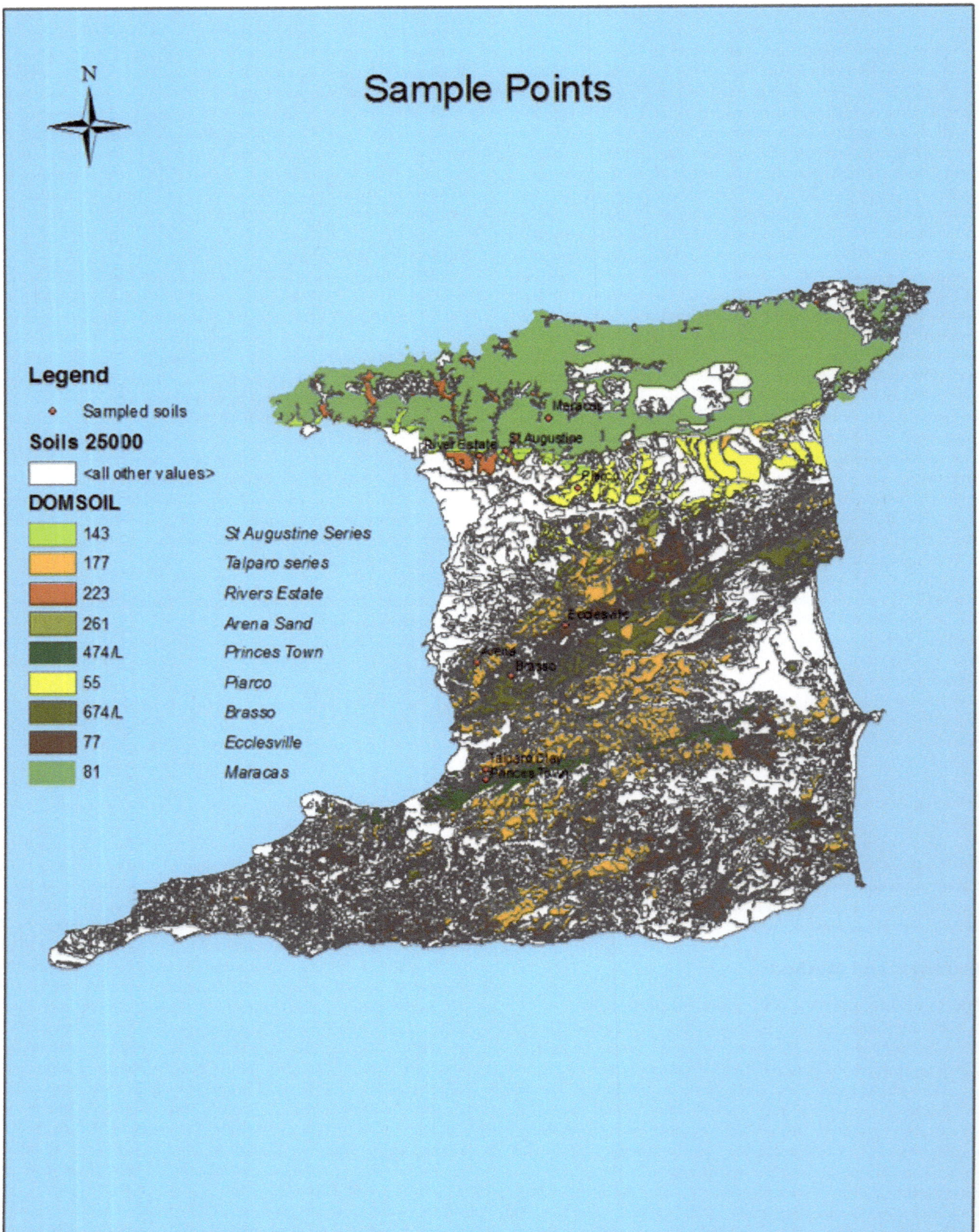

Figure 1. Map of Trinidad showing distribution of the soil series analyzed and the sampling locations.

archaeal *amoA* was done with the primers and conditions of Francis et al. [30]. Triplicate PCR were done for each soil by using an Eppendorf MasterCycler (Eppendorf, Hauppauge, NY), the PCR products were pooled, examined by electrophoresis, purified (QIAquick PCR Purification Kit; QIAGEN, Germantown, MD) and DNA concentrations determined fluorometrically as described above. Aliquots (*ca.* 500 ng) were then digested (3 h, 37°C) with either RsaI or MspI (New England Biolabs). Preliminary tests were done with these soils evaluating fragment generation by RsaI, MspI, HhaI and AluI; RsaI and MspI generated the greatest number of fragments, and were selected for use in the full study. DNA Fragments were precipitated by addition of ice cold 100% ethanol and overnight incubation at −20°C. Analysis of digests was done with an Applied Biosystems 3730xl DNA analyzer (Applied Biosystems, Foster City, CA). A preliminary test was done to determine DNA mass loads that were within the optimal range of quantification for the instrument, and then each sample was analyzed in triplicate at the optimized mass loading. This procedure was adopted from preliminary tests examining variability at each stage of the TRF workflow, which showed that the fragment analysis was the most significant source of error. Fragments of <50 fluorescence units were removed from the analysis, and baseline noise were eliminated by an interactive procedure similar to that of Rees et al. [31] that first excluded fragments <1% of the total peak area, followed by re-normalization and elimination of fragments constituting <5% of the total peak area. The replicate profiles were aligned and TRF profiles of averaged fragment abundance were developed for each sample.

All analyses of TRF data were done with the multivariate statistical software package Primer v. 6 (Primer-E Ltd, Plymouth, UK). A Bray-Curtis coefficient was used to generate similarity matrices, which were ordinated by non-metric multi-dimensional scaling (MDS) computed with 100 random restarts. Analysis of similarity (ANOSIM) was applied to assess the magnitude and statistical significance of dissimilarity between soils in TRF profiles. The BEST (Bioenvironmental Step) routine was applied to identify high rank correlations between the TRF similarity matrices and a matrix of all edaphic factors determined in this study. The BEST analysis was configured to include up to six variables in generating correlations, and run 999 permutations for testing correlation significance.

Construction of AOA and AOB clone libraries and bioinformatics

For all soils, aliquots of each of the replicate DNA extracts were pooled, and an aliquot of the pooled extract was used in PCR Amplification of archaeal *amo*A and bacterial *amoA* was done as described above for TRF analyses, except that the forward primers were not labelled. The PCR products were then purified using the QIAquick PCR Purification Kit (QIAGEN), cloned with the Zero Blunt TOPO system (Invitrogen), transformed into *E.coli* JM 109 competent cells (Promega, USA) and transformants were selected by plating on Luria Bertani medium containing ampicilin and X-gal (each at 100 μg mL^{-1}). Sixteen clones were selected from each soil for sequencing. Plasmid templates in these clones were amplified by using the TempliPhi system (GE HealthcareLife Sciences, Pittsburgh, PA), and then inserts sequenced by using an M13 primer and the BigDye Terminator Cycle Sequencing system (Applied Biosystems). Reactions were analyzed with an Applied Biosystems 3730xl DNA analyzer.

Raw sequence files were imported into Geneious Pro 5.4 (Biomatters Ltd., Auckland 1010, New Zealand) for manual curation and translation. Libraries were examined for potential chimeric sequences with UCHIME (Edgar et al. 2011; http://www.drive5.com/uchime/) and sequence identities were determined by searching GenBank with the BLAST-N web server (http://www.ncbi.nlm.nih.gov/). Sequences were aligned with MUSCLE (Edgar 2004; http://www.drive5.com/uchime/). Phylogenetic trees were constructed by the neighbour-joining method and the Jukes-Cantor distance model, and were created with Geneious Pro v. 5.4.5. The bootstrap re-sampling method was used with 1000 replicates and minimum bootstrap values of 50. Phylogenetic trees were illustrated with FigTree (v. 1.3.1, http://tree.bio.ed.ac.uk/software/figtree/). Distance matrices were created with PHYLIP v. 3.69 (http://evolution.genetics.washington.edu/phylip.htm) and assignment of operational taxonomic units (OTU) and rarefaction analyses were done with Mothur (http://www.mothur.org, [32]). Principle coordinate analyses (PCoA) were done with Fast Unifrac *via* webserver (http://bmf.colorado.edu/fastunifrac/).

Statistical analyses

Analysis of variance (ANOVA) was used to test for significant differences in nitrification potential between soils at the 95% confidence level with Genstat 13 (VSN International, Hemel Hempstead, HPI IESUK). Data from nitrification potential assays was examined for outliers by Grubbs' Test by using an online calculator (www.graphpad.com/quickcalcs/grubbs2). All other analyses were done by using Prism v. 5.4.5 (Graphpad Software, La Jolla, CA).

Sequence accession numbers

Representative sequences of each archaeal *amoA* and bacterial *amoA* OTU are deposited in Genbank under accessions KF888663-KF888698 for archaeal *amoA* and KF888699-KF888724 for bacterial *amoA*).

Results and Discussion

Soil physico-chemical characteristics and nitrification potential

All soils were sampled from non-managed locations vegetated primarily by grasses *viz.* Elephant grass (*Pennisetum purpureum*) and Fowl Foot grass (*Eleusine indica*) as the predominant species. The exception was the Arena sandy loam which was sampled from a preserve of seasonal evergreen forest, which was originally predominant in Trinidad. The soils developed from a variety of different parent materials (Table S1 in File S1), ranged in texture from sandy loam to clay, spanned pH 4.8 to 8.2 and varied in a number of other physico-chemical characteristics (Tables 1, 2). Five acidic soils (pH<6) spanned textural classes, two circum-neutral soils (pH 6.2, 6.6) were loams, and two alkaline soils (pH 7.8, 8.2) were clays.

The three clay soils were vertisols, a defining characteristic of which is a high content of the swelling phylosilicate, montmorillonite. The remainder of the soils either had non-swelling kaolinite as the main phylosilicate mineral or contained relatively little clay (Table 1). Clay content and pH were significantly correlated ($p = 0.0245$) while pH and TC levels were not. Total carbon levels ranged from 8.90 g kg^{-1} soil to 39.13 g kg^{-1} soil, but TC was not correlated with clay content ($p = 0.3317$). The CEC was strongly correlated with clay content ($p = 0.0003$), but not levels of soil OC ($p = 0.3438$). Total N content and TC contents were strongly correlated ($p<0.0001$), but neither parameter was correlated with levels of either ammonium-N or nitrate-N.

Nitrification potential (NP) differed significantly among soils (ANOVA $p<0.001$, Table 3) with the Arena sandy loam being lowest at 0.6 mg N kg^{-1} soil d^{-1} and Brasso being highest at

Table 1. Characteristics of soils used in this study[a].

Soil Name	Texture	pH	Clay	Sand	Silt	TOC	TC	TN	TKN	NH_4-N	NO_3-N	BD	CEC
			g/kg							mg/kg		Mg/m^3	cmol(+)/kg
Arena	Sandy Loam	5.4	4.57	74.00	21.43	6.1	8.90	0.65	0.7	16.4	2.1	1.78	0.3
Ecclesville	Silty Loam	4.8	21.81	34.2	43.99	26.2	3.91	3.46	3.2	46.9	56	1.54	8.8
Piarco	Silty Loam	4.8	17.71	32.00	50.28	13.6	17.87	1.56	1.5	20.5	6.65	1.58	3.1
Maracas	Loam	4.8	10.96	47.32	41.72	13.6	15.89	1.39	1.5	69.5	<0.01	1.56	0.9
River Estate	Loam	6.6	11.79	49.54	38.67	7.9	10.81	0.97	0.9	3.6	0.01	1.56	2.3
St. Augustine	Loam	6.2	19.76	39.16	41.08	24.5	24.75	1.58	1.5	16.4	35.6	1.52	5.6
Brasso	Clay	8.2	57.17	15.21	27.61	16.5	25.46	2.37	1.8	27.1	73.0	1.22	35.3
Princes Town	Clay	7.8	52.52	14.59	32.89	15.5	21.19	1.76	1.1	24.8	24.0	1.24	17.1
Talparo	Clay	5.8	46.71	17.32	35.97	21.3	29.81	2.69	2.3	52.0	46.1	1.28	29.1

[a]TOC, Total Organic Carbon; TC, Total Carbon; TN, Total Nitrogen; TKN, Total Kjeldahl Nitrogen; BD, Bulk Density; CEC, Cation Exchange Capacity.

14 mg N kg^{-1} soil d^{-1}). Rates were linear over the incubation period with $r^2 = 0.952$ to 0.998. Within the clays, there was a clear impact of pH on NP, as the rate in the pH 8.2 Brasso soil was nearly three times greater that in the pH 5.8 Talparo soil (Table 3). The high NP of the Brasso soil was a significant outlier (Grubbs Test $p<0.01$) from that of the other soils, and was thus excluded from the correlation analyses. The NP rates had a significant correlation ($p<0.0001$) with pH (Table 4) and with edaphic characteristics relating to soil texture/structure including: clay content, sand content, bulk density and cation exchange capacity (Table 4). The elemental composition also strongly correlated with NP, as only one element (P) of the 11 determined, lacked a signification relation to NP (Table 4). All of the carbon and nitrogen parameters measured were strongly correlated to NP, except total organic carbon and ammonium-N, which were not significant (Table 4). The significant positive correlation between NP and soil nitrate concentrations could indicate that variation in soil nitrate levels was at least in part attributable to varying rates of nitrification.

The positive correlation of NP rates with soil pH observed in the present study data was consistent with the majority of the literature [33,34,35,36]. But, our study also demonstrated that clay content had a significant factor modulating the pH effect. For example, the Arena sandy loam and Talparo clay were two acidic soils of roughly similar pH (5.4 and 5.8, respectively), yet NP in the montmorrillonic clay soil was *ca.* eight-fold greater than that of the sandy loam. Prior investigators determined that ammonia oxidation by AOB pure cultures was stimulated by expanding clays (montmorillonite) but not non-expanding clays, such as kaolinite [37,38]. Jiang et al. [39] demonstrated that amending a kaolinitic Oxisol with montmorillonite greatly increased NP as well as abundance of both AOA and AOB. For both pure culture and soil studies, the stimulatory effect of montmorrillonite was hypothesized to reflect the ability of this high-CEC mineral to enable localized exchange of ammonium and protons and thereby buffer acidity [37]. Our study builds on prior findings from laboratory experiments, and illustrated with soil field samples the potential impact on nitrification of naturally existing levels of expanding clays like montmorillonite.

Microbial community analysis by qPCR

The majority of the samples (93%) yielded amplification of archaeal *amoA*, which ranged in abundance from 300 copies (g soil)$^{-1}$ in the Maracas loam to 11830 copies g^{-1} soil in the Arena sandy loam (Fig. 2A). The abundance of archaeal *amoA* showed

Table 2. Soil Elemental Properties[a].

Soil Name	P	K	Ca	Mg	S	Zn	B	Mn	Fe	Cu	Al	Na
	g/kg					mg/kg						
Arena	0.03	0.1	0	0.04	0.1	2.77	<2	4.31	1046.9	<0.5	642.9	17.6
Eccelville	1	2.8	5	3.5	0.4	112.2	<2	777.73	24453.3	32.13	22058.6	125.5
Piarco	0.9	1.2	2	0.3	0.2	41.45	<2	48.55	4724.8	14.35	12731.3	160.1
Maracas	0.5	0.6	0	0.5	0.2	23.87	<2	392.56	26648	34.76	11909.5	42.7
River Estate	0.3	0.8	1	0.9	0.1	28.23	<2	127.27	10617.5	11.1	6599.8	110.2
St.Augustine	1	1.9	2	0.4	0.2	126.3	<2	333.66	20695.8	34.49	12277.3	245.4
Brasso	0.9	4.5	10	5	0.5	130	5.48	604.38	34142.7	34.37	46458.4	198.6
Princes Town	0.5	4.1	15	8.2	0.3	165	8.2	977.84	31128.9	44.68	54124	204.5
Talparo	1.9	4.6	9	4.2	0.5	158.7	4.96	624.97	33908.2	28.45	45810.3	207.2

[a]Abbreviations: P = Phosphorus; K = Potassium; Ca = Calcium; Mg = Magnesium; S = Sulphur; Zn = Zinc, B = Boron; Mn = Manganese; Fe = Iron, Cu = Copper; Al = Aluminium; Na = Sodium.

Table 3. Nitrification Potential[a].

Soil	pH	mg N kg^{-1} soil d^{-1}
Arena	5.4	0.7
Ecclesville	4.8	2.1
Piarco	4.8	0.6
Maracas	4.8	1.1
River Estate	6.6	2.4
St. Augustine	6.2	2.7
Brasso	8.2	14.1
Princes Town	7.8	6.6
Talparo	5.8	5.4

[a]N = NO_3-N+NO_2-N, $p<0.001$, LSD (5%) = 0.17, SEM = 0.06.

significant negative correlations with: ammonium concentrations ($r^2 = 0.2504$, $p = 0.0079$), OC levels ($r^2 = 0.2148$, $p = 0.0149$) and clay content ($r^2 = 0.1589$, $p = 0.0394$). Archaeal *amoA* abundance was also inversely related to total N and total carbon ($p = 0.0569$ and 0.0789, respectively). But, there was no significant correlation between archaeal *amoA* abundance and NP ($p = 0.4371$) or pH ($p = 0.8794$). In contrast to archaeal *amoA*, amplification of bacterial *amoA* occurred in only 33% of the samples (Table S2 in File S1). For most soils, amplification occurred in either just one of the three replicates or none of these (Table S2 in File S1). For example, the highest bacterial *amoA* abundance was 12,010 copies g^{-1} soil obtained from one of the three Piarco silty loam replicates, while the other two Piarco soil replicates did not yield bacterial *amoA* signals that were within the quantification limit (Table S2 in File S1). The low frequency of detection for bacterial *amoA* precluded assessing relationships between AOB abundance and soil properties. It should be noted that ammonia-oxidzers could exist that are not detected by currently available PCR primers.

Genes for bacterial 16S rRNA and archaeal 16S rRNA were amplified from all replicates of all soils. Archaeal 16S rRNA gene numbers ranged from 9.24×10^3 copies g^{-1} soil in the Princes Town clay to 2.85×10^6 copies g^{-1} soil in the Piarco silty loam (Fig. 2B). There were no significant correlations with archaeal 16S rRNA gene abundance and any soil properties. Bacterial 16S rRNA copies were *ca.* 10^6-fold greater than archaea. Bacterial 16S rRNA gene numbers ranged from 8.1×10^8 copies g^{-1} soil in the Princes Town clay to 3.5×10^{11} copies g^{-1} soil in the Piarco silty loam (Fig. 2C), the same two soils that showed minimum and maximum archaeal 16S rRNA gene abundance. There were no significant correlations with bacterial 16S rRNA gene abundance and any soil property, and there was no correlation between NP

Table 4. Correlations between nitrification potential and edaphic properties[a].

Property[b]	*p* value	r^2
pH	<0.0001	0.5929
clay	<0.0001	0.8888
sand	<0.0001	0.5756
silt	0.3230	0.0444
TOC	0.1161	0.1085
Total C	0.0031	0.3342
Total N	0.0897	0.1253
TKN	0.6111	0.0119
NH_4-N	0.7743	0.0038
NO_3-N	0.0172	0.2319
BD	<0.0001	0.8612
CEC	<0.0001	0.7205
P	0.1897	0.0822
K	<0.0001	0.7887
Ca	<0.0001	0.8573
Mg	<0.0001	0.7965
S	<0.0019	0.3623
Zn	<0.0001	0.7404
Mn	<0.0001	0.5508
Fe	0.0001	0.4576
Cu	0.0012	0.3850
Al	<0.0001	0.8560
Na	0.0006	0.4206

[a]Correlations exclude the Brasso soil;
[b]See Table 1 for abbreviations.

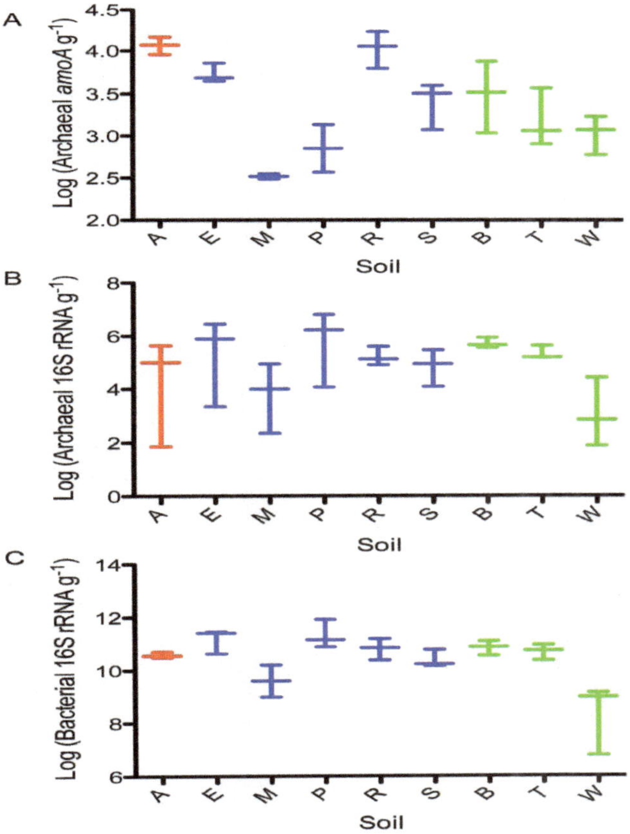

Figure 2. Box and whisker plots of gene abundance determined by qPCR. Panel A, Archaeal *amoA*; Panel B, Archaeal 16S rRNA; Panel C, Bacterial 16S rRNA. Symbol colors correspond to soil types: red, sandy loam; blue, silty loam; green, clay. Soil name abbreviations are: A, Arena; B, Brasso; E, Ecclesville; P, Piarco; R, River Estate; S, St. Augustine; T, Talparo; W, Princes Town.

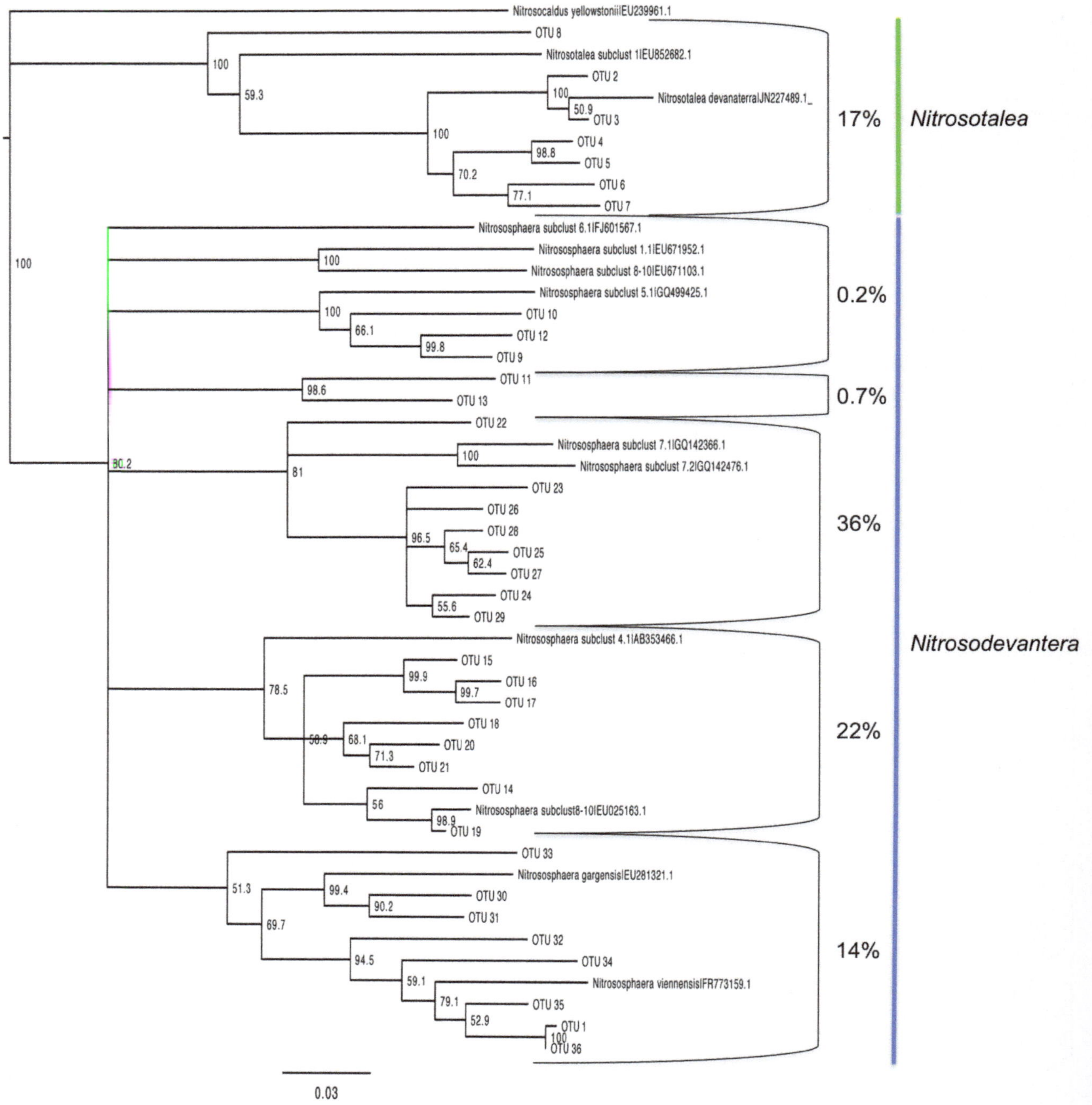

Figure 3. Neighbour joining consensus tree of archaeal *amoA* phylotypes. Colored branches indicate the *Nitrosotalea* cluster (green) and *Nitrososphaera* cluster (blue). Boot strap values are indicated at nodes. Cluster designations follow the classification of Pester et al. [25] based on nucleic acid alignment of archaeal *amoA*.

and the abundance of 16S rRNA genes for either archaea or bacteria.

The negative relation of AOA abundance with ammonium levels was in accordance with similar findings by prior investigations indicating a preference of soil AOA for low ammonium levels [18,21,40,41,42,51]. Since all soils in the present study were from non-managed locations, the soil N dynamics were not impacted by N inputs derived from agricultural practices (*e.g.*, mineral or organic N fertilizers, tillage, grazing). Therefore, differences in AOA abundance that arose from variation in ammonium levels resulted from N cycling processes intrinsic to the soils. Mineralization would be a major mechanism for ammonium input, and the present study thus supported the concept that a key factor favouring soil habitation by AOA *vs.* AOB is the predominance of mineralization as the N input pathway [12,43].

The negative correlation of AOA abundance to OC levels could have reflected the preference of AOA for low organic nutrient conditions [44,45]. But, soil organic matter affects a wide range of

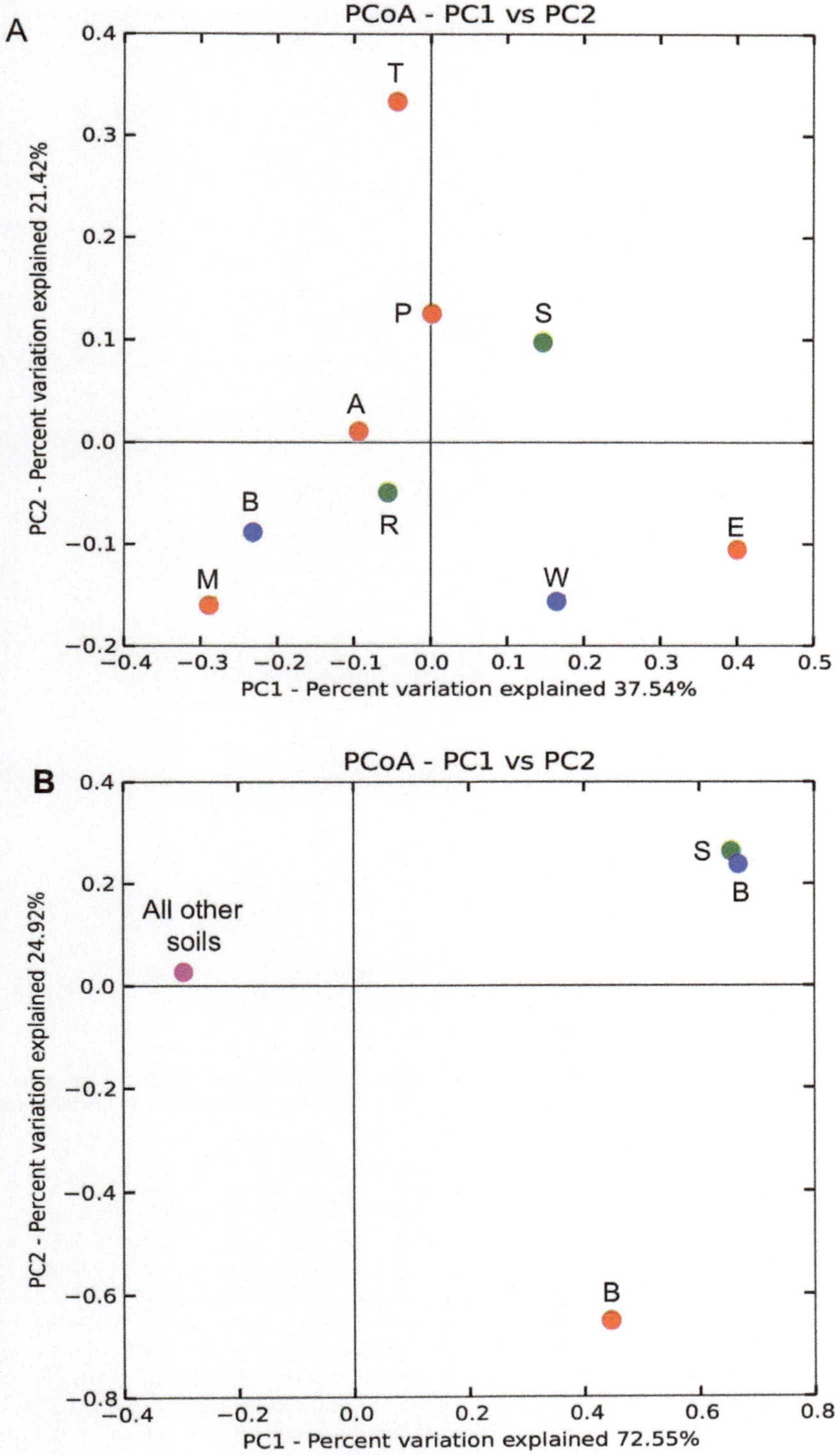

Figure 4. Principle coordinate plot of archaeal *amoA* clone libraries (Panel A) and bacterial *amoA* clone libraries (Panel B). Symbol colors correspond to soil pH: red, acidic (pH<6); green, circum-neutral (pH 6.2, 6.6), blue, alkaline (pH 7.8, 8.2), purple, multiple pH. Letters indicate soil names that are abbreviated as: A, Arena; B, Brasso; E, Ecclesville; P, Piarco; R, River Estate; S, St. Augustine; T, Talparo; W, Princes Town.

soil physico-chemical properties (*e.g.*, bulk density, water-holding capacity, aeration, *etc.*) that also influence microbial communities, and one or more of these broad soil properties cannot be ruled out as exerting effects on AOA. Similarly, clay minerals have many impacts on soil characteristics that affect microbial activity, which complicates interpretation of the negative correlation observed here of clay content with AOA abundance. In a survey of an organic farm, Wessen and colleagues [21] determined a negative

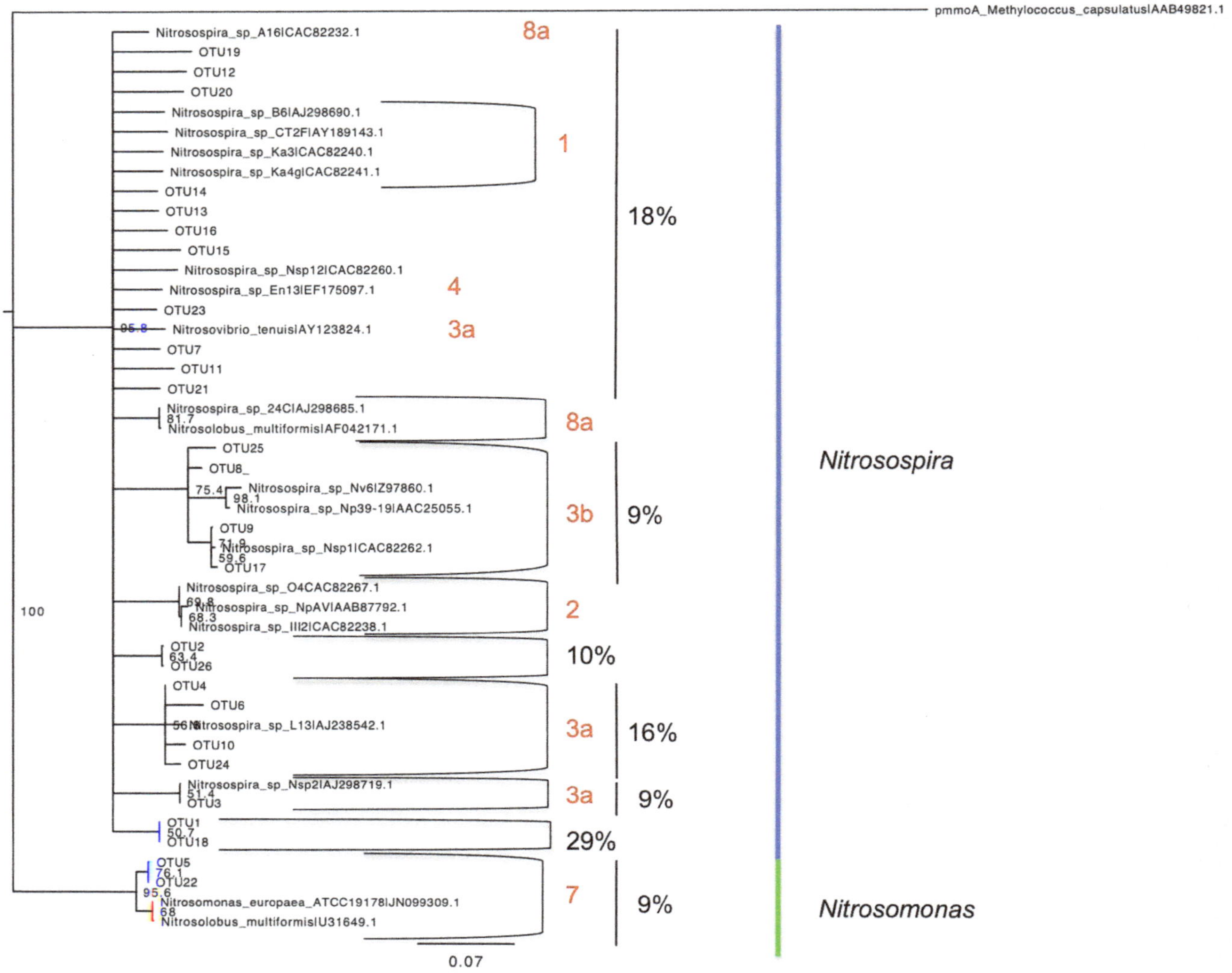

Figure 5. Neighbor joining consensus tree of bacterial AmoA phylotypes. Colored branches indicate the *Nitrosomonas* cluster (green) and *Nitrosospira* cluster (blue). Boot strap values are indicated at nodes. Values in red are cluster designations following the classification of Avrahami et al. [48] based on amino acid alignment of bacterial AmoA.

correlation of AOA with clay content, which they theorized to reflect increased binding of ammonium, implying reduced ammonium bioavailability. But, the potential for clays to impede AOA growth *via* ammonium binding is unclear, as even ammonium tightly bound between interlayers of expanding clays ("fixed ammonium") can be readily accessed by nitrifiers [46]. Also, as described above, expanding clays could have a stimulatory effect on nitrification. Thus, while a clay-associated inhibitory effect cannot be ruled out, we hypothesize that these minerals could exert a negative impact on AOA abundance by their effects on broader physico-chemical characteristics of soil (*e.g.*, bulk density, water-holding capacity, aeration, *etc.*).

The AOB populations in these tropical soils were low, and at levels not readily assayed by qPCR. Similarly low AOB populations have been reported by other investigators [7,43,47]. For example, Pett-Ridge and co-workers [47] were unable to amplify bacterial *amoA* from soils of a tropical forest and, in a survey of soils across Scotland, AOB were below qPCR detection limits in *ca.* 62% of the 184 sites sampled [17]. Thus, for soils having AOB populations below qPCR detection, AOA have been inferred to be the primary group driving nitrification [7,43,47]. By this reasoning, AOA are implicated as dominant drivers of nitrification in the tropical soils studied here. While AOA appeared to play a key role in nitrification, AOA numbers were not correlated with NP. A lack of correspondence between nitrification levels and nitrifier community density (AOB and/or AOA) has been reported by many other investigators [23,43,47], and the reasons for this disconnection are unknown. But, as mentioned above, it's possible that ammonia-oxidzers exist that are divergent from known betaproteobacterial AOB and thaumarchaeal AOA, and thus evade detection by PCR primers developed from currently available sequence data.

Analysis of community composition by *amoA* sequencing and TRF

Sequencing of archaeal *amoA* clone libraries generated 35 OTUs from a total of 121 sequences (Table S3 in File S1, Fig. S1). Most phylotypes (83%) were identified as *Nitrososphaera* and the

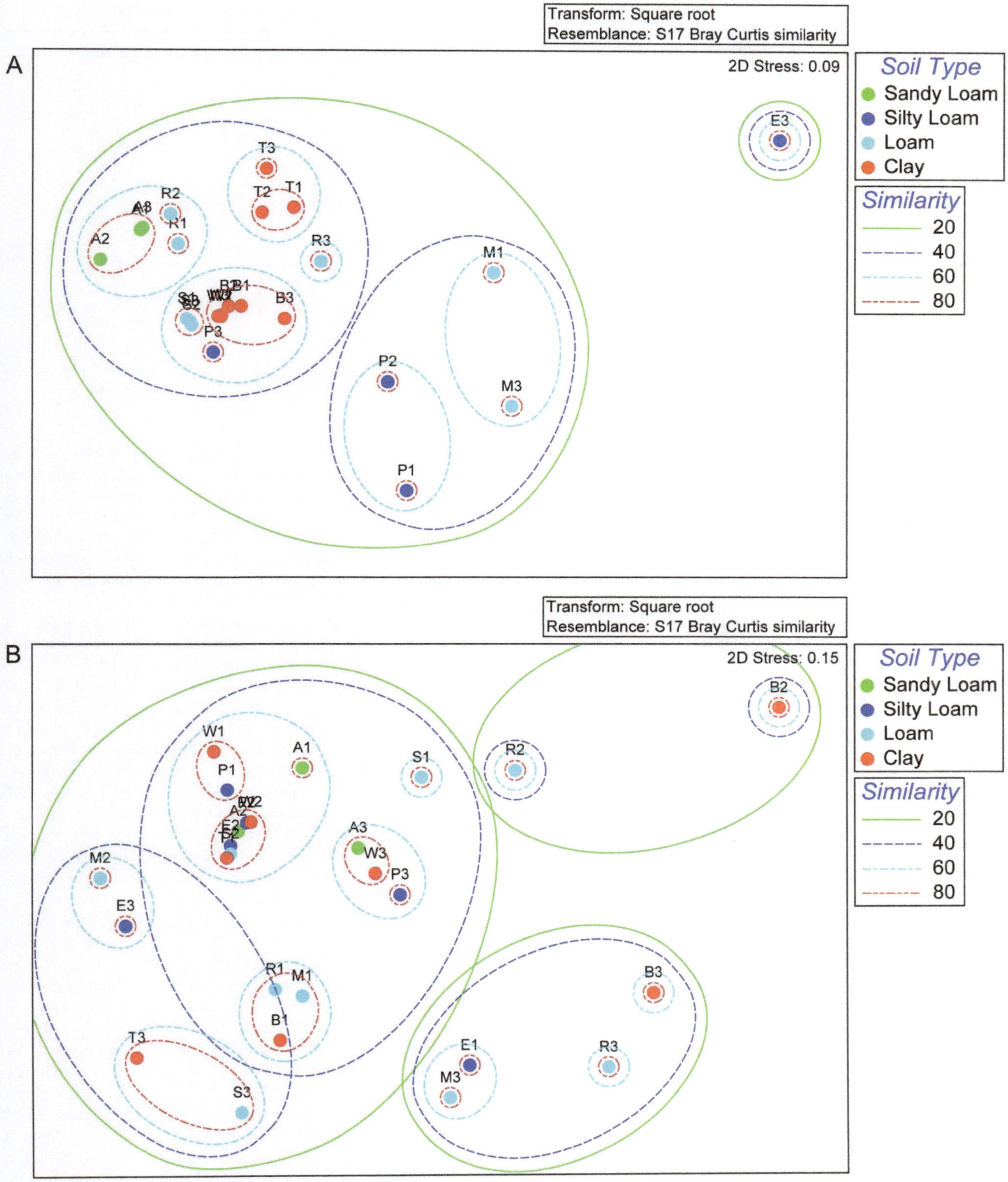

Figure 6. Non-metric multidimensional scaling plots of archaeal *amoA* terminal restriction fragment profiles generated by digestion with MspI (Panel A) or RsaI (Panel B). Symbol colors correspond to soil types, letters indicate soil names and values are the replicate number for the indicated soil. Soil name abbreviations: A, Arena; B, Brasso; E, Ecclesville; P, Piarco; R, River Estate; S, St. Augustine; T, Talparo; W, Princes Town.

Table 5. BEST variables explaining variation between soils in archaeal *amoA* RsaI terminal restriction fragment profiles[a].

No. Variables	Correlation	Variable						
		NH4	TN	TKN	Zn	TC	pH	P
2	0.630	X		X				
2	0.624	X	X					
3	0.620	X		X	X			
2	0.619	X				X		
3	0.618	X	X		X			
3	0.618	X	X					
3	0.613	X	X				X	
5	0.612	X	X			X	X	X
4	0.611	X	X	X	X			
4	0.609	X		X		X	X	

[a]Variables included in correlation indicated by "X". Abbreviations: P, Phosphorus; Zn, Zinc; TC, Total Carbon; TN, Total Nitrogen; TKN, Total Kjeldahl Nitrogen.

remainder as *Nitrosotalea*. The largest group of sequences (36%) were associated with *Nitrososphaera* subclusters 7.1 and 7.2 (Fig. 3; Table S3 in File S1), and identified in all soils except the Princes Town clay (Table S3 in File S1). Sequences grouping wih *Nitrososphaera* subcluster 4 and *Nitrososphaera* subclusters 8–10 ranked second in abundance accounting for 22% of the sequences (Fig. 3; Table S3 in File S1). The *Nitrosotalea* phylotypes were identified in clone libraries from the St. Augustine, Maracas and River Estate loams, but were most abundant in the latter two (Table S3 in File S1). Analysis of the archaeal *amoA* phylotypes by PCoA showed no significant correlation with pH (Fig. 4A). Furthermore, while all *Nitrosotalea* phylotypes were identified from loams, there was no consistent soil characteristic that explained their more frequent detection in these soils.

Sequencing of bacterial *amoA* clone libraries generated 26 OTUs from a total of 128 sequences (Table S4 in File S1). Most OTU (91%) were *Nitrosospira* phylotypes with the remainder associated with the *Nitrosomonas* (Fig. 5). Of the *Nitrosospira* phylotypes, 27% were affiliated with Cluster 3a and Cluster 3b (Fig. 5), and the remainder did not group within AOB clusters defined by Avrahami and colleagues [48]. Among those not grouping with a defined cluster was the largest phylotype (OTU 1), which comprised 29% of the sequences and occurred in all soils except the St. Augustine loam and Talparo clay (Table S3 in File S1). Analysis of bacterial *amoA* phylotypes by PCoA showed no significant corrrelation of phylotypes with pH (Fig. 4B). While one alkaline soil and one cirum-neutral soil clustered apart from one acidic soil, the rest of the soils, representing a range of pH, did not segregate (Fig. 4B).

In TRF analysis of archaeal *amoA*, RsaI profiles were more effective than those of MspI in differentiating soils by MDS (Figs. 6A,B). The ANOSIM global R of the RsaI profiles was 0.752 ($p<0.001$) indicating significant dissimilarity between soils, whereas for the MspI profiles the ANOSIM global R was 0.058 and not significant. Comparison of the observed TRF phylotypes with those predicted from archaeal *amoA* sequences indicated that the majority of TRF mapped to the above-mentioned *Nitrososphaera* phylotypes. One *Nitrosotalea*-affiliated TRF (RsaI 315) was most abundant in the River Estate soil, which also had the highest proportion of *Nitrosotalea*-like archaeal *amoA* clones. For archaeal *amoA*, BEST analysis identified the combination of ammonium levels and total Kjeldahl N (TKN) as most important in explaining variation in the RsaI profiles (correlation 0.630, $p<0.001$, Table 5). Nitrogen-related variables were a dominant factor, as ammonium appeared in all ten of the significant combinations, and was accompanied by TKN and/or TN in all but one. Four other factors (Zn, TC, pH, P) were included in some explanations with these N variables, but none yielded a more significant correlation with TRF similarity patterns than did the combination of ammonium with either TKN or TN (Table 5).

In the TRF analysis of bacterial *amoA*, soil differentiation was more effective with MspI profiles than with those of RsaI (Figs. 7A,B). The former had an ANOSIM global R of 0.819 ($p<0.001$) indicating significant dissimilarity between groups. But, the RsaI ANOSIM Global R of 0.058 was not significant. From BEST analysis of the bacterial *amoA* MspI profiles, TRF phylotype variation across the soils was best explained by phosphorus (P), bulk density (BD) and Fe (correlation 0.648, $p<0.001$, Table 6). These variables were dominant factors explaining MspI TRF profile variation, as P occurred in all ten correlations, BD in nine and Fe in eight (Table 4). Aluminium was also prominent, and was included in five correlations (Table 6). Other significant factors were Cu, Na, clay, S, B, OC, and K. Notably, neither pH nor any N-related variable was included in a significant correlation.

Our analysis of archaeal *amoA* clone libraries was consistent with prior reports establishing *Nitrososphaera* as a cosmopolitan soil inhabitant [5,7,49]. But, our identification of *Nitrosotalea* phylotypes in soils of widely varying pH (and other edpahic factors) contrasted with prior studies, which have indicated *Nitrosotalea* as associated primarily with acidic soils [25]. Furthermore, PCoA analyses of archaeal *amoA* clone libraries did not support the hypothesis that pH was the primary factor in AOA phylotype selection, and contrasted with results from prior investigators [3,7,50]. For bacterial *amoA* clone libraries, predomination by *Nitrsosospira* was congruent with the findings of other investigators [23,48,49]. But, as with AOA, the data did not confirm the hypothesized effect of pH as a major driver of AOB community structure.

Collectively, results from the BEST and qPCR analyses indicated that communities of AOA and AOB were shaped by differing sets of edaphic factors. The study conducted by Wessen et al. [52] revealed a similar conclusion, although specific soil characteristics that differentially affected these communities were not identified. In the present study, soil N characteristics

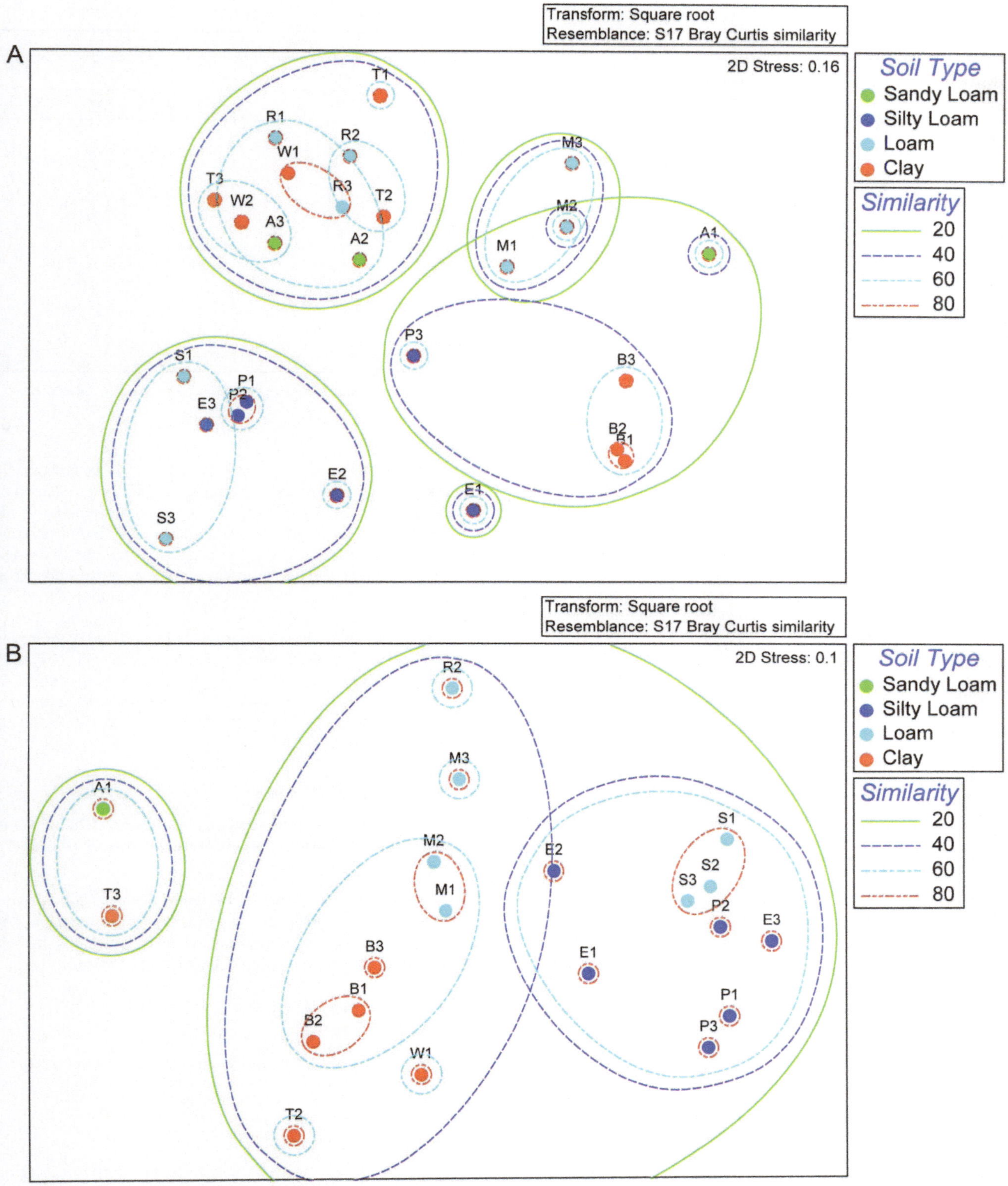

Figure 7. Non-metric multidimensional scaling plots of bacterial *amoA* terminal restriction fragment profiles generated by digestion with MspI (Panel A) or RsaI (Panel B). Symbol colors correspond to soil types, letters indicate soil names and values are the replicate number from the indicated soil. Soil name abbreviations: A, Arena; B, Brasso; E, Ecclesville; P, Piarco; R, River Estate; S, St. Augustine; T, Talparo; W, Princes Town.

Table 6. BEST Variables explaining variation between soils in bacterial *amoA* MspI terminal restriction fragment profiles[a].

No. Variables	Correlation	Variable[a]										
		P	BD	Fe	Al	Cu	Na	Clay	S	B	OC	K
3	0.648	X	X	X								
3	0.632	X		X	X							
4	0.623	X	X	X	X							
2	0.621	X	X									
6	0.619	X	X	X			X	X	X			
5	0.618	X	X	X	X	X						
5	0.617	X	X	X		X				X		
4	0.616	X	X		X	X						
5	0.612	X	X	X	X						X	
6	0.609	X	X	X			X	X				X

[a]Variables included in correlation indicated by "X". Abbreviations: P = Phosphorus; K = Potassium; S = Sulphur; B = Boron; Mn = Manganese; Fe = Iron, Cu = Copper; Al = Aluminium; Na, Sodium; BD, Bulk Density; OC, Organic Carbon.

(ammonium-N and organic N pools) were the major factor affecting AOA communities, but were not identified to affect those of AOB. This finding was significant as it provided field-based evidence that AOA ecophysiology was responsive to natural variation in N levels, data which is limited in the current literature. Also, absence of soil N as a factor affecting AOB would be consistent with adaptation of AOB to relatively high N inputs, which were lacking in the soils studied here. If so, our field-based data would support the concept that soil N levels are one factor important in affecting niche separation of AOA and AOB.

For AOB, the dominant edaphic factors affecting community structure were the combination of P, BD and Fe. Soil P level was the only variable identified to affect communities of both AOA and AOB, but it was much more significant for the AOB. Phosphorus levels have been previously identified as affecting AOB communities [19,20,24]. But, those effects emerged in response to P enrichment through exogenous P application. In contrast, in the present study, the effect of P was associated with variation in P levels endogenous to the soils. The eco-physiological significance of the P effect on AOB community structure is unclear. It may reflect a direct impact on some aspect of AOB physiology and/or indirect effects, such as competition for P with other organisms [19,20,24]. The effect of BD on AOB could have reflected one or more impacts of BD on soil characteristics, perhaps most importantly oxygen diffusion. The strong effect of iron on AOB community structure might have reflected the iron-heme dependent respiration of AOB [53,54], and possibly indicated iron acquisition was a limiting factor. The BEST analyses identified Cu and Al as community structure determinants for AOB. Both metals could have impacted community structure *via* toxicity effects, but the significance of Cu could also have reflected its core role in AOB biology as a component of ammonia mono-oxygenase and other enzymes in N metabolism [53,54]. However, since Cu and Al were intercorrelated with Fe levels, it's difficult to resolve metal-specific effects.

Conclusions

Detectable *amoA* genes of AOA predominated over detectable bacterial *amoA* genes in these tropical soils, and their community structure was affected primarily by soil N characteristics. In contrast, communities of the low abundance AOB were shaped by edaphic factors other than soil N. Contrary to our hypothesis, pH was not a major factor affecting abundance or structure of either AOA or AOB communities. Nitrification levels were strongly affected by soil pH and clay content, but not by the abundance of AOA or AOB, or aspects of AOA or AOB community structure. Thus, edaphic factors other than pH differentially affected evolution and niche differentiation of AOA and AOB communities. But, the impact of pH (and clay content) on nitrification levels appeared to more strongly reflect physico-chemical effects on AOA or AOB activity, rather than alterations in ammonia-oxidizer abundance or selection for AOA or AOB phylotypes differing in nitrifying capacity.

Supporting Information

Figure S1 Rarefaction curves for clone libraries of archaeal *amoA* (Panel A) and bacterial *amoA* (Panel B).

File S1 Supporting tables. Table S1. Soil Taxonomy and Soil Sites Sampled. **Table S2.** Amplification of Bacterial *amoA* by qPCR. **Table S3.** Distribution of archaeal *amoA* OTU by soil. **Table S4.** Distribution of bacterial *amoA* OTU by soil.

Acknowledgments

The authors thank Ronald Roopnarine and David Dyer for their assistance with this study. The authors are also grateful to Lisa Stein for many constructive comments that improved this manuscript.

Author Contributions

Conceived and designed the experiments: WJH VdG GE. Performed the experiments: VdG. Analyzed the data: WJH VdG GE. Contributed reagents/materials/analysis tools: WJH VdG GE. Wrote the paper: VdG WJH.

References

1. de Gannes V, Eudoxie G, Dyer DH, Hickey WJ (2012) Diversity and abundance of ammonia oxidizing archaea in tropical compost systems. Front Microbio 3: 1–16.
2. Di HJ, Cameron KC, Shen JP, Winefield CS, O'Callaghan M, et al. (2009) Nitrification driven by bacteria and not archaea in nitrogen-rich grassland soils. Nat Geosci 2: 621–624.
3. Gubry-Rangin C, Nicol GW, Prosser JI (2010) Archaea rather than bacteria control nitrification in two agricultural acidic soils. FEMS Microbiol Ecol 74: 566–574.
4. Treusch AH, Leininger S, Kletzin A, Schuster SC, Klenk H, et al. (2005) Novel genes for nitrite reductase and amo-related proteins indicate a role of uncultivated mesophilic crenarchaeota in nitrogen cycling. Environ Microbiol 7: 1985–1995.
5. He J-Z, Shen J-P, Zhang L-M, Zhu Y-M, Zheng M-G, et al. (2007) Quantitative analyses of the abundance and composition of ammonia-oxidizing archaea of a Chinese upland red soil under long term fertilization practices. Environ Microbiol 9: 2364–2374.
6. Taylor AE, Zeglin LH, Dooley S, Myrold DD, Bottomley PJ (2011) Evidence for different contributions of archaea and bacteria to the ammonia-oxidizing potential of diverse Oregon soils. Appl Environ Microbiol 76: 7691–7698.
7. Yao H, Gao Y, Nicol GW, Campbell CD, Prosser JI, et al. (2011) Links between ammonia oxidizer community structure, abundance and nitrification potential in acidic soils. Appl Environ Microbiol 77: 4618–4625.
8. Zeglin LH, Taylor AE, Myrold DD, Bottomley PJ (2011) Bacterial and archaeal amoA gene distribution covaries with soil nitrification properties across a range of land uses. Environ Microbiol Rps 3: 717–726.
9. Jia Z, Conrad R (2009) Bacteria rather than archaea dominate microbial ammonia oxidation in an agricultural soil. Environ Microbiol 11: 1658–1671.
10. Shen J-P, Zhang L-M, Zhu Y-G, Zhang J-B, He J-Z (2008) Abundance and composition of ammonia-oxidizing bacteria and ammonia-oxidizing archaea communities of an alkaline sandy loam. Environ Microbiol 10: 1601–1611.
11. Gubry-Rangin C, Hai B, Quince C, Engel M, Thomson BC, et al. (2011) Niche specialization of terrestrial archaeal ammonia oxidizers. Proc Natl Acad Sci (USA) 108: 21206–21211.
12. Levicnik-Hofferle S, Nicol GW, Ausec L, Mandic-Mulec I, Prosser JI (2012) Stimulation of thaumarchaeal ammonia oxidation by ammonia derived from organic nitrogen but not added inorganic nitrogen. FEMS Microbiol Ecol 80: 114–123.
13. Hatzenpichler R, Lebedeva EV, Spieck E, Stoecker K, Richter A, et al. (2008) A moderately thermophilic ammonia-oxidizing crenarchaeota from a hot spring. Proc Natl Acad Sci (USA) 105: 2134–2139.
14. Tourna M, Stieglmeier M, Spang A, Konneke M, Schintlmeister A, et al. (2011) *Nitrososphaera viennensis*, an ammonia oxidizing archaeon from soil. Proc Natl Acad Sci (USA) 108: 8420–8425.
15. Hatzenpichler R (2012) Diversity, physiology and niche differentation of ammonia-oxidizing archaea. Appl Environ Microbiol 78: 7501–7510.
16. Prosser JI, Nicol GW (2008) Relative contributions of archaea and bacteria to aerobic ammonia oxidation in the environment. Environ Microbiol 10: 2931–2941.
17. Yao H, Campbell CD, Chapman SJ, Freitag TE, Nicol GW, et al. (2013) Multi-factorial drivers of ammonia oxidizer communities: Evidence from a national soil survey. Environ Microbiol 15: 2545–2556.
18. Di HJ, Cameron KC, Shen JP, Winefield CS, O'Callaghan M, et al. (2010) Ammonia-oxidizing bacteria and archaea grow under contrasting soil nitrogen conditions. FEMS Microbiol Ecol 72: 386–394.
19. He JZ, Shen JP, Zhang LM, Zhu YG, Zheng YM, et al. (2007) Quantitative analyses of the abundance and composition of ammonia-oxidizing bacteria and ammonia-oxidizing archaea of a Chinese upland red soil under long-term fertilization practices. Environ Microbiol 9: 2364–2374.
20. Klumpp K (2008) Effects of aboveground grazing on coupling among nitrifier activity, abundance and community structure. ISME J 2: 221–232.
21. Wessen E, Soderstrom M, Stenberg M, Bru D, Hellman M, et al. (2011) Spatial distribution of ammonia-oxidizing bacteria and archaea across a 44-hectare farm related to ecosystem functioning. ISME J 5: 1213–1225.
22. Ying J-Y, Zhang L-m, He J-Z (2010) Putative ammonia-oxidizing bacteria and archaea in an acidic red soil with different land utilization patterns. Environ Microbiol Rps 2: 304–312.
23. O'Sullivan CA, Wakelin SA, Fillery RP, Roper MM (2013) Factors affecting ammonia-oxidizing microroganisms and potential nitrification rates in southern Australian agricultural soils. Soil Res 51: 240–252.
24. Fierer N, Carney KM, Horner-Devine MC, Megonigal JP (2009) The biogeography of ammonia-oxidizing bacterial communities in soil. Microb Ecol 58: 434–445.
25. Pester M, Rattei T, Flechl S, Gröngröft A, Richter A, et al. (2012) *AmoA*-based consensus phylogeny of ammonia-oxidizing archaea and deep sequencing of *amoA* genes from soils of four different geographic regions. Environ Microbiol 14: 525–539.
26. Hart SC, Stark JM, Davidson EA, Firestone MK (1994) Nitrogen mineralization, immobilization, and nitrification. In Methods of Soil Analysis, Part 2. Microbiological and Biochemical Properties; Weaver RW, Angle S, Bottomley P, Bezdicek D, Smith S et al., editors. Madison, WI: Soil Science Society of America. 985–1018 p.
27. Shand C, Williams B, Coutts G (2008) Determination of N-species in soil extracts using microplate techniques. Talanta 74: 648–654.
28. Karlsson AE, Johansson T, Per Bengtson P (2012) Archaeal abundance in relation to root and fungal exudation rates. FEMS Microbiol Ecol 80: 305–311.
29. Horz H-P, Rotthauwe J-H, Lukow T, Liesack W (2000) Identification of major subgroups of ammonia-oxidizing bacteria in environmental samples by T-RFLP analysis of *amoA* PCR products. J Microbiol Meth 39: 197–120.
30. Francis CA, Roberts KJ, Beman JM, Santoro AE, Oakley BB (2005) Ubiquity and diversity of ammonia-oxidizing archaea in water columns and sediments of the ocean. Proc Natl Acad Sci (USA) 102: 14683–14688.
31. Rees GN, Baldwin DS, Watson GO, Perryman S, Nielsen DL (2004) Ordination and significance testing of microbial community composition derived from terminal restriction fragment length polymorphisms: application of multivariate statistics. Antonie van Leeuwenhoek 86: 339–347.
32. Schloss PD, Westcott SL, Ryabin T, Hall JR, Hartmann M, et al. (2009) Introducing mothur: Open-source, platform-independent, community-supported software for describing and comparing microbial communities. Appl Environ Microbiol 75: 7537–7541.
33. de Boer W, Kowalchuk GA (2001) Nitrification in acid soils: Microorganisms and mechanisms. Soil Biol & Biochem 28: 203–211.
34. Sahrawat KL (1982) Nitrification in some tropical soils. Plant Soil 65: 281–286.
35. Sahrawat KL (2008) Factors affecting nitrification in soils. Commun Soil Sci Plan 39: 1436–1446.
36. Kyveryga PM, Blackmer AM, Ellsworth JW, Isla R (2004) Soil pH effects on nitrification of fall-applied anhydrous ammonia. Soil Sci Soc Am J 68: 545–551.
37. Kunc F, Stotzky(1980) Acceleration by montmorillonite of nitrification in soil. Folia Microbiol 25: 106–125.
38. Powell SJ, Prosser JI (1991) Protection of *Nitrosomonas europaea* colonizing clay minerals from inhibition by nitrapyrin. J Gen Microbiol 137: 1923–1929.
39. Jang X, Maa Y, Yuan J, Wright AL, Li H (2011) Soil particle surface electrochemical property effects on abundance of ammonia-oxidizing bacteria and ammonia-oxidizing archaea, NH_4 activity, and net nitrification in an acid soil. Soil Biol & Biochem 43: 2215–2221.
40. Hofferle S, Nicol GW, Pal L, Hacin J, Prosser JI, et al. (2010) Ammonium supply rate influences archaeal and bacterial ammonia oxidizers in a wetland soil vertical profile. FEMS Microbiol Ecol 74: 302–315.
41. Zhang LM, Offre PR, He J-Z, Verhamme DT, Nicol GW, et al. (2010) Autotrophic ammonia oxidation by soil thaumarchaea. Proc Natl Acad Sci (USA) 107: 17240–17245.
42. Verhamme DT, Prosser JI, Nicol GW (2011) Ammonia concentration determines differential growth of ammonia-oxidising archaea and bacteria in soil microcosms. ISME J 5: 1067–1071.
43. Stopnisek N, Gubry-Rangin C, Hofferle S, Nicol GW, Mandic-Mulec I, et al. (2010) Thaumarchaeal ammonia oxidation in an acidic forest peat soil is not influenced by ammonium amendment. Appl Environ Microbiol 76: 7626–7634.
44. Erguder TH, Boon N, Wittebolle L, Marzorati M, Verstraete W (2009) Environmental factors shaping the ecological niches of ammonia-oxidizing archaea. FEMS Microbiol Rev 33: 855–869.
45. Valentine DL (2007) Adaptations to energy stress dictate the ecology and evolution of the Archaea. Nat Rev Microbiol 5: 316–323.
46. Green CJ, Blackmer AM, Yang NC (1994) Release of fixed ammonium during nitrification in soils. Soil Sci Soc Am J 58: 1411–1415.
47. Pett-Ridge J, Petersen DG, Nuccio E, Firestone MK (2013) Influence of oxic/anoxic fluctuations on ammonia oxidizers and nitrification potential in a wet tropical soil. FEMS Microbiol Ecol 85: 179–194.
48. Avrahami S, Conrad R, Braker G (2002) Effect of soil ammonium concentration on N_2O release and on the community structure of ammonia oxidizers and denitrifiers. Appl Environ Microbiol 68: 5685–5692.
49. Boyle-Yarwood SA, Bottomley PJ, Myrold DD (2008) Community composition of ammonia-oxidizing archaea in soils under stands of red alder and Douglas fir in Oregon. Environ Microbiol 10: 2956–2965.
50. Nicol GW, Leininger S, Schleper C, Prosser JI (2008) The influence of soil pH on the diversity, abundance and transcriptional activity of ammonia oxidizing archaea and bacteria. Environ Microbiol 10: 2966–2978.
51. Webster G, Embley TM, Freitag TE, Smith Z, Prosser PI (2005) Links between ammonia oxidizer species composition, functional diversity and nitrification kinetics in grassland soils. Environ Microbiol 7: 676–684.
52. Wessen E, Hallin S, Philippot L (2010) Differential responses of bacterial and archaeal groups at high taxonomical ranks to soil management. Soil Biol & Biochem 42: 1759–1765.
53. Glass JB, Orphan VJ (2012) Trace metal requirements for microbial enzymes involved in the production and consumption of methane and nitrous oxide. Front Microbio 3:61.
54. Godfrey LV, Glass JB (2011) The geochemical record of the ancient nitrogen cycle, nitrogen isotopes, and metal cofactors. Methods Enzymol 486: 483–506.

5

Quantitative Field Testing *Heterodera glycines* from Metagenomic DNA Samples Isolated Directly from Soil under Agronomic Production

Yan Li[1], Gary W. Lawrence[1], Shien Lu[1], Clarissa Balbalian[1], Vincent P. Klink[2]*

1 Department of Biochemistry, Molecular Biology, Entomology and Plant Pathology, Mississippi State University, Mississippi State, Mississippi, United States of America,
2 Department of Biological Sciences, Mississippi State University, Mississippi State, Mississippi, United States of America

Abstract

A quantitative PCR procedure targeting the *Heterodera glycines* ortholog of the *Caenorhabditis elegans uncoordinated-78* gene was developed. The procedure estimated the quantity of *H. glycines* from metagenomic DNA samples isolated directly from field soil under agronomic production. The estimation of *H. glycines* quantity was determined in soil samples having other soil dwelling plant parasitic nematodes including *Hoplolaimus*, predatory nematodes including *Mononchus*, free-living nematodes and biomass. The methodology provides a framework for molecular diagnostics of nematodes from metagenomic DNA isolated directly from field soil.

Editor: Vishal Shah, Dowling College, United States of America

Funding: The work was supported by the Mississippi Soybean Promotion Board. The funders had no role in study design, data collection and analysis, decision to publish, or preparation of the manuscript.

Competing Interests: The authors have declared that no competing interests exist.

* E-mail: vklink@biology.msstate.edu

Introduction

Plant parasitic nematodes present a major problem for the cultivation of *Glycine max* (soybean) [1–5]. Among them, the dominant problem is the soybean cyst nematode, *Heterdera glycines* [2–5,6]. *H. glycines* was first detected in North Carolina in 1954 [7] and subsequently spread rapidly to as far away as Mississippi by 1957 [8]. The host range of *H. glycines* is broad, updated to over 399 plant species [9–13]. *H. glycines* are a bisexual, cyst-forming species consisting of six life stages including the egg, four juvenile stages (J1–J4), and the adult stage [6,14,15]. The duration of the *H. glycines* life cycle is 3–4 weeks, but this time frame may be influenced by environmental conditions [15]. *H. glycines* feed from their plant host by producing a nurse cell called a syncytium which forms by the fusion and incorporation of approximately 200 cells [16]. Depending upon the environment, several generations of *H. glycines* can be completed in a typical soybean growing season. Adding to management problems, eggs within cysts can remain dormant for 9 years [17]. Management is further complicated because cultivars under production, while experiencing as much as a 30% decrease in yield loss, may appear symptom-free [18]. Infection of soybean by *H. glycines* results in annual losses of $ 1 billion in yield in the U.S. [19] and about $15 billion worldwide. To mitigate these losses, management of *H. glycines* is accomplished partly by the use of soybean genotypes that are resistant to infection [20–24]. Although the use of resistant genotypes is a component of recommended control measures [25], consistent use of soybean cultivars with the same sources of resistance can lead to adaptation of the existing *H. glycines* population to cultivars with that source of resistance [26]. Along with the use of resistant germplasm, the application of crop rotation and nematicides are also used practices in the management of *H. glycines* [27].

A long-used agricultural management strategy is to determine the types and quantities of nematodes present in the soil before planting and this has been particularly important for managing soybean cultivation in *H. glycines*-infested fields [13,28–31]. While these types of tests are traditionally done by trained nematologists, the analyses can be time consuming (i.e. up to 100 days or longer). Therefore, molecular diagnostic tests which could be done in a shorter time frame (i.e. hours) would be useful in the management of many plant-parasitic nematodes [32–39]. Quantitative PCR (qPCR) is a procedure that specifically assays the PCR reaction by incorporating a molecular tag whose amplification products are measured during the PCR reaction. The qPCR method works quantitatively in pathogen detection and measurement because it detects the pathogen by the using the amount of DNA present in a sample to obtain a cycle threshold (Ct) value which corresponds to the level of infection [40]. The qPCR methodology has been shown to be a very powerful tool in pathogen detection, used to identify and quantify various bacteria, fungi, and viruses [41–44]. Population estimates for nematode species such as *M. javanica*, *Pratylenchus zeae*, and *Xiphinema elongatum* have been determined with qPCR on native soil samples from trial plots [45]. The qPCR procedure has also been used to determine the quantity of the plant parasitic nematode *Rotylenchulus reniformis* from metagenomic DNA isolated directly from field plots [46].

It is hypothesized that conserved gene sequences can be used in determining the number of *H. glycines* from metagenomic DNA isolated directly from soil. In the analysis presented here, highly conserved *H. glycines* homologs of the *C. elegans* uncoordinated gene family genes [47] were screened for their use as biomarkers against

Table 1. PCR primer pairs for *H. glycines* homologs of *uncoordinated* (*unc*) class genes.

H.glycines gene	*H.glycines* est	Product Size	Primer	TM
Hg-unc-9	CB281382	124	F-CCATGGTGCGCTATTTGTCA	62.95
			R-CTGCCCCCAAATTGTTTGAA	63
Hg-unc-9	CB281382	253	F-GCGCGAAAGGACGATGATTTTTC	66.5
			R-CTTACGGCCGGACGAATACCTCTC	66.6
Hg-unc-22	CB378705	120	F-GACGAAATTGTGGCCGAGTC	62.91
			R-AAATTGTCCCGCGTCCTCTT	63.11
Hg-unc-31	CB378080	110	F-CTCCGATGGTTGTCCGCTAC	62.91
			R-GGTTGAGCAACCGTCTTTGC	63.05
Hg-unc-52	CK350534	114	F-ACCGCAGGTGTACGATGGTT	62.69
			R-CCGTAGGCGGTCACTTTGTC	62.92
Hg-unc-78	CB238521	97	F-CGTTTTTGGGACACCACACA	62.79
			R-TGCTGTCCTCAGACCACGAA	63.04
		qPCR probe	GAAGTCGGAGTTCGCTCTTCTTTCG	70.98
Hg-unc-78	CB238521	313	F-GTGGAGACCAATCGGGCAAAATC	66.4
			R-GAAAGGAGGGCCTTCGAAAATGG	66.5
Hg-unc-89	CB379143	120	F-GCGCGGTACTGACGAAAGTC	63.19
			R-GCAGGACAGTTTCCGCATTC	63.04
Hg-unc-89	CB379143	289	F-CCCGTACACACATTTCCGCAGTC	66.3
			R-CAGCCGACCATCGAGTTCCATAC	66.5
Hg-unc-101	CB379764	94	F-CATGCAAGGCAACAGATTCG	62.7
			R-TAACAGCGCACATCCAAACG	63.09
Hg-unc-115	CK350435	103	F-ACGGAAGTCGCGCTATTCAA	63.06
			R-GTCGTTGTCCACGGAAGAGG	63.01
Hg-dys-1	CB934909	124	F-GCTATTTGCCGGTCGAACAA	63.28
			R-TTGTCCAATCTCGCGGCTAT	62.92
Hg-dys-1	CB934909	303	F-GTTTCCGATCGTTGGACTTCG	66.5
			R-GCTGGTGCATTGCCTCTGTTTC	66.4
Hg-nep-1	CB824545	114	F-TATTCGGGCGTCAAAAATGC	63.04
			R-GCCAATCACTGCTCCAATCC	62.9
Hg-MRCK-1 (control)	CB380016	125	F-CCACCGACACGTCCAACTTT	63.27
			R-GAAGGTGAAGCCGATGAACG	63.02

other phylogenetically related and unrelated nematodes [48]. The molecular diagnostic was able to specifically identify *H. glycines* from metagenomic DNA samples isolated directly from soil under agricultural production. Furthermore, a single, *H. glycines* could be identified from the metagenomic DNA sample. The methodology is rapid, and can be performed *en masse* in a single afternoon.

Results

Probe Identification

Alkharouf et al. [49] analyzed 8,334 *H. glycines* unigenes that resulted in the identification of numerous genes involved in essential aspects of their biology. From those sequences, 11 genes were targeted for their use as molecular probes to determine the level of soil infestation by *H. glycines* (**Table 1**). The 11 primer pairs, designed from these gene sequences, were used in a series of PCR reactions to identify genes that would robustly amplify *H. glycines* DNA isolated from pure single cysts as compared to a control gene (**Figure 1**). As demonstrated by an increase in the amount of amplification product, semiquantitative PCR tests demonstrated Hg-unc-78 primer pairs were able to detect increasing amounts of DNA from increasing numbers of *H. glycines* J2 (**Figure 2**). Gene sequencing confirmed the PCR amplicon was Hg-unc-78 (**data not presented**).

Detecting *H. glycines* using Metagenomic DNA Isolated Directly from Soil

The experiments then focused on determining if *H. glycines* DNA could be isolated and detected directly from soil samples. Metagenomic DNA isolation, followed by PCR, demonstrated that the Hg-unc 78 amplification product could be detected as a band in samples with as few as 1 added J2 (**Figure 3**). To further confirm the specificity of Hg-unc 78 primers, experiments on agriculturally significant off-target plant-parasitic nematodes

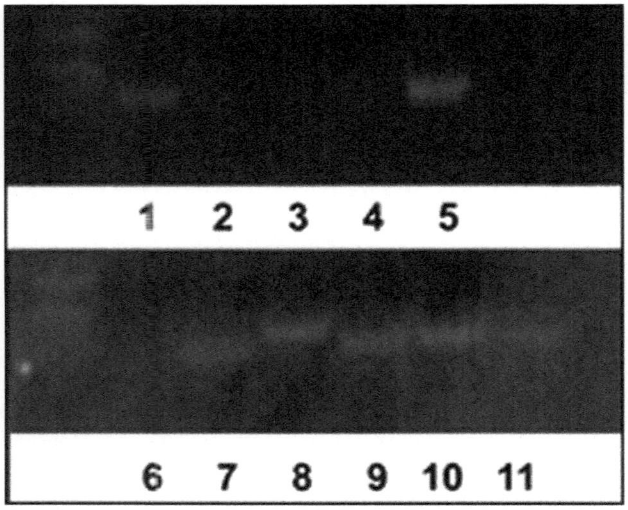

Figure 1. Screening of Hg-unc primer pairs. Lane 1, Hg-unc 9; Lane 2, Hg-unc 22; Lane 3, Hg-unc 31; Lane 4, Hg-unc 52; Lane 5, Hg-unc 115; Lane 6, Hg-unc 101; Lane 7, Hg-dys-1; Lane 8, Hg-nep-1; Lane 9, Hg-unc 89; Lane 10, Hg-unc 78; Lane 11, Hg-MRCK-1.

including *R. reniformis* [50] (Hoplolaimidae) and *M. incognita* (root knot nematodes, RKN) (Heteroderidae), along with *H. glycines*, DNA were used in PCR experiments. No amplification was achieved in reactions containing *R. reniformis* or *M. incognita* DNA, but amplification was successful on reactions containing *H. glycines* DNA (**Figure 4**). This outcome indicates that the primers designed to amplify Hg-unc 78 is specific to *H. glycines*. Since it was possible to amplify target *H. glycines* DNA from soil containing as little as a single *H. glycines* J2 nematode, experiments were done whereby metagenomic DNA was isolated directly from field samples containing numerous parasitic and free-living nematodes as well as other living organisms. Soil samples were collected in triplicate and a visual assessment of the nematodes present within the soil sample were determined (**Table 2**). This procedure determined that the soil contained other species of nematodes that are more distantly related to *H. glycines*, including *Hoplolaimus* (sp.) [51] (Hoplolaimidae) and *Mononchus* (sp.) [52] (Mononchidae). Known numbers of *H. glycines* were then added to the soil samples and the metagenomic DNA was isolated directly from the soil as described previously. The amplification profile of the Hg-unc-78 DNA is shown (**Figure 5**). The demonstration of increasing amounts of amplification product in samples having increasing amount of added *H. glycines* indicated that it would be possible to adapt qPCR to the detection method.

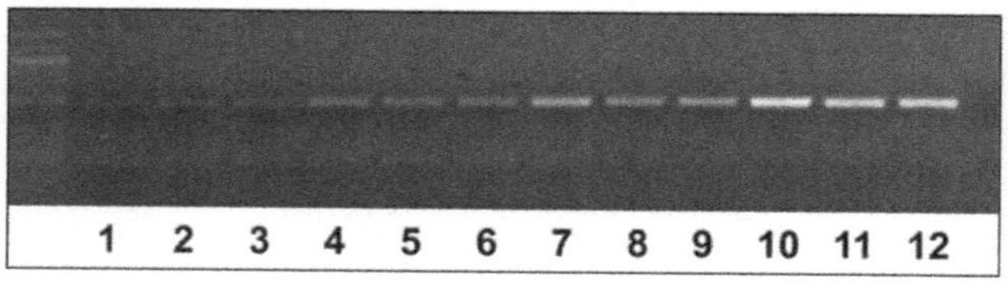

Figure 2. DNA amplification of Hg-unc-78 from different numbers of pure, extracted *H. glycines* (SCN) J2 cultured in soil in greenhouse. Lanes 1–3, 1 SCN; Lanes 4–6, 10 SCN; Lanes 7–9, 100 SCN; Lanes 10–12, 1,000 SCN.

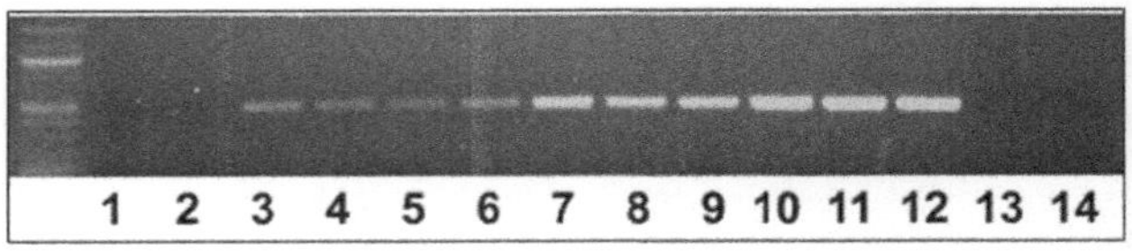

Figure 3. DNA amplification of Hg-unc-78 from metagenomic DNA isolated directly from soil containing different numbers of *H. glycines* (SCN). Lanes 1–3, 1 SCN; Lanes 4–6, 10 SCN; Lanes 7–9, 100 SCN; Lanes 10–12, 1,000 SCN; Lane 13, without primer (control); Lane 14, without DNA (control).

qPCR Estimate of *H. glycines*

The qPCR experiments began with pure, greenhouse cultured *H. glycines* J2s (**Figure 6**). From this standard curve, it is demonstrated that the Hg-unc-78 qPCR primers are able to measure, quantitatively, the different amount of *H. glycines* DNA within the samples. This measurement allows for the correlation of the quantity of the amplified DNA amount to estimate number of *H. glycines* J2s. The same procedure then was used to estimate the number of *H. glycines* J2s directly from metagenomic DNA isolated directly from soil samples (**Figure 7**). Subsequently, actual soil samples under agricultural production were assayed using qPCR for *H. glycines* (**Table 3**). Metagenomic DNA was isolated from those samples and used in qPCR experiments. Since the number of *H. glycines* is a pool of both cysts and J2s, it was necessary to estimate the number of J2s per cyst. Using that estimate, a correlation between the counted *H. glycines* as compared to the qPCR estimate is provided (**Table 3**). DNA sequencing of the qPCR product demonstrated the specificity of the reaction by revealing the amplification product was Hg-unc-78 (**data not presented**).

Discussion

The determination of the numbers of specific types of plant parasitic nematodes in the soil is important in agriculture because it allows for understanding threshold populations that could detrimentally affect plant productivity. This knowledge, in turn, will allow growers to make informed choices in terms of cultivar and/or crop selection as nematode populations can vary from year to year. The analysis presented here resulted in the development of a simple and reliable molecular diagnostic technique that determines the number of *H. glycines* from metagenomic DNA isolated directly from field soil under agricultural production. During the course of the analysis, gene sequences for *H. glycines* homologs of *C. elegans unc* genes were identified, tested and analyzed for their reliability in molecular diagnostics. During the analysis, qPCR primers were designed and used on metagenomic

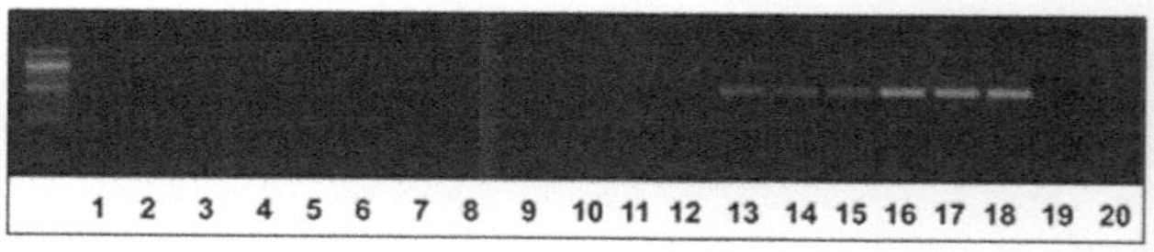

Figure 4. The specificity of the Hg-unc78 PCR reaction from DNA isolated from different numbers of pure, extracted off-target nematodes as compared to DNA isolated from pure, extracted *H. glycines*. Lanes 1–3; 100 Rr; Lanes 4–6, 1,000 Rr; Lanes 7–9, 100 RKN; Lanes 10–12, 1,000 RKN; Lanes 13–15, 100 SCN; Lanes 16–18, 1,000 SCN; Lane 19, without primers (control); Lane 20, without DNA (control). Rr, *Rotylenchulus reniformis*; RKN, *Meloidogyne incognita*.

Table 2. Quantification of major nematode fauna from field soil.

Microplots	*Hoplolaimus* spp	*Mononchus* spp	Free living
1	1287.5	772.5	772.5
2	772.5	2317.5	257.5
3	257.5	0	515
4	257.5	257.5	515

DNA isolated directly from actual soil samples from real agricultural sites. While the focus of the analysis was low numbers of *H. glycines* (1–1,000 individual J2s), it is expected that the methodology would also be reliable for analyzing field samples that are more highly infested.

Hg-unc-78 as a Molecular Probe

Gene sequences analyzed in Alkharouf et al. [49] was mined for genes that could be used as reliable molecular probes for the presence of *H. glycines* in actual field samples under agricultural production. Genes that function in nematode movement, an important aspect of parasitism, and were also conserved in their primary structure to *C. elegans* were identified and used for PCR primer design. Many of these genes were *unc* genes, first described in *C. elegans* [53]. The experiments presented here show that it is possible to use highly conserved DNA sequences that are homologous to *unc* genes from *H. glycines* and use them to specifically amplify gene fragments from as few as 1 J2. The specificity was shown when attempts in amplification reactions of off-target nematodes and, in particular, metagenomic DNA isolated directly from soil samples. DNA sequence analyses of the DNA amplification products from the PCR reactions show that the amplification product perfectly matches its target DNA from *H. glycines*. Lastly, using the same primers to amplify target DNA from metagenomic DNA isolated directly from soil samples reveals that it is possible to amplify the target as shown by sequencing reactions of that amplification product to as low as a single *H. glycines* J2. Furthermore, in regard to the major nematode species found in Mississippi soils, the procedure appears to distinguish *H. glycines*. However, all nematodes have not been examined.

Quantitative PCR

The qPCR experiments were designed with the goal of determining the numbers of *H. glycines* in soil samples collected from actual field sites under agricultural production. From the described PCR studies, it was determined that it would be possible

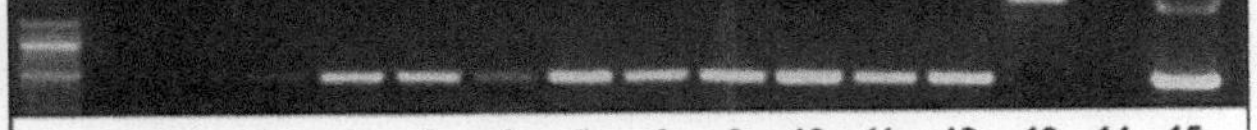

Figure 5. DNA amplification of the Hg-unc78 from metagenomic DNA samples. Lanes 1–3, metagenomic DNA including 1 SCN J2; Lanes 4–6, metagenomic DNA including 10 SCN J2; Lanes 7–9, metagenomic DNA including 100 SCN J2; Lanes 10–12, metagenomic DNA including 1,000 SCN J2; Lane 13, cloned gene without primers (control); Lane 14, without DNA (control); Lane 15, positive control, amplification product is Hg-unc 78 amplified from the cloned gene.

to identify a single *H. glycines* from metagenomic DNA. Similar observations have been made for *R. reniformis* [46]. The DNA samples isolated from pure cultures of *H. glycines* could then be used in the development of a standard curve from which it was possible to estimate the number of *H. glycines*. With this capability, it was then tested whether it would be possible to take metagenomic DNA samples isolated from soil having predefined numbers of *H. glycines* to quantify the number of *H. glycines*. As shown, the Hg-unc-78 primer set could be used in experiments to accurately estimate the number of J2s in the sample. The quantitative analysis, using qPCR, demonstrates that low numbers of *H. glycines* can be identified reliably from the metagenomic samples. Notably, the quantity of *H. glycines* has been determined through visual analysis from actual soil samples under agricultural production, serving as experimental samples. Metagenomic DNA then was isolated from those samples and used for subsequent qPCR analyses. The analysis demonstrated that the qPCR did not perfectly determine the exact number of *H. glycines* in the agricultural samples. The same observation was made for *R. reniformis* [46]. However, it is well known that *H. glycines* are not evenly distributed within soil samples [27]. Thus, the limits of detection may have been met. Further experimentation is required to explore this problem further.

A key question remaining is whether different compositions of soil will affect the DNA isolation efficiency. In Mississippi where the experiments were carried out, there are 5 major soil types. The soil types vary among alfisols, entisols, inceptisols, ultisols and vertisols [54]. It is possible that these different soil compositions could affect the efficiency of the DNA isolation and amplification. While no major influence was observed on amplification from the different sample types used in the various analyses, a thorough analysis on different soil types from the obtained unknown samples requires further testing.

Materials and Methods

Ethics Statement

Since the focal point of the research was plant parasitic nematodes, IACUC regulations were not relevant to this study.

Materials Statement

No specific permits were required for the described field studies because the soil samples were collected on the grounds owned by Mississippi State University and contained no endangered or protected species as determined by a stereomicroscopic analysis of soil.

Plant Inoculation

Prior to nematode inoculation, *G. max* seedlings are grown in a sterilized sand: soil mix (1:1) in clay pots for a period of 7 to 14 days. The nematodes of second stage juveniles were counted, concentrated and diluted to a final concentration of 2,000 J2 per 3 ml which were then added to each root. This meant that 3 ml of inoculum that contained 2,000 nematodes were added to each root system on each plant by inoculating on two spots evenly.

Plant and Nematode Culture

The target nematode *H. glycines* and off-target *M. incognita* (root knot nematode, RKN) [55–56] (Heteroderidae) and *R. reniformis* (reniform nematode) [48] (Hoplolaimidae) were cultured under ambient conditions in a greenhouse at the Mississippi Agriculture and Forestry Experiment Station (MAFES), RR Foil Plant Science Research Center, North Farm, Mississippi State University. Supplemental fluorescent light was provided to bring the day

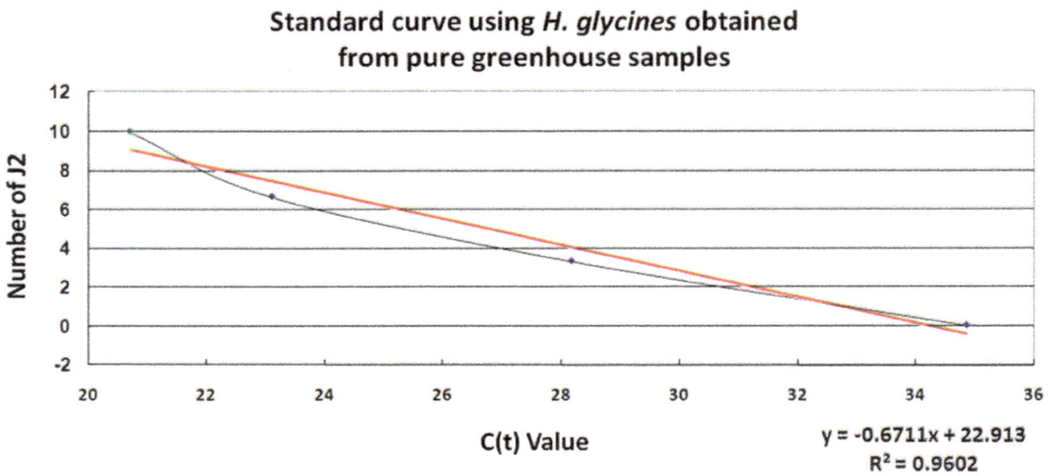

Figure 6. Estimation of *H. glycines* population size by qPCR. Standard curve obtained using pure greenhouse samples, (b) soil samples. The statistical significance between the actual number of SCN and the estimated number are provided.

length to a 16 hour day/8 hour night cycle. Temperatures were kept in a constant temperature range between 28.9–34.4°C. Nematodes were cultured in 500 cm^3 diameter clay pots for a period of 2–6 months in a 50:50 mixture of a fine sandy loam (46.25% sand, 46.50% silt, and 7.25% clay) and sand (93.00% sand, 5.75% silt, and 1.25% clay). Harvesting cysts was accomplished by massaging the infected roots in water. Massaging was achieved by placing the infected root mass between the index finger and thumb. The index finger and thumb were gently rubbed together with the root mass between them. This gentle rubbing activity dislodged the cysts so that they could be collected on a 150 μm pore sieve under a constant water steam. The cysts are not damaged because of their protective hardened nature. Cysts were collected by rinsing them through nested 850 μm pore sieve (debris) and a 150 μm sieve (cyst). Then cysts were rinsed out of the 150 μm pore sieve into a beaker with 100 ml of water. Juveniles were collected by rinsing them through 250 μm pore sieve (debris) and 45 μm pore sieve (J2) and transferred to a beaker with 100 ml water in it for counting. Cysts were crushed by using Janke & Kunkel IKA-WERK crushing machine (IKA-WORKS, INC), for 30 seconds to 1 minute with a timer. Removal of debris smaller than the eggs was done by washing the slurry though a 25 μm pore sieve. The eggs were transferred to a beaker with 100 ml water in it for counting [57].

Soybean Cyst Nematode DNA Isolation

To isolate DNA from a single cyst, a single cyst was picked up under a stereomicroscope and placed into a 1.5 ml microcentrifuge tube, with 14 μl sterile distilled water. The tube was placed in −20°C for 1 hour to freeze the cyst (or liquid nitrogen for 1 min). A glass rod sterilized with 75% ethanol was used to grind the cyst containing mixture until it is thawed. A 3 μl 10×PCR buffer and 3 μl proteinase K was then added. The material was incubated at −20°C for at least 2 hours. The solution was incubated for 90 min at 65°C to denature the protein, and 15 min at 9°C to unwind the DNA. The solution was centrifuged for 1 min at 8,000 rpm/h with the supernatant transferred to a new 1.5 ml microcentrifuge tube. The DNA was stored at −20°C for later PCR analysis.

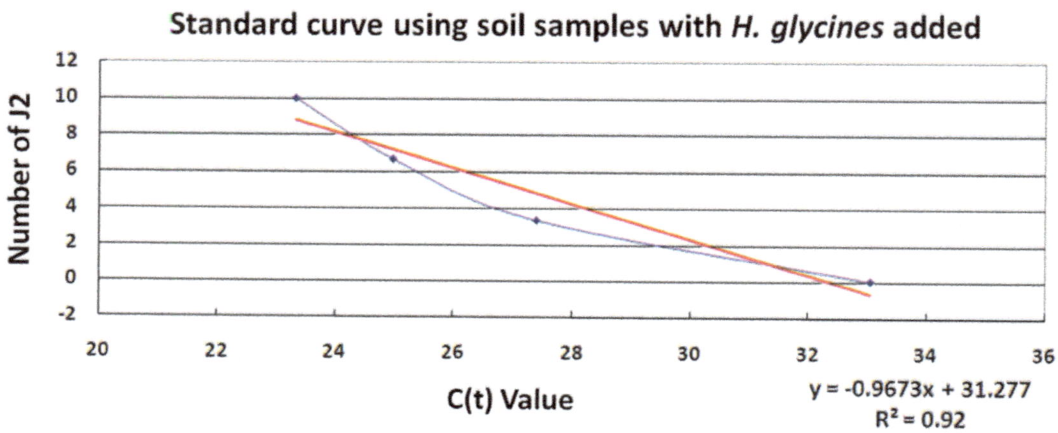

Figure 7. Estimation of *H. glycines* population size by qPCR. Standard curve obtained from metagenomic DNA isolated directly from soil samples. The statistical significance between the actual number of SCN and the estimated number are provided.

Table 3. qPCR estimation of *H. glycines* from field soil samples.

Sample	J2/pint	Cysts/pint	Cyst to J2*	Actual NO. J2/pint	qPCR J2 estimate/pint	fold difference
1	0	8	1,600	1,600	8,394.38921	0.19
2	0	16	3,200	3,200	5,599.6761	0.57
3	0	24	4,800	4,800	7,692.002288	0.62
4	24	16	3,200	3,224	4,928.816234	0.65
5	0	16	3,200	3,200	2,369.447	1.35
6	63	24	4,800	4,863	1,857.50756	2.62
7	8	8	1,600	1,608	421.9439991	3.81
8	0	8	1,600	1,600	370.9181429	4.32
9	16	47	9,400	9,416	1,557.579433	6.05
10	32	32	6,400	6,432	747.7392119	8.62

Gene Screening

H. glycines genes identified in Alkharouf et al. [49] were used for primer design according to Showmaker et al. [46] who used the identical methodology for quantitative molecular detection of *Rotylenchulus reniformis* DNA from metagenomic samples isolated from soil. Highly conserved genes from Group 1 [49], having 266 genes with *E*-values between 0 and 1*E*-100, were the focus of the analysis because no other *Heterodera* spp. agricultural important species has been published to be found in Mississippi soils at this time. Many of these genes included the *uncoordinated* (*unc*) class were first described in *C. elegans*. The genetically-defined *unc* genes group contains 114 different members and function in muscle focal adhesion, architecture and stimulation [53].

Metagenomic DNA Isolation

Metagenomic DNA from soil isolations were conducted by using the Powersoil DNA extraction kit (MO BIO Laboratories, Inc; Carlsbad, CA) using the manufacturer's protocol with modifications. The modifications included using 0.3 grams of soil instead of 0.25 grams of soil in step for developing the quantitative PCR standard curve and adding 0.1 ml of nematode suspension (water containing the numbers of J2 extracted from greenhouse cultured soil samples). In its place, 0.1 ml of the nematode suspension, extracted from either greenhouse pots or field soil, was pipetted into the bead beating tube. Secondly, in the step when first instructed to remove the supernatant, a standard volume of 550 ml of supernatant was removed from each tube. This ensured for downstream applications. The DNA was eluted from the spin column in 50 µl of nuclease free water (Promega; Madison, WI) and stored at –20°C.

PCR Reaction Conditions

For PCR, a 50 µl PCR reaction consisting of 2 µl DNA template, 5.0 µl 10×PCR buffer, 1.0 µl dNTP, 3.0 µl $MgCl_2$, 0.25 µl recombinant Taq Polymerase, 1.25 µl of 100 nM forward and reverse primers each, 36. µl nuclease free water (Ambion) (Promega) was used. The reaction conditions, as reported by Agudelo et al. [39] were modified to include a 2 minute pre-denaturation step at 94°C. The procedure then followed the Agudelo et al. [39] protocol that included a denaturation at 94°C for 45 sec, annealing at 59°C for 45 sec and primer extension at 72°C for 60 sec for 40 cycles. The PCR reaction products were run out by gel electrophoresis on a 1% agarose gel with 0.01% SYBR-Green incorporated into the gel. The DNA amplification products were visualized and recorded with digital imagery using a FOTO/Analyst Apprentice System (FOTODYNE Inc.; Hartland, WI).

Cloning and Sequencing

After using Accuprime polymerase for the PCR, the DNA amplification products were extracted on the gel by using the gel extraction kit (QIAQuick Gel Extraction Kit), mixing 4 µl of the gel purified DNA with 1 µl of salt solution and 1 µl of TOPO vector for the TOPO cloning reaction (pENTR Directional TOPO Cloning Kits, Invitrogen) in a 1.5 ml microcentrifuge tube. The tube was incubated at room temperature for 5 min and then placed on ice. Chemically competent *E. coli* stored at −80°C was thawed on ice. Transformation began by distributing the *E. coli* cells in a separate 1.5 ml microcentrifuge tube, at a volume of 20 µl each, which is proceeded on ice. Then, 3 µl of TOPO cloning reaction was added to the *E. coli* cells, tapping the tube gently. The *E. coli* solution were incubated on ice for 15 min followed by heat shocking the *E. coli* cells at 42°C for 30 seconds on the heat block. The *E. coli* solution then was placed back on ice immediately. Subsequently, 100 µl of SOC media was added to the tube and shaken at 37°C for 1 hour. After incubation, 70 µl of the solution were plated out on LB-kanamycin plates. The plates were incubated at 37°C overnight for 12–24 hours. The next day, single colonies were picked and placed in a 50 ml centrifuge tube containing LB-kanamycin culture agar. The tubes were cultured for 37°C for no more than 16 hours, usually around 15 hours. A plasmid prep was done on the culture (QIAPrep, Miniprep, QIAGEN).

To confirm that the DNA amplification in both PCR and qPCR reactions were products of *H. glycines* DNA and not spurious amplification of off-target DNA, DNA amplification products were run out electrophoretically on and then isolated from the 1% agarose gels. The DNA was purified using the Qiaquick Gel Extraction Kit (Qiagen; Valencia, CA) as according to the manufacturer's specifications. The isolated DNA was ligated into the pGEM-T Vector System II (Promega). The ligation reaction was shuttled into competent JM109 *E. coli* cells and selected on 50 µg/ml ampicillin on LB-agar plates. Colonies were selected and grown in liquid culture in LB media containing 50 µg/ml ampicillin. Plasmid DNA was isolated from the bacteria using the Qiaprep kit (Qiagen). The DNA from the plasmid preps was sequenced to determine if the DNA amplification product was correctly amplifying the proper target. The DNA sequence was trimmed using the Crimson Editing freeware (http://www.

crimsoneditor.com/). In this procedure, the pGEM-T Vector DNA sequence was trimmed leaving the qPCR-generated sequence. The trimmed sequence was blasted in GenBank using the blastn query option. This additional quality control step demonstrated the accuracy of the qPCR reaction conditions.

Quantitative PCR

Taqman 6-carboxyfluorescein (6-FAM) qPCR probes (MWG Operon; Birmingham, AL) were used. The 6-FAM probes have a maximum excitation at 495 nm and maximum emission at 520 nm. The qPCR quencher was the Black Hole Quencher (BHQ1) (MWG Operon), with maximum excitation at 534 nm. Assays were conducted for primers that produced a single amplicon and had no off target amplification. The qPCR reaction conditions included a 20 µl Taqman Gene Expression Master Mix (Applied Biosystems; Foster City, CA), 0.9 µl 100 µM forward primer, 0.9 µl 100 µM reverse primer, 2 µl 2.5 µM 6-FAM (MWG Operon) probe and 4.4 µl metagenomic template DNA. The conditions were a denaturation at 94°C for 45 sec, annealing at 54°C for 45 sec and primer extension at 72°C for 60 sec for 40 cycles. The qPCR reactions were performed on an ABI 7300 (Applied Biosystems).

To generate a standard curve for the amount of *H. glycines* in a soil sample, estimates of approximately 1,000 nematodes in 0.1 ml of water were placed into the Powersoil DNA isolation kit bead beating tubes and extracted as described previously, replicated this by 2 more times. A 1:10 serial dilution series of DNA extracted from approximately 1,000 nematodes was created and used for generation of the standard curve by qPCR. To evaluate the accuracy of the standard curve, samples containing 0, 1, 10 and 100 *H. glycines* nematodes were generated by carefully hand collecting them under a stereoscope, 3 replicates of each, and isolated the DNA by the Powersoil DNA isolation kit as described previously. The qPCR methodology works quantitatively because it detects pathogens by using the amount of DNA present in a sample to obtain a cycle threshold (C(t)) value which corresponds to the amount of target DNA (Livak and Schmittgen, 2001). The lower the C(t), the greater the amount of the corresponding DNA (target organism) is present in a sample.

H. glycines Estimation

50 g of soil was weighed and put into a cylinder. Subsequently, 50 ml of water was put in the cylinder. Because of the volume of the soil, the total volume increased to 70 ml, meaning the 50 g of soil was equal to a volume of 20 ml. With 50 g taking up 20 ml, 0.3 g of soil takes (0.3 g×20 ml)/50 g = 0.12 ml. So the estimated number of J2 in 1 pint = 473.176 ml×C(t)/0.12.

Acknowledgments

The authors thank Dr. Larry Heatherly and the Mississippi Soybean Promotion Board and the Mississippi Agricultural and Forestry Experimental Station (MAFES). Dr. Scott Willard, Department of Biochemistry, Molecular Biology, Entomology and Plant Pathology (BMEP), Mississippi State University provided greenhouse space. Yixiu Pinnix (BMEP) provided technical support. The authors are thankful to Dr. Tom Allen, Delta Research and Extension Center, Stoneville, MS.

Author Contributions

Conceived and designed the experiments: YL GL VK. Performed the experiments: YL. Analyzed the data: YL GL SL CB VK. Contributed reagents/materials/analysis tools: YL GL CB VK. Wrote the paper: YL GL VK.

References

1. Wrather JA, Chambers AY, Fox JA, Moore WF, Sciumbato GL (1995) Soybean disease loss estimates for the southern United States, 1974–1994. Plant Disease 79: 1076–1079.
2. Wrather JA, Anderson TR, Arsyad DM, Tan Y, Ploper LD, et al. (2001a) Soybean disease loss estimates for the top ten soybean producing countries in 1998. Canadian Journal of Plant Pathology 23: 115–121.
3. Wrather JA, Steinstra WC, Koenning SR (2001b) Soybean disease loss estimates for the United States from 1996–1998. Canadian Journal of Plant Pathology 23: 122–131.
4. Wrather JA, Koenning SR, Anderson TR (2003) Effect of diseases on soybean yields in the United States and Ontario (1999–2002). Online. Plant Health Progress doi: 10.1094/PHP-2003-0325-01-RV,http://www.plantmanagement network.org/sub/php/review/2003/soybean/>(19 November 2003).
5. Wrather JA, Koenning SR (2006) Estimates of disease effects on soybean yields in the United States 2003–2005. Journal of Nematology 38: 173–180.
6. Ichinohe M (1952) On the soybean nematode, *Heterodera glycines* n. sp., from Japan. Magazine of Applied Zoology 17: 1–4.
7. Winstead NN, Skotland CB, Sasser JN (1955) Soybean cyst nematodes in North Carolina. Plant Disease Reporter 39: 9–11.
8. Spears JF (1957) Review of soybean cyst nematode situation for presentation at public hearing on the need for Federal Domestic Plant Quarantine, July 24, 1957.
9. Epps JM, Chambers AY (1958) New host records for *Heterodera glycines* including one in the Labiate. Plant Disease Reporter 42: 194.
10. Riggs RD, Hamblen ML (1962) Soybean-cyst nematode host studies in the Leguminosae. Arkansas Agriculture Experimental Station Reporter Series 110 Fayetteville AR 17p.
11. Riggs RD, Hamblen ML (1966a) Additional weed hosts of *Heterodera glycines*. Plant Disease Reporter 50: 15–16.
12. Riggs RD, Hamblen ML (1966b) Further studies on the host range of the soybean-cyst nematode. Arkansas Agriculture Experimental Station Bulletin 718 Fayetteville AR 19p.
13. Niblack TL, Arelli PR, Noel GR, Opperman CH, Orf JH, et al. (2002) A revised classification scheme for genetically diverse populations of Heterodera glycines. Journal of Nematology 34: 279–288.
14. Lauritis JA, Rebois RV, Graney LS (1983) Development of *Heterodera glycines* Ichinohe on soybean, *Glycine max*. Journal of Nematology 15: 272–280.
15. Davis EL, Tylka GL (2000) Soybean cyst nematode disease. The Plant Health Instructor. DOI: 10.1094/PHI-I-2000-0725-01.
16. Jones MGK, Northcote DH (1972) Nematode-induced syncytium-a multinucleate transfer cell. J Cell Sci 10: 789–809.
17. Inagaki H, Tsutsumi M (1971) Survival of the soybean cyst nematode, *Heterodera glycines* Ichinohe (Tylenchida: Heteroderidae) under certain storage conditions. Applied Entomology and Zoology (Jpn) 8: 53–63.
18. Wang J, Meyers D, Yan Y, Baum T, Smant G, et al. (2003) The soybean cyst nematode reduces soybean yield without causing obvious symptoms. Plant Disease 87: 623–628.
19. Lawton K (2010) Defeating soybean cyst nematode requires planting non-host crops and varieties with different sources of resistance. The Progressive Farmer: 28.
20. Ross JP, Brim CA (1957) Resistance of soybeans to the soybean cyst nematode as determined by a double-row method. Plant Dis Rep 41: 923–924.
21. Ross JP (1958) Host-Parasite relationship of the soybean cyst nematode in resistant soybean roots. Phytopathology 48: 578–579.
22. Epps JM, Hartwig EE (1972) Reaction of soybean varieties and strains to soybean cyst nematode. Journal of Nematology 4: 222.
23. Concibido VC, Diers BW, Arelli PR (2004) A decade of QTL mapping for cyst nematode resistance in soybean. Crop Sci. 44: 1121–1131.
24. Shannon JG, Arelli PR, Young LD (2004) Breeding for resistance and tolerance. Pp. 155–180 *in* D. P. Schmitt, J. A. Wrather, and R. D. Riggs, eds. Biology and management of soybean cyst nematode, 2nd ed. Marceline, MO: Schmitt & Associates of Marceline.
25. Riggs RD, Niblack TL (1999) Soybean cyst nematode. In: Hartman GL, Sinclair JB, Rupe JC, editors. Compendium of Soybean Diseases. Fourth Edition. St. Paul: APS Press; 52–53.
26. Luedders VD, Anand SC (1989) Attempt to select a cyst nematode population on soybean plant introduction 437654. Journal of Nematology 21: 264–267.
27. Niblack TL (2005) Soybean cyst nematode management reconsidered. Plant Disease Vol. 89 No.10.
28. Cobb NA (1918) Estimating the Nema population of the soil. Agricultural Technology Circular Bureau of Plant Industry, US Department of Agriculture, 1: 1–48.
29. Golden AM, Epps JM, Riggs RD, Duclos LA, Fox JA, et al. (1970) Terminology and identity of infraspecific forms of the soybean cyst nematode (Heterodera glycines). Plant Dis Rep 54: 544–546.
30. Riggs RD, Schmitt DP (1988) Complete characterization of the race scheme for Heterodera glycines. Journal of Nematology 20: 392–395.
31. Riggs RD, Schmitt DP (1991) Optimization of the *Heterodera glycines* race test procedure Journal of Nematology 23: 149–154.

32. Harris TS, Sandall LJ, Powers TO (1990) Identification of single Meloidogyne juveniles by polymerase chain reaction amplification of mitochondrial DNA. Journal of Nematology 22: 518–524.
33. Roosien J, Van Zandvoort PM, Folkertsma RT, Van der Voort JN, Goverse A, et al. (1993) Single juveniles of the potato cyst nematodes Globodera rostochiensis and G. pallida differentiated by randomly amplified polymorphic DNA. Parasitology 107: 567–572.
34. Gasser RB, Monti JR (1997) Identification of parasitic nematodes by PCR-SSCP of ITS-2 rDNA. Molecular Cell Probes 11: 201–209.
35. Powers TO, Harris TS (1993) A polymerase chain reaction method for identification of five major Meloidogyne species. Journal of Nematology 25: 1–6.
36. Cenis JL (1993) Identification of four major *Meloidogyne* spp. by random amplified polymorphic DNA (RAPD-PCR). Phytopathology 83: 76–80.
37. Semblat JP, Wajnberg E, Dalmasso A, Abad P, Castagnone-Sereno P (1998) High-resolution DNA fingerprinting of parthenogenetic root-knot nematodes using AFLP analysis. Molecular Ecology 7: 119–125.
38. Blok VC (2005) Achievements in and future prospects for molecular diagnostics of plant-parasitic nematodes. Canadian journal of plant pathology-revue canadienne de phytopathologie 27: 176–185.
39. Agudelo P, Robbins RT, Stewart JMcD, Szalanski AL (2005) Intraspecific variability of *Rotylenchulus reniformis* from cotton-growing regions in the United States. Journal of Nematology 37: 105–114.
40. Livak KJ, Schmittgen TD (2001) Analysis of relative gene expression data using real-time quantitative PCR and the 2(−Delta Delta C(T)) method. Methods 25: 402–408.
41. Gao X, Hartman GL, Lambert KN, Li S (2004) Detection and quantification of Fusarium solani f. sp. glycines in soybean roots with real-time quantitative polymerase chain reaction. Plant Disease 88: 1372–1380.
42. Finetti-Sialer MM, Ciancio A (2005) Isolate-Specific Detection of Grapevine fanleaf virus from Xiphinema index through DNA-Based Molecular Probes. Phytopathology 95: 262–268.
43. Atallah ZK, Stevenson WR (2006) A methodology to detect and quantify five pathogens causing potato tuber decay using real-time quantitative polymerase chain reaction [electronic resource]. Phytopathology 96: 1037–1045.
44. Okubara PA, Schroeder KL, Paulitz TC (2008) Identification and Quantification of *Rhizoctonia solani* and *R. oryzae* Using Real-Time Polymerase Chain Reaction [electronic resource]. Phytopathology 98: 837–847.
45. Berry SD, Fargette M, Spaull VW, Morand S, Cadet P (2008) Detection and quantification of root-knot nematode (*Meloidogyne javanica*), lesion nematode (*Pratylenchus zeae*) and dagger nematode (*Xiphinema elongatum*) parasites of sugarcane using real-time PCR. Molecular Cell Probes 22: 168–176.
46. Showmaker K, Lawrence GW, Lu S, Balbalian C, Klink VP (2011) Quantitative field testing *Rotylenchulus reniformis* DNA from metagenomic samples isolated directly from soil. PLoS ONE 6 (12): e28954. doi:10.1371/journal.pone.0028954.
47. Klink VP, Martins VE, Overall CC, Alkharouf NW, MacDonald MH, et al. (2007) A decline in transcript abundance for *Heterodera glycines* homologs of *Caenorhabditis elegans* uncoordinated genes accompanies its sedentary parasitic phase. BMC Developmental Biology 7: 35.
48. van Megen H, van den Elsen S, Holterman M, Karssen G, Mooyman P, et al. (2009) A phylogenetic tree of nematodes based on about 1200 full-length small subunit ribosomal DNA sequences. Nematology 11: 927–950.
49. Alkharouf N, Klink VP, Matthews BF (2007) Identification of *Heterodera glycines* (soybean cyst nematode [SCN]) DNA sequences with high similarity to those of *Caenorhabditis elegans* having lethal mutant or RNAi phenotypes. Experimental Parasitology 115: 247–258.
50. Yik C-P, Birchfield W (1982) Reactions of sweet potato root tissue to the reniform nematode. Plant Disease 66: 707–709.
51. Daday E von (1905) Untersuchungen fiber die Siisserwasser-Mikrofauna Paraguays. Zoologica, Stuttgart 18: 1–349.
52. Bastian HC (1865) Monograph on the Anguillulidae or free nematodes, marine, land and freshwater, with description of 100 new species. Transactions of the Linnaean Society of London, 25: 73–184.
53. Zengel JM, Epstein HF (1980) Identification of genetic elements associated with muscle structure in the Nematode Caenorhabditis elegans. Cell Motility 1: 73–97.
54. Ritter ME (2009) The Physical Environment: an Introduction to Physical Geography.
55. Kofoid CA, White WA (1919) A new nematode infection of man. Journal of the American Medical Association 72: 567–569.
56. Chitwood BG (1949) Root-knot nematodes. I. A revision of the genus Meloidogyne Goeldi, 1887. Proc. Helminthol. Soc. Wash. 16: 90–104.
57. Matsye PD, Lawrence GW, Youssef RM, Kim K-H, Matthews BF, et al. (2012) The expression of a naturally occurring, truncated allele of an α-SNAP gene suppresses plant parasitic nematode infection. Plant Molecular Biology 80: 131–155.

6

The Characteristics of Extreme Erosion Events in a Small Mountainous Watershed

Nu-Fang Fang[1,2], Zhi-Hua Shi[1,2,3]*, Ben-Jiang Yue[2], Ling Wang[2]

1 State Key Laboratory of Soil Erosion and Dryland Farming on the Loess Plateau, Northwest A & F University, Yangling, PR China, **2** Institute of Soil and Water Conservation of Chinese Academy of Sciences and Ministry of Water Resources, Yangling, PR China, **3** College of Resources and Environment, Huazhong Agricultural University, Wuhan, China

Abstract

A large amount of soil loss is caused by a small number of extreme events that are mainly responsible for the time compression of geomorphic processes. The aim of this study was to analyze suspended sediment transport during extreme erosion events in a mountainous watershed. Field measurements were conducted in Wangjiaqiao, a small agricultural watershed (16.7 km^2) in the Three Gorges Area (TGA) of China. Continuous records were used to analyze suspended sediment transport regimes and assess the sediment loads of 205 rainfall–runoff events during a period of 16 hydrological years (1989–2004). Extreme events were defined as the largest events, ranked in order of their absolute magnitude (representing the 95th percentile). Ten extreme erosion events from 205 erosion events, representing 83.8% of the total suspended sediment load, were selected for study. The results of canonical discriminant analysis indicated that extreme erosion events are characterized by high maximum flood-suspended sediment concentrations, high runoff coefficients, and high flood peak discharge, which could possibly be explained by the transport of deposited sediment within the stream bed during previous events or bank collapses.

Editor: Vanesa Magar, Plymouth University, United Kingdom

Funding: Financial support for this research was provided by the National Natural Science Foundation of China (41071190) and the Program for New Century Excellent Talents in University (NCET-10-0423). The funders had no role in study design, data collection and analysis, decision to publish, or preparation of the manuscript.

Competing Interests: The authors have declared that no competing interests exist.

* E-mail: pengshi@mail.hzau.edu.cn

Introduction

Soil erosion poses a serious problem for sustainable agriculture and the environment [1,2]. The soil and water conservationist is typically more interested in flood or erosion damage caused by large events than in the damage caused by smaller, more common hydrological events. Many studies have clearly emphasized that large-magnitude, low-frequency events are assumed to be dominant with respect to soil erosion [3–9]. Estimation of the role of sediment transport produced by extreme events is necessary for the calculation of sediment yields from basins, as a single event may represent the transport of several "normal" years [10]. However, soil erosion is a temporally compressed process; the effect of one event must be isolated when designing erosion control technologies and conservation planning [11,12].

Previous projects have documented the effects of extreme events on total soil erosion and sediment transport on catchment scales [13–17]. These studies have mainly focused on extreme rainfall events and have discussed erosion based on those events. Analysis of the relationships between suspended sediment transport and rainfall characteristics during erosion events can help in understanding the factors and processes that determine sediment responses [18,19]. However, runoff and erosion processes are strongly affected by many other factors in addition to rainfall [20–22], and few systematic attempts have been made to distinguish extreme rainfall events and erosion events.

The Three Gorges Project (TGP) on the Yangtze River in China is the world's largest hydropower project. Following construction of the Three Gorges Dam, many farmers resettled in surrounding mountain areas and cultivated marginal lands, which are mostly on steep slopes with soil of poor structure. The TGP is controversial for several reasons, including the likely impact of sedimentation on the operation and useful life of the reservoir [23,24]. The TGA refers to the riparian counties along the Yangtze valley between Yichang and Chongqing (Fig. 1). This area is periodically impacted by catastrophic floods, and soil erosion is a major environmental problem [25]. High levels of suspended sediment may result in high sediment deposition rates, thereby reducing the useful life of the Three Gorges Reservoir. Thus, understanding the temporal variations of suspended sediment transport and sediment loads during extreme events is essential for future watershed management plans.

In this study, we investigated the runoff and erosion processes of the Wangjiaqiao watershed, which is a representative watershed within the TGA [26,27]. The main objectives of this work were to (1) characterize extreme erosion events and the hydrological and sediment responses in a small watershed in the TGA and (2) improve our understanding of the factors that control sediment transport patterns and loads during extreme erosion events.

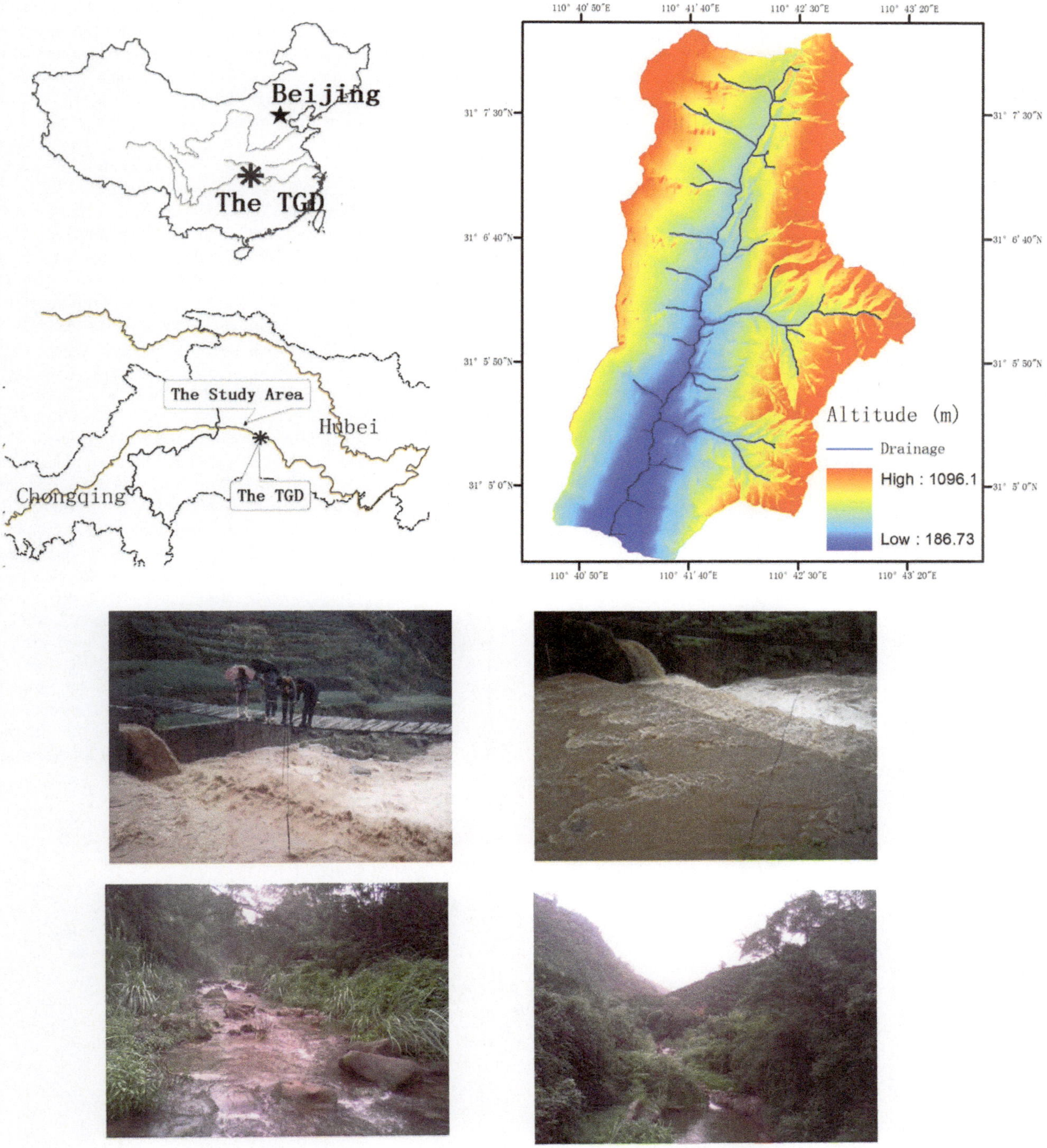

Figure 1. Location of the study watershed in the Three Gorges Area, China. Photos are of an overall view, the main channel, and the outlet of the Wangjiaqiao watershed. Photos are of an overall view, the main channel, and the outlet of the Wangjiaqiao watershed during an extreme event.

Study Area and Methods

Study Area

The study was conducted in the Wangjiaqiao watershed (31°5′N–31°9′N, 110°40′E–110°43′E), which lies in Zigui County in Hubei Province, China. It is approximately 50 km northwest of the Three Gorges Dam and covers an area of 16.7 km^2 (Fig. 1). Elevations within the watershed range from 184 to 1,180 m and slopes range from 2° to 58°, with an average slope of 23°. The parent materials of this area are mostly Cretaceous and Tertiary purple shale, sandy shale, and sandstone, which contain large quantities of iron and manganese oxides in specific mineralogical forms. Two main soil great groups occur in the study watershed:

purple soils derived from purple sandy shale and paddy soils developed from the purple soil. According to the USDA Soil Taxonomy, purple soils and paddy soils are classified as entisols and aquepts, respectively. The climate is subtropical, with mean temperatures between 11°C and 18°C. The annual precipitation averages 1,015 mm, 70% of which occurs between May and September. Land use is mainly a function of elevation and topography. Remnant forest patches exist primarily on steep, inaccessible peaks and slopes. Little natural vegetation is observed, and most areas are covered by secondary vegetation under human influence. A field survey was carried out in 1995 using a topographic map (scale 1:10,000) and aerial photos. The results indicated that forest covered 44.5% of the study area, whereas cropland covered 23.3% (389.4 ha) and paddy fields covered 19.8% (330.7 ha). The other land use types were relatively minor and consisted of shrubland (3.2%), orchards (4.2%), rural residential land (3.9%), and water bodies (0.7%). The main agricultural crops are rice (*Oryza sativa* L.), maize (*Zea mays* L.), and wheat (*Triticum aestivum* L.). The streams in this watershed have a pinnate drainage pattern, and the length of the main channel is approximately 3,500 m [26].

Field and Lab Methods

Suspended sediment yields represent the sum of the erosion produced by all active sources within a watershed, although yields cannot be used as reliable indicators of total hill slope erosion due to the difficulty of quantifying sediment storage and identifying the mixture of sources [3]. A set of instruments consisting of a continuous recording rain gauge, water-level stage recorder, and bottle-type silt sampler were used to record rainfall, stream flow, and sediment load, respectively. The water stage was measured every 15 min and transformed into discharge using a calibrated rating curve obtained from periodic flow measurements. Suspended sediment concentrations (SSCs) were determined by the gravimetric method. Water samples were vacuum filtered through a 0.45-μm filter, and the residue was oven dried at 105°C for 24 h. Suspended sediment samples were taken only during rainfall–runoff events, and more than 10 samples were required during each event based on the "Technical code of practice on water and soil conservation monitoring" [28]. In practice, samples were collected based on the variation of the discharge and the magnitude of the SSC. Generally, samples were collected frequently during events with high SSC values and were collected infrequently during events with relatively small SSC values. The weight of each dried sample of residue and the sample volume were used to determine the SSC (g m^{-3}). The suspended sediment load was then calculated from the SSC and water discharge data. Watershed runoff and rainfall data have been collected since 1989. No specific permissions were required for these sampling activities because the location is not privately-owned or protected in any way and the field activities did not involve endangered or protected species.

Data Processes

Hydrograph separation was conducted for 205 events during the 1989–2004 period, for which we have fairly complete records of sediment concentrations or hydrographs. Floods were identified in cases in which the increase in stream discharge exceeded 1.5 times the base flow recorded at the beginning of the rainfall event [29,30]. Runoff was separated between storm flow and base flow using the classical hydrograph separation method of Hewlett and Hibbert (1967) [31]. Some continuous events were excluded because they were hard to separate as "an event" using our separation method. Some other events with complete hydrograph records were included even though their SSC values were below the threshold for our monitoring method. Total suspended sediment load of an event is the accumulation of sediment yield during each time step. The combined SS load of the 205 events represented 68% of the total suspended sediment load during the study years.

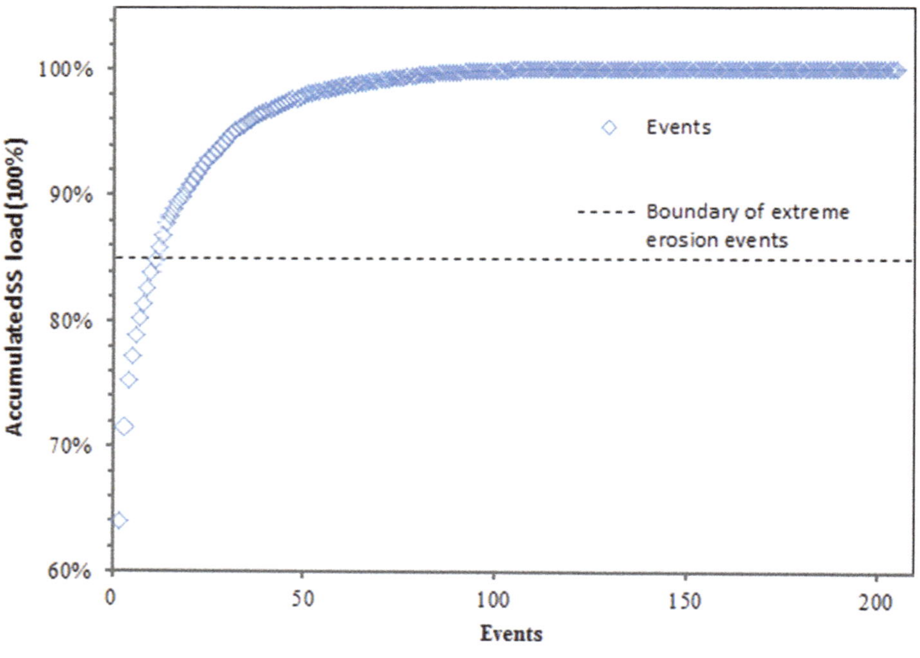

Figure 2. The percentage of accumulated suspended sediment yield in relation to the number of events. Events were independently ranked from largest to smallest for sediment.

Table 1. Flood variables and associated abbreviations used in the statistical analysis of the rainfall–runoff-suspended sediment transport relations.

Rainfall-related variables	Runoff-related variables	Suspended sediment-related variables
Total precipitation	Runoff	Maximum flood suspended sediment concentration
(*P*, mm)	(*R*)	(SSC_{max}, g m^{-3})
Duration of the event	Runoff coefficient	Total suspended sediment load
(*D*, h)	(*RC*)	(*TL*, t)
Maximum 30-min rainfall intensity	Flood peak discharge	
(I_{30}, mm)	(Q_{max}, m^3 s^{-1})	
Antecedent precipitation one day before		
(*AP1d*, mm)		
Antecedent precipitation three days before		
(*AP3d*, mm)		
Antecedent precipitation seven days before		
(*AP7d*, mm)		

Statistical Analysis

The flood events were characterized using several variables (Table 1).

$$RC = R/P$$

$$R = Q - BF$$

where *RC*, *R*, *P*, *Q*, and *BF* are the runoff coefficient, surface runoff, total discharge, precipitation, and base flow, respectively.

$$SY_i = SSC_i \times Q_i$$

$$TL = \sum_{i=1}^{n} SY_i$$

$$SSC_{mean} = TL/Q_{mean}$$

Where SY_i, SSC_i and Q_i are the suspended sediment yield, Suspended sediment concentrations, and discharge during a time term *i*. *TL*, Q_{mean}, and SSC_{mean} are the total suspended sediment load, mean discharge, and mean flood-suspended sediment concentration, respectively.

After evaluating the descriptive statistics, canonical discriminant analysis (CDA) was performed with the SPSS13.0 statistical software package using the variables mentioned above.

Table 2. Statistical features of different event regimes.

Regimes	Variables	Mean	SD	Variation	Frequency (time)
	P (mm)	82.8	36.9	45%	10
I	I_{30} (mm)	15.4	7.0	46%	
	D (min)	1,408.3	899.0	64%	
	TL (kg)	7,261,812	11,308,672	156%	
	P (mm)	33.2	19.5	59%	132
II	I_{30} (mm)	7.4	7.0	95%	
	D (min)	1,034.7	776.8	75%	
	TL (kg)	106,207	179,681	169%	
	P (mm)	19.6	9.9	51%	63
III	I30 (mm)	4.8	5.2	108%	
	D (min)	834.3	583.2	70%	
	TL (kg)	0.0	0.0	/	

P, *D*, and I_{30} represent precipitation depth, duration, and maximum 30-min intensity, respectively.

Results

Events Characteristics and Arrangement

An extreme event was defined classically as a rare, low-probability event, typically in relation to the exceedance of certain threshold values (e.g., means, percentiles). For this study, extreme erosion events are events that cause large amounts of SS load; no quantitative definition, e.g., in kg, was used. A partial definition might refer to an SS amount of a different order of magnitude than that caused by normal events. The definition of an extreme event in terms of damage caused by the event is discussed in [32]. A total of 10 events (approximately representing the 95th percentile) were identified as extreme erosion events based on the SS load caused by those 10 events having been clearly greater than that caused by the other events.

Based on the ranking of SS loads, 205 events were classified into three regimes; extreme erosion events, normal events, and no-SS load events. Four variables were compared: *P*, *D*, I_{30}, and *TL* (Table 2). An extreme erosion event has a destructive effect on the soil surface. The mean *SS* load of the extreme erosion events that occurred during the study period was almost 68 times that of normal events, and the mean *P* and mean I_{30} were 2.5 and 2.1 times greater, respectively. Non-SS load events had observable hydrograph processes with a mean P of 19.6 mm but did not

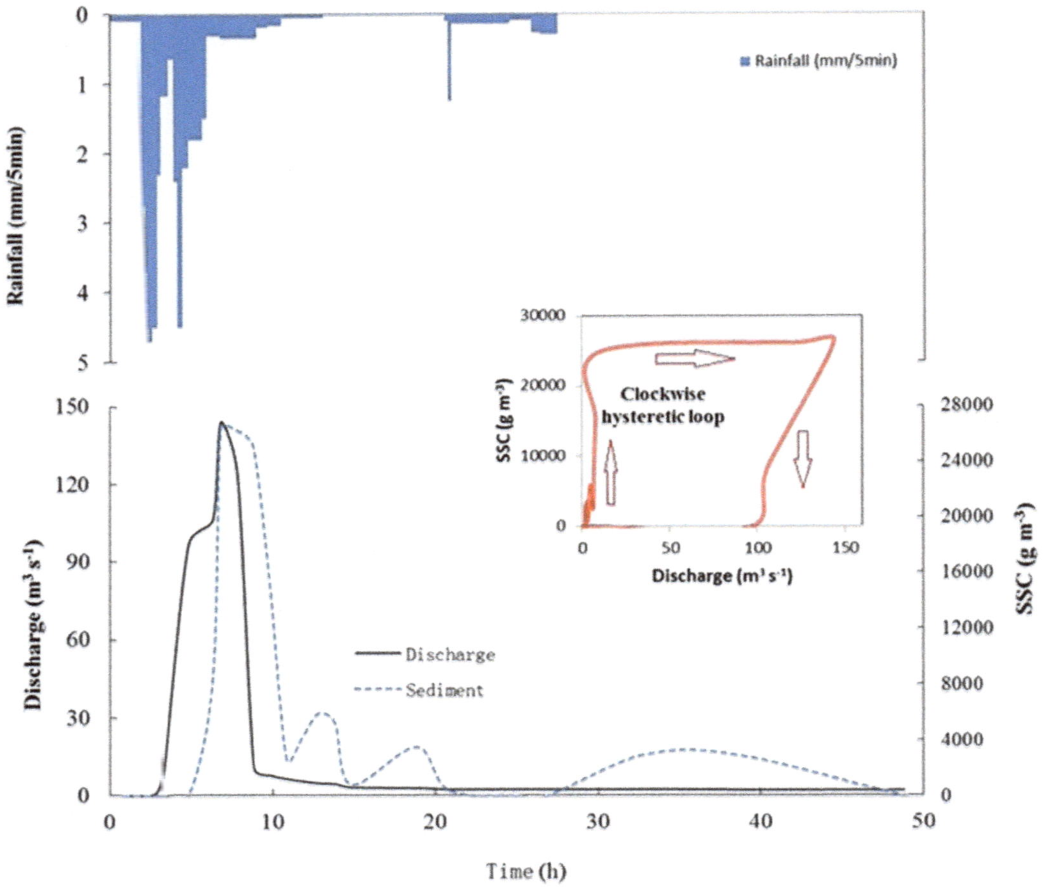

Figure 3. Hyetographs, hydrographs, sedigraphs, and hysteretic loop during the event of 6/7/1999.

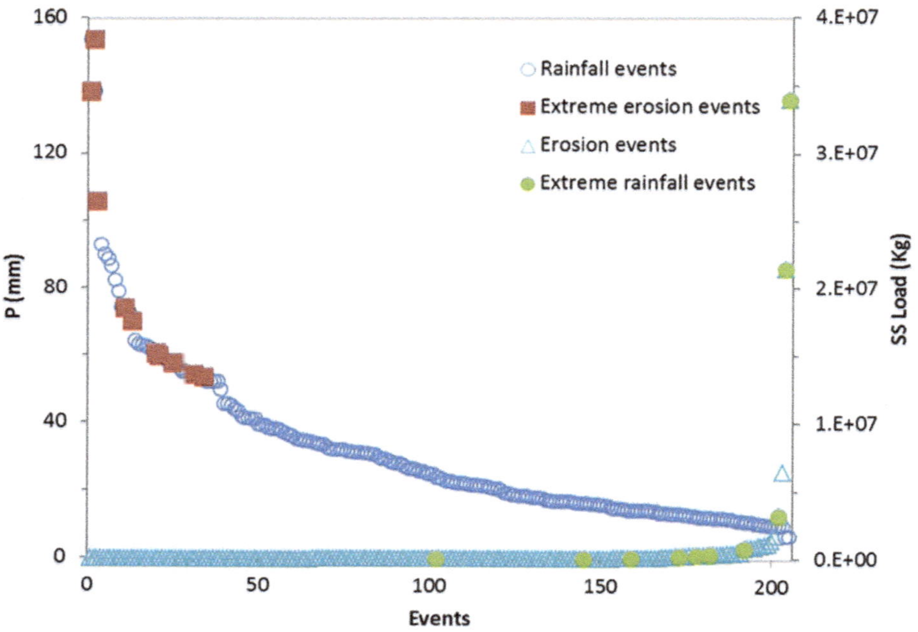

Figure 4. Distributions of extreme erosion events and extreme rainfall events.

Table 3. Characteristics of extreme erosion and rainfall (1), extreme erosion (2), and extreme rainfall events (3).

Date	Regime	D(min)	P(mm)	I30(mm)	SS (kg)	AP1d(mm)	AP3d(mm)	AP7d(mm)	RC	SSmax(g/m³)	Qmax(m³/s)
30/6/1991	1	973.2	153.7	20.1	21,467,789	1.2	1.2	10	0.55	62,138	87.9
6/7/1999	1	1,610	138.5	18.2	33,987,159	0	0	0	0.94	25,060	144
30/4/1990	1	2,890.2	105.9	18.9	3,224,476	0	0	8.9	0.5	42,820	61.2
1/8/1997	2	320	59.8	24.7	6,505,964	0	0	0	0.38	34,820	30.1
19/5/1993	2	1,669.8	54.3	14.9	1,738,339	7.1	8.6	28.4	0.52	40,745	19.6
22/7/1993	2	919.8	74	9.6	1,385,369	7.1	33.2	45.8	0.54	11,970	17.3
2/8/1998	2	1,400	53.4	6.7	1,175,040	12.2	23.3	34.7	0.54	2,289	24.1
28/6/1998	2	1,225	70.1	7.7	1,056,682	0	0	6.8	0.36	38,680	48.1
6/6/1996	2	2,820	57.7	8.2	1,033,247	2.9	65	71.8	0.84	3,539	16.2
6/7/1996	2	255	60.6	25.3	1,044,054	0	0	139.9	0.43	1,260	21.9
19/6/1990	3	1,596	92.8	8.9	329,701	0	0	0	0.39	3,007	11.9
25/7/1989	3	2,565	90	11.6	831,571	0.5	0.5	0.5	0.35	7,558	15.2
6/7/1995	3	3,400	88.8	6.4	7,483	0	2.1	2.1	0.25	1,030	3.08
2/8/1991	3	2,119.8	86.6	12.3	287,019	8	17.2	20.3	0.44	4,517	6.6
18/9/1996	3	2,345	82.3	15.9	215,205	35	35	37.8	0.39	1,333	14.3
12/6/2004	3	2,700	79	5.9	103,077	0.1	3.9	3.9	0.33	2,366	7.3
9/4/1994	3	610.2	74.3	9.1	53,428	0.2	2.7	8.4	0.1	5,070	5.6

produce *SS* loads or their SSCs were below the threshold of our monitoring method. These results emphasize the importance of the rainfall type as a major cause of erosion. From this standpoint, the rainfall depth is an important factor in the degree of soil erosion in the study area. Meanwhile, the variance of *TL* is much higher than the variance of *P* for both extreme erosion events and normal events, which confirms the complex nature of sediment response in the study area.

Characterization of Extreme Erosion Events

The greatest proportion of the SS load was produced by the extreme erosion events. This finding is illustrated in Figure 2, which illustrates the percentage of the accumulated sediment load as a function of the percentage of events. Extreme erosion events were responsible for 83.8% of the *SS* loads. For each extreme erosion event, the erosive characteristics of individual storms were evaluated. The I_{30} ranged from 6.7 to 24.7 mm h^{-1}. Antecedent rainfall amounts were highly variable, ranging from 0 to 12.2, 0 to 33.2, and 0 to 15.8 mm of precipitation during the previous one-, five-, and seven-day periods, respectively. The maximum *SS* load was 33,987,157 kg on 7/6/1999, which was 4.2 times greater than the mean annual load during the study years (Fig. 3). This event was caused by a Q_{max} of 144 m^3/s. The curve representing sediment concentration vs. discharge over time exhibited a clockwise hysteretic loop [33]. The maximum *SS* occurred on 30/6/1991 and was 62,138 g/m^3. This flood load was 2.68 times greater than the mean annual load and was generated by the highest precipitation recorded during the study period (P = 153.7 mm), which created a flood peak discharge of 87.9 m^3 s^{-1} and a total flood runoff depth of 84.8 mm. The minimum *SS* load of an extreme erosion event was 1,033,247 kg. The *RC* of extreme erosion events ranged from 0.94 to 0.36. The peak discharge ranged from 16.2 to 144 m^3 s^{-1}. The direct SSC_{max} ranged from 3,539 to 62,138 g m^{-3}. I_{30} ranged from 6.7 to 24.7 mm h^{-1}. The antecedent rainfall amounts were highly variable, ranging from 0 to 12.2, 0 to 33.2, and 0 to 15.8 mm of precipitation during the previous one-, five-, and seven-day periods, respectively.

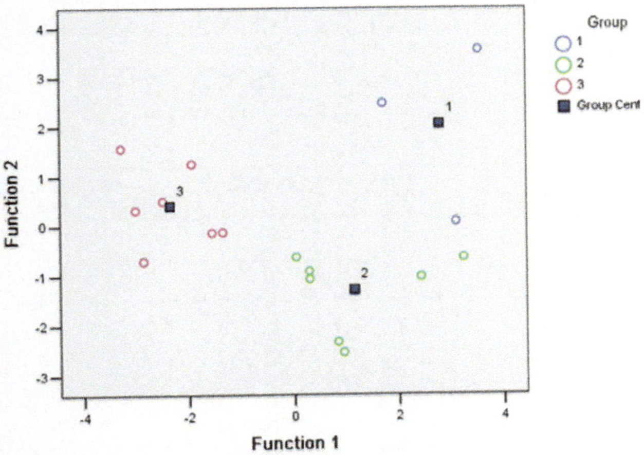

Figure 5. Distribution of cases with discriminant functions.

Large Cumulative Rainfalls

The study of extreme erosion events is almost always associated with the study of precipitation [34]. To examine the relationship between rainfall and extreme erosion events, we identified the 10 events with the highest *P* values as large cumulative rainfalls (using the same method that was used to identify extreme erosion events). We then compared large cumulative rainfalls with extreme erosion events and evaluated the descriptive statistics of the extreme erosion and rainfall events, extreme erosion events, and large cumulative rainfalls shown in Table 3. Some studies have found that the most large cumulative rainfalls do not necessarily produce extreme fluvial discharge or the maximum soil erosion [7,35,36]. Our results confirm that the most extreme erosion events were not

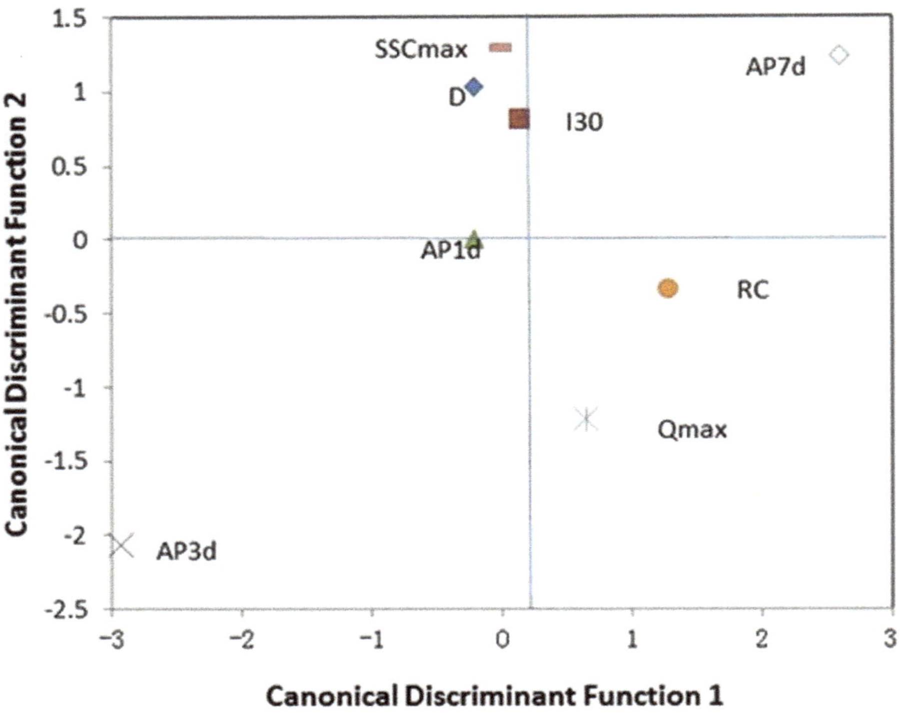

Figure 6. Standardized discriminant function coefficients of the variables included in the discriminant functions; the values are shown in Table 3.

necessarily produced by the most large cumulative rainfalls. Only three extreme erosion events were found to be associated with large cumulative rainfalls. Fig. 3 illustrates the distributions of extreme erosion events and large cumulative rainfalls. A rainfall of 53.4 mm caused extreme erosion (117,504 kg) on 2/8/1998, whereas a rainfall of 74.3 mm caused only 53,428 kg of *SS* load on 9/4/1994. Extreme erosion events were greater than 16.6% if ranked by *P*, and large cumulative rainfalls were greater than 50% if ranked by *SS* load (Fig. 4).

Canonical Discriminant Analysis

To identify factors (excluded *TL* and *P*) that might explain the measured hydrological and sedimentological responses of extreme events, we conducted discriminant analyses using the data shown

Table 4. Data on the canonical discriminant functions (cdf).

Eigenvalues of cdf							
				Group centroids			
Eigenvalue	**Variance (%)**	**Cumulative variance (%)**	**Canonical correlation**	**1**	**2**		**3**
5.093	73.7	73.7	0.914	2.783	1.158		−2.351
1.819	26.3	100	0.803	2.056	−1.289		0.408

Standardized cdf coefficients

	D	**I_{30}**	**AP1d**	**AP3d**	**AP7d**	**RC**	**Qmax**	**SSCmax**
F1	−0.214	0.136	−0.217	−2.940	2.599	1.271	0.645	−0.084
F2	1.025	0.815	0.003	−2.068	1.235	−0.341	−1.220	1.289

Structure matrix of cdf

	SSC_{max}	**RC**	**Q_{max}**	**I_{30}**	**AP1d**	**AP7d**	**AP3d**	**D**
F1	0.411*	0.409*	0.393*	0.260*	−0.101*	0.111	0.041	−0.170
F2	0.272	0.054	0.004	0.143	−0.076	−0.481*	−0.417*	0.280*

The highest correlations of the variables with the functions are indicated with *.

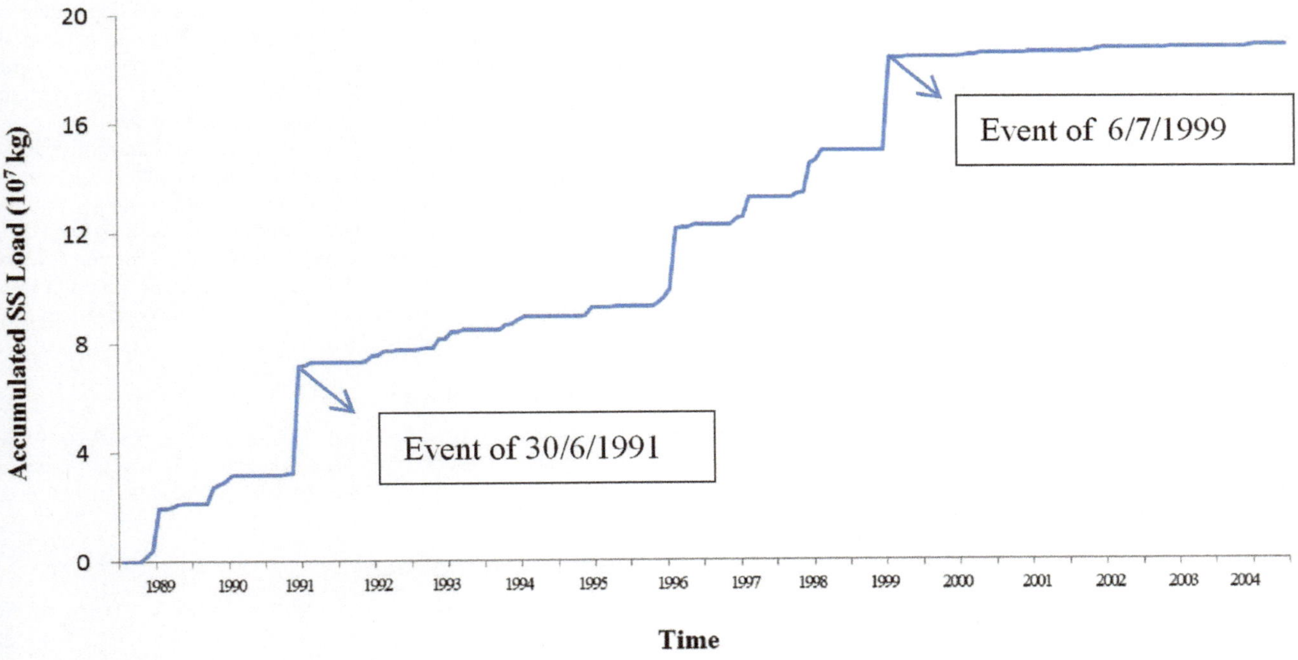

Figure 7. Curve of accumulated SS load during the study period. (The time step is one month.).

in Table 3. We created two canonical discriminant functions that included eight of the input variables to separate the groups (Table 4 and Fig. 5). The first canonical discriminant function (f1) explained 73.7% of the variance (Table 4), was highly positively correlated with SSC_{max} (0.411), RC (0.409), and Q_{max} (0.393), and was more weakly correlated with I_{30} and $AP1d$. For this reason, f1 expresses increasing SSC_{max}, RC, and Q_{max} in the watershed at the time of the floods with higher function values. The second discriminant function (f2) correlated with the remaining parameters (−0.481 for AP7d and −0.417 for SSC_{max}), as expected, given the lower proportion of variance explained. In addition, f2 was found to increase with increasing SSC_{max}, although with low intensity. Table 3 also presents the discriminant function values at the centroids of the three event types. The distribution of the flood events in the two dimensions created by the functions is shown in Fig. 6 (discriminant function coefficients are shown in Table 4). The centroids of the extreme erosion and rainfall group were high for both functions (f1 = 2.783; f2 = 2.056). The centroids of the extreme erosion events had relatively small values for f2 but high values for f1. These events were generated for very high values of RC, SSC_{max}, and Q_{max}. The large cumulative rainfalls have very low values for f1 and values close to zero for f2 (0.408). That is, the large cumulative rainfalls were generated for particularly low values of SSC_{max}, RC, and Q_{max} but with no special antecedent rainfall characteristics.

Discussion

A comparison of the accumulated suspended sediment transport with the accumulated rainfall and the number of single events indicates that 83.3% of sediment transport occurred during 5% of the erosion events and was caused by 13% of the precipitation. Fig. 7 illustrates the importance of the SS loads of extreme events to the total accumulated SS load. Two sharp rises in the curve were caused by the events of 1991/06/30 and 1999/07/06. The soil thickness of the Wangjiaqiao watershed ranges predominantly from 30 to 50 cm. The most extreme erosion event caused a lot of soil loss in the watershed. In the TGA, the sediment delivery rate on a small watershed scale has been thought to be very low in "common events," ranging from 0.1 to 0.4 [37,38], but mean sediment delivery radios are inappropriate for characterizing extreme erosion events. A study by Zhang et al. [39] reported a long-term natural sediment delivery radios close to 1 for a 10.88-km^2 watershed.

The results confirm the complexity of erosion processes and the relationships between TL and P in particular. The most cumulative rainfalls do not necessarily produce the most extreme erosion. Events of low rainfall depth and short duration typically cause very limited hydrological responses and almost no sediment transport. Other studies [17,40] have suggested that sediment yield rates may change in response to changes in rainfall, including changes in both total precipitation and intensity.

The geomorphologic impacts of an extreme event are closely related to such factors as the erodibility of the parent material, topography, vegetation cover, and land use [41–43]. Moreover, soil erosion is largely determined by on-site sediment production and the connectivity of sediment sources and streams [44]. Lu and Higgitt [45] found that 60% of sediment is contributed from arable land in 32 catchments in the TGA. Slopes with gradients in excess of 30% comprise 76% of the area of the Wangjiaqiao watershed. Cultivated sloping lands are major contributors to sediment yield. Tillage activities are generally carried out between April and September. Soil is eroded and then transported to the stream networks [27]. The soil parent material of the watershed is predominantly purple sandy shale, and bedrock is typically exposed in the channel bed; thus, channel erosion is rare. Sediments stored in the channel and distributed within tributaries are transported after flood events with sufficient transport capacity [46]. Therefore, the extreme erosion events characterized by the largest values of SSC_{max}, RC, and Q_{max} could be understood to result from the transport of deposited sediment within the stream bed during previous events. The clockwise hysteretic loop of the largest erosion event confirms this explanation [47]. Moreover, bank collapse has been identified as an important source of extreme erosion. Investigations in the Wangjiaqiao watershed have indicated that bank collapse along the stream results in bank erosion that contributes to sediment transport, particularly during major floods of high intensity [48]. The soils of the watershed are mainly purple soils developed from purple sandstone with a content of rock fragments and are rich in macrospores [49]. The macrospores of shallow soils become saturated easily during large rainfall events, the soil bulk density increases rapidly, and the combination of poor drainage in the soil and high-intensity rainfall easily causes bank collapse [50].

Conclusions

This study investigated the important issue of extreme erosion events in a typical watershed in the TGA. These events, which occur primarily during the summer season, can result in serious soil loss and great damage. The results confirm the complex and heterogeneous nature of hydrological and sediment responses in the watershed. A total of 83.3% of the sediment transport occurred during 5% of the erosion events that occurred in the TGA and was caused by 13% of the precipitation that the TGA received. The maximum TL of an event was 4.2 times the mean annual load during the study period. Large rainfall depth does not necessarily produce large fluvial discharge or large soil erosion. Extreme erosion events were generated under very high *RC*, SSC_{max}, and Q_{max}. The results indicate that the use of event average values or mean index values may not be suitable for analyzing soil erosion processes in the TGA. The effects of a small number of events appear to determine the magnitudes of soil loss; catastrophism may be more suitable for explaining soil erosion processes. Although our work provides a considerable database of extreme sediment load during rainfall events, we provide a speculative explanation concerning extreme erosion events, subject to the constraints of the field methods. Further investigations should be performed to assess the sediment yield during extreme events, which is expected to be of significant value in environmental management and development of strategies to control sediment dynamics at the catchment scale.

Acknowledgments

We would like to acknowledge members of the Soil and Water Conservation Bureau of Zigui County in Hubei Province for assistance in the field and in the laboratory. We thank Cheng Jiang, who provided valuable help in data collection.

Author Contributions

Conceived and designed the experiments: NFF ZHS. Performed the experiments: NFF ZHS. Analyzed the data: NFF ZHS BJY LW. Contributed reagents/materials/analysis tools: NFF BJY LW. Wrote the paper: NFF ZHS.

References

1. Ng SL, Cai QG, Ding SW, Chau KC, Qin J (2008) Effects of contour hedgerows on water and soil conservation, crop productivity and nutrient budget for slope farmland in the Three Gorges Region (TGR) of China. Agroforestry Systems 74: 279–291.
2. Shi ZH, Chen LD, Fang NF, Qin DF, Cai CF (2009) Research on the SCS-CN initial abstraction ratio using rainfall-runoff event analysis in the Three Gorges Area, China. Catena 77: 1–7.
3. Estrany J, Garcia C, Batalla RJ (2010) Hydrological response of a small mediterranean agricultural catchment. Journal of Hydrology 380: 180–190.
4. Hooke JM, Mant JM (2000) Geomorphological impacts of a flood event on ephemeral channels in SE Spain. Geomorphology 34: 163–180.
5. Lana-Renault N, Regues D, Marti-Bono C, Begueria S, Latron J, et al. (2007) Temporal variability in the relationships between precipitation, discharge and suspended sediment concentration in a small Mediterranean mountain catchment. Nordic Hydrology 38: 139–150.
6. Mano V, Nemery J, Belleudy P, Poirel A (2009) Assessment of suspended sediment transport in four alpine watersheds (France): influence of the climatic regime. Hydrological Processes 23: 777–792.
7. Osterkamp WR, Friedman JM (2000) The disparity between extreme rainfall events and rare floods - with emphasis on the semi-arid American West. Hydrological Processes 14: 2817–2829.
8. Poesen JWA, Hooke JM (1997) Erosion, flooding and channel management in Mediterranean environments of southern Europe. Progress in Physical Geography 21: 157–199.
9. White S, Garcia-Ruiz JM (1998) Extreme erosional events and their role in mountain areas of northern Spain. Ambio 27: 300–305.
10. Wolman MG, Miller JP (1960) Magnitude and frequency of forces in geomorphic processes. The Journal of Geology: 54–74.
11. Seeger M, Errea MP, Begueria S, Arnaez J, Marti C, et al. (2004) Catchment soil moisture and rainfall characteristics as determinant factors for discharge/suspended sediment hysteretic loops in a small headwater catchment in the Spanish pyrenees. Journal of Hydrology 288: 299–311.
12. Stroosnijder L (2005) Measurement of erosion: Is it possible? Catena 64: 162–173.
13. Gonzalez-Hidalgo JC, Batalla RJ, Cerda A, de Luis M (2012) A regional analysis of the effects of largest events on soil erosion. Catena 95: 85–90.
14. Larson WE, Lindstrom MJ, Schumacher TE (1997) The role of severe storms in soil erosion: A problem needing consideration. Journal of Soil and Water Conservation 52: 90–95.
15. Lenzi MA, Mao L, Comiti F (2003) Interannual variation of suspended sediment load and sediment yield in an alpine catchment. Hydrological Sciences Journal-Journal Des Sciences Hydrologiques 48: 899–915.
16. McBroom MW, Beasley RS, Chang M, Ice GG (2008) Water quality effects of clearcut harvesting and forest fertilization with best management practices. Journal of Environmental Quality 37: 114–124.
17. Nearing MA, Jetten V, Baffaut C, Cerdan O, Couturier A, et al. (2005) Modeling response of soil erosion and runoff to changes in precipitation and cover. Catena 61: 131–154.
18. Nadal-Romero E, Regues D, Latron J (2008) Relationships among rainfall, runoff, and suspended sediment in a small catchment with badlands. Catena 74: 127–136.
19. Zabaleta A, Martinez M, Uriarte JA, Antiguedad I (2007) Factors controlling suspended sediment yield during runoff events in small headwater catchments of the Basque Country. Catena 71: 179–190.
20. Chen LD, Wang J, Fu BJ, Qiu Y (2001) Land-use change in a small catchment of northern Loess Plateau, China. Agriculture Ecosystems & Environment 86: 163–172.
21. Kirkby MJ, Bracken LJ, Shannon J (2005) The influence of rainfall distribution and morphological factors on runoff delivery from dryland catchments in SE Spain. Catena 62: 136–156.
22. Siriwardena L, Finlayson BL, McMahon TA (2006) The impact of land use change on catchment hydrology in large catchments: The Comet River, Central Queensland, Australia. Journal of Hydrology 326: 199–214.
23. Chu ZX, Zhai SK (2008) Yangtze River sediment: In response to Three Gorges Reservoir (TGR) water impoundment in June 2003. Journal of Coastal Research 24: 30–39.
24. Dai ZJ, Chu A, Stive M, Zhang XL, Yan H (2011) Unusual Salinity Conditions in the Yangtze Estuary in 2006: Impacts of an Extreme Drought or of the Three Gorges Dam? Ambio 40: 496–505.
25. Yang YS, Shi DM (1994) Study on soil erosion in the Three Gorge area of the Changjiang River. Southeast University Press, Nanjing (in Chinese).
26. Fang NF, Shi ZH, Li L, Jiang C (2011) Rainfall, runoff, and suspended sediment delivery relationships in a small agricultural watershed of the Three Gorges area, China. Geomorphology 135: 158–166.
27. Shi ZH, Cai CF, Ding SW, Wang TW, Chow TL (2004) Soil conservation planning at the small watershed level using RUSLE with GIS: a case study in the Three Gorge Area of China. Catena 55: 33–48.
28. Ministry of Water Resources of the People's Republic of China (MWRPRC) (2002) Technical code of practice on water and soil conservation monitoring. 7 p. (in Chinese).
29. Garcia-Ruiz JM, Arnaez J, Begueria S, Seeger M, Marti-Bono C, et al. (2005) Runoff generation in an intensively disturbed, abandoned farmland catchment, Central Spanish Pyrenees. Catena 59: 79–92.
30. Lana-Renault N, Regues D (2009) Seasonal patterns of suspended sediment transport in an abandoned farmland catchment in the Central Spanish Pyrenees. Earth Surface Processes and Landforms 34: 1291–1301.
31. Hewlett JD, Hibbert AR (1967) Factors affecting the response of small watersheds to precipitation in humid areas. Forest hydrology: 275–290.

32. Bathurst JC, Amezaga J, Cisneros F, Novillo MG, Iroume A, et al. (2010) Forests and floods in Latin America: science, management, policy and the EPIC FORCE project. Water International 35: 114–131.
33. Williams GP (1989) Sediment concentration versus water discharge during single hydrologic events in rivers. Journal of Hydrology 111: 89–106.
34. Soler M, Regüés D, Latron J, Gallart F (2007) Frequency–magnitude relationships for precipitation, stream flow and sediment load events in a small Mediterranean basin (Vallcebre basin, Eastern Pyrenees). Catena 71: 164–171.
35. Gonzalez-Hidalgo JC, Pena-Monne JL, de Luis M (2007) A review of daily soil erosion in Western Mediterranean areas. Catena 71: 193–199.
36. Nunes JP, Vieira GN, Seixas J, Gonclaves P, Carvalhais N (2005) Evaluating the MEFIDIS model for runoff and soil erosion prediction during rainfall events. Catena 61: 210–228.
37. Shi DM (1992) Soil and water loss and its control in the Yangtze river basin. Resources and Environment in the Yangtze Valley 1: 62–70. (in Chinese).
38. Tu EQ, Liao XY, Chen ZJ, Wang HM, (2010) Effects of rainfall on runoff from a small watershed in three gorges reservoir area. Bulletin of Soil and Water Conservation 30: 7–11. (in Chinese).
39. Zhang XB, Wen AB, Zhang YQ, Wang YC, Qi YQ, et al. (2006) Perliminary study on natural erosion rates in small watershed of Sichuan hilly basin. Joural of Soil and Water Conservation 20:1–5. (in Chinese).
40. Old GH, Leeks GJL, Packman JC, Smith BPG, Lewis S, et al. (2003) The impact of a convectional summer rainfall event on river flow and fine sediment transport in a highly urbanised catchment: Bradford, West Yorkshire. Science of the Total Environment 314: 495–512.
41. Coppus R, Imeson AC (2002) Extreme events controlling erosion and sediment transport in a semi-arid sub-andean valley. Earth Surface Processes and Landforms 27: 1365–1375.
42. Belmonte AMC, Beltran FS (2001) Flood events in Mediterranean ephemeral streams (ramblas) in Valencia region, Spain. Catena 45: 229–249.
43. Bull LJ, Kirkby MJ, Shannon J, Hooke JM (2000) The impact of rainstorms on floods in ephemeral channels in southeast Spain. Catena 38: 191–209.
44. Bakker MM, Govers G, van Doom A, Quetier F, Chouvardas D, et al. (2008) The response of soil erosion and sediment export to land-use change in four areas of Europe: The importance of landscape pattern. Geomorphology 98: 213–226.
45. Lu XX, Higgitt DL (2001) Sediment delivery to the Three Gorges 2: Local response. Geomorphology 41: 157–169.
46. Oeurng C, Sauvage S, Sanchez-Perez JM (2010) Dynamics of suspended sediment transport and yield in a large agricultural catchment, southwest France. Earth Surface Processes and Landforms 35: 1289–1301.
47. Klein M (1984) Anti clockwise hysteresis in suspended sediment concentration during individual storms: Holbeck Catchment; Yorkshire, England. Catena 11: 251–257.
48. Si JJ (2009) Research of man–made accelerated erosion and farm households' behavior of Water&Soil conservation in small watershed. Master dissertation of Huazhong agricultural university. China academic journal (cd-rom version) electronic magazine,Wuhan. (in Chinese).
49. Fu ZY, Li ZX, Cai CF, Shi ZH, Xu QX, et al. (2011) Soil thickness effect on hydrological and erosion characteristics under sloping lands: A hydropedological perspective. Geoderma 167–68: 41–53.
50. Wu CL, Zhang PC (2002) primary study on soil erosion factors of the wangjiaqiao watershed in three gorges reservoir area. Resources and Env ironment in the Yangtze Basin 11: 165–170. (in Chinese).

7

Is the Inherent Potential of Maize Roots Efficient for Soil Phosphorus Acquisition?

Yan Deng[1], Keru Chen[1], Wan Teng[2], Ai Zhan[3], Yiping Tong[2], Gu Feng[1], Zhenling Cui[1], Fusuo Zhang[1], Xinping Chen[1]*

1 Center for Resources, Environment and Food Security, China Agricultural University, Beijing, China, **2** State Key Laboratory for Plant Cell and Chromosome Engineering, Institute of Genetics and Developmental Biology, Chinese Academy of Sciences, Beijing, China, **3** State Key Laboratory of Soil Erosion and Dryland Farming on the Loess Plateau, Institute of Soil and Water Conservation, Chinese Academy of Sciences and Ministry of Water Resource, Yangling, China

Abstract

Sustainable agriculture requires improved phosphorus (P) management to reduce the overreliance on P fertilization. Despite intensive research of root adaptive mechanisms for improving P acquisition, the inherent potential of roots for efficient P acquisition remains unfulfilled, especially in intensive agriculture, while current P management generally focuses on agronomic and environmental concerns. Here, we investigated how levels of soil P affect the inherent potential of maize (*Zea mays* L.) roots to obtain P from soil. Responses of root morphology, arbuscular mycorrhizal colonization, and phosphate transporters were characterized and related to agronomic traits in pot and field experiments with soil P supply from deficiency to excess. Critical soil Olsen-P level for maize growth approximated 3.2 mg kg^{-1}, and the threshold indicating a significant environmental risk was about 15 mg kg^{-1}, which represented the lower and upper levels of soil P recommended in current P management. However, most root adaptations involved with P acquisition were triggered when soil Olsen-P was below 10 mg kg^{-1}, indicating a threshold for maximum root inherent potential. Therefore, to maintain efficient inherent potential of roots for P acquisition, we suggest that the target upper level of soil P in intensive agriculture should be reduced from the environmental risk threshold to the point maximizing the inherent potential of roots.

Editor: Malcolm Bennett, University of Nottingham, United Kingdom

Funding: This work was supported by the National Natural Science Foundation of China (30890133, 30971872), National Maize Production System in China (CARS-02-24), the Innovative Group Grant of NSFC (30971892), and the Innovation Fund for Graduate Student of China Agricultural University (KYCX2011032). The funders had no role in study design, data collection and analysis, decision to publish or preparation of the manuscript.

Competing Interests: The authors have declared that no competing interests exist.

* E-mail: chenxp@cau.edu.cn

Introduction

Modern high-intensity agriculture strongly relies on phosphorus (P) fertilization [1], [2], but sustainable agriculture requires improved P management. On one hand, there is increasing concern about P scarcity because the world's main source of P fertilizers, phosphate rock, is a limited and non-renewable resource [3], [4], [5], [6]. On the other hand, P fertilization can harm the environment by contributing to the eutrophication of water bodies [7], [8]. Inappropriate P fertilization accelerates the soil P imbalance in croplands worldwide [9].

P is readily fixed in most soils and has low availability to plants [10]. To enhance P acquisition, plants and their root-associated microbes have evolved a series of strategies that include modified root growth and functioning. Common strategies about root growth are increased root/shoot ratio [11], [12], modified root architecture [13], [14], decreased root diameter [15], enhanced specific root length (root length per unit root mass) [16], higher root hair length and/or density [17], [18], and production of aerenchyma [13], [19]. These morphological adaptations can greatly enhance the volume of soil root will exploit, and/or benefit exploitation of P-rich patches [20]. Also associations with arbuscular mycorrhizal (AM) fungi greatly extend the soil exploration space beyond the roots for many higher plant species [21]. Besides increasing soil volume exploited, roots and associated microbes can increase P availability from touched inorganic and organic sources by enhancing synthesis and exudation of organic acids and phosphatases [15], [22]. Increased P-uptake capacity by enhancing expression of high-affinity phosphate (Pi) transporters is another typical response of root functioning to facilitate P acquisition [23].

Current P management in intensive agriculture focuses on agronomic and environmental concerns, aiming to maintain soil P level between critical values that maximize crop yield but minimize P loss [24]. Most research, which explores the inherent potential of roots for efficient P acquisition, has focused on the adaptive mechanisms of P-efficient plants under P deficiency in natural ecosystems or low-yielding agricultural systems [13], [25], [26], [27]. In intensive agriculture, where soil P supply is increased by fertilization, the inherent potential of roots for efficient P acquisition is unfulfilled. In addition, interactions of P between soil and plants have often been studied in controlled and short term experiments not representative of field cropping systems.

Here, we hypothesized that the inherent potential of roots can be manipulated by managing soil P level to achieve a soil P supply that maximizes root uptake of P, optimizes crop yield and minimizes P loss. In pot and field experiments, we investigated root morphology, AM colonization, and expression of Pi transporter genes in maize (*Zea mays* L.) with different levels of P supply. The responses of these root traits were related to agronomic traits, and the optimum soil P supply was estimated.

Materials and Methods

Pot experiment

A pot experiment was carried out in the glasshouse at China Agricultural University from April to June, 2011. Maize (NE15, a test-cross variety) plants were planted in a calcareous silt loam soil which was collected from the same experimental site that was used for the field study (see next section). The initial soil properties were: pH 8.35 (1 : 5 soil : water ratio), organic matter 7.09 g kg^{-1}, total N 0.51 g kg^{-1}, Olsen-P 1.19 mg kg^{-1}, and exchangeable K 90 mg kg^{-1}. Seven P application rates (0, 12.5, 15, 50, 75, 100 and 300 mg P kg^{-1} soil) were used with P added as calcium superphosphate. In addition, N and K (as urea and potassium sulfate, respectively) were each applied at 150 mg kg^{-1} soil. Soil was air-dried and ground to pass through a 2-mm sieve. The added nutrients were mixed well with 8 kg of soil and filled into 4.5-L pots, with six replicate pots per treatment. Maize seeds were surface sterilized (30 min in a 10% H_2O_2 solution), rinsed, imbibed (8 h in a saturated $CaSO_4$ solution), and germinated in a dark and humid environment for 24 h. Two germinated seeds were planted per pot, and the seedlings were thinned to one per pot at the 3-leaf stage.

Plants were harvested 56 days after planting (DAP) at the 8-leaf stage. In each treatment, three pots were randomly chosen for shoot, root and soil sampling. Shoots were removed, oven-dried at 60°C for three days, weighed, and ground for nutrient analysis. All visible roots in each pot were carefully picked out by hand and stored in an ice box before transferring to the laboratory, after which soil samples were taken. In the laboratory root samples were carefully cleaned with tap water and frozen at −20°C before measurement of root morphology and AM colonization. Soil samples were air-dried and ground to pass through a 1-mm sieve for analysis of soil Olsen-P and $CaCl_2$-P. The remaining three pots in each treatment (except pots treated with 75 mg P kg^{-1} soil) were sampled for analysis of expression of Pi transporters in roots. Plant roots were gently taken out and immediately washed and then stored in liquid nitrogen for later RNA extraction.

Field experiment

The field experiment was conducted at the Shangzhuang Experimental Station of China Agricultural University (40°8′27″N, 116°10′39″E) in Beijing. This site is located in the northern North China Plain and has a typical semi-humid monsoon climate of the warm temperate zone. The annual average temperature ranges from 11 to 13°C, and annual rainfall ranges from 480 to 580 mm with precipitation mainly occurring from June to August. Annual mean sunshine is 2750 h, and 180–200 days are frost-free. Similar to the pot experiment, the soil in the field experiment was silt loam and with these properties at 0–30 cm depth: pH 8.00 (1 : 5 soil : water ratio), organic matter 8.02 g kg^{-1}, total N 0.37 g kg^{-1}, Olsen-P 1.82 mg kg^{-1}, and exchangeable K 82 mg kg^{-1}.

The treatments consisted of eight P application rates (0, 12.5, 25, 50, 75, 100, 150 and 300 kg P ha^{-1}) with four replicate plots per rate in a randomized complete block design. Each plot had an area of 17.5 m^2 (5 m×3.5 m). Before sowing, the entire quantity of P (as calcium superphosphate), 75 kg N ha^{-1} (as urea), and 62 kg K ha^{-1} (as potassium sulfate) were broadcast and mixed into top 20 cm of soil by disking. At 6-leaf and 13-leaf stages, an additional 150 kg N ha^{-1} as urea was top-dressed to each plot, with 75 kg N ha^{-1} each time. The same maize variety as that in pot experiment was used and seeds were planted at 67,500 plants ha^{-1} on 6 June, 2011.

At the flowering stage (57 DAP), three plants per plot were harvested from all plots and separated into shoots and roots. The shoot samples were used for determination of shoot dry weight and P content. Root samples were washed for assessment of AM colonization. Root samples for analysis of morphological traits were collected using the monolith method (see below) in all replicates of the following six P treatments: 0, 12.5, 25, 50, 100 and 300 kg P ha^{-1}. With the shoot at the center, each monolith measured 40 cm perpendicular to the rows (row spacing was 60 cm)×20 cm parallel with the rows (plant spacing was 25 cm)×30 cm deep. Each monolith was subdivided into three depths (0–10, 10–20, and 20–30 cm). All visible roots in each soil layer were carefully collected by hand and stored in an ice box before transport to the laboratory for washing. Additional roots for analysis of Pi transporter genes were taken from the same plots for root morphology assessment. Roots of three plants per plot were carefully excavated and washed with water, separated from stems, and combined into one sample. The samples were preserved in liquid nitrogen for transport to laboratory. At the grain maturity stage (111 DAP), three plants per plot were harvested for determination of total shoot dry weight and P content. Grain yield was determined by manually harvesting and drying (at 60°C) ears from two rows per plot. At flowering and grain maturity stages, topsoil (0–20 cm) was sampled from each plot for analysis of soil Olsen-P and $CaCl_2$-P.

Sample analysis

We used Olsen-P as the indicator of plant available P in the calcareous soils used here [28] and $CaCl_2$-P as the indicator of P loss risk [29]. Soil Olsen-P level was determined by the molybdo-vanadophosphate method based on extraction of air-dried soil with 0.5 M $NaHCO_3$ (pH 8.5) at 25°C [28]. Soil $CaCl_2$-P was measured by extracting air-dried soil with 0.01 M $CaCl_2$ according to Hesketh and Brookes [29]. Plant P concentration was measured by the molybdo-vanadophosphate method after samples were digested with concentrated H_2SO_4 and H_2O_2 [30]. Plant P uptake was then calculated from plant dry weight and P concentration.

To measure root morphological traits, cleaned root samples were dispersed in water in a transparent array (30 cm×20 cm×2 cm) and imaged with a scanner (Epson Expression 1600, Seiko Epson, Nagano, Japan) at a resolution of 800 dpi. The images were analyzed by WinRhizo software (Regent Instrument Inc., Quebec, QC, Canada) to determine root length. Root dry weight was determined by weighing the oven-dried samples after scanning. Specific root length was calculated from root length and root dry weight, and root/shoot ratio was assessed from root dry weight and shoot dry weight, respectively.

AM colonization was measured in 1-cm fine root segments that had been thoroughly mixed. Roots were cleared with 10% KOH at 90°C for 1 h and stained with Trypan blue [31]. The percent root colonization by AM fungi was assessed by examining 30 randomly selected stained root segments at 100–400 × magnification with a light microscope according to Trouvelot *et al.* [32].

Six Pi transporters of the Pht1 family have been reported for maize: from *ZEAma;Pht1;1* to *ZEAma;Pht1;6* [33]. For simplification, these genes are presented as *ZmPht1;1* to *ZmPht1;6* here. Expression of these Pi transporter genes was analyzed in maize root samples by real-time quantitative RT-PCR (qRT-PCR) [34]. Total RNA was isolated from frozen root tissue using the Trizol reagent (cat. no. 15596018, Invitrogen, USA). The isolated RNA was treated with the RNase-Free DNase Set (cat. no. 79254,

Table 1. Agronomic traits of pot maize plants in response to different P application rates.

P rate	Olsen-P	$CaCl_2$-P	Shoot dry weight	Shoot P concentration	P uptake
(mg P kg^{-1} soil)	(mg kg^{-1})	(mg kg^{-1})	(g $plant^{-1}$)	(g kg^{-1})	(mg $plant^{-1}$)
0	1.21 (0.11)^{d}	0.15 (0.00)^{b}	1.9 (0.1)^{d}	1.06 (0.12)^{e}	2.1 (0.2)^{d}
12.5	1.71 (0.12)^{d}	0.15 (0.00)^{b}	8.3 (0.5)^{c}	1.14 (0.12)^{de}	9.4 (0.9)^{cd}
25	2.39 (0.24)^{d}	0.14 (0.01)^{b}	12.4 (0.7)^{b}	1.42 (0.04)^{d}	17.6 (1.4)^{c}
50	4.80 (0.88)^{c}	0.13 (0.01)^{b}	17.8 (1.3)^{a}	1.94 (0.11)^{c}	34.6 (2.5)^{b}
75	6.86 (0.15)^{b}	0.13 (0.01)^{b}	18.1 (1.0)^{a}	2.24 (0.05)^{bc}	40.7 (3.0)^{ab}
100	7.65 (0.87)^{b}	0.15 (0.01)^{b}	18.4 (2.2)^{a}	2.43 (0.18)^{b}	45.2 (8.3)^{ab}
300	35.77 (0.88)^{a}	0.83 (0.07)^{b}	16.2 (0.7)^{a}	3.18 (0.13)^{a}	51.7 (4.3)^{a}

Plants were sampled at the 8-leaf stage (56 days after planting).
Each value is the mean (± SE) of three replicates.
Values in a column followed by different letters are significantly different at $p < 0.05$.

Qigen, Germany) to eliminate genomic DNA contamination before it was cleaned further with the RNeasy Plant Mini Kit (cat. no. 74904, Qigen, Germany). The first-strand cDNA was synthesized using the PrimeScript® RT reagent Kit Perfect Real Time (cat. no. DRR037A, Takara, Dalian) according to the manufacturer's protocol. Then the qRT-PCR was performed on a Mastercycler Realplex4 Real Time PCR System (Eppendorf, Germany) based on the protocol of the SYBR® Premix EX TaqTM (cat. no. DRR041A, Takara) in 20 μl reaction volume, which contained 10 μl of SYBR Green PCR mix, 0.4 μM of each forward and reverse primer, 0.4 μg of diluted cDNA template, and the appropriate amounts of sterile double distilled water. The applied program was set as initial polymerase activation at 95°C, 30 s, then 40 cycles at 95°C, 5 s; 60°C, 35 s. The specificity of the PCR amplification was evaluated with a melt curve analysis from 60°C to 95°C following the final cycle of the PCR. All reactions were set up using four biological replicates. We used UBQ2 as the internal control gene as reported by Calderon-Vazquez *et al.* [34], and the transcription levels of each gene were normalized to that of UBQ2 by the $2^{-\Delta\Delta Ct}$ method. The sequences of the gene-specific primers used for the six transporters and UBQ2 were provided by L. Z. Long, pers. comm.:

ZmPht1;1 primers: - 5′-GACCCAGATGGTGTAGAATCGAA-CAT-3′, and

- 5′-TCACTTACTTTCCCGCCTATAACACACA-3′.

Table 2. Agronomic traits of field maize plants in response to different phosphorus application.

DAP	P rate	Olsen-P	$CaCl_2$-P	Shoot dry weight	Grain yield	Shoot P concentration	Grain P concentration	P uptake
	(kg P ha^{-1})	(mg kg^{-1})	(mg kg^{-1})	(Mg ha^{-1})	(Mg ha^{-1})	(g kg^{-1})	(g kg^{-1})	(kg ha^{-1})
57	0	1.02 (0.11)^{e}	0.03 (0.00)^{c}	4.01 (0.02)^{c}		1.57 (0.14)^{c}		6.29 (0.60)^{d}
	12.5	1.39 (0.03)^{e}	0.03 (0.00)^{c}	4.43 (0.13)^{c}		1.83 (0.21)^{bc}		8.16 (1.05)^{cd}
	25	1.67 (0.11)^{e}	0.04 (0.00)^{c}	6.02 (0.16)^{b}		1.89 (0.11)^{bc}		11.39 (0.69)^{bc}
	50	3.04 (0.45)^{de}	0.05 (0.00)^{c}	6.15 (0.11)^{b}		2.04 (0.15)^{bc}		12.57 (1.09)^{ab}
	75	5.97 (0.14)^{cd}	0.06 (0.01)^{c}	6.69 (0.29)^{a}		2.12 (0.13)^{b}		14.21 (1.20)^{ab}
	100	7.03 (0.49)^{c}	0.04 (0.00)^{c}	6.62 (0.17)^{a}		2.15 (0.30)^{ab}		14.26 (2.14)^{ab}
	150	11.57 (2.61)^{b}	0.13 (0.04)^{b}	6.70 (0.11)^{a}		2.23 (0.23)^{ab}		15.01 (1.77)^{ab}
	300	18.00 (2.50)^{a}	0.19 (0.03)^{a}	5.87 (0.14)^{b}		2.68 (0.13)^{a}		15.75 (0.83)^{a}
111	0	1.21 (0.01)^{d}	0.02 (0.00)^{c}	13.44 (0.68)^{e}	2.97 (0.04)^{d}	0.47 (0.02)^{cd}	1.89 (0.05)^{c}	10.57 (0.56)^{e}
	12.5	1.53 (0.25)^{d}	0.03 (0.00)^{bc}	13.95 (1.19)^{de}	5.06 (0.01)^{c}	0.42 (0.02)^{d}	1.95 (0.18)^{c}	12.20 (0.52)^{e}
	25	1.54 (0.25)^{d}	0.02 (0.00)^{c}	14.26 (0.50)^{cde}	4.66 (0.04)^{c}	0.41 (0.00)^{d}	1.85 (0.01)^{c}	12.60 (0.18)^{e}
	50	2.12 (0.23)^{d}	0.03 (0.01)^{bc}	16.20 (0.20)^{bcd}	6.08 (0.30)^{b}	0.40 (0.02)^{d}	2.25 (0.06)^{b}	17.69 (0.61)^{d}
	75	4.43 (0.17)^{c}	0.03 (0.01)^{bc}	16.36 (1.01)^{bc}	7.12 (0.25)^{a}	0.56 (0.03)^{bc}	2.45 (0.11)^{a}	22.73 (1.83)^{c}
	100	5.41 (0.46)^{c}	0.06 (0.01)^{bc}	16.59 (0.95)^{b}	7.41 (0.22)^{a}	0.62 (0.06)^{b}	2.41 (0.03)^{ab}	23.54 (0.66)^{bc}
	150	9.82 (0.67)^{b}	0.09 (0.02)^{ab}	18.93 (0.42)^{a}	7.35 (0.58)^{a}	0.62 (0.02)^{b}	2.50 (0.07)^{a}	25.48 (1.05)^{b}
	300	14.12 (0.80)^{a}	0.13 (0.06)^{a}	19.29 (0.88)^{a}	7.74 (0.20)^{a}	0.75 (0.04)^{a}	2.66 (0.07)^{a}	29.07 (0.80)^{a}

Each value is the mean (± SE) of four replicates.
Within each column and for each sampling time, values followed by different letters are significantly different at $p < 0.05$.
Abbreviation: DAP: days after planting.

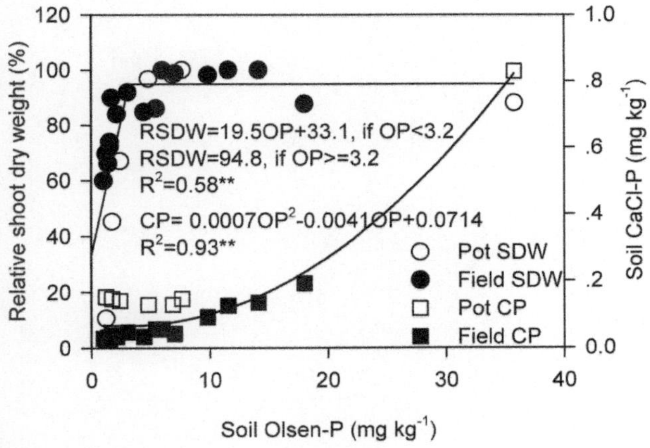

Figure 1. Maize growth and P loss risk in response to increasing soil P supply. Maize growth was presented as relative shoot dry weight, which was expressed relative to the highest mean value at each sampling time in each experiment. P loss risk was presented as soil $CaCl_2$-P level. Abbreviations: RSDW: relative shoot dry weight; OP: Olsen-P; CP: $CaCl_2$-P.

ZmPht1;2 primers: - 5′-GTCTGGTGAGGCTGAAGACTCA-GAGG-3′, and
- 5′-ACATGATAGCCCACCATGTGCAGTGC-3′.

ZmPht1;3 primers: - 5′-TGTTTCCGTTCTGTCTGGTGCTT-GTG-3′, and
- 5′-TCCCGACGGTGACCTCCGATTATTTA-3′.

ZmPht1;4 primers:- 5′-GAGACCCAGATGGTGTAGAGA-ATCG-3′, and
- 5′-CATCAAAACACAGCCAGGGTTGACT-3′.

ZmPht1;5 primers: - 5′-CCAAAGGTAAGTCGCTGGAAGA-GAT-3′, and
- 5′-CCATTGCGTGCAACAAACAGTGAC-3′.

ZmPht1;6 primers: - 5′-CGGACGTGAGCAAGGATGACAA-3′, and
- 5′-GGATTCCACACCCCCTGTGTAGT-3′.

ZmUBQ2 primers: -5'- CTTTGCTGCTGCACGGGAG-GAATG- 3', and
-5'- ATGGACGCACGCTGGCTGACTA-3'.

Statistical analysis

Data are presented as means and standard errors (SE). One-way ANOVA (SPSS 13.0, USA) was conducted, and significant differences among means were determined by LSD at the $p < 0.05$ probability level. To explore the relationship between shoot dry weight and soil Olsen-P, we used relative shoot dry weight by normalizing shoot dry weight data according to the maximum value obtained at each sampling time for each experiment; the data were then fitted to a linear-plateau model using SAS statistical software (SAS 8.1, USA) [35]. The relationships between soil $CaCl_2$-P, root morphological traits, AM colonization, expression of the six Pi transporter genes and soil Olsen-P were plotted using SigmaPlot statistical software (SigmaPlot 10.0, USA).

Results

Plant growth, P uptake, and risk of soil P loss

P application significantly increased soil P supply in terms of Olsen-P in both experiments, with the highest Olsen-P level in pot experiment being almost twice of that in the field (Table 1 and 2). Accordingly, shoot dry weight of 8-leaf plants in pot experiment increased fast with increasing P application up to 50 mg kg^{-1} soil, then increased only slightly and were reduced with further P supply (Table 1). Similar results were obtained for field plants at the flowering stage, with the critical P application rate for best shoot growth being 75 kg P ha^{-1} (Table 2). At maturity stage in the field, total shoot dry weight increased with increasing P supply, while grain yield plateaued when P application rate exceeded 75 kg P ha^{-1} (Table 2). Taken together, the result show that shoot dry weight was positively correlated with soil Olsen-P up to a threshold of 3.2 mg kg^{-1}, at which value almost 95% of the relative shoot dry weight had been achieved (Figure 1).

In the pot experiment, shoot P concentration were similar until P application rate reached 50 mg kg^{-1} soil, above which it increased significantly with further P supply (Table 1). Contrarily, shoot P uptake in aboveground parts increased rapidly below P application rate of 50 mg kg^{-1} soil and then more slowly with higher P applications (Table 1). Similar responses were observed for field plants at both sampling stages, with a critical P application rate of 50 kg P ha^{-1} at flowering stage and 75 kg P ha^{-1} at maturity stage, respectively (Table 2).

The risk of soil P loss was indicated by the level of soil $CaCl_2$-P. In the pot experiment, soil $CaCl_2$-P level remained unchanged with P application rates between 0 and 100 mg P kg^{-1} soil and significantly increased only with the highest application rate (Table 1). Similarly, soil $CaCl_2$-P in the field did not increase until P was applied at 150 kg P ha^{-1} at both sampling stages (Table 2). Overall, soil $CaCl_2$-P level remained unchanged with soil Olsen-P level < 15 mg kg^{-1} (Figure 1).

Responses of root morphology, AM colonization and expression of Pi transporter genes to soil P supply

In the pot experiment, root dry weight increased as soil P supply increased, but plateaued once soil Olsen-P level reached 5 mg kg^{-1} (Figure 2a). In the field experiment, root dry weight also initially increased with increasing soil P supply, peaked when soil Olsen-P was about 2.5 mg kg^{-1}, and then gradually declined to a plateau at an Olsen-P level around 10 mg kg^{-1} (Figure 2a). The responses of root length to P supply in both experiments were very similar to that of root dry weight of field plants, with the critical Olsen-P level indicating a plateau at 8 mg kg^{-1} (Figure 2b). As soil Olsen-P increased from very low levels, specific root length and root/shoot ratio of pot plants declined substantially at first, and then gradually reached a plateau when Olsen-P exceeded 8 mg kg^{-1}; field plants showed similar responses, but decreases were not so pronounced compared with those of pot plants, and both traits reached the plateau at a lower critical Olsen-P level about 5 mg kg^{-1} (Figure 2c, d). Generally, specific root length and root/shoot ratio were much higher for pot-grown plants than field-grown plants.

In the pot experiment, root AM colonization of plants at the 8-leaf stage initially increased with increasing soil P supply, peaked at an Olsen-P level about 2.5 mg kg^{-1}, and then tended to gradually decrease until Olsen-P level reached 10 mg kg^{-1}, above which AM colonization remained stable at an average value of 70% with further P supply (Figure 3). In the field at the flowering stage, with an increase of soil Olsen-P root AM colonization declined rapidly at first, and gradually plateaued at 40% when Olsen-P level exceeded 5 mg kg^{-1} (Figure 3). AM colonization rates were generally higher for pot plants than field plants.

We detected expressions of all the six Pht1 Pi transporter genes in both experiments. For pot plants, the six genes responded similarly to soil P supply, i.e. their transcript levels initially decreased rapidly with increasing P supply until soil Olsen-P level reached about 10 mg kg^{-1}, above which their expression kept very

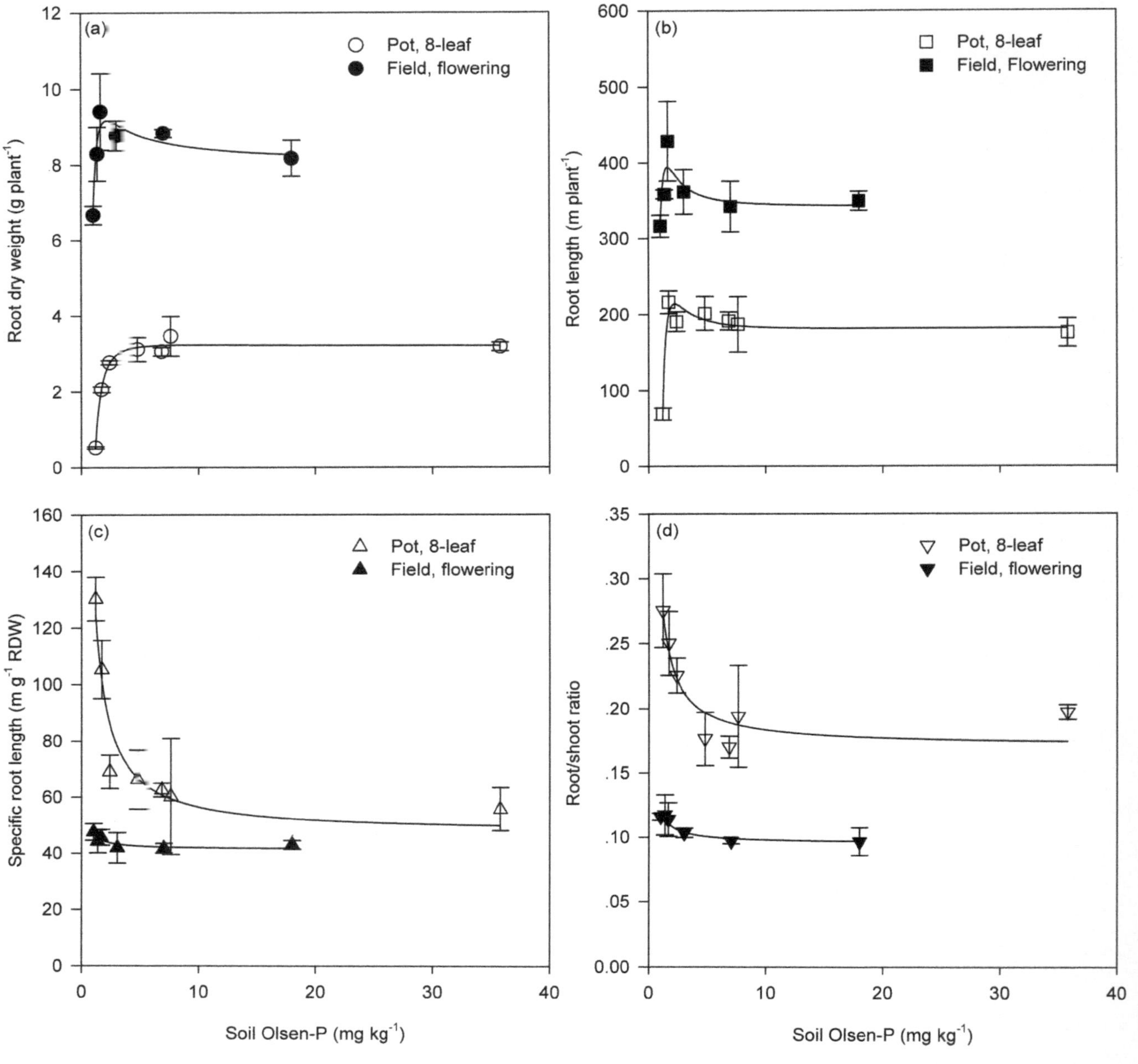

Figure 2. Root morphological traits in response to increasing soil P supply. In the pot experiment (open symbols), data were collected when plants were at the 8-leaf stage, and all visible roots in each pot were collected. In the field experiment (closed symbols), data were collected at the flowering stage, and roots were collected in a soil volume of 40 cm (row spacing) × 20 cm (plant spacing) × 30 cm (depth). Each symbol represents the mean (± SE) of three replicates for the pot experiment and four replicates for the field experiment, respectively. Abbreviation: RDW: root dry weight.

low and stable (Figure 4). Results were more complex in the field experiment. Expressions of *ZmPht1;1* to *ZmPht1;4* had similar responses to increase of soil Olsen-P, which were in accord with their expressions in pot plants, with the critical Olsen-P level indicating up-regulation approximating 5 mg kg^{-1} (Figure 4a–c). The transcript level of *ZmPht1;5* increased with increasing soil Olsen-P, peaked around 2.5 mg kg^{-1}, and then gradually declined to a stable status at Olsen-P level about 8 mg kg^{-1} (Figure 4e). Expression response of *ZmPht1;6* was similar to that of *ZmPht1;5* when soil Olsen-P was below 8 mg kg^{-1}, but declined further above this level. Generally, the expression responses of the six genes to soil P supply were more significant in the pot experiment than those in the field experiment.

Discussion

Many studies have reported that as soil P supply increased from initially low levels, crop yield increases quickly at first and then more slowly to an asymptote [36], [37], [38]. Based on this typical response, it has been proposed that there is a critical level of soil P for optimum crop yield [38], [39]. Under the conditions of our study, the critical level of soil P for maize production was 3.2 mg kg^{-1} (Olsen-P) (Figure 1). This value is relatively low as the critical

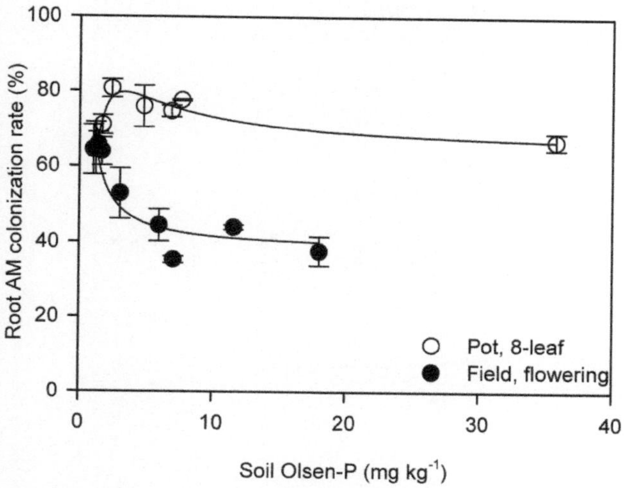

Figure 3. Root AM colonization in response to increasing soil P supply. In the pot experiment (open symbols), data were collected when plants were at the 8-leaf stage. In the field experiment (closed symbols), data were collected at the flowering stage. Each symbol represents the mean (± SE) of three replicates for the pot experiment and four replicates for the field experiment, respectively. Abbreviation: AM: arbuscular mycorrhizal.

P values reported in other studies have ranged from 3.9 to 17.3 mg kg^{-1} for maize production [35], [39], [40], [41].

To maximize crop yield, farmers tend to increase soil P and maintain it at a level that is greater than the critical value required for optimum yield [24]. Soil, however, cannot retain unlimited quantities of P. When soil P rise to a certain point, the environmental risk threshold, the risk of P loss increases significantly; and P loss may cause water pollution [7], [42]. As a result, P management has developed to embraces both agronomic and environmental goals, with the environmental risk threshold used as the building-up threshold for soil P [43], [44], [45]. Using $CaCl_2$-P as an indicator of P loss risk, we found that the environmental risk threshold in terms of Olsen-P, which resulted in a significant increase in $CaCl_2$-P, was about 15 mg kg^{-1} in this study (Figure 1). This value is in the range of 10–119 mg kg^{-1} previously reported for a range of soils [29]. Therefore, based on the current P management goals of maximizing crop yield and minimizing adverse environmental effects, our results indicate that soil Olsen-P level should be maintained between 3.2 and 15 mg kg^{-1} for maize production under our experimental conditions.

When soil Olsen-P decreased from 10 to 3.2 mg kg^{-1}, shoot growth was maintained but root length (and probably root dry weight) gradually increased (Figure 2a, b), so did root surface area (data not shown), whilst specific root length and root/shoot ratio remained relatively low and stable. This suggests that maize root growth was more sensitive than shoot growth to the reduction in soil P supply, and that maize root morphology acclimated to a reduced P supply firstly by increasing root length (and probably root dry weight and root surface area). When soil P supply was further reduced below 3.2 mg kg^{-1}, increases in specific root length and in the root/shoot ratio appeared to be the main morphological adaptation (Figure 2c, d). These results are consistent with commonly observed responses of plants to P limitation: plants allocate more biomass to root and produce more root length per unit of metabolic investment in root tissue in order to improve the capacity of roots to explore soil [22], [46]. Soil Olsen-P level higher than 10 mg kg^{-1} showed an inhibitory effect on all the morphological traits we measured.

For AM plants like maize, there are two P uptake pathways. One is the direct P uptake via root epidermis and root hairs. Among the six Pi transporters in Pht1 family reported for maize, *ZmPht1;1*–*ZmPht1;4* are considered to facilitate the direct uptake pathway [33], although the affinity for Pi of these four transporters remains uncertain. Phylogenetic analysis shows that *ZmPht1;1*, *ZmPht1;2*, and *ZmPht1;4* are closely related with rice *OsPT8* while *ZmPht1;3* closely clusters with rice *OsPT6* [33]. Both *OsPT6* and *OsPT8* are identified high-affinity Pi transporters [47]. Thus we speculate that *ZmPht1;1* to *ZmPht1;4* may also play the role of high-affinity Pi transporters. In the current study we found that these four genes responded similarly to soil P supply, and the up-regulation of their expressions was induced at an Olsen-P level (10 and 5 mg kg^{-1} in pot and field experiments, respectively) higher than the critical level for shoot growth (Figure 1, 4a–d), suggesting that root response on the molecular level to reduced P supply occurred at a higher P level than shoot growth. The up-regulation of the expressions for the four genes coincided with the increase of root length until soil Olsen-P was reduced to 2.5 mg kg^{-1}, from which root length began to decrease with further P supply reduction (Figure 2b, 4a–d). Such responses probably indicate the cooperation between root morphology and physiology for efficient P acquisition in the direct pathway when soil P supply reduced from sufficiency to deficiency. We noticed the overall up-regulation levels of these four genes were significantly higher in pot-grown plants than field-grown plants. One reason may be that the pot plants suffered relatively more P deficiency at lower P supply as indicated by the lower shoot P concentrations of pot plants than those of field plants (Table 1 and 2).

The other P uptake pathway is through AM symbioses. The rate of root colonization by AM fungi is considered an important factor for the extent to which AM symbioses contribute to P uptake and growth response of AM plants [48]. The soil P level plays an important role in influencing AM colonization [49], [50], [51]. In the current study, we found the influence of soil P on AM colonization was strongest when soil Olsen-P level was below 10 mg kg^{-1} (Figure 3), indicating a threshold of soil P supply for AM colonization as reported by others [51], [52]. As Olsen-P reduced from 10 to 3.2 mg kg^{-1}, the increasing root AM colonization suggests that the mycorrhizal pathway was enhanced to increase P acquisition without loss of shoot growth.

There are two types of Pht1 transporter genes whose expression can be induced in mycorrhizal roots: AM-specific Pi transporter genes which are expressed strictly in response to AM symbioses, and AM-inducible Pi transporter genes which can be strongly induced by AM symbioses but have a basal expression in non-mycorrhizal roots [53]. *ZmPht1;6* has been identified as an AM-inducible Pi transporter gene for maize [33]. Currently there is no report regarding an association between *ZmPht1;5* and AM symbioses. *ZmPht1;5*, however, clusters with *OsPT13* on the phylogenetic tree [33], [53], and *OsPT13* is an AM-inducible gene in rice [54], possibly indicating a role of *ZmPht1;5* in the mycorrhizal pathway. Coinciding with the response of root AM colonization, the expression levels of the two genes were gradually up-regulated when soil Olsen-P was reduced from 10 to 3.2 mg kg^{-1} in both experiments (Figure 4e, f). When soil P supply further decreased below the critical level for shoot growth, we noticed that both AM colonization and expressions of the two genes in the pot experiment were different from those in the field study. This may be partly in relation to the dependency on mycorrhizal pathway for P uptake under P deficiency when maize plants were at different growth stages. In the early growth season, root growth

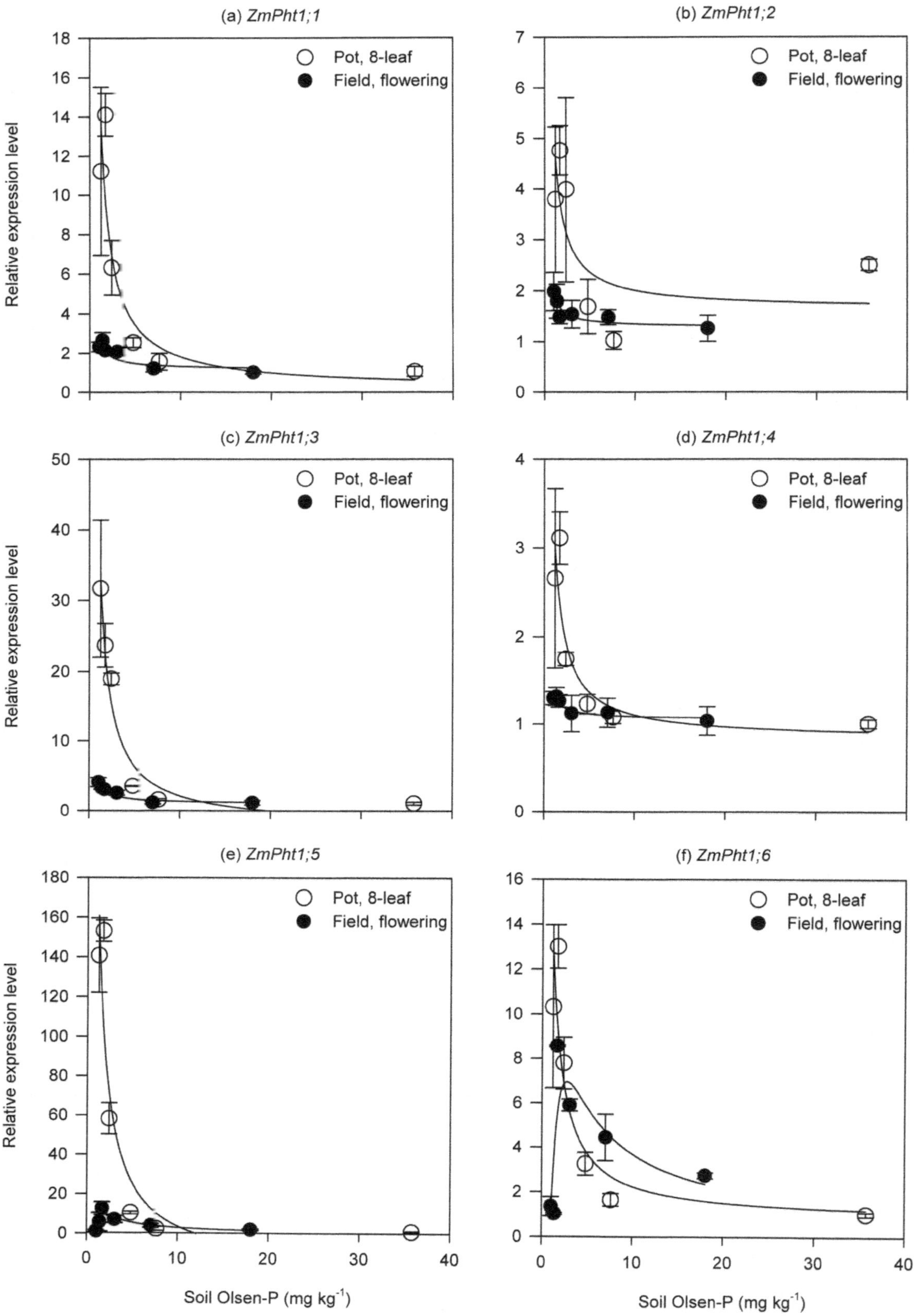

Figure 4. Expression of root Pht1 transporter genes in response to increasing soil P supply. Roots were sampled at the 8-leaf stage for pot plants (open symbols) and at the flowering stage for field plants (closed symbols). Gene relative expression level was measured by real-time quantitative RT-PCR. UBQ2 was used as the internal control. For each gene, the lowest expression level was set equal to 1.0. Each symbol represents the mean (± SE) of three replicates for the pot experiment and four replicates for the field experiment, respectively.

and development is relatively poor, but plant requirement of P is high to support fast growth [55], [56]. Thus an effective AM association may help to improve P nutrition for young plants. Our data showed that pot plants at the 8-leaf stage had much less root length (Figure 2b) and root surface area (data not shown) for P acquisition compared with field plants at the flowering stage. To compensate the direct pathway, the younger plants in the pot experiment might depend more on mycorrhizal pathway through enhancing expression of AM-inducible Pi transporter genes but limiting fungal growth as represented by reduced AM colonization, possibly a strategy to save carbon for efficient P acquisition. Besides, the much smaller soil volume for root growth and exploration in the pot experiment compared with field conditions may trigger pot plants to depend more on the mycorrhizal pathway under P deficiency, as indicated by the overall higher AM colonization rates and up-regulation levels of AM-inducible Pi transporter genes.

In summary, under the conditions of our study, most root adaptations (root morphology, AM colonization and expression of Pi transporter genes) involved with P acquisition were triggered at Olsen-P level < 10 mg kg^{-1}. Therefore, to maximize the inherent potential of maize roots to obtain P, we suggest that the upper level of optimum soil P supply in present study should be reduced to 10 mg kg^{-1}, rather than the environmental risk threshold of 15 mg kg^{-1}.

Soil P status varies across the world's croplands, and P deficits occur on 30–40% of the world's arable land [57], especially in Africa and South Asia. Sustainable crop production on these soils requires some level of P fertilization, a level that can optimize agronomic goals while maximizing inherent potential of roots to obtain P efficiently. In contrast to areas with P-depleted soils, most areas with intensive agriculture (Western Europe, North America, and East Asia) have the opposite problem in that levels of soil P have reached or exceeded the levels required by crops. In these soils, management should prevent further increases in P level or reduce P level not only to reduce environmental risk but also to better use the inherent potential of roots to acquire P. Although the importance of developing cultivars with an increased ability to obtain and utilize P is now recognized, our study illustrates that improvement of agronomic soil P management will also be important to explore the inherent potential of roots for efficient P acquisition.

Acknowledgments

We are very grateful to Prof. Fanjun Chen for supplying the plant material. We also thank Prof. Chunqin Zou and Prof. Susanne Schmidt for comments on an earlier version of the manuscript.

Author Contributions

Conceived and designed the experiments: YD YT FZ XC. Performed the experiments: YD KC WT AZ YT GF ZC XC. Analyzed the data: YD. Wrote the paper: YD XC.

References

1. Tilman D (1999) Global environmental impacts of agricultural expansion: The need for sustainable and efficient practices. Proceedings of the National Academy of Sciences of the United States of America 96: 5995–6000.
2. Tilman D, Cassman KG, Matson PA, Naylor R, Polasky S (2002) Agricultural sustainability and intensive production practices. Nature 418: 671–677.
3. Cordell D, Drangert J-O, White S (2009) The story of phosphorus: Global food security and food for thought. Global Environmental Change 19: 292–305.
4. Gilbert N (2009) Environment: The disappearing nutrient. Nature 461: 716.
5. Vaccari D (2009) Phosphorus: A looming crisis. Scientific American Magazine 300: 54–59.
6. Cordell D, Rosemarin A, Schröder J, Smit A (2011) Towards global phosphorus security: A systems framework for phosphorus recovery and reuse options. Chemosphere 84: 747–758.
7. Sharpley AN, Chapra S, Wedepohl R, Sims J, Daniel TC, et al. (1994) Managing agricultural phosphorus for protection of surface waters: Issues and options. Journal of Environmental Quality 23: 437–451.
8. Conley DJ, Paerl HW, Howarth RW, Boesch DF, Seitzinger SP, et al. (2009) Controlling eutrophication: Nitrogen and phosphorus. Science 323: 1014–1015.
9. MacDonald GK, Bennett EM, Potter PA, Ramankutty N (2011) Agronomic phosphorus imbalances across the world's croplands. Proceedings of the National Academy of Sciences of the United States of America 108: 3086–3091.
10. Hinsinger P (2001) Bioavailability of soil inorganic P in the rhizosphere as affected by root-induced chemical changes: A review. Plant and Soil 237: 173–195.
11. Hermans C, Hammond JP, White PJ, Verbruggen N (2006) How do plants respond to nutrient shortage by biomass allocation? Trends in Plant Science 11: 610–617.
12. Hill J, Simpson R, Moore A, Chapman D (2006) Morphology and response of roots of pasture species to phosphorus and nitrogen nutrition. Plant and Soil 286: 7–19.
13. Lynch JP (2007) Roots of the second green revolution. Australian Journal of Botany 55: 493–512.
14. Vance CP, Uhde-Stone C, Allan DL (2003) Phosphorus acquisition and use: Critical adaptations by plants for securing a nonrenewable resource. New Phytologist 157: 423–447.
15. Raghothama K (1999) Phosphate acquisition. Annual Review of Plant Biology 50: 665–693.
16. Richardson AE, Hocking PJ, Simpson RJ, George TS (2009) Plant mechanisms to optimise access to soil phosphorus. Crop and Pasture Science 60: 124–143.
17. Bates T, Lynch J (1996) Stimulation of root hair elongation in *Arabidopsis thaliana* by low phosphorus availability. Plant, Cell & Environment 19: 529–538.
18. Ma Z, Bielenberg D, Brown K, Lynch J (2001) Regulation of root hair density by phosphorus availability in *Arabidopsis thaliana*. Plant, Cell & Environment 24: 459–467.
19. Fan M, Zhu J, Richards C, Brown KM, Lynch JP (2003) Physiological roles for aerenchyma in phosphorus-stressed roots. Functional Plant Biology 30: 493–506.
20. Brown L, George T, Dupuy L, White P (2013) A conceptual model of root hair ideotypes for future agricultural environments: What combination of traits should be targeted to cope with limited P availability? Annals of Botany 112 (2): 317–330.
21. Bolan N (1991) A critical review on the role of mycorrhizal fungi in the uptake of phosphorus by plants. Plant and soil 134: 189–207.
22. Lambers H, Shane MW, Cramer MD, Pearse SJ, Veneklaas EJ (2006) Root structure and functioning for efficient acquisition of phosphorus: Matching morphological and physiological traits. Annals of Botany 98: 693–713.
23. Ramaekers L, Remans R, Rao IM, Blair MW, Vanderleyden J (2010) Strategies for improving phosphorus acquisition efficiency of crop plants. Field Crops Research 117: 169–176.
24. Li H, Huang G, Meng Q, Ma L, Yuan L, et al. (2011) Integrated soil and plant phosphorus management for crop and environment in China: A review. Plant and Soil 349: 157–167.
25. Coelho GT, Carneiro NP, Karthikeyan AS, Raghothama KG, Schaffert RE, et al. (2010) A phosphate transporter promoter from *Arabidopsis thaliana* AtPHT1; 4 gene drives preferential gene expression in transgenic maize roots under phosphorus starvation. Plant Molecular Biology Reporter 28: 717–723.
26. Chin JH, Gamuyao R, Dalid C, Bustamam M, Prasetiyono J, et al. (2011) Developing rice with high yield under phosphorus deficiency: Pup1 sequence to application. Plant Physiology 156: 1202–1216.
27. Gamuyao R, Chin JH, Pariasca-Tanaka J, Pesaresi P, Catausan S, et al. (2012) The protein kinase Pstol1 from traditional rice confers tolerance of phosphorus deficiency. Nature 488: 535–539.
28. Sims JT (2000) Soil test phosphorus: Olsen P. In: Pierzynski GM, editor. Methods of phosphorus analysis for soils, sediments, residuals, and waters. Raleigh, nc: North Carolina State University. pp. 20–21.
29. Hesketh N, Brookes P (2000) Development of an indicator for risk of phosphorus leaching. Journal of Environmental Quality 29: 105–110.
30. Shi R (1986) Soil and agricultural chemistry analysis. Beijing: China Agricultural Press.
31. Phillips J, Hayman D (1970) Improved procedures for clearing roots and staining parasitic and vesicular-arbuscular mycorrhizal fungi for rapid assessment of infection. Transactions of British Mycological Society 55: 158–161.
32. Trouvelot A, Kough JL, Gianinazzi-Pearson V (1986) Mesure du taux de mycorhization VA d'un système radiculaire. Recherche de méthodes d'estimation ayant une signification fonctionnelle. In: Gianinazzi-Pearson V, Gianinazzi S, editors. Physiological and genetical aspects of mycorrhizae. Paris: Institut geNational de la Recherche Agronomique. pp. 217–221.
33. Nagy R, Vasconcelos M, Zhao S, McElver J, Bruce W, et al. (2006) Differential regulation of five Pht1 phosphate transporters from maize (*Zea mays* L.). Plant Biology 8: 186–197.

34. Calderon-Vazquez C, Ibarra-Laclette E, Caballero-Perez J, Herrera-Estrella L (2008) Transcript profiling of *Zea mays* roots reveals gene responses to phosphate deficiency at the plant-and species-specific levels. Journal of Experimental Botany 59: 2479–2497.
35. Tang X, Ma Y, Hao X, Li X, Li J, et al. (2009) Determining critical values of soil Olsen-P for maize and winter wheat from long-term experiments in China. Plant and Soil 323: 143–151.
36. Johnston A, Dawson C (2005) Phosphorus in agriculture and in relation to water quality. Peterborough: Agricultural Industries Confederation.
37. Howard AE (2006) Agronomic thresholds for soil phosphorus in Alberta: A review. In: Alberta soil phosphorus limits project volume 5: Background information and reviews. Alberta, Canada: Alberta Agriculture, Food and Rural Development. 42 p.
38. Kirkby EA, Johnston AEJ (2008) Soil and fertilizer phosphorus in relation to crop nutrition. In: White PJ, Hammond JP, editors. The ecophysiology of plant-phosphorus interactions. Dordrecht, the Netherlands: Springer. pp. 177–223.
39. Mallarino A, Blackmer A (1992) Comparison of methods for determining critical concentrations of soil test phosphorus for corn. Agronomy Journal 84: 850–856.
40. Mallarino AP, Atia AM (2005) Correlation of a resin membrane soil phosphorus test with corn yield and routine soil tests. Soil Science Society of America Journal 69: 266–272.
41. Colomb B, Debaeke P, Jouany C, Nolot J (2007) Phosphorus management in low input stockless cropping systems: Crop and soil responses to contrasting P regimes in a 36-year experiment in southern France. European Journal of Agronomy 26: 154–165.
42. Heckrath G, Brookes P, Poulton P, Goulding K (1995) Phosphorus leaching from soils containing different phosphorus concentrations in the Broadbalk experiment. Journal of Environmental Quality 24: 904–910.
43. Higgs B, Johnston A, Salter J, Dawson C (2000) Some aspects of achieving sustainable phosphorus use in agriculture. Journal of Environmental Quality 29: 80–87.
44. Sharpley AN, Daniel T, Sims T, Lemunyon J, Stevens R, et al. (2003) Agricultural phosphorus and eutrophication, 2nd ed. Washington, DC: U.S. Department of Agriculture, Agricultural Research Service.
45. Elliott H, O'Connor G (2007) Phosphorus management for sustainable biosolids recycling in the United States. Soil Biology and Biochemistry 39: 1318–1327.
46. Lynch JP, Brown KM (2008) Root strategies for phosphorus acquisition. In: White PJ, Hammond JP, editors. The ecophysiology of plant-phosphorus interactions. Dordrecht, the Netherlands: Springer. pp. 83–116.
47. Nussaume L, Kanno S, Javot H, Marin E, Pochon N, et al. (2011) Phosphate import in plants: Focus on the PHT1 transporters. Frontiers in Plant Science 2: 1–12.
48. Smith SE, Smith FA (2011) Roles of arbuscular mycorrhizas in plant nutrition and growth: New paradigms from cellular to ecosystem scales. Annual Review of Plant Biology 62: 227–250.
49. Abbott L, Robson A (1991) Factors influencing the occurrence of vesicular-arbuscular mycorrhizas. Agriculture, Ecosystems & Environment 35: 121–150.
50. Grant C, Bittman S, Montreal M, Plenchette C, Morel C (2005) Soil and fertilizer phosphorus: Effects on plant P supply and mycorrhizal development. Canadian Journal of Plant Science 85: 3–14.
51. Covacevich F, Echeverría HE, Aguirrezabal LA (2007) Soil available phosphorus status determines indigenous mycorrhizal colonization of field and glasshouse-grown spring wheat from Argentina. Applied Soil Ecology 35: 1–9.
52. Fernandez M, Gutierrez B, Flavio H, Rubio F (2009) Arbuscular mycorrhizal colonization and mycorrhizal dependency: A comparison among soybean, sunflower and maize. Proceedings of the International Plant Nutrition Colloquium XVI. Department of Plant Science, Univ. Calf., Davis, USA.
53. Javot H, Pumplin N, Harrison MJ (2007) Phosphate in the arbuscular mycorrhizal symbiosis: Transport properties and regulatory roles. Plant, Cell & Environment 30: 310–322.
54. Glassop D, Godwin RM, Smith SE, Smith FW (2007) Rice phosphate transporters associated with phosphate uptake in rice roots colonized with arbuscular mycorrhizal fungi. Botany 85: 644–651.
55. Römer W, Schilling G (1986) Phosphorus requirements of the wheat plant in various stages of its life cycle. Plant and soil 91: 221–229.
56. Grant C, Flaten D, Tomasiewicz D, Sheppard S (2001) The importance of early season phosphorus nutrition. Canadian Journal of Plant Science 81: 211–224.
57. Von Uexküll, HR, Mutert E (1995) Global extent, development and economic impact of acid soils. Plant and Soil 171: 1–15.

8

Efficacy Evaluation of Fungus *Syncephalastrum racemosum* and Nematicide Avermectin against the Root-Knot Nematode *Meloidogyne incognita* on Cucumber

Wen-Kun Huang[1], Jian-Hua Sun[2], Jiang-Kuan Cui[1], Gang-Feng Wang[1], Ling-An Kong[1], Huan Peng[1], Shu-Long Chen[3], De-Liang Peng[1]*

1 State Key Laboratory for Biology of Plant Diseases and Insect Pests, Institute of Plant Protection, Chinese Academy of Agricultural Sciences, Beijing, P. R. China, **2** Tianjin Key Laboratory of Animals and Plants Resistance, College of Life Sciences, Tianjin Normal University, Tianjin, P. R. China, **3** Institute of Plant Protection, Hebei Academy of Agricultural and Forestry, Baoding, Hebei Province, P. R. China

Abstract

The root-knot nematode (RKN) is one of the most damaging agricultural pests.Effective biological control is need for controlling this destructive pathogen in organic farming system. During October 2010 to 2011, the nematicidal effects of the *Syncephalastrum racemosum* fungus and the nematicide, avermectin, alone or combined were tested against the RKN (*Meloidogyne incognita*) on cucumber under pot and field condition in China. Under pot conditions, the application of *S. racemosum* alone or combined with avermectin significantly increased the plant vigor index by 31.4% and 10.9%, respectively compared to the *M. incognita*-inoculated control. However, treatment with avermectin alone did not significantly affect the plant vigor index. All treatments reduced the number of root galls and juvenile nematodes compared to the untreated control. Under greenhouse conditions, all treatments reduced the disease severity and enhanced fruit yield compared to the untreated control. Fewer nematodes infecting plant roots were observed after treatment with avermectin alone, *S. racemosum* alone or their combination compared to the *M. incognita*-inoculated control. Among all the treatments, application of avermectin or *S. racemosum* combined with avermectin was more effective than the *S. racemosum* treatment. Our results showed that application of *S. racemosum* combined with avermectin not only reduced the nematode number and plant disease severity but also enhanced plant vigor and yield. The results indicated that the combination of *S. racemosum* with avermectin could be an effective biological component in integrated management of RKN on cucumber.

Editor: Martin Heil, Centro de Investigación y de Estudios Avanzados, Mexico

Funding: This research work were supported by the Special Project for Public Agriculture Research and Development of Ministry of Agriculture (Project No. 201103018), the National Basic Research Program of China (973 Program, Project No. 2013CB127502) and the National Science Foundation Project (Projects No. 31272022 and No. 31272019). The funders had no role in study design, data collection and analysis, decision to publish of the manuscript.

Competing Interests: The authors have declared that no competing interests exist.

* E-mail: dlpeng@ippcaas.cn

Introduction

The root-knot nematode (RKN), *Meloidogyne* spp., is one of the most damaging agricultural pests attacking a wide range of crops [1,2], and it can cause dramatic yield losses, mainly in tropical and sub-tropical agriculture [3,4]. In addition to RKNs infecting economically important plants, namely cucumber and tomato in China, RKN is the most common species and perhaps the most damaging pathogen in greenhouses [5]. Approximately 50% of greenhouse-grown vegetables are infected by RKN with an annual loss caused by RKN that is estimated to be more than $400 million. In intensive cropping systems, RKN has usually been managed with chemical nematicides for decades. However, the potential negative impact on the environment and humans has led to a total ban or restricted use of most chemical nematicides [6]. With increased pressure on growers to reduce nematicide usage and without effective alternatives, there is great interest in biological control as a management tool for this destructive pathogen.

Scientists throughout the world are engaged in research to develop nonchemical and eco-friendly RKN management strategies, such as sanitation, organic soil amendments, fertilization, heat-based methods, and cultivation of resistant cultivars and transgenic plants [5,7,8]. Many biological antagonists, such as *Pochonia chlamydosporia* [9], *Pseudomonas fluorescens*, *Paecilomyces lilacinus* [10], *Muscodor albus* [11], and *Trichoderma harzianum* [1], have been reported to be effective in controlling RKN. Application of antagonistic microorganisms or compounds produced by theses antagonists to RKN might provide an additional option for managing RKN. However, the use of exotic antagonists that are less adapted to local climates, conditions, or target species has led to limited success in RKN control and variability of results. Therefore, it is desirable to use local species and isolates in nematode management programs [12].

The metabolites produced by the filamentous fungus *Syncephalastrum racemosum* (isolated from soil samples infected by RKN in Tianjin Province, China are potential nematode biocontrol agents against RKN, soybean cyst nematodes (*Heterodera glycines*) and the pine wood nematode (*Bursaphelenchus xylophilus*) [13,14]. These metabolites cause more than 90% mortality of *B. xylophilus* juveniles after 120 h of exposure [14]. In greenhouse experiments, *S. racemosum* reduces multiplication rates of RKN in soil and the gall index of cucumber by 94.7 and 53.5%, respectively [15]. However, the reproductive capacity of female RKNs in roots treated with *S. racemosum* is yet to be measured under field conditions. The nematicide avermectin was first discovered in the 20th century, and it has been recently used in controlling plant parasitic nematodes [16,17]. Qiao et al. [18] reported that avermectin treatment has significant efficacy in lowering population levels of RKN compared to the untreated control. Avermectin is also effective in increasing plant height and yield as well as in reducing the incidence of *M. incognita* in potato and cucumber in field experiments [17]. However, few studies have focused on the combined effects of bioagents and avermectin.

The aim of our research was to test and evaluate the efficacy of *S. racemosum* as a biological control agent in controlling RKN on cucumber and its effect on the growth of cucumber over two successive years under field conditions. We also compared the efficacies of *S. racemosum* (fungus), avermectin (biological nematicide), and their combination to expect an increase in plant production or vigor with the combined treatment.

Materials and Methods

Nematode Inoculum Preparation

Eggs of RKN (*Meloidogyne incognita*) were extracted from infected roots of tomato (*Solanum lycopersicum*) using 1% sodium hypochlorite (NaClO), and the eggs were gently washed with tap water to remove the NaClO [19]. The nematode was identified on the basis of the perineal pattern of mature females [20].

Syncephalastrum Racemosum Inoculum Preparation

The *S. racemosum* Sr18 isolate was used for all experiments and was maintained on potato sucrose malt extract agar (PSMA) plate slants. Potato sucrose (PS) liquid medium was inoculated with a spore suspension to a final concentration of 10^7 spores per liter in a shaking flask according to the procedure described by Sun et al. [14]. Fermentations were carried out in a 30 L fermenter containing PS liquid medium with aeration (1:0.5 vvm) and stirring (100–400 rpm) at 27°C for 42 h. Culture filtrate and mycelium were separated by filtration and adjusted to pH 7.0 with NaOH.

Pot Experiment

The experiment was conducted in plastic pots (30 cm in diameter) filled with steam-sterilized sandy loam soil amended with organic peat and sand (field soil: organic peat: sand = 1:3:4). All pots (three plants per pot and 10 replicates) were arranged in a completely randomized block design on a bench in the greenhouse at 25–30°C. The treatments were as follows: (1) untreated control without RKN; (2) untreated control inoculated with *M. incognita*; (3) *S. racemosum* (2 ml); (4) 1.8% avermectin emulsifiable concentrate (EC) (0.6 ml) and (5) *S. racemosum* (1 ml) +1.8% avermectin EC (0.4 ml) (Table 1). The concentrations of *S. racemosum* and avermectin in treatments 3–5 were suspended in 400 ml of tap water and applied as a soil drench before sowing. The treatment 1 and 2 pots were irrigated with only 400 ml of tap water. Cucumber (*Cucumis sativus*) seeds were surface sterilized with 1% sodium hypochlorite (5 min) and sown directly in the pots (5 seeds per pot). The seed germination rate was calculated to be 10 d. When the cucumber seedlings reached the two-leaf stage, each pot in treatments 2–5 was inoculated with approximately 2000 second-stage juveniles (J2) of *M. incognita* by pouring 10 ml of the nematode suspension into holes with a 2–4 cm depth around the bases of the plants. Forty-five days after nematode inoculation, the cucumber plants were uprooted and washed free of adhering soil. The shoot length, root length, and number of galls per root system were measured and counted, as appropriate. The vigor of the seedlings was evaluated by calculating the vigor index (VI) [21] as follows:

$$\text{Vigor index (VI)} = (\text{mean of shoot length} + \text{mean of root length}) \times \text{percentage of germination}$$

Meanwhile, nematodes were extracted from the soil using a Baermann pan [22]. The cucumber roots were stained with acid fuchsin [23]. Nematodes extracted from the soil or stained in the host plant roots were examined and counted under a dissecting microscope. Nematode density in the soil was expressed as the number of nematodes per 200 g of soil, and nematode density in the roots was expressed as the number of nematodes per gram of wet root.

Field Experiment

The study was performed in a commercial greenhouse on private land located in Baodi County, Tianjin Province, China (E117°17′5″ and N39°44′7″; elevation of 10.6 m) in October 2010 and 2011. The field studies did not involve endangered or protected species, and no specific permissions were required for these locations/activities. Cucumber had been cultivated in this greenhouse for 10 years before our experiments. In July 2010, the disease incidence was 80–95%, and the galling index varied between 46 and 67. The soil texture was sandy loam with the following properties: pH = 7.4; electrical conductivity = 4.2 dS m^{-1}; NH_4^+-N content = 0.7 mg kg^{-1}; NO_3^-N content = 59.4 mg kg^{-1}; total N = 1.35 g kg^{-1}; extractable P = 264.5 mg kg^{-1}; extractable K = 167.0 mg kg^{-1}; and organic C = 8.2 mg kg^{-1}.

Four experiments were designed in randomized blocks with four replicates as follows: (1) untreated control; (2) *S. racemosum* filtrate (10 ml m^{-2}); (3) 1.8% avermectin EC (3 ml m^{-2}); and (4) *S. racemosum* filtrate (5 ml m^{-2}) +1.8% avermectin EC (2 ml m^{-2}) (Table 2). Plots were 5 m long and 3 m wide (4 plant rows). Seedlings of cucumber cv. Jinglv (Tianjin Kernel Cucumber Research Institute) were produced on a field without RKN, and 15 plants were transplanted in each row. Two days before transplanting, *S. racemosum* and avermectin were dissolved in 10 L of water and used for irrigation on the planting furrow in each replicate. The untreated control was irrigated with 10 L of water only.

Prior to the application of biological agents, 10 soil cores (10 cm in diameter and 25 cm in depth) were collected from each plot for calculating nematode population density. Soil samples were also collected at 30, 60 and 180 d after transplanting (DAT). Second-stage juveniles of *M. incognita* were extracted from a subsample of soil (200 g) by the Baermann funnel method [24] and counted under a stereoscopic microscope. At 30, 60 and 180 DAT, 10 plants from each plot were uprooted, and the soil adhering to the roots was removed by gentle agitation in water. Root gall severity was assessed using a 0–5 rating scale according to the percentage of galled tissue (0 = 0%; 1 = 1–15%; 2 = 16–25%; 3 = 26–50%; 4 = 51–75%; and 5 = 76–100%). Roots were then carefully washed and stained with acid fuchsin [23]. Females and juveniles per gram

Table 1. Effect of biocontrol agents on seed germination and seedling vigor of cucumber plants in pots.

Treatments	Growth indices			
	Germination (%)	MRL[c] (cm)	MSL[c] (cm)	VI[c]
1. *M. incognita*-free control	89.3 a	19.5 a	48.3 b	60.5±5.8 b
2. *M. incognita*-inoculated control[a]	87.8 ab	15.8 b	43.3 c	51.3±2.7 c
3. *Syncephalastrum racemosum* (2 ml[b])	86.3 ab	21.5 a	53.5 a	67.4±3.5 a
4. Avermectin (0.6 ml[b])	81.5 b	19.0 ab	50.3 ab	56.5±5.7 b
5. *S. racemosum* (1 ml)+avermectin (0.4 ml[b])	86.8 ab	20.8 a	49.8 ab	60.8±2.8 b
LSD 0.05	6.6	3.7	4.3	6.5

[a]Treatments 2–5 were inoculated with approximately 2000 second-stage juveniles at the two-leaf stage of cucumber seedlings.
[b]Treatments 1–2 were irrigated with 400 ml of tap water. The concentrations of *S. racemosum* and avermectin in treatments 3–5 were suspended in 400 ml of tap water and applied as a soil drench before sowing.
[c]MRL = mean root length, MSL = mean shoot length, VI = vigor index. The values in columns followed by the same letter(s) are not significantly different according to LSD (P = 0.05).

of root were counted. Mature fruits were weighed every 4 d from all of the plants per plot until the end of the harvest.

Statistical Analysis

Growth indices of plants, galls and Second-stage juveniles of RKN in pot experiment were used for statistical analysis. Data for total cumulative yield per month, galls and nematode number under field condition were used for statistical analysis. Analysis of variance (ANOVA) was performed for the pot experiment and the field experiment. The significant differences among the treatments were determined according to Least Significant Difference (LSD) test using SAS software version 8.0 (SAS Institute,Cary, NC) ($P < 0.05$) [25].

Results

Pot Experiment

Effects of biological agents on seedling vigor. To determine the effect of biological agents on the vigor index of cucumber, we cultivated cucumber in the presence or absence of nematodes and biological agents. The mean root length (MRL), mean shoot length (MSL) and vigor index (VI) of the control plants inoculated with J2 of *M. incognita* were significantly reduced ($p < 0.05$) (Table 1). The germination of the seeds was not influenced in any of the treatments. However, all treatments, except avermectin, increased the VI and MRL. All treatments increased the percentage of MSL compared to the untreated inoculated control. Compared with the non-inoculated control, only the *S. racemosum* treatment induced an increase of these growth parameters. The highest VI (67.4) and MRL (24.3 cm) occurred in the treatment of *S. racemosum*.

Effect of biological control agents on *M. incognita*. To determine the effect of biological agents on the gall index and nematode number, we cultivated cucumber in the presence or absence of nematodes and biological agents. Cucumber in the nematode-inoculated control showed the largest number of galls and nematodes both in soil and in roots (Table 2). All of the biocontrol treatments significantly reduced root galling compared to the nematode-inoculated control. The avermectin and *S. racemosum*+avermectin treatments yielded the smallest number of root galls. The J2 in the soil or plant roots showed a similar trend to root galls. In the soil, the highest reduction of J2 was observed in the avermectin and *S. racemosum*+avermectin treatments. The avermectin and *S. racemosum*+avermectin treatments also significantly reduced the number of J2 and females in the plant roots. However, a smaller reduction of nematodes and root galls resulted from treatment with *S. racemosum* compared to treatments with avermectin and avermectin+*S. racemosum*.

Field Experiment

To determine the effect of biological agents on nematode number under field condition, soil samples were collected from different time after cucumber transplanting. At the beginning of the experiment, the field was uniformly infested with RKN and varied between 221 and 246 J2 per 200 g of soil. All of the treatments significantly decreased the number of nematodes in both soil and roots at 30, 60, and 180 DAT (Table 3). At 30 DAT, the avermectin and *S. racemosum*+avermectin treatments showed greater efficacy at reducing nematodes in both soil and roots compared to *S. racemosum*. At 60 DAT, the *S. racemosum* treatment showed lower efficacy at reducing nematodes in roots compared to the *S. racemosum*+avermectin treatment. No significant difference in the number of nematodes in the soil was observed among *S. racemosum*, Avermectin and *S. racemosum*+avermectin treatments. At 180 DAT, the avermectin and *S. racemosum*+avermectin treatments showed greater efficacy at reducing nematodes in roots compared to the *S. racemosum* treatment. However, the *S. racemosum* and *S. racemosum*+avermectin treatments showed lower efficacy at reducing nematodes in soil compared to the avermectin treatment.

The reduction in root galling followed the same trend as the nematode number in soil and roots. All of the treatments significantly decreased the gall index compared to the untreated control at all observation times. At 30 DAT, no significant difference was observed among the treatments. At 60 and 180 DAT, the lowest gall index was registered after application of the avermectin and *S. racemosum*+avermectin treatments, and this index was significantly different from the *S. racemosum* treatment ($P < 0.05$).

Cucumber Yield

To determine the effect of biological agents on the yield of cucumber, cumulative cucumber yields per month were recorded from December 2010 to July 2011 (Figure 1). At the initial stage (December 2010 to March 2011), the yield did not differ significantly between the treatments and the untreated control. From midseason until the end of harvest time (April to July 2011), the yields in all treated plots were higher than in the untreated control. The highest overall yield was recorded in

Table 2. Effect of biocontrol agents on the number of galls and second-stage juveniles in soil and roots of cucumber infected with *Meloidogyne incognita* in the pot experiment.

Treatments	Galls			Second-stage juveniles (J2) in soil			Second-stage juveniles (J2) in roots		
	Galls (x)/root system	$\sqrt{x+1}$	Decrease over control (%)	J2 (y)/200 g soil	$\sqrt{y+1}$	Decrease over control (%)	J2 (z)/g root	$\sqrt{z+1}$	Decrease over control (%)
1. Non-inoculated control									
2. Untreated inoculated control	547	23.4±0.74 a[c]		418	20.5±0.36 a		31.4	5.7±0.31 a	
3. *S. racemosum* (2 ml)	74	8.7±0.36 b	62.8	98	9.9±0.42 b	51.7	7.6	2.6±0.30 b	54.4
4. Avermectin (0.6 ml)	24	5.0±0.44 c	78.6	27	5.3±0.49 c	74.2	2.6	1.9±0.13 c	66.7
5. *S. racemosum* (1 ml)+avermectin (0.4 ml)	29	5.5±0.54 c	76.5	24	5.0±0.48 c	75.6	3.2	2.1±0.05 c	63.2
LSD 0.05		0.83			0.68			0.35	

[a]Treatments 2–5 were inoculated with approximately 2000 second-stage juveniles at the two-leaf stage of cucumber seedlings.
[b]Treatments 1–2 were irrigated with 400 ml of tap water. The concentrations of *S. racemosum* and avermectin in treatments 3–5 were suspended in 400 ml of tap water and applied as a soil drench before sowing.
[c]MSL = mean root length, MSL = mean shoot length, VI = vigor index. The values in columns followed by the same letter are not significantly different according to LSD (P = 0.05).
[c]Each transformed figure is the average of four replicates±SD. The mean in columns followed by the same letter(s) did not significantly different according to LSD (p = 0.05).

Table 3. Effect of different biocontrol agents alone or in combination on the number of *Meloidogyne incognita* in soil and roots as well as symptom expression on cucumber plants in the field experiment.

Treatments	30 days after transplanting			60 days after transplanting			180 days after transplanting		
	Nematode density per 200 g soil	Nematode density per g root	Root gall index	Nematode density per 200 g soil	Nematode density per g root	Root gall index	Nematode density per 200 g soil	Nematode density per g root	Root gall index
1. Untreated inoculated control[a]	331±18.2 a	24.7±6.0 a	21.3±2.6 a	584±40.5 a	91.3±4.5 a	40.3±5.6 a	762±23.5 a	157.5±13.1 a	56.1±6.8 a
2. *S. racemosum* (10 ml m^{-2})	68±9.8 b	15.9±1.9 b	5.7±0.7 b	94±7.5 b	36.8±4.9 b	16.7±0.7 b	348±16.8 b	42.1±3.3 b	29.5±6.5 b
3. Avermectin (3 ml m^{-2})	42±7.1 c	5.3±0.9 c	4.8±1.2 b	65±10.4 b	8.6±0.8 c	11.4±1.1 c	297±17.2 c	9.3±1.5 c	22.5±1.9 c
4. *S. racemosum* (5 ml m^{-2})+ avermectin (2 ml)	31±8.6 c	4.9±1.3 c	4.2±1.3 b	74±6.7 b	8.1±1.0 c	10.9±1.5 c	341±11.8 b	11.7±0.7 c	21.8±3.3 c
LSD 0.05	18.1	5.0	2.5	33.1	5.2	4.5	27.4	10.5	7.9

[a]The *S. racemosum* and avermectin products in treatments 2–4 were dissolved in 10 L of water and used for irrigation on the planting furrow in one replicate. The untreated control was irrigated with 10 L of water only.

plots treated with *S. racemosum*+avermectin, with a mean of 756.6 t ha^{-1}, followed by the avermectin treatment, with a total mean of 731.6 t ha^{-1}. The lowest overall yield was recorded in the untreated control with a total mean of 508.5 t ha^{-1}. The total yield after the *S. racemosum*, avermectin and *S. racemosum*+ avermectin treatments was increased by 23.2, 43.9, and 48.8%, respectively. A significant yield increase during the entire season was also observed in the *S. racemosum*+avermectin and avermectin treatments compared to the *S. racemosum* treatment ($p<0.05$). No significant difference in the cumulative yield was observed between the treatments of *S. racemosum*+avermectin and avermectin($p>0.05$).

Discussion

Research on the use of antagonists to control plant parasitic nematodes is receiving increasing attention [26], and *S. racemosum* has been studied very little. The results from pot and field experiments in this study showed that the combination of *S. racemosum* and avermectin can cause the paralysis of RKN nematodes and reduce the application of pesticides. The decrease of final cost and the increase of plant yield may help to increase the farmers' income. Therefore, *S. racemosum* and avermectin are promising biological agents in the integrated control of plant parasitic nematodes.

The combination of *S. racemosum* and avermectin reduced the number of galls and developmental stages in cucumber roots, which was consistent with previous field results [15,27,28,29]. Sun et al. [27] reported that *S. racemosum* reduces the population density of *M. incognita* by 82.7% on tomato plants in a greenhouse. Secondary metabolites produced by *S. racemosum* were shown to cause a significant reduction of the reproduction of *M. incognita* on cucumber in a field experiment [15]. A previous *in vitro* experiment showed that *S. racemosum* metabolites inhibit hatching and reduce development of *M. incognita, H. glycines* and *B. xylophilus* [14,28]. More recently, Huang et al. [29] reported that a combination of *S. racemosum*, mustard biofumigation and solarization is effective in reducing *M. incognita*, which subsequently enhances cucumber growth. However, little attention had been paid to the effects of bioagents on plant vigor and yield. Present findings have confirmed that the highest VI and MRL were obtained after application of *S. racemosum*, which suggests that *S. racemosum* may have stimulative effects on both shoot growth and plant vigor. The efficacy of *S. racemosum* on the yield also demonstrated that *S. racemosum* had a significant promoting effect on the growth of cucumber. The widely recognized mechanisms of bioagents include the production of toxins, enzymes and other metabolic products, as well as the promotion of plant growth and induction of systemic resistance of host plants to pathogens [30]. In this study, the reduction of the gall index caused by the strain may be attributed to secondary metabolites produced by the fungus in the soil and direct parasitism on nematode eggs, which may have also induced the plant defense mechanism leading to systemic resistance [1,10]. The beneficial effects induced by *S. racemosum* combined with avermectin on the seedling vigor index and fruit weight may be attributed to the combined effect of production of growth-promoting substances and antibiotic metabolites effective against the microbial community [31].

Many fungi and their metabolites are known to possess nematicidal activity against a wide range of plant-parasitic nematodes including RKNs [3,26,32]. In the present studies, the antibiotic metabolites of *S. racemosum* showed good nematicidal activity against *M. incognita*. However, little is known about the nematicidal mechanisms of *S. racemosum* against RKNs despite excellent toxicologic actions. For RKNs, J2 move freely in the soil before infecting host plants, and their movement is regulated by active compounds [33,34]. Avermectin is a streptomycete-derived macrocyclic lactone biological nematicide that blocks the transmittance of electrical activity in nerves and muscle cells by stimulating the release and binding of gamma-aminobutyric acid at nerve endings [16], which causes paralysis of the neuromuscular systems that adversely affect nematode hatching and movement in soil, subsequently reducing the extent of root invasion [32]. Secondary metabolites of *S. racemosum* may inhibit the incubation of eggs after the RKN attach to the roots of cucumber plants, resulting in reduced

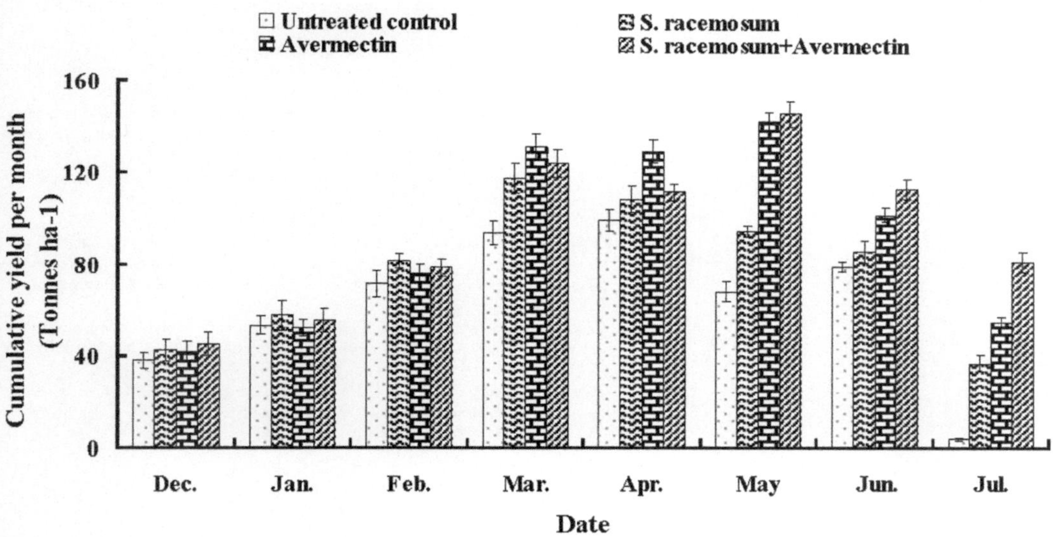

Figure 1. Final cumulative fruit yield per month of cucumber from December 2010 to July 2011 in the field experiment. Four experiments were designed in randomized blocks with untreated control, *S. racemosum* filtrate 10 ml m-2 (*S. racemosum*), 1.8% avermectin EC 3 ml m-2 (Avermectin) and *S. racemosum* filtrate 5 ml m-2+1.8% avermectin EC 2 ml m-2 (*S. racemosum*+Avermectin).

density of nematode populations in the soil. Therefore, the combination of *S. racemosum* and avermectin can not only cause the paralysis of nematodes in soil but can also reduce the infection of nematodes in roots. However, the exact mechanism of disease defense by this strain needs to be verified in future research.

Controlling root-knot nematodes is of major economic importance to the agricultural industry of northern China. The current study provides evidence that soil drench with the tested bio-products,avermectin and *S. racemosum* effectively reduced the number of galls and juveniles of RKN and consequently enhanced cucumber growth. The obtained results are highly encouraging, demonstrating their promising candidates as an alternative for the control of RKN in cucumber under glasshouse conditions. Further research is required to identify nematicidal components of *S. racemosum* and their mechanism of action that may be used as a part of an integrated strategy with other control methods, such as biofumigation, solarization, soil amendments, trap cropping and chemical nematicides.

Acknowledgments

The authors thank Prof. Maurice Moens at Ghent University and Prof. Shiming Liu at Southern Illinois University for their critical revision of the manuscript.

Author Contributions

Conceived and designed the experiments: DLP. Performed the experiments: JKC GFW LAK HP. Analyzed the data: WKH. Contributed reagents/materials/analysis tools: JHS. Wrote the paper: WKH DLP SLC.

References

1. Sahebani N, Hadavi N (2008) Biological control of the root-knot nematode *Meloidogyne javanica* by *Trichoderma harzianum*. Soil Biol Biochem 40: 2016–2020.
2. Wesemael WML, Viaene N, Moens M (2011) Root-knot nematodes (*Meloidogyne* spp.) in Europe. Nematology 13: 3–16.
3. Kiewnick S, Sikora RA (2006) Biological control of the root-knot nematode *Meloidogyne incognita* by *Paecilomyces lilacinus* strain 251. Biol Control 38: 179–187.
4. Moens M, Perry RN, Starr JL (2009) *Meloidogyne* species - A diverse group of novel and important plant parasites. In: Perry RN, Moens M, Starr JL, editors. Root-knot Nematodes. Wallingford, UK: CAB International. 1–17.
5. Dong L, Huang C, Huang L, Li X, Zuo Y (2012) Screening plants resistant against *Meloidogyne incognita* and integrated management of plant resources for nematode control. Crop Prot 33: 34–39.
6. Zukerman BM, Esnard J (1994) Biological control of plant nematodes current status and hypothesis. Jpn J Nematol 24: 1–13.
7. Li XQ, Tan A, Voegtline M, Bekele S, Chen CS, et al. (2008) Expression of Cry5B protein from *Bacillus thuringiensis* in plant roots confers resistance to root-knot nematode. Biol Control 47: 97–102.
8. Collange B, Navarrete M, Peyre G, Mateille T, Tchamitchian M (2011) Root-knot nematode (*Meloidogyne*) management in vegetable crop production: the challenge of an agronomic system analysis. Crop Protection 30, 1251–1262.
9. Atkins SD, Hidalgo-Diaz L, Kalisz H, Mauchline TH, Hirsh PR, et al. (2003) Development of a new management strategy for the control of root-knot nematodes (*Meloidogyne* spp.) in organic vegetable production. Pest Manag Sci 59: 183–189.
10. Hashem M, Abo-Elyousr Kamal A (2011) Management of the root-knot nematode *Meloidogyne incognita* on tomato with combinations of different biocontrol organisms. Crop Prot 30: 285–292.
11. Riga E, Lacey LA, Guerra N (2008) Muscodor albus, a potential biocontrol agent against plant-parasitic nematodes of economically important vegetable crops in Washington State, USA. Biol Control 45: 380–385.
12. Stirling GR (199) Biological control of plant parasitic nematodes: Progress, problems and prospects. Wallingford, UK: CAB International. 282 p.
13. Sun JH, Yu KL, Bi P, Peng DL (1999) Study on biological control of plant parasitic nematodes by using fungal metabolites. J Yunnan Agric Univ 14: 122–123.
14. Sun JH, Wang HK, Lu FP, Du LX, Wang GF (2008) The efficacy of nematicidal strain *Syncephalastrum racemosum*. Ann Microbiol 58: 369–373.
15. Sun JH, Qi JS, Feng X, Lu FP, Du LX (2005) Study on efficacy of bio-nematicide from Sr18 fungus for biological control of cucumber root-knot nematode. Hua Bei Nong Xue Bao 20: 74–78.
16. Putter JG, Preiser FA, Haidri AA, Ristich SS, Dybas RA (1981) Avermectins: Novel insecticides, acaricides and nematocides from a soil microorganism. Experientia 37: 963–965.
17. Qiao K, Dong S, Wang H, Xia X, Ji X, et al. (2012) Effectiveness of 1, 3-dichloropropene as an alternative to methyl bromide in rotations of tomato (*Solanum lycopersicum*) and cucumber (*Cucumis sativus*) in China. Crop Prot 38: 30–34.
18. Qiao K, Zhang H, Wang H, Ji X, Wang K (2011) Efficacy of aluminium phosphide as a soil fumigant against nematode and weed in tomato crop. Sci Hortic 130: 570–574.
19. Hussey RS, Barker RK (1973) A comparison of methods of collecting inocula of *Meloidogyne* spp., including a new technique. Plant Dis Rep 57: 1025–1028.
20. Eisenback JD (1985) Detailed morphology and anatomy of second-stage juveniles, males, and females of the genus *Meloidogyne* (root-knot nematode). In: Sasser JN, Carter CC, editors. An advanced treatise on meloidogyne, vol. 1: Biology and control. Raleigh, NC: North Carolina State University Graphics. 95–112.
21. Abdul Baki AA, Anderson JD (1973) Vigour determination in soybean seed by multiple criteria. Crop Sci 13: 630–633.
22. Townshend JL (1963) A modification and evaluation of the apparatus for the Oostenbrink direct cotton-wool filter extraction method. Nematologica 9: 106–110.
23. Byrd DWT, Kirkpatrick T, Barker KR (1983) An improved technique for clearing and staining plant tissues for detection of nematodes. J Nematol 15: 142–143.
24. Graeff-Teixeira C, Medeiros E, Zanini GM, Brasil CAA, Cardozo BL, et al. (1997) Inexpensive alternative material for the isolation of larvae with the Baermann method. Mem Inst Oswaldo Cruz 92: 399–400.
25. Gomez KA, Gomez AA (1984) Statistical procedures for agricultural research. New York: John Wiley and Sons.
26. Hallman J, Davies KG, Sikora R (2009) Biological control using microbial pathogens, endophytes and antagonists. In: Perry RN, Moens M, Starr JL, editors. Root-knot Nematodes. Wallingford, UK: CAB International. 380–411.
27. Sun JH, Yu KL, Chen H, Peng DL (2002) Effect of *Syncephalastrum racemosum* on the control of root knot nematode in tomato. Hua Bei Nong Xue Bao 17: 119–123.
28. Li P, Tian Y, Gao BL, Sun JH (2012) The nematicidal spectrum of bio-nematicides from filamentous fungus Sr18. Chin Agric Sci Bull 28: 238–245. (in Chinese).
29. Huang WK, Zhang GJ, Zhang C, Zhang DS, Wang BY, et al. (2010) Management technology of *Meloidogyne incognita* on greenhouse vegetables using biofumigation combined with solarization. Plant Prot 36: 139–142.
30. Sikora RA, Schäfer K, Dababat AA (2007) Modes of action associated with microbially induce in planta suppression of plant-parasitic nematodes. Australas Plant Pathol 36: 124–134.
31. Wu SC, Cao ZH, Li ZG, Cheung KC, Wong MH (2005) Effects of biofertilizer containing N-fixer, P and K solubilizers and AM fungi on maize growth: a greenhouse trial. Geoderma 125: 155–166.
32. Tobin JD, Haydock PPJ, Hare MC, Woods SR, Crump DH (2008) Effect of the fungus *Pochonia chlamydosporia* and fosthiazate on the multiplication rate of potato cyst nematodes (*Globodera pallida* and *G. rostochiensis*) in potato crops grown under UK field conditions. Biol Control 46: 194–201.
33. L'Etoile ND, Bargmann CI (2000) Olfaction and odor discrimination are mediated by the *Caenorhabditis elegans* guanylyl cyclase ODR-1. Neuron 25: 575–586.
34. Abad P, Favery B, Rosso MN, Castagnone-Sereno P (2003) Root-knot nematode parasitism and host response: molecular basis of a sophisticated interaction. Mol Plant Pathol 4: 217–224.

9

Influence of Environmental Factors on the Germination of *Urena lobata* L. and Its Response to Herbicides

Tahir Hussain Awan[1]*, Bhagirath Singh Chauhan[1], Pompe C. Sta. Cruz[2]

1 Weed Science, Crop and Environmental Sciences Division, International Rice Research Institute (IRRI), Los Baños, Laguna, Philippines, **2** Crop Science Cluster, College of Agriculture, University of Philippines Los Baños, Laguna, Philippines

Abstract

Urena lobata is becoming a noxious and invasive weed in rangelands, pastures, and undisturbed areas in the Philippines. This study determined the effects of seed scarification, light, salt and water stress, amount of rice residue, and seed burial depth on seed germination and emergence of *U. lobata*; and evaluated the weed's response to post-emergence herbicides. Germination was stimulated by both mechanical and chemical seed scarifications. The combination of the two scarification methods provided maximum (99%) seed germination. Germination was slightly stimulated when seeds were placed in light (65%) compared with when seeds were kept in the dark (46%). Sodium chloride concentrations ranging from 0 to 200 mM and osmotic potential ranging from 0 to −1.6 MPa affected the germination of *U. lobata* seeds significantly. The osmotic potential required for 50% inhibition of the maximum germination was −0.1 MPa; however, some seeds germinated at −0.8 MPa, but none germinated at −1.6 MPa. Seedling emergence and biomass increased with increase in rice residue amount up to 4 t ha^{-1}, but declined beyond this amount. Soil surface placement of weed seeds resulted in the highest seedling emergence (84%), which declined with increase in burial depth. The burial depth required for 50% inhibition of maximum emergence was 2 cm; emergence was greatly reduced (93%) at burial depth of 4 cm or more. Weed seedling biomass also decreased with increase in burial depth. Bispyribac-sodium, a commonly used herbicide in rice, sprayed at the 4-leaf stage of the weed, provided 100% control, which did not differ much with 2,4-D (98%), glyphosate (97%), and thiobencarb + 2,4-D (98%). These herbicides reduced shoot and root biomass by 99–100%.

Editor: Manuel Reigosa, University of Vigo, Spain

Funding: Co author Dr Bhagirath Singh Chauhan supervised this study and provided funding for this study. The funders had role in study design, analysis, and preparation of the manuscript.

Competing Interests: The authors have declared that no competing interests exist.

* E-mail: t.awan@irri.org

Introduction

Urena lobata L., an annual weed, belongs to the family Malvaceae. This weed is known by many local names in different countries, for example, Afulut Gad (Philippines), China pink flowered Chinese Burr or Xiao fan tian hua (China), Africain Jute (Africa), Kongojute (Germen), Bachita (India), Aguaxima (Brazil), Aramina or Cadillo (Spanish), Ke hoa dao (Vietnam), and *Urena lopastnaia* (Russia) [1]. *U. lobata* originated in Asia and has invaded most of the tropical and subtropical area of the world [2]. According to the United States Department of Agriculture (USDA) plant profile database, *U. lobata* has infested areas in Puerto Rico, Florida, Louisiana, the U.S. Virgin Islands, and New South Wales [2]. At present, *U lobata* is a common intruder in natural open areas, pastures, and gardens. It can grow over a wide altitude range; from near sea-level to about 1000-m above sea level. This weed usually grows in rainforests along roads, monsoon forests, and in disturbed areas. From these observations, the Exotic Pest Plant Council of Florida promptly placed *U. lobata* on the list of nuisance plants under Category II in 1999, which implies that the plant is increasing in number but has not yet been established as ecologically harmful [3].

During the 1970s and 1980s, the new *U. lobata* cultivar Ex-Mokwa was introduced and began to be grown as a fiber crop in Sierra Leone [4] and experimentally grown for many years [5]. This species is also used for its medicinal properties in Malaysia and Java. The leaves and barks are used as a contraceptive, and leaf and root extracts are used to prepare herbal medicines [6]. In some parts of Africa, its leaves and flowers are used as food during famine. *U. lobata* had limited success as a crop, but its cultivation resulted in its rigorous infestation as a weed in many areas. Rangelands, poorly managed disturbed areas, pastures, road sides, and eroded areas are easily infested by *U. lobata.* It can thrive on a wide range of soil types and can grow up to a height of 3 m with a basal diameter of 7 cm [7]. Throughout the year, *U. lobata* produces flowers and fruits with up to 600 seeds per plant [8]. *U. lobata* fruits are produced in about 1 cm-diameter capsules covered by straight trichomes that end in about 4–6 small hooks. These hooks cause the fruits to stick on horses' manes and tails, sheep's hair, clothes, and other similar objects that come into contact with plants.

Under field conditions, *U. lobata* seeds revealed dormancy due to an impermeable seed coat (testa) [9]. This is the main cause of poor germination under field conditions [5]. Low or irregular germination, owing to seed dormancy, has been reported by Kirkby [10]. To improve germination, mechanical scarification through removal of carpels and by rubbing the seeds in sand, have been suggested [11]. Sulfuric acid scarification has also been found successful in overcoming seed dormancy [9]. The focus of most of the research work done on *U. lobata* in the past was to improve plant characteristics including growth, seed yield, and fiber yield [5,8,12]. From research conducted on seed germination, it has been speculated that scarification of seeds is necessary to break seed dormancy [5]. The cause of poor germination is the high levels of dormancy due to an impermeable seed coat. However, the role of light, water stress, salt stress, and soil burial depth on seed germination is not well understood for the species. *U. lobata* is becoming a noteworthy weed in many parts of the world, especially in rice fields of Bangladesh, India, Indonesia, Philippines, Thailand, Vietnam [13] and Florida pastures [14]. Research is needed to better understand the seed biology of *U. lobata* and its control with herbicides. A better understanding of the environmental factors that affect the germination of *U. lobata* seeds may help in explaining its fast proliferation in various ecosystems.

Seed germination is usually influenced, directly or indirectly, by many environmental factors such as temperature, moisture, and light [15]. For example, temperature can affect germination by regulating the enzymes activities involved in germination and by promoting or inhibiting the synthesis of hormones that affect seed dormancy [15]. Germination response of seed to fluctuating temperatures is a mechanism by which seeds detect gaps in vegetation canopies and depth of burial in soil [16]. Requirement of light for germination of a weed species means that the species will germinate only when its seeds are on or near the soil surface. Similarly, the ability of a species to germinate under moisture- and salt-stressed conditions will allow that species to take advantage of conditions that suppress the germination of other species. Information on the effect of seed burial depth on seedling emergence may indicate the effectiveness of tillage operations in weed control. Crop residue on the soil surface may be helpful in weed suppression [17] and can be incorporated as a tool in integrated weed management plans. Seedling emergence may be inhibited or stimulated by crop residue, and will depend on the quantity and quality of crop residue and the biology of the weed species involved [18].

The herbicide 2,4-D is used for broadleaf weed control in agricultural and nonagricultural situations. Glyphosate is usually used as a non-selective herbicide to control all weeds before crop-sowing. However, glyphosate is marginally effective on some weeds, such as the *Ipomoea* species [19]. Little is known about the effect of herbicides on *U. lobata*, and, therefore, there is a need to evaluate the performance of available herbicides against it.

The objectives of this study were to determine the optimum time for soaking seeds in sulfuric acid to break seed dormancy; assess the effect of light, water stress, salt stress, soil burial depth, and quantity of rice straw on *U. lobata* seed germination and seedling emergence; and evaluate the performance of commonly available post-emergence herbicides on *U. lobata*.

Materials and Methods

All experiments in the laboratory and screenhouse were laid out in a randomized complete block design with four replications. Each experiment was conducted twice.

Seed collection and germination test

Seeds of *U. lobata* were collected in March 2011 at the International Rice Research Institute (IRRI) farm in Los Baños, Laguna, Philippines. The first two authors work at this institute, and thus no permission was needed to collect the weed seeds. They also confirmed that the study did not involve endangered or protected species.

Seeds collected from at least 150 plants were bulked and cleaned. Experiments were conducted at laboratory and screenhouse facilities of IRRI. Seed germination of *U. lobata* was determined by placing 25 seeds evenly spaced in a 9 cm-diameter Petri dish containing two layers of Whatman No. 1 filter paper (Whatman International Ltd., Maidstone, U.K.) moistened with 5 ml of distilled water or a treatment solution. Parafilm was used to wrap the Petri dishes to reduce water loss. Seeds were scarified with concentrated sulfuric acid (98% H_2SO_4) for 30 minutes (min). Seeds were washed with running tap water for 5 min after treating with sulfuric acid and before being subjected to various treatments. Seeds were placed in an incubator at fluctuating day/night temperatures of 30/20°C in a light/dark regime, unless otherwise specified. This temperature regime was selected because it has been found appropriate for several weed species in the Philippines. The photoperiod was set at 12 h to coincide with the high-temperature photoperiod. Fluorescent lamps were used to produce a photosynthetic photon flux density (PPFD) of 88 micro mol m^{-2} s^{-1}.

Seed germination was counted daily up to 15 d after placement in the incubator except for the continuous dark treatment, in which germination was counted once at 15 DAS. Germinated seeds (with the radicle visible) were removed from the Petri dishes at each counting. Water was added to the filter paper as needed. Non-germinated seeds were subjected to a simple pressure test to evaluate viability. White firm embryos were considered viable while brown soft embryos were considered nonviable [15].

Effect of seed scarification on germination

To establish whether germination in *U. lobata* is inhibited by an impermeable seed coat, seeds were placed in concentrated sulfuric acid (98%) for 5, 10, and 30 min. In another treatment, seeds were mechanically scarified first by using two rubber stoppers, and then soaking the seeds in sulfuric acid for 30 min. Non-treated seeds were used as a control treatment. Seeds were incubated as described above, and germination was determined.

Effect of light on germination

The effect of light on germination was determined by incubating scarified seeds of *U. lobata* at 30/20°C temperatures (alternating day/night) in both light/dark (12 h/12 h) and continuous (24 h) dark regimes. To maintain the conditions of complete darkness, Petri dishes were wrapped with a double layer of aluminum foil. Other experimental conditions were constant, as previously stated in the germination protocol. As weed seeds do not undergo continuous light and constant temperature conditions under normal conditions in nature [18], these conditions were not included in the experimental variables.

Effect of salt stress on germination

Seed germination as affected by salt stress was investigated by placing scarified seeds of *U. lobata* in dishes each containing a 5 ml of solution of 0, 25, 50, 100, and 200 mM sodium chloride (NaCl) (Mallinckrodt Baker Inc., Phillipsburg, NJ). These ranges of NaCl reflect salinity levels occurring in the tropics [20].

Effect of osmotic stress on germination

The effect of osmotic stress on seed germination was determined by incubating seeds in solutions with osmotic potentials of 0, −0.1, −0.2, −0.4, −0.8, and −1.6 MPa, which were prepared by dissolving 0, 28.9, 40.9, 57.8, 81.8, and 115.6 g of polyethylene glycol 8000 (Sigma-Aldrich Co., St. Louis, MO) in 300 ml of distilled water [21].

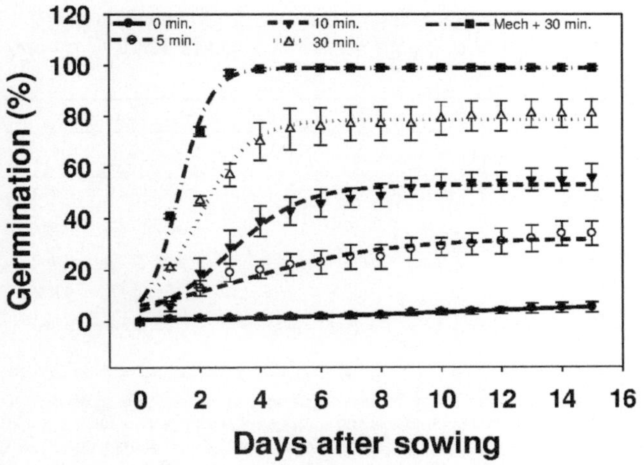

Figure 1. Effect of seed-soaking time in concentrated sulfuric acid (0, 5, 10, and 30 min) and mechanical scarification + soaking time (Mech + 30 min) on germination of *Urena lobata* seed, modeled with the use of equation $G = G_{max}/(1+e[-(x - T_{50})/G_{rate}]$. Estimated parameters are given in Table 1. Vertical bars represent standard error of the mean.

Effect of amount of rice residue on emergence and biomass of seedlings

Fifty scarified seeds of *U. lobata* were sown on the soil surface in plastic pots. Rice residue (leaves and stems of the rice variety NSICRc222) was spread on the soil surface at rates equivalent to 0, 1, 2, 4, and 6 t ha^{-1}. The amounts of rice straw used in this study reflect the amount of straw produced in low-yield rainfed environments and high-yield irrigated environments. The condition of the soil and the pots, emergence counting, and harvesting in this experiment were done as described above for the seed burial experiment.

Effect of seed burial depth on emergence and biomass of seedlings

The effect of seed burial depth on seedling emergence was investigated in a screenhouse. Fifty scarified seeds of *U. lobata* were covered with soil to depths of 2, 4, 6, and 8 cm or placed on the soil surface in plastic pots (15 cm in diameter). The soil had 22% sand, 38% silt, 40% clay, 35% organic carbon, 0.107% N, 0.121% Kjeldahl N, 43 mg kg^{-1} soil available P, 1.26 meq 100^{-1} g soil available K, and a pH of 6.0. Soil used for this experiment was collected from rice fields, autoclaved, and passed through a 3-mm sieve. Pots were watered initially with an overhead mist sprinkler and later sub-irrigated. Plants were watered throughout the study as needed to maintain the optimal moisture level for seed germination. Seedlings were considered emerged when a cotyledon was visible on the soil surface. Emergence was counted at 3-day intervals up to 30 days after sowing (DAS). Emerged seedlings were counted and harvested at 30 DAS. After harvesting, sample plants were oven-dried at 70°C for 72 h to obtain dry biomass. After oven-drying, the biomass of plants was measured.

Efficacy of herbicides

To evaluate the effect of recommended rates of herbicides on survival, shoot biomass, root biomass, and control of *U. lobata*, 20 scarified seeds were planted on the soil surface in plastic pots and covered with a thin layer of soil. The condition of the soil and pots used in this experiment were the same as that described for the seed burial experiment. Seedlings were thinned to 8 plants per pot at 5 DAS and later maintained at this density. Post-emergence herbicides 2,4-D ester at 0.5 kg ai ha^{-1}; glyphosate at 1 kg ai ha^{-1}; bispyribac-sodium at 0.03 kg ai ha^{-1}; thiobencarb + 2,4-D at 0.8 kg ai ha^{-1}; and fenoxaprop-p-ethyl + ethoxysulfuron at 0.045 kg ai ha^{-1} were applied using a Research Track Sprayer (DeVries Manufacturing, Hollandale, MN) that delivered 210 L ha^{-1} spray solution at a spray pressure of 140 kPa. Flat-fan nozzles (Teejet 80015) were used in the sprayer. Herbicides were sprayed on the seedlings of *U. lobata* at the 4- and 6-leaf stages. The difference between the two leaf stages was 9 days. Herbicides were not sprayed in the control treatments for each leaf stage. Seedling survival was determined at 25 days after herbicide application, with the criterion being the appearance of a new sprouting leaf. After counting the surviving plants, these were removed from the pots. Soil was washed off through a steel strainer. Plants were separated into shoots and roots and oven-dried at 70°C for 72 h to obtain dry biomass weight.

Table 1. Effect of seed-soaking time in concentrated sulfuric acid (0, 5, 10, and 30 min) and mechanical scarification + soaking time in sulfuric acid (Mech + 30 min) on germination of *Urena lobata* seed.

Soaking time in sulfuric acid (min)	G_{max}	G_{rate}	T_{50}	R^2
0	3.0 (1.5)	4.67 (0.73)	10.90 (1.96)	0.98
5	32.0 (1.4)	2.39 (0.44)	3.36 (0.43)	0.94
10	53.1 (0.9)	1.23 (0.14)	2.95 (0.15)	0.98
30	78.6 (1.1)	0.87 (0.10)	1.92 (0.11)	0.98
Mechanical + 30	99.9 (0.8)	0.53 (0.05)	1.30 (0.05)	0.99

Parameter estimates [G_{max}, maximum germination (%); T_{50}, time to reach 50% of maximum germination (days); and G_{rate}, slope] of a three-parameter sigmoid model, $G = G\text{max}/(1 + e[-(x - T50)/Grate]$, fitted to seed germination data in Figure 1. Values in parentheses represent standard error of the mean.

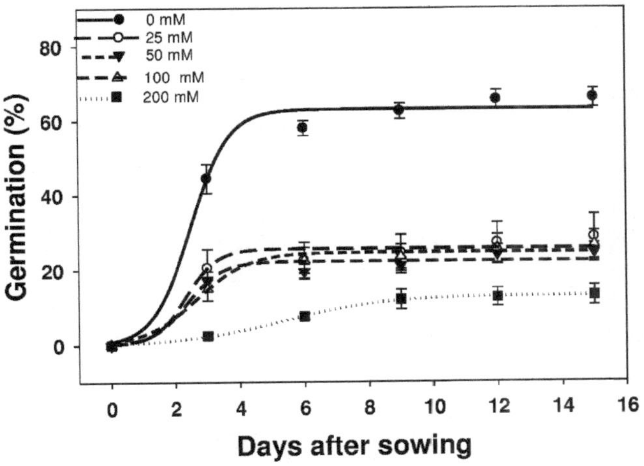

Figure 2. Effect of sodium chloride concentrations (mM) on the germination of *Urena lobata* seeds, incubated at 30/20°C in light/dark, modeled with the use of equation $G = G_{max}/(1+e[-(x - T_{50})/G_{rate}]$. Estimated parameters are given in Table 2. Vertical bars represent standard error of the mean.

Table 2. Effect of sodium chloride concentrations (mM) on the germination of *Urena lobata* seeds, incubated at 30/20°C in light/dark.

NaCl (mM)	G_{max}	G_{rate}	T_{50}	R^2
0	63.1 (1.8)	0.59 (0.47)	2.49 (0.43)	0.99
25	25.6 (1.3)	0.48 (1.03)	2.34 (1.45)	0.98
50	24.6 (0.9)	0.80 (0.44)	2.68 (0.28)	0.98
100	22.3 (1.2)	0.54 (0.86)	2.37 (1.03)	0.96
200	12.9 (0.3)	1.55 (0.15)	5.42 (0.17)	0.998

Parameter estimates [G_{max}, maximum germination (%); T_{50}, time to reach 50% of maximum germination (days); and G_{rate}, slope] of a three-parameter sigmoid model, $G = G\max/(1+e[-(x-T50)/Grate]$, fitted to seed germination data in Figure 2. Values in parentheses represent standard error of the mean.

Statistical analyses

In the laboratory and screenhouse, each replication was arranged on different shelves in the incubator or on different benches in the screenhouse. Different shelves in the incubator or different benches in the screenhouse may experience yet varying conditions, and therefore each replication was considered as a block. Data were subjected to ANOVA and means were separated using least significant difference (LSD) at 5% level of significance. No interaction was detected between the treatments and experimental runs; therefore, data were pooled over the runs for analysis (GenStat 8.0). Before statistical analysis, homogeneity of variance was visually confirmed by inspecting the residuals.

Germination, emergence, and biomass data were analyzed using regression analysis. Germination resulting from scarification and salt- and water-stress treatments, and emergence resulting from different burial depths and residue amounts were modeled using a functional three-parameter sigmoid function. The fitted model was:

$$G = G\max/(1+e[-(x-T50)/Grate]$$

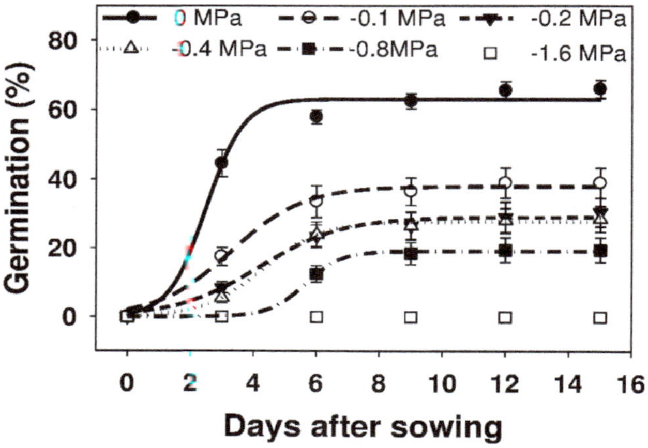

Figure 3. Effect of different osmotic potential (MPa) on germination of *U. lobata* seed, in 3-day intervals, incubated at 30/20°C in light/dark, modeled with the use of equation $G = G_{max}/(1+e[-(x-T_{50})/G_{rate}]$. Estimated parameters are given in Table 3. Vertical bars represent standard error of the mean.

Table 3. Effect of different osmotic potential (MPa) on germination of *U. lobata* seed, incubated at 30/20°C in light/dark conditions.

Osmotic Potential (MPa)	G_{max}	G_{rate}	T_{50}	R^2
0	63.1 (1.8)	0.59 (0.47)	2.49 (0.43)	0.988
−0.1	38.0 (1.1)	1.16 (0.25)	3.29 (0.24)	0.987
−0.2	29.0 (1.0)	1.30 (0.23)	4.19 (0.29)	0.990
−0.4	27.8 (0.5)	0.93 (0.09)	4.30 (0.15)	0.998
−0.8	19.2 (0.3)	0.57 (0.27)	5.65 (0.17)	0.998
−1.6	0	0	0	0

Parameter estimates [G_{max}, maximum germination (%); T_{50}, time to reach 50% of maximum germination (days); and G_{rate}, slope] of a three-parameter sigmoid model, $G = G\max/(1+e[-(x-T50)/Grate]$, fitted to seed germination data in Figure 3. Values in parentheses represent standard error of the mean. No germination occurred at −1.6 MPa.

where G is the cumulative percentage germination at time x, G_{max} is the maximum germination (%), T_{50} is the time (d) required for 50% of maximum germination, and G_{rate} indicates the slope. An exponential decay curve in the form of

$$E = E\max.e(-Erate.x)$$

was fitted to seedling emergence (%) or seedling biomass (g) obtained at different depths, where E represents cumulative emergence (%) and biomass (g) at depth x, E_{max} is the maximum emergence or biomass, and E_{rate} indicates the slope. Parameter estimates for each model were compared using their standard errors [22–24]. Survival and weed control data were transformed before statistical analysis using arcsine transformation. The transformation did not improve the results; therefore, the nontransformed values were used for analysis.

Results and Discussion

Effect of seed scarification on germination

This study included the effect of mechanical and chemical scarifications on the germination of *U. lobata*. Increased soaking time in concentrated sulfuric acid was found to improve the germination percentage (Figure 1). Maximum germination (80%) was achieved at 30 min of soaking in sulfuric acid compared to the control (3%). Mechanical scarification plus 30 min of soaking in sulfuric acid resulted in the highest germination rate (99%) (Table 1). Similar observations were reported in an earlier study [14], in which mechanical and chemical scarification enhanced germination rate of *U. lobata* by 42 and 78%, respectively.

Mechanical scarification was laborious and time consuming, and may not be done uniformly on a large quantity of seeds. Thus, *U. lobata* seeds were chemically scarified (30 min of soaking) for all the subsequent experiments. Non-germinated seeds in all treatments were tested through the pressing method as described earlier [15] and were found to be viable. These results are consistent with earlier findings that sulfuric acid scarification was effective in breaking seed dormancy in *U. lobata* [17] and *Ipomoea triloba* L. [25].

Seed dormancy of *U. lobata* was mainly caused by its hard seed coat. Physical dormancy is widespread among Malvaceae species [5] and is the most ubiquitous form of dormancy among plant species [27]. Mechanical or chemical scarification can break,

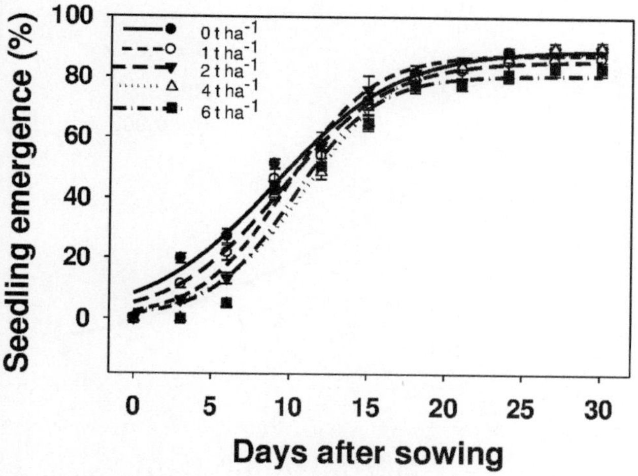

Figure 4. Effect of rice residue amount (t ha^{-1}) on emergence of *Urena lobata* at various days after sowing, modeled with the use of equation $G = G_{max}/(1 + e\ [-(x - T_{50})/G_{rate}]$. Parameter estimates are given in Table 4. Vertical bars represent standard error of the mean.

Table 4. Effect of rice residue amount (t ha^{-1}) on seedling emergence of *Urena lobata.*

Residue quantity (t ha^{-1})	G_{max}	G_{rate}	T_{50}	R^2
0	89.5 (2.7)	3.97 (0.49)	9.04 (0.55)	0.98
1	85.9 (1.8)	3.44 (0.32)	9.47 (0.37)	0.99
2	87.9 (1.7)	2.76 (0.28)	9.79 (0.32)	0.99
4	88.5 (3.4)	3.03 (0.52)	11.17 (0.61)	0.98
6	81.0 (3.0)	2.65 (0.50)	10.25 (0.58)	0.98

Parameter estimates [G_{max}, maximum germination (%); T_{50}, time to reach 50% of maximum germination (days); and G_{rate}, slope] of a three-parameter sigmoid model, $G = G\max/(1 + e[-(x - T50)/Grate]$, fitted to seedling emergence data in Figure 4. Values in parentheses represent standard error of the mean.

scratch, or soften seed coats, which effectively release dormancy in *U. lobata* seeds. Based on previous research on *U. lobata* [14], it was pragmatic that the germination percentage of intact seeds (without removing husk) was less than 2%. The low percentage of germination was attributed to the enclosure of the seed inside a bur, which resulted in low germination rates [28]. Removal of the husk increased germination slightly (10%), while water leaching inside resulted in a nearly 20% germination.

Research suggests that seed coat-imposed dormancy may be a significant cause for low germination. The hard seed coat helps *U. lobata* persist for a long period of time in the soil. When dormancy is released, seeds start to germinate. The results of the present study suggest that *U. lobata* seeds are less likely to germinate in the field unless scarified. It may thus persist in the soil for a long period of time, causing a setback in future crops, if scarification does not take place. Under natural environmental conditions, scarification usually occurs due to soil burial of seed [28,29], microbial attack [29], acute changes in temperature which cause contraction and expansion of the seed coat) [30,31], humid atmospheric conditions [32,33], passage through the digestive system of animals [34,35], flora and fauna actions, burning of vegetation and fire release the dormancy of hard seed especially if fires coincide with wet and hot conditions [36,37,38], and scratching by soil particles probably during tillage operations [28]. However, the effects of these factors are not yet known in the case of *U. lobata*.

Effect of light on germination

The average germination percentage of *U. lobata* seed was 65% (±2.5%) in light/dark and 46% (±6.1%) in continuous dark conditions (data not shown). The results suggest that light is not a prerequisite for the germination of *U. lobata* and that its seeds can germinate without light. This may explain why *U. lobata* is often found under tree canopies within pastures, rangelands, and natural areas. Other plant species of the Malvaceae family, such as *Sida spinosa* L., do not require light for germination [26].

Effect of salt stress on germination

Salt concentration affected the germination of *U. lobata* greatly (Figure 2). Seeds had the highest cumulative germination (63%) in the nonstressed treatment (0 mM NaCl); cumulative germination declined with increasing salt concentration. Germination of *U. lobata* was at 22% at NaCl concentration of 100 mM, and decreased drastically to 13% at 200 mM concentration. It is evident from the data that salt stresses significantly delayed the start of germination and reduce germination percentage. The time to inhibit 50% of the maximum germination (T_{50}) increased with increase in the level of salt stress (Table 2).

Similar results on the effect of varying NaCl concentrations were reported for *Amaranthus spinosus* L. and *A. viridis* L. [24], *I. triloba* [25], *Chromolaena odorata* (L.) King and H.E. Robins and *Tridax procumbens* L. [22], and *Digitaria ciliaris* (Retz.) Koel. and *D. longiflora* (Retz.) Pers. [23]. In Asia, a significant area (21.5 M ha) has been documented to have salt-affected soil [39]. Soil with more than 100 mM concentration of NaCl is considered as having a high level of salinity. Rice, a staple food of people in the tropics, is considered a salt-sensitive crop [40]. Salinity stress, even at 40 mM NaCl, can affect flooded rice [41]. Some species in the Asteraceae, Fabaceae, and Poaceae families were reported as able to adapt to saline conditions [20,22,23]. In this study, *U. lobata* germinated even at 200 mM NaCl. This characteristic may allow

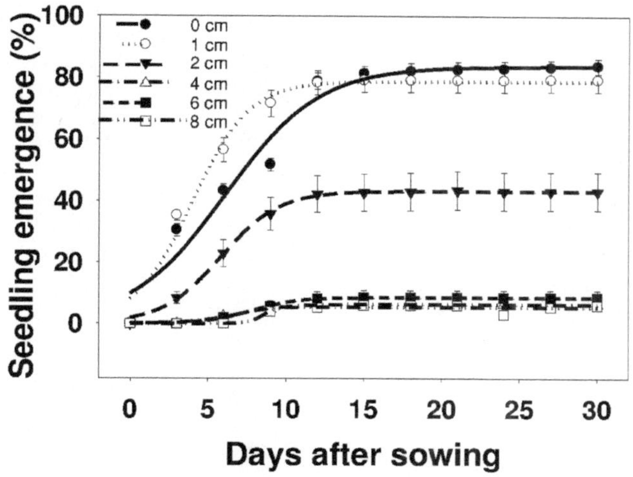

Figure 5. Effect of seed burial depth (cm) on seedling emergence of *Urena lobata* at various days after sowing, modeled with the use of equation $G = G_{max}/(1 + e\ [-(x - T_{50})/G_{rate}]$. Estimated parameters are given in Table 5. Vertical bars represent standard error of the mean.

Table 5. Effect of seed burial depth (cm) on seedling emergence of *Urena lobata* Parameter estimates [G_{max}, maximum germination (%); T_{50}, time to reach 50% of maximum germination (days); and G_{rate}, slope] of a three-parameter sigmoid model, $G = G\max/(1+e[-(x-T50)/Grate]$, fitted to seedling emergence data in Figure 5.

Soil depth (cm)	G_{max}	G_{rate}	T_{50}	R^2
0	83.8 (2.7)	3.02 (0.65)	6.09 (0.60)	0.97
1	78.7 (1.5)	1.84 (0.27)	3.96 (0.30)	0.98
2	43.2 (0.3)	1.87 (0.10)	5.82 (0.11)	0.99
4	6.3 (0.1)	1.53 (0.17)	7.05 (0.20)	0.99
6	8.5 (0.1)	1.45 (0.08)	7.87 (0.09)	0.99
8	5.3 (0.4)	0.45 (3.38)	8.52 (3.65)	0.89

Values in parentheses represent standard error of the mean.

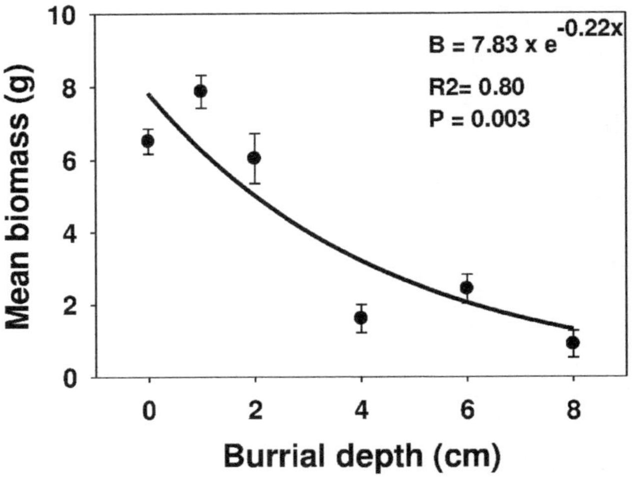

Figure 6. Effect of different seed burial depth (cm) on seedling biomass (g pot^{-1}) of *Urena lobata* at 30 days after sowing, modeled with the use of equation $E = E_{max} \times e^{(-Erate.x)}$. Vertical bars represent standard error of the mean.

U. lobata to have a distinct advantage over other crops, such as rice, in these saline environments.

Effect of osmotic stress on germination

Germination of *U. lobata* was affected greatly by increasing water stress (Figure 3). The maximum germination of *U. lobata* at 0 MPa was 63%. An osmotic potential of −0.1, −0.2, −0.4, −0.8, and −1.6 MPa reduced germination of *U. lobata* to 38, 29, 28, 19, and 0%, respectively (Figure 3; Table 3). Time to 50% germination (T_{50}) increased with increase in degree of osmotic stress.

The results of our study suggest that germination of *U. lobata* is favored by a moist environment and that seeds can remain ungerminated in drought conditions. This could be a valuable tactic of *U. lobata*, allowing for an increased ability to delay germination until favorable conditions occur, which thus prolongs the existence of these seeds in the seed bank. Similar results were reported by Wang and colleagues, wherein *U. lobata* seed germination was significantly affected by increasing water stress. Germination percentage was at 86% in deionized water and, as water stress is increased, germination declined [14].

Seed germination is a process of growth of a previously dormant or inactive seed starting with the imbibitions of water [42]. Seed imbibition rate and germination normally decrease as the water potential of the surrounding decreases, and the critical hydration level for seed germination is species-specific [43].

Based on the present study, germination of *U. lobata* cannot occur when water potential is equal or lower than −1.6 MPa, which is the case for soils that are dry or have high salt levels. The tolerance level of a weed species to water stress is related to its ecological habitat. Species, such as *Eupatorium compositifolium* Walt., when in dry or well-drained terrestrial habitats, can usually germinate at low moisture or water potential [44]. Conversely, species that require moist soils for germination, such as *A. spinosus* and *A. viridis*, can germinate only when water potential is at −0.6 MPa or higher [24,45]. Our data support the findings reported for *I. triloba* [25], *A. spinosus* and *A. viridis* [24,45], in which germination decreased with decrease in water potential. At 30/20°C alternating temperatures, no seeds germinated at water potential of −0.6 MPa and below. Germination of *A. retroflexus* L. was also sensitive to water availability, with final germination being reduced by 50% at −0.4 MPa compared to 0 MPa [46]. *U. lobata* seeds are sensitive to water stress and may not germinate during the dry season in well-drained soils. Rainfall is typically low during April and May in the Philippines, which may delay the germination of *U. lobata* until the rainy season starts in June.

Effect of amount of rice residue on emergence and biomass of seedlings

Seedling emergence continued until 27 DAS and subsequently became constant (Figure 4). The emergence of *U. lobata* was not influenced by the amount of residue tested, and it varied from 81 to 90% (Figure 4; Table 4). Seedling biomass was not influenced by the amount of rice residue, and was in the range of 5.4–6.2 g pot^{-1} (data not shown). Similar results were reported for *A. viridis* [24]. Contrary to our findings on *U. Lobata*, however, the emergence of several weed species (e.g., *C. odorata*, *Tridax procumbens*, *A. spinosus*, and *A. viridis*) were reduced by the addition of crop residue [22,24].

Effect of seed burial depth on emergence and biomass of seedlings

The emergence of *U. lobata* was markedly affected by seed burial depth (Figure 5). Seeds placed on the soil surface had an emergence percentage of 84% that decreased with increase in burial depth, i.e., 79% at 1 cm and 50% at 2 cm. Increase in burial depth to 6 and 8 cm resulted in 9 and 5% emergence, respectively. With increase in soil burial depth, there was a resulting increase in time (d) to reach 50% of the maximum emergence, except in the treatment where seeds were placed on the soil surface. For example, time to reach 50% emergence was 4, 6, 7, 8, and 9 days at seed burial depths of 1, 2, 4, 6, and 8 cm, respectively (Table 5).

Biomass was lower (6.5 g pot^{-1}) for seedlings emerging from the soil surface compared with those emerging from 1 cm burial depth (7.9 g pot^{-1}) (Figure 6). However, seedling biomass decreased with burial depth beyond 1 cm. Biomass was only 0.9 g pot^{-1} for seedlings emerging from a depth of 8 cm.

These results are in agreement with reported findings on *I. triloba* [25]. *U. lobata* seeds are most likely to germinate within 2 cm of the soil surface, but low germination rate (6%) was observed at

Table 6. Effect of post-emergence herbicides on survival (%), shoot and root biomass (g pot^{-1}), and control (%) of *Urena lobata* when sprayed at 4- and 6-leaf stages of the weed.

Herbicide	Rate	4-leaf stage				6-leaf stage			
		Survival	Shoot biomass	Root biomass	Control	Survival	Shoot biomass	Root biomass	Control
	kg ha^{-1}	%	g pot^{-1}	g pot^{-1}	%	%	g pot^{-1}	g pot^{-1}	%
Control	-	100	4.342	1.647	0	100	6.971	2.053	0
2,4-D ester	0.50	1.6	0.006	0.001	98.4	17.2	0.229	0.051	82.8
Bispyribac-sodium	0.03	0	0	0	100	89.1	1.336	0.329	10.9
Fenoxaprop+ ethoxysulfuron	0.045	90.6	1.014	0.307	9.4	100	2.447	0.809	0
Glyphosate	1.00	3.1	0.014	0.003	96.9	12.5	0.117	0.021	87.5
Thiobencarb + 2,4-D	0.80	1.6	0.008	0.005	98.4	18.8	0.218	0.072	81.2
$LSD_{0.05}$		7.88	0.1727	0.0748	7.88	10.74	0.3376	0.1309	10.74
P		<0.001	<0.001	<0.001	<0.001	<0.001	<0.001	<0.001	<0.001

burial depth of 8 cm. A similar germination percentage (84%) was reported by Wang and workers when *U. lobata* seeds were placed on the soil surface [14]. Probably at deeper depths, light and seed size are usually the limiting factors for seedling emergence. Light penetration is generally limited to the first few millimeters below the soil surface [47]. However, the present study found that light is not a prerequisite for *U. lobata* seed germination. In previous studies, reduced emergence of several weed species due to increased burial depth were reported [22,24].

Fluctuating temperatures are more likely to promote germination of small-seeded species than of large-seeded species [16]. Seeds of small weed species are more likely to be buried [48,49], and their seedlings are less likely to emerge from deeper soil, from where seedlings of large seeds can emerge [50]. Thus, depth sensing is much more imperative for small seeds than for large seeds. Small-seeded species, such as *A. spinosus* and *A. viridis*, may have limited reserves to support germination and seedling emergence, thus limiting the depth from which their seedlings can emerge [24]. Conversely, larger seeds usually have higher reserves of carbohydrate and are capable of emerging from greater burial depths [17], such as in the case of *Sicyos angulatus* L. that could emerge even from a depth of 16 cm [51]. The seed diameter of *U. lobata* is about 2 mm, compared to that of *S. angulatus* that is about 5 mm. *U. lobata* seedlings thus failed to emerge when buried too deep. *U. lobata* seedlings were able to emerge from a depth of 6 cm, suggesting the possibility of an escape mechanism when treated with pre-emergence herbicides. Another possible reason for germination inhibition caused by burial depth may be the secondary dormancy imposed by the interaction between soil gases and seed metabolism [52,53]. The lack of emergence from deeply buried seeds could be attributed to hypoxia and low rates of gaseous diffusion deep in the soil [47,54].

The results of the present study suggest that farming practices that enable shallow burial of weed seeds will only suppress the emergence of *U. lobata* seedlings but may not provide complete control of *U. lobata* emergence. On the one hand, deep-tillage operations may be needed to overturn the soil and bury *U. lobata* seeds very deep, from where its seedlings are unable to emerge. On the other hand, succeeding tillage operations may bring viable weed seeds near the soil surface and cause these to germinate and re-infest the succeeding crop. To cope with weed infestation, it is necessary to understand seed viability in the soil seed bank.

Efficacy of herbicides

Urena lobata seedlings did not survive when sprayed with bispyribac-sodium at the 4-leaf stage, resulting in 100% weed control. Similar weed control was achieved with 2,4-D, glyphosate, and thiobencarb + 2,4-D (Table 6). These herbicides reduced the shoot and root biomass by 99–100%. At the 6-leaf stage, the above-mentioned herbicides, except bispyribac-sodium, had similar effects but with little decline in weed control (80%). Bispyribac-sodium at this leaf stage provided poor control (11%). As far as root and shoot biomass is concerned, other herbicides reduced biomass by 97% while bispyribac-sodium reduced biomass by 82%. Similar results from the use of 2,4-D and glyphosate were reported for *I. triloba* [25]. Percentage control and survival data suggest that fenoxaprop-p-ethyl + ethoxysulfuron provided poor control (9 and 0% at 4- and 6-leaf stages, respectively).

Manual control of a weed plant that has a fibrous root system, such as *U. lobata*, is very difficult, time consuming, and expensive. Therefore, farmers have to rely on chemicals to control *U. lobata*. Knowledge of herbicides that are effective against *U. lobata* is very important for its control. Based on the results of this study, it can be concluded that if *U. lobata* infests upland rice, farmers can easily control this weed through the application of 2,4-D or thiobencarb + 2,4-D. These herbicides can provide excellent control of *U. lobata* and the control is comparable to glyphosate application. Glyphosate application remains as the ideal control option for *U. lobata* when there is no rice crop in the field.

Author Contributions

Conceived and designed the experiments: THA BSC PCSC. Performed the experiments: THA BSC. Analyzed the data: THA BSC. Contributed reagents/materials/analysis tools: THA BSC. Wrote the paper: THA BSC PCSC.

References

1. Philippine Medicinal plants. Available: http://www.stuartxchange.com/Dalupang.html. Accessed 2013 Nov 27.
2. Leisure services department exotic management plan 2009. Available: http://seminolecountyfl.gov/parksrec/pdf/scnl_exoticmgmtplan_FINAL.pdf. Accessed 2013 Sept 23.
3. FLERPC (2008) Florida Exotic Pest Plant Council's 2008 list of invasive plant species. Wildland Weeds 10: 13–16.
4. Dasgupta DK (1973) Studies on the crop physiology and cultural practices of *Urena lobata* L. for fibre production in Sierra Leone. 2 Cultural practices. Sols Afr 18: 33–46.
5. Harris PJC (1981) Seed viability, dormancy, and field emergence of *Urena lobata* L. in Sierra Leone. Trop Agric 58: 205–213.
6. Forest Research Institute of Malaysia (2003) Plant Information: *Urena lobata* Griff., pulut-pulut, Malvaceae. Forest Research Institute of Malaysia, Kuala Lumpur, Malaysia. http://www.frim.gov.my/tu/Urena.htm.
7. Francis JK (2003) *Urena lobata* L. Malvaceae. US Department of Agriculture, Forest Service, International Institute of Tropical Forestry, Jardin Botanico Sur, 1201 Calle Ceiba, San Juan PR 00926-1119, in cooperation with the University of Puerto Rico, Rio Piedras, PR 00936-4984. http://www.fs.fed.us/global/iitf/pdf/shrubs/Urena%20lobata.pdf.
8. Harris PJC, Brewah Y (1986) Seed production of *Urena lobata* in Sierra Leone: effect of sowing date. Trop Agric 63: 30–32.
9. Horn CL, Natal Colon JE (1942) Acid scarification of the seed of two Cuban fiber plants. J Am Soc Agron 34: 1137–1138.
10. Kirby RH (1963) Vegetable fibres, pp. 135–136. London: Leonard Hill (Books) Ltd.
11. Dempsey JM (1975) Fiber crops, p 382. Gainsville: University Presses of Florida.
12. Harris PJC (1985) Seed production of *Urena lobata* in Sierra Leone: effect of harvest date on yield. Trop Agric 62: 229–232.
13. Moody K (1989) Weeds Reported In Rice in South And Southeast Asia. International Rice Research Institute, Los Baños, Laguna, Philippines. 83 pp.
14. Wang J, Jason F, Gregory MD, Brent S (2009) Factors affecting seed germination of Cadillo (*Urena lobata*). Weed Sci 57: 31–35.
15. Baskin CC, Baskin JM (1998) Seeds: ecology, biogeography, and evolution of dormancy and germination. San Diego, California, USA: Academic Press. 666 p.
16. Liu K, Baskin JM, Baskin CC, Bu H, Du G, et al. (2013) Effect of diurnal fluctuating versus constant temperatures on germination of 445 species from the Eastern Tibet Plateau. PLOS ONE 8(7): 1–7. e69364. doi:10.1371/journal.pone.0069364
17. Chauhan BS, Johnson DE (2010) The role of seed ecology in improving weed management strategies in the tropics. Adv Agron. 105: 221–262.
18. Bolfrey AGEK, Chauhan BS, Johnson DE (2011) Seed germination ecology of itchgrass (*Rottboellia cochinchinensis*). Weed Sci 59: 182–187.
19. Culpepper AS, Gimenez AE, York AC, Batts RB, Wilcut JW (2001) Morningglory (*Ipomoea* spp.) and large crabgrass (*Digitaria sanguinalis*) control with glyphosate and 2,4-DB mixtures in glyphosate-resistant soybean (*Glycine max*). Weed Technol 15: 56–61.
20. Chauhan BS, Johnson DE (2008a) Influence of environmental factors on seed germination and seedling emergence of eclipta (*Eclipta prostrata*) in a tropical environment. Weed Sci 56: 383–388.
21. Michel BE (1983) Evaluation of the water potentials of solutions of polyethylene glycol 8000 both in the absence and presence of other solutes. Plant Physiol 72: 66–70.
22. Chauhan BS, Johnson DE (2008b) Germination ecology of two troublesome Asteraceae species of rainfed rice: Siam weed (*Chromolaena odorata*) and coat buttons (*Tridax procumbens*). Weed Sci 56: 567–573.
23. Chauhan BS, Johnson DE (2008c) Germination ecology of southern crabgrass (*Digitaria ciliaris*) and India crabgrass (*Digitaria longiflora*): two important weeds of rice in tropics. Weed Sci 56: 722–728.
24. Chauhan BS, Johnson DE (2009) Germination ecology of spiny (*Amaranthus spinosus*) and Slender Amaranth (*A. viridis*): troublesome weeds of direct-seeded rice. Weed Sci 57: 379–385.
25. Chauhan BS, Abugho SB (2012) Threelobe morningglory (*Ipomoea triloba*) germination and response to herbicides. Weed Sci 60: 199–204.
26. Smith CA, Shaw DR, Newsom LJ (1992) Arrow leaf sida (*Sida rhombifolia*) and prickly sida (*Sida spinosa*): germination and emergence. Weed Res 32: 103–109.
27. Foley ME (2001) Seed dormancy: an update on terminology, physiological genetics, and quantitative trait loci regulating germinability. Weed Sci 49: 305–317.
28. Crocker W (1906) Role of seed coats in delayed germination. Bot Gaz 44: 375–380.
29. Taylor GB (1985) Effect of tillage practices on the fate of hard seeds of subterranean clover in a ley farming system. Aust J Exp Agri. 25: 568–573.
30. Taylor GB, Ewing MA (1988) Effect of depth of seed burial on the longevity of hard seeds of subterranean clover and annual medics. Aust J Exp Agri 28: 77–81.
31. Hagon MW (1971) The action of temperature fluctuating on hard seed of subterranean clover Aust J Exp Agri Ani Husb 11: 440–443.
32. Scott KA (2006) Seed coat morphology of *Parkinsonia aculeata* L. & pattern of heat-induced fractures. In: Proceedings of the 15th Australian Weeds Conference (Adelaide, SA, Australia, 24–28 September 2006). Weed Management Society of South Australia, Adelaide, SA, p. 268–271.
33. Taylor GB (1982) Longevity of subterranean clover. Annual Report, CSIRO Division of Land Resources management, CSIRO.
34. Fairbrother TE (1991) Effect of fluctuating temperature and humidity on the softening rate of hard seed of subterranean clover (*Trifolium subterraneum* L.). Seed Sci Tech 19: 93–105.
35. Carter Ed, Challis S, Knowles RC, Bahrani J (1989) Pasture legume seed survival following ingestion by sheep. In 'Proceedings of the 5th Australian Agronomy Conference'. Perth, WA, p. 437. (Australian Society of Agronomy: Carlton, Vic.).
36. Squella F, Carter ED (1993) Significance of seed size and level of hard seededness on survival of medic seeds ingested by sheep. In 'Proceedings of the XVII International Grassland Congress Palmerston North, New Zealand Vol. 1. Pp. 419–420. (Keely and Mundy: Palmerston North, New Zealand). Perth, WA, p. 437. (Australian Society of Agronomy: Carlton, Vic.).
37. Auld TD, O'Connel MA (1991) Predicting patterns of post-fire germination in eastern Australian Fabaceae. Aust J Ecol 16: 53–70.
38. Tieu A, Dixon KW, Meney KA, Sivasitijamparam K (2001) The interaction of heat and smoke in the release of seed dormancy in seven species from southwestern Western Australia. Ann Bot 88: 259–265.
39. Lafitte HR, Ismail A, Bennett J (2006) Abiotic stress tolerance in tropical rice: progress and future prospects. Oryza 43: 171–186.
40. Tanji KK, Kielen NC (2002) Agricultural Drainage Water Management in Arid and Semi-Arid Areas. FAO Irrigation and Drainage Paper 61. Rome: Food and Agriculture Organization of the United Nations. 202 p.
41. Sanchez PA (1976) Soil management in rice cultivation systems. Pages 413–477 in Properties and Management of Soils in the Tropics. Raleigh, NC: John Wiley and Sons.
42. Bradford KJ (1990) A water relations analysis of seed germination rates. Plant Physiol 94: 840–849.
43. Evans CE, Eitherington JR (1990) The effect of soil water potential on seed germination of some British plants. New Phytol 115: 539–548.
44. MacDonald GE, Brecke BJ, Shilling DG (1992) Factors affecting germination of dogfennel (*Euparotium capillifolium*) and yankeeweed (*Eupatorium compositifolium*). Weed Sci 40: 424–428.
45. Thomas WE, Burke IC, Spears JF, Wilcut JW (2006) Influence of environmental factors on slender amaranth (*Amaranthus viridis*) germination. Weed Sci 54: 316–320.
46. Ghorbani R, Seel W, Leifert C (1999) Effects of environmental factors on germination and emergence of *Amaranthus retroflexus*. Weed Sci 47: 505–510.
47. Woolley JT, Stoller E (1978) Light penetration and light-induced seed germination in soil. Plant Physiol 61: 597–600.
48. Peart MH (1984) The effects of morphology, orientation and position of grass diaspores on seedling survival. J Ecol 72: 437–453. doi:10.2307/2260057
49. Thompson K, Green A, Jewels AM (1994) Seeds in soil and worm casts from a neutral grassland. Funct Ecol 8: 29–35. doi: 10.2307/2390108
50. Pearson TRH, Burslem DFRP, Mullins CE, Dalling JW (2002) Germination ecology of neotropical pioneers: interacting effects of environmental conditions and seed size. Ecology 83: 2798–2807.
51. Mann RK, Rieck CE, Witt WW (1981) Germination and emergence of burcucumber (*Sicyos angulatus*). Weed Sci 29: 83–86.
52. Benvenuti S, Macchia M (1995) Effects of hypoxia on buried weed seeds germination. Weed Res. 35: 343–351.
53. Benvenuti S, Macchia M (1997) Phytochrome-meditated germination of *Datura stramonium* L. seeds after seed burial. Weed Res 38: 199–205.
54. Benvenuti S (2003) Soil texture involvement in germination and emergence of buried weed seeds. Agron J 95: 191–198.

10

Application of Organic Amendments to a Coastal Saline Soil in North China: Effects on Soil Physical and Chemical Properties and Tree Growth

Linlin Wang[1,2], Xiangyang Sun[1*], Suyan Li[1*], Tao Zhang[1], Wei Zhang[1], Penghui Zhai[1]

1 College of Forestry, Beijing Forestry University, Beijing, China, **2** Forestry Research Institute of Hulun Buir, Hulun Buir, Inner Mongolia, China

Abstract

The ability of the following four organic amendments to ameliorate saline soil in coastal northern China was investigated from April 2010 to October 2012 in a field experiment: green waste compost (GWC), sedge peat (SP), furfural residue (FR), and a mixture of GWC, SP and FR (1:1:1 by volume) (GSF). Compared to a non-amended control (CK), the amendments, which were applied at 4.5 kg organic matter m^{-3}, dramatically promoted plant growth; improved soil structure; increased the cation exchange capacity (CEC), organic carbon, and available nutrients; and reduced the salt content, electrical conductivity (EC), and exchangeable sodium percentage (ESP). At the end of the experiment in soil amended with GSF, bulk density, EC, and ESP had decreased by 11, 87, and 71%, respectively, and total porosity and organic carbon had increased by 25 and 96% respectively, relative to the CK. The GSF treatment resulted in a significantly lower Na^{+}+K^{+} content than the other treatments. CEC and the contents of available N, P, and K were significantly higher in the GSF-treated soil than in the CK and were the highest in all treatments. The FR treatment resulted in the lowest pH value and Ca^{2+} concentration, which decreased by 8% and 39%, respectively, relative to the CK. Overall, the results indicate that a combination of green waste compost, sedge peat and furfural residue (GSF treatment) has substantial potential for ameliorating saline soils in the coastal areas of northern China, and it works better than each amendment alone. Utilization of GWC and FR can be an alternative organic amendment to substitute the nonrenewable SP in saline soil amelioration.

Editor: Ben Bond-Lamberty, DOE Pacific Northwest National Laboratory, United States of America

Funding: This research was supported by the Eleventh Five-Year Plan of the National Scientific and Technical Supporting Programs (No. 2009BADB2B0504), and the Fundamental Research Funds for the Central Universities. The funders had no role in study design, data collection and analysis, decision to publish, or preparation of the manuscript.

Competing Interests: The authors have declared that no competing interests exist.

* E-mail: sunxy1965@126.com (XS); lisuyan@bjfu.edu.cn (SL)

Introduction

Soil salinization is a major obstacle to the optimal utilization of land resources. Salt-affected soils are widely distributed throughout the world, and about 20% of the world's cultivated land is salt-affected [1]. In the coastal regions of China, approximately 10^6 hm^2 of land is salt-affected because of sea water intrusion and the dry climate. The salt-affected soils in the coastal areas of northern China have several unique characteristics including a high water table (saline groundwater is 0.5–1.0 m below the soil surface), a high groundwater mineralization degree (7–10 $g{\cdot}L^{-1}$), serious secondary salinization, and a low nutrient content [2]. The present extent of salt-affected soils substantially restricts plant growth in these areas.

Sodium (Na^+) is the predominant soluble cation in most saline soils, particularly in the coastal areas. Most plants are hypersensitive to saline environments [3]. The tissues of plants growing in saline media generally exhibit an accumulation of Na^+ and Cl^- and/or the reduced uptake of mineral nutrients, especially Ca^{2+}, K^+, N, and P [4]. Previous studies have reported on the adverse effects of excessive amounts of salts on the physical and chemical properties of soils and on plant growth and yield [5,6]. Clark et al. (2007) [7] demonstrated that salt-affected soils exhibit poor soil structure resulting from soil physical processes such as slaking, swelling, and dispersion of clay. Tejada and Gonzalez (2005) [8] showed that an increase in electrical conductivity (EC) adversely affects soil total porosity, bulk density, and structural stability. Salinity also affects soil chemical properties such as pH, cation exchange capacity (CEC), exchangeable sodium percentage (ESP), soil organic carbon, and available nutrients [1]. Salinity also alters the osmotic and matric potential of the soil solution, thereby adversely affecting soil microbial communities and their activity [9]. Moreover, the changes in the proportions of exchangeable ions in the soil solution have osmotic and ion-specific effects that can produce imbalances in plant nutrients, including deficiencies of several nutrients or excessive levels of Na^+ [4]. Reclaiming these salt-affected soils will require new amelioration methods and improved management practices [10].

Leaching with water, chemical amendment, and phytoremediation are among the methods used to ameliorate saline soils [10]. The use of gypsum, calcite, calcium chloride, and other chemical agents that provide Ca, which tends to replace exchangeable Na, is effective for saline soil amelioration [11]. Another important practice is the application of organic matter conditioners, which can both ameliorate and increase the fertility of saline soils [12]. Salt-affected soils generally exhibit poor structural stability due to low organic matter content. Many researchers have suggested that the structural stability of soil can be improved by the addition of organic materials (e.g. green manures, farmyard and poultry manures, compost, and food processing wastes) [13,14]. Barzegar et al. (1997) [15] found that the application of organic matter to saline soils can accelerate Na^+ leaching, increase the percentage of water-stable aggregates, and decrease the ESP, EC, and soil salinity. Wichern et al. (2006) [16] reported that the addition of maize straw to a saline soil reduced the negative effects of salinity on the microbial community and mineralization.

In China, green waste is mostly produced by urban landscape maintenance and represents a significant portion of municipal solid waste [17]. In view of both environmental protection and economic development, composting is the best method to manage and utilize green waste. Compost has a unique ability to improve the chemical, physical, and biological characteristics of soils. It improves water retention and soil structure by increasing the stability of soil aggregates [18]. Green waste compost (GWC) is a biologically stable, humus-like substance that can be used for saline soil amelioration. Peat is a partly-decomposed plant material that slowly accumulates in pond and lake bottoms and swamp areas. Sedge peat (SP) and sphagnum peat moss are the most common types of peat used for soil remediation, and the former is a fine-textured and more decomposed type of peat. Sedge peat can hold water and nutrients ten to fifteen times their own weight when fully saturated, yet still can hold 40% air. They have high soil CEC and pH buffering capacity owing to their great soil specific surface and chelation capacity [18]. The utilization of sedge peat as a soil amendment in agriculture represents a potentially large market for peat.

However, the use of peat must be reduced, because it is considered as a nonrenewable resource [19]. Thus, an alternative to peat must be developed to achieve the function of peat amendment and meet consumer's environmental concern. Furfural residue (FR) is the by-product of furfural production; furfural is mainly obtained from corn cobs by acid catalysis at high temperature. Among organic amendments, FR has the highest potential for low-cost, high-volume production [20]. Because the cellulose and lignin in corn cobs are relatively stable, FR is enriched in cellulose and lignin [21]. The advantage of organic amendments (GWC, SP, and FR) is that it is unlikely to contain microorganisms that are pathogenic to humans, which is an important concern of such soil amendment [18,22].

Even though there are many studies dealing with organic amendments as degraded soil conditioners, very little is known about their effects on coastal saline soil. In view of the above, three organic materials (GWC, SP, and FR) and five treatments: no amendment, GWC alone, SP alone, FR alone and a mixture of GWC, SP, and FR (abbreviated as GSF) were applied in this study. The aim of the work was to evaluate the effects of different organic amendments on soil salinity and fertility of saline soil as well as plant growth in the coastal areas of northern China. The information obtained from this study will help provide guidance on selection of organic matters in ameliorating coastal saline soil while considering their environmental concerns.

Materials and Methods

Ethics Statement

All necessary permits were obtained for the described field studies. The study was approved by the Dagang Agriculture and Forestry Administration Bureau, Tianjin, China.

Experimental Site and Properties of the Organic Amendments

The research was conducted from April 2010 to October 2012 at the Coastal Salt-tolerant Plant Science and Technology Park (38°46′ N, 117°13′ E, 1.3 m a.s.l.), Dagang, Tianjin, China. The study site has a semi-humid continental monsoon climate. The mean annual temperature is 12.3°C, the highest monthly temperature is in July (26°C), and the lowest monthly temperature is in January (−4°C). On average, there are 2618 h of sunshine per year. The frost-free period is 211 d, and the global radiation is 121.1 $MJ\cdot m^{-2}$. The annual precipitation and evaporation are 593.6 mm and 1979.4 mm, respectively, and the mean relative humidity is 65%. Rainfall from June to September accounts for 84% of the total annual precipitation. The general properties of soil samples (0–25 cm depth) taken from the study site are shown in Table 1.

The organic amendments applied in this study were GWC, SP, FR, and GSF. The general properties of the organic amendments are shown in Table 2. The green waste used to prepare GWC consisted of fallen leaves and branch cuttings collected during greening maintenance. The waste was reduced to about 1 cm

Table 1. Properties of the saline soil before organic amendments were applied.

Property	Value
pH	7.75 (0.22)
Electrical conductivity ($dS\cdot m^{-1}$)	3.69 (0.17)
Clay ($g\cdot kg^{-1}$)[a]	96 (9)
Silt ($g\cdot kg^{-1}$) [a]	680 (22)
Sand ($g\cdot kg^{-1}$) [a]	224 (15)
Bulk density ($mg\cdot m^{-3}$)	1.54 (0.01)
Total porosity (%)	43.46 (0.18)
Total N ($g\cdot kg^{-1}$)	0.69 (0.03)
Total C ($g\cdot kg^{-1}$)	3.69 (0.13)
C/N ratio	5.35 (2.00)
Available N ($mg\cdot kg^{-1}$)	44.33 (1.17)
Available P ($mg\cdot kg^{-1}$)	5.38 (0.59)
Available K ($mg\cdot kg^{-1}$)	143.33 (12.02)
Na^+ ($g\cdot kg^{-1}$)[b]	2.40 (0.03)
Ca^{2+} ($g\cdot kg^{-1}$)[b]	1.25 (0.04)
Mg^{2+} ($g\cdot kg^{-1}$)[b]	0.38 (0.01)
K^+ ($g\cdot kg^{-1}$)[b]	0.06 (0.01)
Cl^- ($g\cdot kg^{-1}$)[b]	2.81 (0.02)
SO_4^{2-} ($g\cdot kg^{-1}$)[b]	2.38 (0.26)
HCO_3^- ($g\cdot kg^{-1}$)[b]	0.33 (0.01)
Total salt ($g\cdot kg^{-1}$)[c]	8.74 (0.30)
Cation exchange capacity ($cmol\cdot kg^{-1}$)	15.70 (0.05)
Exchangeable Na ($cmol\cdot kg^{-1}$)	2.5 (0.2)
Exchangeable sodium percentage (%)	15.8 (0.2)

Values are means of four samples and standard errors are presented in parentheses.
[a]Determined using the methods described by Bao (2005) [25].
[b]Soluble salt ions.
[c]Total salt = $Na^+ + K^+ + Ca^{2+} + Mg^{2+} + Cl^- + SO_4^{2-} + HCO_3^-$.

particle size with a grinder. Before composting, the moisture content of the raw material was adjusted to 60%, and the C/N ratio was adjusted to between 25 and 30 to optimize microbial activity [23]. In this experiment, the GWC was used when the primary fermentation was completed [24]. In SP treatment, sedge peat was obtained from The Garden Waste Consumption Centre of Chaoyang District (Beijing, China). FR was supplied by the Kain Furfural Corporation (Beijing, China). In GSF treatment, the GWC, SP, and FR were mixed at a ratio of 1:1:1 by volume.

Experiment

The experiment was conducted with five treatments : (1) a non-amended control (CK), (2) addition of 14.2 kg·m^{-3} of GWC, (3) addition of 9.1 kg·m^{-3} of SP, (4) addition of 5.5 kg·m^{-3} of FR, and (5) addition of 8.1 kg·m^{-3} of GSF. The quantities of amendments applied per m^3 were selected so that the same mass of organic matter (4.5 kg·m^{-3}) was applied for each amendment. All treatments were arranged in a randomized complete block design with four blocks (78 m×12 m). Each block was again divided into five plots (12 m×12 m) and hence there were 20 (4×5) unit plots. The distance between two adjacent blocks and plots were 6 m and 3 m, respectively. The five treatments were randomly assigned to each plot within the individual blocks with a separate randomization for each block. The amendments were added to nine planting sites in each plot. The nine sites were evenly distributed based on a planting spacing of 3 m×3 m. Each planting site was 1 m×1 m×1 m. On April 25, 2010, the amendments were manually mixed with the soil in each site, and one 2-year-old Pagoda tree (*Sophora japonica* Linn) was planted in each site. Seedlings with similar height, basal diameter, and growth potential were obtained from the Coastal Salt-tolerant Plant Science and Technology Park. Each planting site was irrigated with 5 liters of water every other week from April 26, 2010 to June 25, 2010. All plots and planting sites were managed in the same manner. We measured the height and basal diameter of each Pagoda tree on October 10, 2010, October 12, 2011, and October 14, 2012.

Soil Sampling and Analyses

An auger was used to collect one soil core (0–25 cm depth, 5 cm diameter) from the amended zone of each of three randomly selected trees in each plot on October 11, 2010, October 13, 2011, and October 15, 2012. After drying, the soil samples were ground to pass through a 0.25-mm sieve and stored in sealed polyethylene bags at 4°C until chemical analysis.

Soil bulk density and total porosity were determined using the methods described by Bao (2005) [25]. pH and EC were determined with an MP521 pH/EC meter (Shanghai, China) for a slurry consisting of 1:5 (W/V) soil/distilled water. Soil organic carbon content was determined by the potassium dichromate wet-combustion method. Total N was determined using the micro-Kjeldahl method. Total P was determined with a colorimetric method [25]. Total K was measured with flame photometric detectors. Available N was determined with a micro-diffusion technique after alkaline hydrolysis. Available P was extracted using 0.5 M $NaHCO_3$ and was measured spectrometrically. Available K was measured by flame photometry after NH_4OAc neutral extraction [26].

Soluble Na^+ and K^+ were determined with a flame photometer, and Ca^{2+} and Mg^{2+} were measured by titration with EDTA [27]. The CEC was determined with a 1 M ammonium chloride solution in ethanol/water (60:40, v/v) at pH 8.2 [28]. Extracted Na was determined by flame photometry.

The ESP was determined by the formula,

$$ESP = 100 \times \left(\frac{Na_x}{CEC}\right)$$

Table 2. Properties of the organic amendments applied to the saline soil.

	Organic amendment			
Property†	**GWC**	**SP**	**FR**	**GSF**
pH	8.88 (0.03)	4.93 (0.08)	2.54 (0.03)	7.20 (0.03)
Electrical conductivity (dS· m^{-1})	6.82 (0.04)	1.54 (0.02)	2.99 (0.03)	3.36 (0.06)
Organic carbon (g·kg^{-1})	184 (5)	286 (10)	477 (15)	312 (8)
Total N (g·kg^{-1})	17.5 (0.04)	10.5 (0.01)	8.8 (0.12)	12.6 (0.22)
C/N ratio	10.5 (2.3)	27.2 (12.0)	54.2 (1.8)	30.0 (0.4)
Total P (g·kg^{-1})	3.20 (0.01)	0.07(0.01)	0.47 (0.03)	1.38 (0.03)
Total K (g·kg^{-1})	6.20 (0.03)	1.33 (0.02)	1.71 (0.02)	3.28 (0.11)
Na^+ (g·kg^{-1})	7.38 (0.07)	0.34 (0.01)	0.29 (0.02)	2.96 (0.07)
K^+ (g·kg^{-1})	11.38 (0.07)	0.03 (0.01)	1.04 (0.01)	4.63 (0.09)
Ca^{2+} (g·kg^{-1})	1.10 (0.07)	1.50 (0.05)	0.65 (0.03)	1.06 (0.09)
Mg^{2+} (g·kg^{-1})	2.26 (0.05)	0.98 (0.04)	7.81 (0.07)	3.76 (1.90)
Cl^- (g·kg^{-1})[a]	20.2 (4.2)	1.1 (0.1)	2.5 (0.6)	8.7 (0.3)
SO_4^{2-} (g·kg^{-1})[a]	13.90 (4.00)	5.90 (0.05)	35.90 (7.10)	19.00 (0.09)

Values are means of six samples and standard errors are presented in parentheses.
GWC, green waste compost; SP, sedge peat; FR, furfural residue; GSF, green waste compost plus sedge peat plus furfural residue (1:1:1 by volume).
†Based on a dry weight except for the moisture content; [a] Determined by titration method [25].

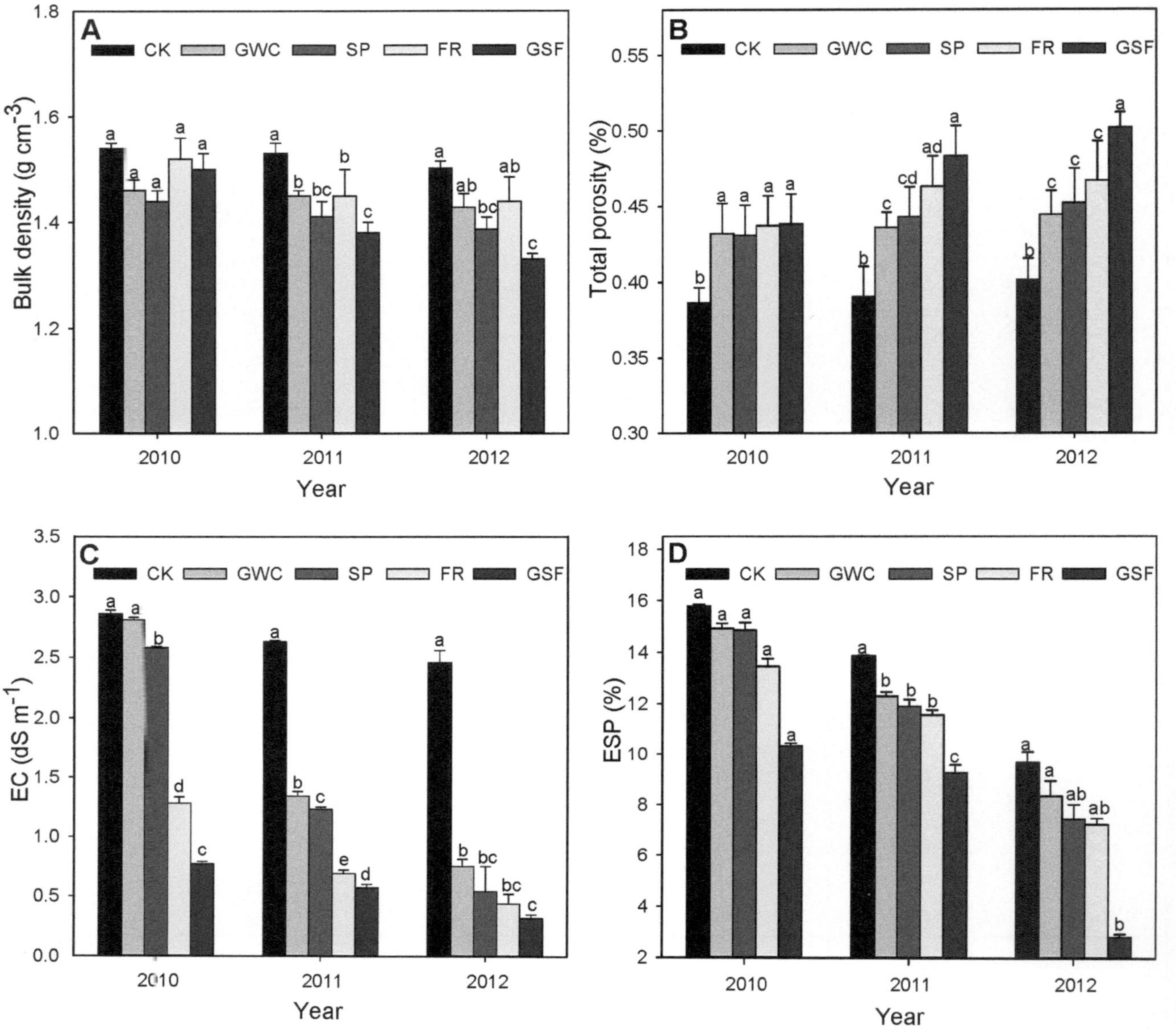

Figure 1. Effects of organic amendments on (A) bulk density, (B) total porosity, (C) electrical conductivity (EC), and (D) exchangeable sodium percentage (ESP) in saline soil. CK, non-amended control; GWC, green waste compost; SP, sedge peat; FR, furfural residue; GSF, green waste compost plus sedge peat plus furfural residue. Rates are indicated in the text but all amendments provided 4.5 kg of organic matter m^{-3}. Treatments were applied in April 2010, and samples were collected in October 2010, October 2011, and October 2012. Values are means+SEMs. Means in the same year followed by the same letter are not significantly different at $p \leq 0.05$ according to LSD.

where Na_x is the exchangeable sodium (cmol·kg^{-1}) and CEC is the cation exchange capacity (cmol·kg^{-1}). Exchangeable sodium (Na_x) was determined with 1 M ammonium acetate at pH 7 [29].

Data Analysis

To test the differences among the treatments, randomized block split-plot analysis of variance (ANOVA) was used with soil amendment treatments as the fixed factor and experimental blocks as the random factor. The probability level used to determine significance was $p<0.05$. If treatment effects were significant, means were compared with an LSD test at $p \leq 0.05$. All statistical analyses were performed using SPSS software package (version 20.0) (IBM Corporation, Armonk, New York).

Results

Bulk Density, Total Porosity, EC, and ESP as Affected by Organic Amendments

In October 2010, about 5.5 months after amendments were applied, there was no significant difference in soil bulk density among the treatments (Fig. 1A). In October 2011, bulk density was significantly lower in all amended plots than in the control (CK). In October 2012, bulk density was significantly lower in SP and GSF plots than in the control. In general, bulk densities were lowest in plots treated with GSF. All four amendments significantly increased the total soil porosity relative to the control (Fig. 1B). Total soil porosity was highest in GSF plots.

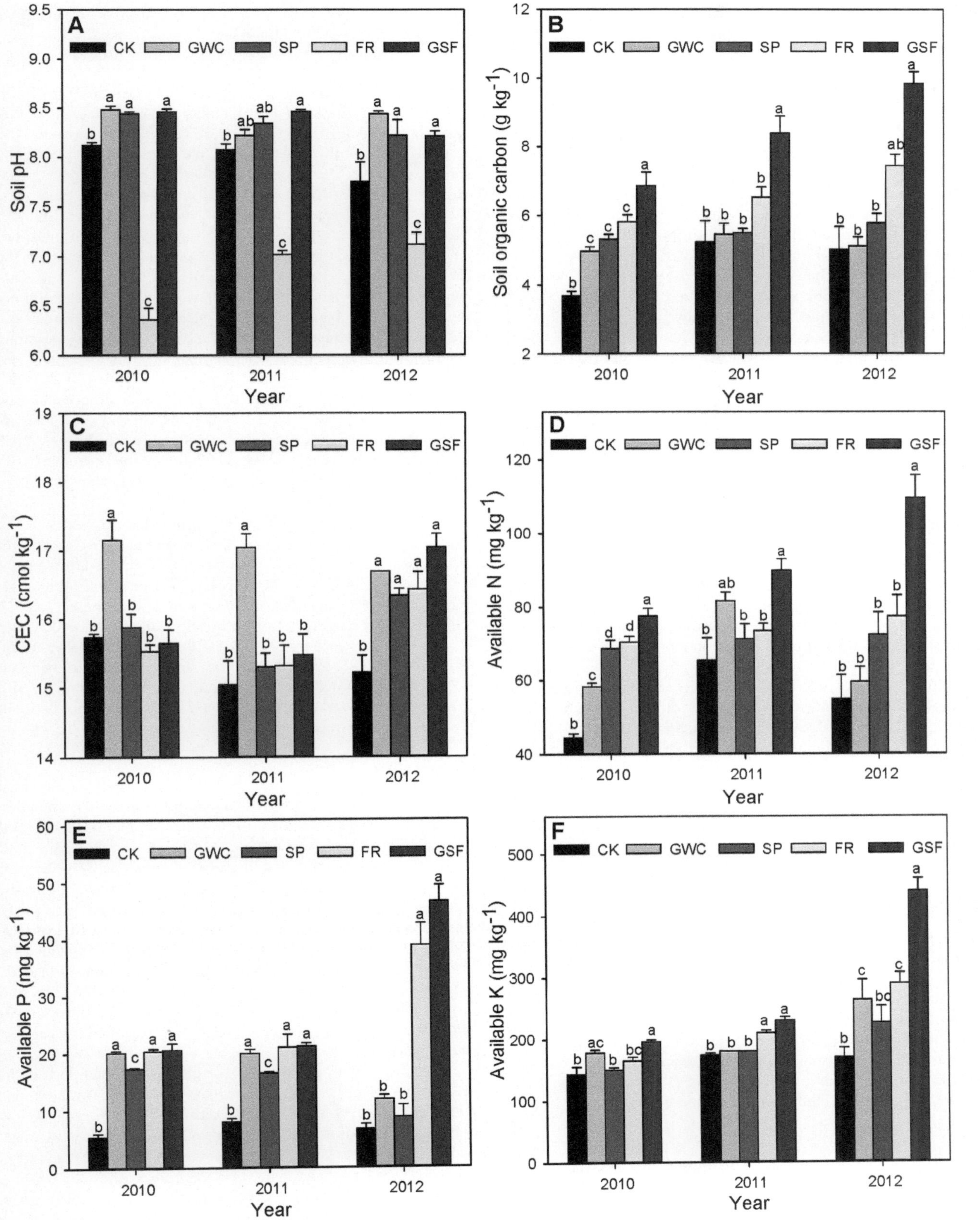

Figure 2. Effects of organic amendments on (A) soil pH, (B) soil organic carbon, (C) cation exchange capacity (CEC), (D) available N, (E) available P, and (F) available K in saline soil. CK, non-amended control; GWC, green waste compost; SP, sedge peat; FR, furfural residue; GSF, green waste compost plus sedge peat plus furfural residue. Rates are indicated in the text but all amendments provided 4.5 kg of organic matter

m^{-3}. Treatments were applied in April 2010, and samples were collected in October 2010, October 2011, and October 2012. Values are means+SEMs. Means in the same year followed by the same letter are not significantly different at $p \leq 0.05$ according to LSD.

By the second year of the study, all four amendments had significantly reduced soil EC values relative to the control (Fig. 1C). EC values dropped more slowly in GWC and SP plots than in FR and GSF plots, but by 2012, the EC values had dropped to a lower level in all plots treated with amendments.

The initial ESP value in the CK plots was 15.8%, which exceeded the critical sodicity value (≈15%) mentioned by Richards (1954) [29]. ESP tended to drop in all plots including the CK plots over time (Fig. 1D). In 2011, ESP values were lower in all amended plots than in the CK plots and were lowest in the GSF plots. In 2012, ESP values were lowest in the GSF plots and did not significantly differ among the other plots.

Soil pH, Organic Carbon, and CEC as Affected by Organic Amendments

In 2010 and 2012, FR decreased soil pH while the other three organic amendments increased pH relative to the control (Fig. 2A). The trends were similar in 2011 but were not always statistically significant. The effect of FR on pH tended to lessen from 2011 to 2012 and especially from 2010 to 2011. Soil pH did not significantly differ among GWC, SP, and GSF plots throughout the experiment.

In 2010, 5.5 months after the amendments were applied, soil organic carbon contents were significantly higher in all amended plots than in the CK plots and were highest in GSF plots (Fig. 2B). In 2011 and 2012, soil organic carbon was higher in GSF plots than in the other plots and did not significantly differ among the other plots, although soil organic carbon tended to also be high in FR plots.

In 2010 and 2011, soil CEC was higher in GWC plots than in the other plots and did not significantly differ among the other plots (Fig. 2C). In 2012, soil CEC was higher in all amended plots than in CK plots and did not significantly differ among the amended plots.

Available Soil Nutrients as Affected by Organic Amendments

The content of available N was significantly higher in the GSF plots than in the other plots, except in 2011, when the level of available N was higher in GSF plots than in all other plots except GWC plots (Fig. 2D). The content of available P was increased by all amendments in 2010 and 2011 (Fig. 2E); in 2012, available P increased to very high levels in FR and GSF plots but did not differ from that in CK plots for the other two amendments. In 2010, the content of available K was greater in GWC and GSF plots than in CK plots but did not differ among SP, FR, and CK plots (Fig. 2F). In 2011, available K was highest in GSF and FR plots and did not differ among the other plots. In 2012, available K increased much more in GSF plots than in the other plots. The level of available K was also significantly higher in GWC and FR plots than in CK plots.

Soluble Cations in the Soil as Affected by Organic Amendments

Table 3 shows the concentration of soluble cations in the soil on October 11, 2010 (about 5.5 months after amendments were applied) and at the end of the experiment on October 15, 2012 (about 29.5 months after treatments were applied). For all treatments including CK, the concentration of Na^{+}+K^{+} in October 2012 was lower than that in October 2010 ($p<0.05$). The Na^{+}+K^{+} concentration in October 2012 was highest in CK plots; intermediate in GWC, SP, and FR plots; and lowest in GSF plots. In addition, the proportion of soluble cations represented by Na^{+}+K^{+} (Na^{+}+K^{+}/total) tended to decline in plots treated with organic amendments but tended to increase in CK plots. The Na^{+}+K^{+}/total value in October 2012 was lowest in the GSF plots. Like Na^{+}+K^{+} concentrations, Ca^{2+} and Mg^{2+} concentrations tended to be lower in October 2012 than in October 2010; the drop in Ca^{2+} concentration was greatest in CK plots. The Ca^{2+} concentration in October 2012 was lowest in FR plots but did not statistically differ between the CK plots and those treated with organic amendments. Mg^{2+} concentrations did not significantly differ among treatments in October 2010 or in October 2012. Unlike the proportion of soluble cations represented by Na^{+}+K^{+} (Na^{+}+K^{+}/total), the proportion of soluble cations represented by Ca^{2+} (Ca^{2+}/total) tended to decrease in the CK plots and increase in the plots treated with organic amendments. In October 2012, the Ca^{2+}/total value was lowest in CK plots; intermediate in SP, FR, and GSF plots; and highest in GWC plots (Table 3).

Tree Growth as Affected by Organic Amendments

In our study, all amendments tended to enhance tree growth. In 2010, about 5.5 months after treatments were applied, tree height was greatest in GSF and FR plots but did not significantly differ among the other plots (Fig. 3A). The trends were similar in 2011 and 2012. Tree height did not statistically differ between GSF and FR treatments except in 2010. Basal diameter was higher in the GSF and FR plots than in CK plots, and did not differ between GWC, SP, and CK plots throughout the experiment (Fig. 3B). Like tree height, basal diameter did not statistically differ between GSF and FR treatments.

Discussion

The GWC, SP, FR, and GSF amendments in the current study decreased soil bulk density and EC value and increased soil total porosity. This agrees with Oo et al. (2013) [14] and Tejada et al. (2006) [13], who found that organic amendments improve soil structure because they promote the flocculation of clay minerals, which is essential for the aggregation of soil particles. The GSF-treated soil had the lowest soil bulk density, the lowest EC, and the highest soil total porosity. These results have been explained by Oo et al. (2013) [14], who reported that combinations of organic amendments resulted in substantial flocculation and in the formation of a large number of soil aggregates. Improvements in soil structure can explain decreases in bulk density and increases in soil porosity [22,30].

Soil EC values were negatively correlated with total porosity in the current study and in a previous study [14]. In addition, Kahlown et al. (2003) [6] and Barzegar et al. (1997) [15] found that the addition of organic matter to soil could increases leaching. This is probably related to the increased soil porosity resulting from the organic amendments [31]. Moreover, the four organic amendments used in our research, especially GSF and FR, contain calcium and magnesium (Table 2), which act as bases because they exist as oxides, hydroxides, and carbonates when applied to soil. As reported by Qadir et al. (2004) [32], an increase in the Ca^{2+} concentration in the soil solution results in the replacement of Na^{+} by Ca^{2+} at the cation exchange sites on the soil particles.

Table 3. Soluble cation contents of the soil in 2010 (about 5.5 months after amendments were applied) and in 2012 (about 29.5 months after amendments were applied).

	Na^{+}+K^{+} ($g·kg^{-1}$)		Ca^{2+} ($g·kg^{-1}$)		Mg^{2+} ($g·kg^{-1}$)		Na^{+}+K^{+}/Total[a] (%)		Ca^{2+}/Total[a] (%)	
Treatment	**2010**	**2012**	**2010**	**2012**	**2010**	**2012**	**2010**	**2012**	**2010**	**2012**
CK	2.46(0.03)a	2.17(0.11)a	1.25(0.13)a	0.28(0.02)ab	0.38(0.05)a	0.08(0.02)a	0.60(0.01)e	0.86(0.03)a	0.31(0.02)a	0.11(0.01)c
GWC	1.66(0.03)b	0.38(0.04)b	1.12(0.04)a	0.45(0.01)a	0.36(0.03)a	0.09(0.01)a	0.53(0.01)d	0.41(0.02)b	0.36(0.03)a	0.51(0.01)a
SP	1.61(0.02)b	0.36(0.03)b	0.98(0.02)ab	0.41(0.05)ab	0.29(0.02)a	0.06(0.01)a	0.56(0.02)c	0.43(0.02)b	0.34(0.02)a	0.49(0.03)a
FR	1.48(0.03)b	0.37(0.02)b	0.43(0.01)b	0.17(0.03)b	0.30(0.01)a	0.06(0.02)a	0.67(0.02)a	0.62(0.02)c	0.19(0.01)b	0.28(0.01)b
GSF	0.89(0.02)c	0.17(0.01)c	0.82(0.03)ab	0.20(0.04)ab	0.32(0.02)a	0.05(0.01)a	0.44(0.02)b	0.40(0.03)b	0.40(0.01)a	0.47(0.02)a

CK, non-amended control; GWC, green waste compost; SP, sedge peat; FR, furfural residue; GSF, green waste compost plus sedge peat plus furfural residue (1:1:1 by volume).
Values are means (SE). Means in a column followed by the same letter are not significantly different at $p \leq 0.05$ according to LSD.
[a]Total = Na^{+}+ K^{+}+ Ca^{2+}+ Mg^{2+}.

Meanwhile, the greater the total porosity, the greater the leaching of the exchanged Na^{+} and the greater the subsequent reduction in soil salinity (as indicated by the EC value).

According to Mamedov et al. (2002) [33], ESP reflects soil aggregation. In the current study, the ESP value in the non-amended soil in 2010 was 15.8%, which indicates that swelling was the main mechanism determining its physical properties. In the amended soils, in contrast, ESP values were always <15%, suggesting that dispersion was probably the main mechanism determining their soil properties. Because ESP is inversely related to organic carbon content [13], it was not surprising that the plots with the lowest ESP values (the GSF plots) had the highest organic carbon contents.

Addition of GWC, SP, and GSF increased the pH of the saline soil, probably because these amendments had a high content of basic cations. Basic cations act in a similar manner as mineral lime, and they increase soil pH [34]. The FR plots had the lowest pH values probably because the FR was quite acidic (pH 2.54). The pH in the FR plots tended to increase over time, perhaps because the degradation of an organic amendment like FR that is rich in basic cations releases inorganic bases into the soil solution and thereby increases the soil pH [34].

Although the same mass of organic matter was added to the soils, the resulting increase in soil organic carbon differed depending on the amendment. This is consistent with Tejada et al. (2006) [30], who reported that the effect of organic amendments on soil organic carbon depended on the chemical nature of the amendments. On the one hand, the chemical nature of the organic amendment is likely to affect the rate at which it is decomposed by the microbial community [35]. On the other hand, the effects of the organic amendments on soil properties will affect plant growth and thus the input of plant residues into the soil [22,36]. The organic carbon content of soil amended with GSF and FR increased gradually over time, and we assume that this

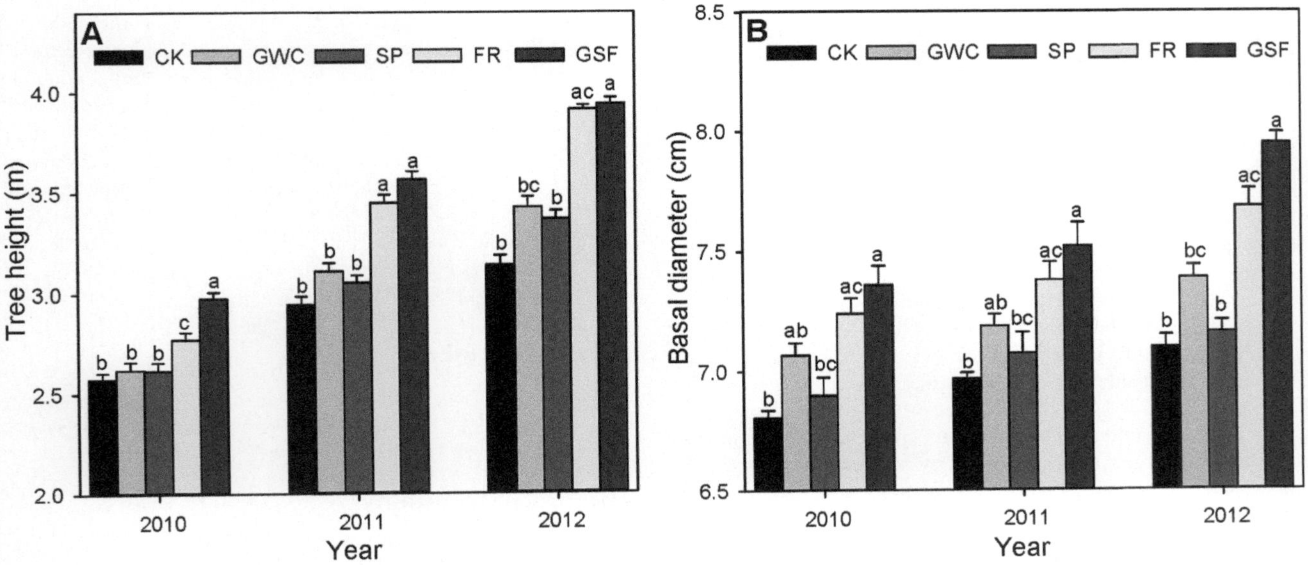

Figure 3. Effects of organic amendments on Pagoda tree (*Sophora japonica* Linn) (A) tree height, and (B) basal diameter in saline soil. CK, non-amended control; GWC, green waste compost; SP, sedge peat; FR, furfural residue; GSF, green waste compost plus sedge peat plus furfural residue. Rates are indicated in the text but all amendments provided 4.5 kg of organic matter m^{-3}. Treatments were applied in April 2010, and data was determined in October 2010, October 2011, and October 2012. Values are means+SEMs. Means in the same year followed by the same letter are not significantly different at $p \leq 0.05$ according to LSD.

increase reflected increased input from the trees (and perhaps reduced decomposition of that input).

CEC provides buffering against changes in pH, available nutrients, calcium levels, and soil structure [37]. The organic amendments used in our research led to an increase in soil CEC, especially in the GWC plots. This might have resulted from a high degree of organic matter oxidation in the GWC-amended soil, which could increase CEC values [38]. We also noted that, from 2010 to 2012, the CEC tended to increase in the SP, FR, and GSF plots. This may primarily be due to the long-lasting decomposition process of organic amendments [22].

Our findings showed that the organic amendments greatly increased the available N content. This agrees with previous reports that organic amendments increase available N [18,39]. However, the effects of organic amendments on nutrient availability depend on their chemical composition and decomposition rates [7]. In the current study, GSF resulted in the highest level of available N, because the mixed organic amendments contain higher percentage of macronutrients and considerable amounts of micronutrients [40]. The C/N ratio was lower in the GWC amendment than in the other organic amendments (Table 2), and a low C/N ratio (less than about 20:1) would result in intense microbial activity, and an increase in available N levels [41]. The available N content was relatively constant in the SP plots, this was probably because the highly-decomposed sedge peat decomposed slowly when added to soil [18]. The FR plots had the second highest level of available N at the end of the experiment, perhaps because the FR amendment had the highest organic matter content, and the organic matter in furfural is mineralized slowly over time [42].

The organic amendments significantly affected the levels of available P in the current study. Like available N content, available P content was highest in the GSF plots, perhaps because the mixed organic amendment has the ability to solubilize and mobilize P [43]. Moreover, P can be adsorbed by soil colloids, and the mixed organic amendment is able to bind large quantities of macronutrients and thereby to reduce their removal from soil by leaching [44]. P availability was also high in the FR plots, probably because the furfural residue was extremely acidic and thereby helped solubilize phosphate. In contrast, available P decreased in the GWC and SP plots probably because these amendments increased the pH.

The organic amendments also increased available K in the soil, which agrees with some researchers [45–47], who found relatively high levels of available K in soils treated with organic amendments. As was the case with available N and available P, available K was highest in the GSF plots and next highest in the FR plots. The difference in available K content among all amendments can be attributed to their chemical compositions. Thus, some K is released as a result of organic matter solubility and decomposition. Available K slightly increased in GWC and SP plots, perhaps because these amendments slowly release nutrients as they slowly decompose [18].

The organic amendments differed in their effects on the concentration of soluble cations. The application of organic amendments would probably increase the amount of Ca^{2+} derived from $CaCO_3$ because of the formation of organic acids [48]. Furthermore, potassium and sodium ions on the soil colloids would be replaced by calcium ions and leached from the soil. The results revealed that, by the end of the experiment, the dominant ions changed from Na^+ and K^+ to Ca^{2+} in GWC, SP, and GSF plots. Mg^{2+} content also tended to decrease in all plots over time. Previous studies have reported that Ca^{2+} could improve soil structure by formed cationic bridges between clay particles and soil organic matter [49]. In addition, Ca^{2+} can inhibit clay dispersion and the associated disruption of aggregates by replacing Na^+ and Mg^{2+} in clay and aggregates, thereby promoting aggregate stability [50]. These results indicate that GWC, SP, and GSF organic amendments could help reduce the salt content and otherwise improve saline soil. FR resulted in the lowest concentration of Ca^{2+} in 2010 and 2012, probably because the high SO_4^{2-} concentration in the furfural residue promoted the precipitation of calcium sulphate [51].

The increased tree height and basal diameter observed in the GSF and FR treatments might be due to better physiological growth of plants. GSF and FR treatments increased the availability of macronutrients as well as micronutrients and improved soil physical properties of saline soil. Increased net primary productivity would have returned higher amounts of plant residues to the soil, perhaps acting as a carbon and nutrient source to maintain a higher microbial biomass and activity [36]. This, in conjunction with the positive effects of organic amendments on soil biota (e.g. increased biodiversity and microbial activity), would lead to continued nutrients release from organic amendments and soil amelioration [22], and help meet the amelioration goal of establishment of self-sustaining ecosystem on saline soil.

Conclusions

The results of this study show that application of GWC, SP, FR, or GSF can improve the physical and chemical properties of saline soil and promote tree growth in the coastal areas of northern China. Application of GSF and FR substantially increased the organic carbon content of the soil and the availability of N, P, and K. The FR amendment effectively reduced the soil pH value. The GWC amendment increased the soil CEC and provided a long-term supply of nutrients. SP had a smaller effect on soil fertility than the other three amendments. Overall, the combination of GWC, SP and FR (the GSF treatment) provides better remediation results than each applied singly and has substantial potential for ameliorating coastal saline soils. The authors believe that SP could be fully or partially replaced by GWC and FR or their combinations in ameliorating coastal saline soil and that additional research should be conducted to optimize the mix proportion of GWC, SP, and FR. Considering the strong correlation between soil biological activities and soil quality [9,16], further studies are also needed to assess the effects of organic amendments on soil microbial community and enzymatic activities.

Acknowledgments

We thank Long Cheng and Donghua Zhang in the Coastal Salt-tolerant Plant Science and Technology Park of Dagang Agriculture and Forestry Administration for many sampling and analysis work.

Author Contributions

Conceived and designed the experiments: LW XS SL. Performed the experiments: LW. Analyzed the data: LW. Contributed reagents/materials/analysis tools: XS SL TZ WZ PZ. Wrote the paper: LW.

References

1. Sumner M (2000) Handbook of soil science. Boca Raton: CRC Press. 2148 p.
2. Wang Z, Zhu S, Yu R (1993) The saline soil in China. Beijing: Science Press. 573p.

3. Qadir M, Oster JD, Schubert S, Noble AD, Sahrawat KL (2007) Phytoremediation of sodic and saline-sodic soils. Advances in Agronomy 96: 197–247.
4. Kaya C, Kirnak H, Higgs D (2001) Enhancement of growth and normal growth parameters by foliar application of potassium and phosphorus in tomato cultivars grown at high (NaCl) salinity. Journal of Plant Nutrition 24: 357–367.
5. Liang Y, Yang Y, Yang C, Shen Q, Zhou J, et al. (2003) Soil enzymatic activity and growth of rice and barley as influenced by organic manure in an anthropogenic soil. Geoderma 115: 149–160.
6. Kahlown MA, Azam M (2003) Effect of saline drainage effluent on soil health and crop yield. Agricultural Water Management 62: 127–138.
7. Clark GJ, Dodgshun N, Sale PWG, Tang C (2007) Changes in chemical and biological properties of a sodic clay subsoil with addition of organic amendments. Soil Biology and Biochemistry 39: 2806–2817.
8. Tejada M, Gonzalez J (2005) Beet vinasse applied to wheat under dryland conditions affects soil properties and yield. European Journal of Agronomy 23: 336–347.
9. Rietz DN, Haynes RJ (2003) Effects of irrigation-induced salinity and sodicity on soil microbial activity. Soil Biology and Biochemistry 35: 845–854.
10. Qadir M, Ghafoor A, Murtaza G (2000) Amelioration strategies for saline soils: a review. Land Degradation and Development 11: 501–521.
11. Hanay A, Büyüksönmez F, Kiziloglu FM, Canbolat MY (2004) Reclamation of saline-sodic soils with gypsum and MSW compost. Compost Science and Utilization 12: 175–179.
12. Melero S, Madejón E, Ruiz JC, Herencia JF (2007) Chemical and biochemical properties of a clay soil under dryland agriculture system as affected by organic fertilization. European Journal of Agronomy 26: 327–334.
13. Tejada M, Garcia C, Gonzalez JL, Hernandez MT (2006) Use of organic amendment as a strategy for saline soil remediation: influence on the physical, chemical and biological properties of soil. Soil Biology and Biochemistry 38: 1413–1421.
14. Oo AN, Iwai CB, Saenjan P (2013) Soil properties and maize growth in saline and nonsaline soils using cassava-industrial waste compost and vermicompost with or without earthworms. Land Degradation and Development DOI: 101002/ldr2208.
15. Barzegar AR, Nelson PN, Oades JM, Rengasamy P (1997) Organic matter, sodicity, and clay type: influence on soil aggregation. Soil Science Society of America Journal 61: 1131–1137.
16. Wichern J, Wichern F, Joergensen RG (2006) Impact of salinity on soil microbial communities and the decomposition of maize in acidic soils. Geoderma 137: 100–108.
17. Liu G, Zhang R, El-Mashad HM, Dong R (2009) Effect of feed to inoculum ratios on biogas yields of food and green wastes. Bioresource Technology 100: 5103–5108.
18. Cooperband L (2002) Building soil organic matter with organic amendments. Center for Integrated Agricultural Systems. University of Wisconsin–Madison. USA.
19. Barkham JP (1993) For peat's sake: conservation or exploitation? Biodiversity and Conservation 2: 556–566.
20. Sun R, Song X, Sun R, Jiang J (2011) Effect of lignin content on enzymatic hydrolysis of furfural residues. Bioresources 6: 317–328.
21. Xing Y, Bu LX, Wang K, Jiang JX (2012) Pretreatment of furfural residues with alkaline peroxide to improve cellulose hydrolysis. Characterization of isolated lignin. Cellulose Chemistry and Technology 46: 249–260.
22. Larney FJ, Angers DA (2012) The role of organic amendments in soil reclamation: A review. Canadian Journal of Soil Science 92: 19–38.
23. Roca-Pérez L, Martinez C, Marcilla P, Boluda R (2009) Composting rice straw with sewage sludge and compost effects on the soil–plant system. Chemosphere 75: 781–787.
24. Zhang L, Sun X, Tian Y, Gong X (2013) Effects of brown sugar and calcium superphosphate on the secondary fermentation of green waste. Bioresource Technology 131: 68–75.
25. Bao SD (2005) Analysis of soil agrochemistry (3rd edition). Beijing, China: China Agriculture Press. 495 p.
26. Page AL, Miller RH, Keeney DR (1982) Methods of soil analysis. Part 2-Chemical and Microbiological Properties (2nd edition). Madison, Wisconsin: American Society of Agronomy and Soil Science. 1159 p.
27. Jackson ML (1962) Soil chemistry of analysis. Englewood Cliffe, New Jersey: Prentice Hall. 78 p.
28. Tucker BM (1954) The determination of exchangeable calcium and magnesium in carbonate soils. Australian Journal of Agricultural Research 5: 706–715.
29. Richards L (1954) Diagnosis and improvement of saline and alkali soils. In: Richards LA, editor. USA Department of Agriculture Handbook 60. Washington, DC: US Government Printing Office. 98–105.
30. Tejada M, Garcia C, Gonzalez JL, Hernandez MT (2006) Organic amendment based on fresh and composted beet vinasse: influence on soil properties and wheat yield. Soil Science Society of America Journal 70: 900–908.
31. Tejada M, Gonzalez JL, García-Martínez AM, Parrado J (2008) Effects of different green manures on soil biological properties and maize yield. Bioresource Technology 99: 1758–1767.
32. Qadir M, Oster JD (2004) Crop and irrigation management strategies for saline-sodic soils and waters aimed at environmentally sustainable agriculture. Science of the Total Environment 323: 1–19.
33. Mamedov AI, Shainberg I, Levy GJ (2002) Wetting rate and sodicity effects on interrill erosion from semi-arid Israeli soils. Soil and Tillage Research 68: 121–132.
34. Pocknee S, Sumner ME (1997) Cation and nitrogen contents of organic matter determine its soil liming potential. Soil Science Society of America Journal 61: 86–92.
35. Hahn AS, Quideau SA (2013) Long-term effects of organic amendments on the recovery of plant and soil microbial communities following disturbance in the Canadian boreal forest. Plant and Soil 363: 331–344.
36. Hyvönen R, Persson T, Andersson S, Olsson B, Ågren GI, et al. (2008) Impact of long-term nitrogen addition on carbon stocks in trees and soils in northern Europe. Biogeochemistry 89: 121–137.
37. Brady AC, Veil RR (2002) The nature and properties of soils, 13th edition. New Jersey: Prentice Hall. 960 p.
38. Liang B, Lehmann J, Solomon D, Kinyangi J, Grossman J, et al. (2006) Black carbon increases cation exchange capacity in soils. Soil Science Society of America Journal 70: 1719–1730.
39. Abbasi MK, Hina M, Khalique A, Khan SR (2007) Mineralization of three organic manures used as nitrogen source in a soil incubated under laboratory conditions. Communications in Soil Science and Plant Analysis 38: 1691–1711.
40. Ros M, Hernandez MT, García C (2003) Soil microbial activity after restoration of a semiarid soil by organic amendments. Soil Biology and Biochemistry 35: 463–469.
41. Yazdanpanah N, Pazira E, Neshat A, Mahmoodabadi M, Sinobas LR (2013) Reclamation of calcareous saline sodic soil with different amendments (II): Impact on nitrogen, phosphorous and potassium redistribution and on microbial respiration. Agricultural Water Management 120: 39–45.
42. Feng YJ, Li F, Wang XL, Liu XM, Zhang LN (2006) Principal chemical properties of artificial soil composed of fly ash and furfural residue. Pedosphere 16: 668–672.
43. Qadir M, Qureshi RH, Ahmad N (1998) Horizontal flushing: a promising ameliorative technology for hard saline-sodic and sodic soils. Soil and Tillage Research 45: 119–131.
44. Withers PJA, Bailey GA (2003) Sediment and phosphorus transfer in overland flow from a maize field receiving manure. Soil Use and Management 19: 28–35.
45. Warman PR, Termeer WC (2005) Evaluation of sewage sludge, septic waste and sludge compost applications to corn and forage: yields and N, P and K content of crops and soils. Bioresource Technology 96: 955–961.
46. Bhattacharyya P, Chakrabarti K, Chakraborty A, Nayak DC, Tripathy S, et al. (2007) Municipal waste compost as an alternative to cattle manure for supplying potassium to lowland rice. Chemosphere 66: 1789–1793.
47. Antil RS, Singh M (2007) Effects of organic manures and fertilizers on organic matter and nutrients status of the soil. Archives of Agronomy and Soil Science 53: 519–528.
48. Wong VNL, Dalal RC, Greene RSB (2009) Carbon dynamics of sodic and saline soils following gypsum and organic material additions: a laboratory incubation. Applied Soil Ecology 41: 29–40.
49. David R, Dimitrios P (2002) Diffusion and cation exchange during the reclamation of saline-structured soils. Geoderma 107: 271–279.
50. Zhang XC, Norton LD (2002) Effect of exchangeable Mg on saturated hydraulic conductivity, disaggregation and clay dispersion of disturbed soils. Journal of Hydrology 260: 194–205.
51. Amezketa E, Aragüés R, Gazol R (2005) Efficiency of sulfuric acid, mined gypsum, and two gypsum by-products in soil crusting prevention and sodic soil reclamation. Agronomy Journal 97: 983–989.

Greenhouse Gas Flux and Crop Productivity after 10 Years of Reduced and No Tillage in a Wheat-Maize Cropping System

Shenzhong Tian[1], Yu Wang[2], Tangyuan Ning[1]*, Hongxiang Zhao[1], Bingwen Wang[1], Na Li[1], Zengjia Li[1], Shuyun Chi[3]*

1 State Key Laboratory of Crop Biology, Shandong Key Laboratory of Crop Biology, National Engineering Laboratory for Efficient Utilization of Soil and Fertilizer Resources, Shandong Agricultural University, Taian, Shandong, China, **2** Shandong Rice Research Institute, Jinan, Shandong, China, **3** College of Mechanical and Electronic Engineering, Shandong Agricultural University, Taian, Shandong, China

Abstract

Appropriate tillage plays an important role in mitigating the emissions of greenhouse gases (GHG) in regions with higher crop yields, but the emission situations of some reduced tillage systems such as subsoiling, harrow tillage and rotary tillage are not comprehensively studied. The objective of this study was to evaluate the emission characteristics of GHG (CH_4 and N_2O) under four reduced tillage systems from October 2007 to August 2009 based on a 10-yr tillage experiment in the North China Plain, which included no-tillage (NT) and three reduced tillage systems of subsoil tillage (ST), harrow tillage (HT) and rotary tillage (RT), with the conventional tillage (CT) as the control. The soil under the five tillage systems was an absorption sink for CH_4 and an emission source for N_2O. The soil temperature positive impacted on the CH_4 absorption by the soils of different tillage systems, while a significant negative correlation was observed between the absorption and soil moisture. The main driving factor for increased N_2O emission was not the soil temperature but the soil moisture and the content of nitrate. In the two rotation cycle of wheat-maize system (10/2007–10/2008 and 10/2008–10/2009), averaged cumulative uptake fluxes of CH_4 under CT, ST, HT, RT and NT systems were approximately 1.67, 1.72, 1.63, 1.77 and 1.17 t ha^{-1} $year^{-1}$, respectively, and meanwhile, approximately 4.43, 4.38, 4.47, 4.30 and 4.61 t ha^{-1} $year^{-1}$ of N_2O were emitted from soil of these systems, respectively. Moreover, they also gained 33.73, 34.63, 32.62, 34.56 and 27.54 t ha^{-1} yields during two crop-rotation periods, respectively. Based on these comparisons, the rotary tillage and subsoiling mitigated the emissions of CH_4 and N_2O as well as improving crop productivity of a wheat-maize cropping system.

Editor: Gil Bohrer, The Ohio State University, United States of America

Funding: This work was financial supported by Financial supported by the National Science and Technology Research Projects of China (2012BAD14B07), the Special Research Funding for Public Benefit Industries (Agriculture) of China (201103001), and the Nature Science Fund of China (30900876 and 31101127). The funders had no role in study design, data collection and analysis, decision to publish, or preparation of the manuscript.

Competing Interests: The authors have declared that no competing interests exist.

* E-mail: ningty@163.com (TN); chishujun1955@163.com (SC)

Introduction

Methane (CH_4) and nitrous oxide (N_2O) play a key role in global climate change [1]. The global warming potential of these gases are respectively 25 and 298 times that of carbon dioxide (CO_2) [2]; thus, the release of these gases is a crucial contributory factor to increasing loads of greenhouse gases (GHG). According to estimations of the IPCC [3], the fluxes of CH_4 and N_2O from agricultural sources account for 50% and 80% of the total emission of these gases, respectively. There have been many studies on CO_2 emission in different ecosystems [4–6]; however, emissions of CH_4 and N_2O emission have been researched incompletely, especially in agricultural ecosystems [7,8].

In general, appropriate soil tilling may reduce GHG emissions because the emissions from soil are strongly affected by tilling, results that have been found by many studies [8–10], most of which, have reported the emissions of CH_4 and N_2O under conventional tillage (CT) and no-tillage (NT) systems in different sites [5,9]. However, the both generally revealed two extremes in maintenance the soil organic carbon stock and crop productivity, agricultural environment protection, the results in these aspects showed the regional character and sometimes they did not suit for agricultural sustainable development [8,9,11]. In which case, some reduced tillage systems such as subsoiling (ST), harrow tillage (HT) and rotary tillage (RT) have been introduced [11], and sometimes they as more important tillage practices combination with no tillage were used to in rotation-tillage systems, which changed some soil environment factors and crop yield [12]. Although these reduced systems were frequently used and developed rapidly due to they are not only advantageous to improve crop yield but also increase the utilization efficiency of soil, water and fertilizer [13,14], the emissions of CH_4 and N_2O under these systems were remain unclear.

The production, consumption and transport of CH_4 and N_2O in soil are strongly influenced by some soil factors. Many studies demonstrated that CH_4 uptake by soil is correlated with soil temperature [15,16] and the N_2O emission. The conditions of soil moisture and N concentration were also shown as two major driving factors of the emission of N_2O: the emission generally peaked during N fertilizer application and irrigation [17,18]. However, the effects of those factors on the emissions of CH_4 and N_2O under different tillage systems in the North China Plain are still not fully understood among available results.

Therefore, the aim of the present study was to quantify the emissions of CH_4 and N_2O under no tillage and three reduced tillage systems in the wheat-maize cropping system and to analyze the correlations between the two gas emissions and the soil temperature, moisture and nitrate content. The crop productivity of the wheat-maize cropping system during two crop-rotation periods was also analyzed.

Materials and Methods

Ethics Statement

This experiment was established in a long-term tillage and residue-management experiment site of Shandong Agricultural University. The farming operations of this experiment were similar to the rural farmers' operations and did not involve endangered or protected species; the operations were approved by the State Key Laboratory of Crop Biology, Shandong Key Laboratory of Crop Biology and National Engineering Laboratory for Efficient Utilization of Soil and Fertilizer Resources, Shandong Agricultural University.

Experimental Site

The study site was located at Tai'an (Northern China, 36°09′N, 117°09′E), which has the typical characteristics of the North China Plain. The average annual precipitation is 697 mm, and the average annual temperature is 13.0°C, with the minimum (−2.6°C) and maximum (26.4°C) monthly temperatures in January and July, respectively. The annual frost-free period is approximately 170–196 d in duration, and the annual sunlight time is 2627.1 hours. The soil is a loam with 40% sand, 44% silt, 16% clay. The characteristics of the surface soil (0–20 cm) were measured as follows: pH 6.8; soil bulk density 1.43g cm^{-3}; soil organic matter 1.36%; soil total nitrogen 0.13%; and soil total phosphorous 0.13%. The meteorological data during the experiment is shown in Figure 1.

Experimental Design

The study based on a 10-year tillage experiment, began in 2002, included no tillage (NT) and three reduced tillage systems involving subsoiling (ST), harrow tillage (HT), rotary tillage (RT), with the conventional tillage (CT) as the control. The treatments were arranged in a randomized block design with three replications. Each plot was 35 m long and 4 m wide.

The experimental site was cropped with a rotation of winter wheat (*Triticum aestivum* Linn.) and maize (*Zea mays L.*). Winter wheat (Jimai-22) was sown at a rate of 90 kg ha^{-1}, on 12 October 2007 and 15 October 2008, and was harvested 6 June 2008 and 10 June 2009. The basal fertilizer was added before sowing and contained with 225 kg N ha^{-1}, 150 kg P_2O_5 ha^{-1} and 105 kg K_2O ha^{-1}, and 100 kg N ha^{-1} was used as topdressing at the jointing stage with 160 mm of irrigation water. The maize was sown on 20 June 2008 and 22 June 2009, which included 66600 plants ha^{-1} and was harvested 8 October 2008 and 10 October 2009. For the maize, 120 kg N ha^{-1}, 120 kg P_2O_5 ha^{-1} and 100 kg K_2O ha^{-1} were used as a basal fertilizer, and 120 kg N ha^{-1} was used as topdressing at the jointing stage.

When the wheat and maize were harvested, the amount of residue retuned to each plot at an equal level according to the crop biomass and water content of residue, in order to ensure their amount and C content of the residue among treatment had no significant difference. The residue equal quality returned to the field, then pulverized using a residue chopper and mixed with the soil in the tillage operations. The operations of tillage and residue-management are shown in Table 1.

Gas Sampling and Analysis

The emission measurements of CH_4 and N_2O for each treatment were conducted using the static-chamber method [19]. According to some studies, there is optimal gas-sample collection duration in a day during which the sample can show the mean gas flux of the day [20–22]. Our previous study have showed that the ratios of the CH_4 flux between 9 a.m. and 10 a.m. and N_2O flux between 9 a.m. and 12 p.m. to the daily mean flux, respectively, verged 1 by the correction coefficient and regression analysis [23]. So, the CH_4 and N_2O were collected between 9 a.m. and 10 a.m. and between 9 a.m. and 12 p.m. respectively from October 2007 to August 2009 at approximately 1-month intervals [12,23]. Both

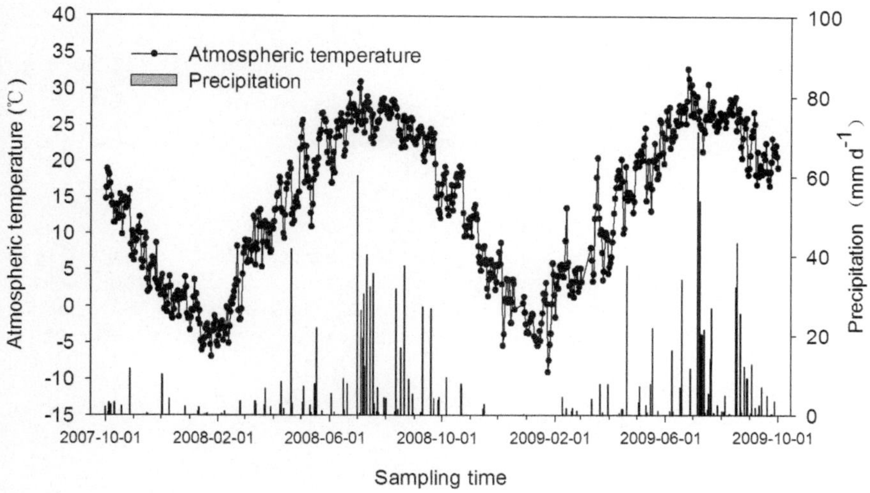

Figure 1. The atmospheric temperature and precipitation at the experimental site. The data were collected by the agricultural meteorological station approximately 500 m from the experiment field.

Table 1. The residue-management and tillage systems in the experimental plots.

Item	Tillage systems				
	CT	ST	HT	RT	NT
Wheat-residue					
Retention rate (t ha^{-1})	11.02	11.02	11.02	11.02	11.02
Residue-C (g kg^{-1})	51.27	51.27	51.27	51.27	51.27
Maize-residue					
Retention rate (t ha^{-1})	10.05	10.05	10.05	10.05	10.05
Residue-C (g kg^{-1})	52.15	52.15	52.15	52.15	52.15
Tillage					
Depth (cm)	25~35	40~45	12~15	10~15	-
Machine	moldboard	vibrating subsoil shovel	disc harrow	rototiller	-

CH_4 and N_2O were sampled at 5 minutes, 20 minutes and 35 minutes using a needle tube. The diurnal variations of CH_4 and N_2O in this study were collected from 2nd to 4th of May at 2-hour interval in 2009. All gases samples were collected at same time in order to avoid the order difference of sampling time. The atmospheric temperature, the temperature in the static chamber, the soil surface temperature and the soil temperature at a depth of 5 cm were determined simultaneously.

The samples were measured using a Shimadzu GC-2010 gas chromatograph. CH_4 was measured using a flame ionization detector with a stainless-steel chromatography column packed with a 5A molecular sieve (2 m long); the carrier gas was N_2. The temperatures of the column, injector and detector were 80°C, 100°C and 200°C, respectively. The total flow of the carrier gas was 30 ml min^{-1}, the H_2 flow was 40 ml min^{-1}, and the airflow was 400 ml min^{-1}. N_2O was measured using an electron-capture detector with a Porapak-Q chromatography column (4 m long); the carrier gas was also N_2. The temperatures of the column, injector and detector were 45°C, 100°C and 300°C, respectively. The total flow of the carrier gas was 40 ml min^{-1}, and the tail-blowing flow was 40 ml min^{-1}. The gas fluctuations were calculated based on the gas-concentration change over time per unit area.

The emission fluxes of CH_4 and N_2O were calculated using the following formula [19]:

$$F = \frac{60HMP}{8.314(273+T)}\frac{dc}{dt}$$

where F is the gas emission flux or uptake flux (μg m^{-2} hour^{-1}), 60 is the conversion coefficient of minutes and hours, H is the height of the static chamber (m), M is the molar mass of gas (g mol^{-1}), P is the atmospheric pressure (Pa), 8.314 is the ideal gas constant (J mol^{-1} K^{-1}), T is the average temperature in the static chamber (°C), and dc/dt is the slope of the line of the gas-concentration change over time.

The cumulative fluxes of CH_4 and N_2O were calculated by summing the products of the daily mean flux of two neighboring observations multiplying the days [6,24], the calculation formula as follows:

$$F_{cumulative} = \sum \left[\frac{(F_i + F_{i+1})}{2} \times d\right]$$

$$F_i = F_{avg} \times 24$$

where, $F_{cumulative}$ is the cumulative flux of CH_4 or N_2O (t ha^{-1} year^{-1}); F_i is the daily flux of observation i (t ha^{-1} d^{-1}); F_{i+1} is the daily flux of observation $i+1$ (t ha^{-1} d^{-1}); d is the days; F_{avg} is the mean flux of CH_4 or N_2O, which able to represent daily mean flux of CH_4 or N_2O according to the correction coefficient and regression analysis between the gas flux in observation duration and the daily total fluxes in diurnal variation.

Soil Sampling and Analysis

The meteorological data during the experiment (10/2007–10/2009) were by the agricultural meteorological station approximately 500 m from the experiment field. We measured soil temperature at a depth of 5 cm and the soil moisture in the 0–20 cm soil layers using a WET Sensor (WET brand, made in the UK). Soil samples (0–10 cm, 10–20 cm and 20–30 cm) were collected at five random positions of each plot and air dried to a constant weight, triturated and passed through a 2 mm sieve after thorough incorporation; they were then used to determine NO_3^-N using the UV colorimetric method [25].

Crop Yield

Winter wheat and maize were harvested at maturity, and the both harvest area was 9 m^2 in the central area of each plot to exclude edge effects. After air-drying, the grains were separated from the plants and oven-dried at 65°C for 48 h, and the dry weight was determined.

Statistical Analyses

The data were mapped using Sigma Plot 10.0, and all of the statistical analyses were performed using SPSS statistical software (SPSS Inc., Chicago, IL). The standard deviation (S.D.) and least significant difference (LSD) were calculated to compare the treatment means.

Results

Seasonal Variation of CH_4 Uptake

The soil acted an absorption sink for CH_4 in all tillage systems, which significantly varied by different soil tillage (Figure 2). The CH_4 flux during the sampling period (10/2007~10/2009) ranged from 6.31 to 41.36 μg m^{-2} h^{-1} under CT, from 4.03 to 33.45 μg

m^{-2} h^{-1} under ST, from 5.30 to 35.21 μg m^{-2} h^{-1} under HT, from 6.78 to 32.54 μg m^{-2} h^{-1} under RT, and from 1.10 to 26.21 μg m^{-2} h^{-1} under NT. Moreover, the fluxes tended to respond to the change of atmospheric temperature during the sampling time (Figure 2), with higher uptake fluxes in the summer and lower in the winter in the different tillage systems.

Seasonal Variation of N_2O Emission

The emitting of N_2O from the soil was observed under the different treatments (Figure 3), but the differences were small among the treatments in the same sampling time. However, the peak N_2O emission coincided with the irrigation and fertilization periods, which had the highest emission fluxes of all of the periods ($P<0.01$), and the flux did not accord with the trend of atmospheric temperature. Meanwhile, the flux in all of the samples of N_2O ranged from 14.07 to 130.39 μg m^{-2} h^{-1} under CT, from 14.20 to 126.43 μg m^{-2} h^{-1} under ST, from 12.68 to 134.93 μg m^{-2} h^{-1} under HT, from 10.81 to 126.42 μg m^{-2} h^{-1} under RT, and from 13.04 to 128.29 μg m^{-2} h^{-1} under NT.

Diurnal Variations of CH_4 and N_2O Under CT and NT

The diurnal flux variations of CH_4 uptake and N_2O emission significantly differed between the CT and NT treatments (Figure 4). In both of the treatments, the CH_4 uptake exhibited the lowest and highest fluxes at 6 a.m. and 2 p.m., respectively, but the lowest and highest values of CH_4 absorption were 12.99 and 23.77 μg m^{-2} h^{-1}, respectively, under CT and 12.23 and 23.03 μg m^{-2} h^{-1}, respectively, under NT (Figure 4a). The emission troughs and peaks of N_2O under CT and NT were observed at 4 a.m. and 4 p.m., respectively, with flux values of 16.90 and 41.89 μg m^{-2} h^{-1} under CT and 13.57 and 34.38 μg m^{-2} h^{-1} under NT, respectively (Figure 4b).

Cumulative Emissions of CH_4 and N_2O

The cumulative fluxes of CH_4 and N_2O under the different tillage systems in the two rotation cycle of wheat-maize systems were shown in Table 2. The highest cumulative uptake flux of CH_4 presented in the RT treatment with 1.85 t ha^{-1} $year^{-1}$ in the first rotation cycle of wheat-maize system (10/2007–10/2008), which was higher 8.3%, 13.7%, 22.5% and 60.9% than CT, ST, HT and NT treatments, respectively. But in the second rotation cycle (10/2008–10/2009), the highest one was the ST treatment. The order of total cumulative uptake fluxes of CH_4 under five tillage systems in two years was RT>ST>CT>HT>NT. The cumulative emission flux of N_2O ranged from 4.18 to 4.63 t ha^{-1} $year^{-1}$ in the first year and from 4.15 to 4.67 t ha^{-1} $year^{-1}$ in the second year, the total cumulative emission flux of N_2O ordered of NT>HT>CT>ST>RT, the flux under NT was only higher 7.2% than that of under RT.

Seasonal Variations of the Soil Temperature, Moisture and NO_3^-N

A significant difference of the soil temperature at 5 cm depth was measured in the different periods. The changes under the different tillage systems were related to the atmospheric temperature (Figure 5a). The averaged soil temperature in all of the periods under RT was higher than under the other tillage methods. The soil moisture of the 0–20 cm layer varied among the different treatments and was related to precipitation or irrigation (Figure 5b). The averaged moisture level of the 0–20 cm layer was highest in the NT treatment. Similarly, higher nitrate contents

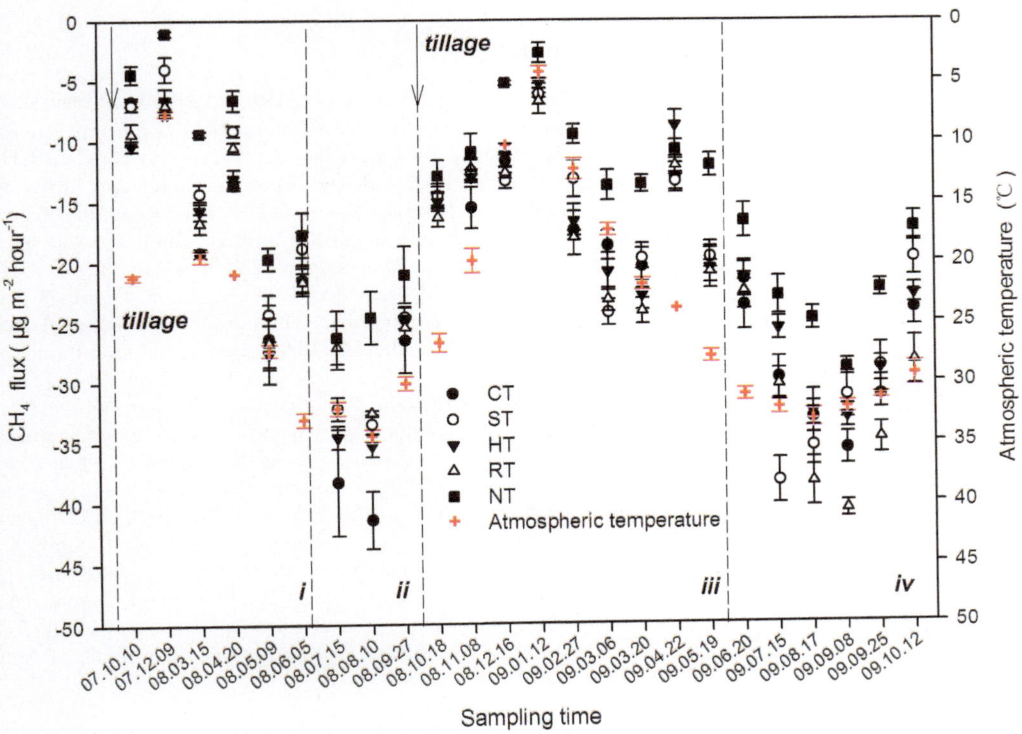

Figure 2. The seasonal characteristics of CH_4 flux under the different tillage systems. i and iii were the periods of wheat growth in 2007–2008 and 2008–2009, respectively; ii and iv were the periods of maize growth in 2008 and 2009, respectively. The arrows indicate the time of tilling, and the dotted lines distinguish the different periods of crop growth. The data are means ± SD (n = 3).

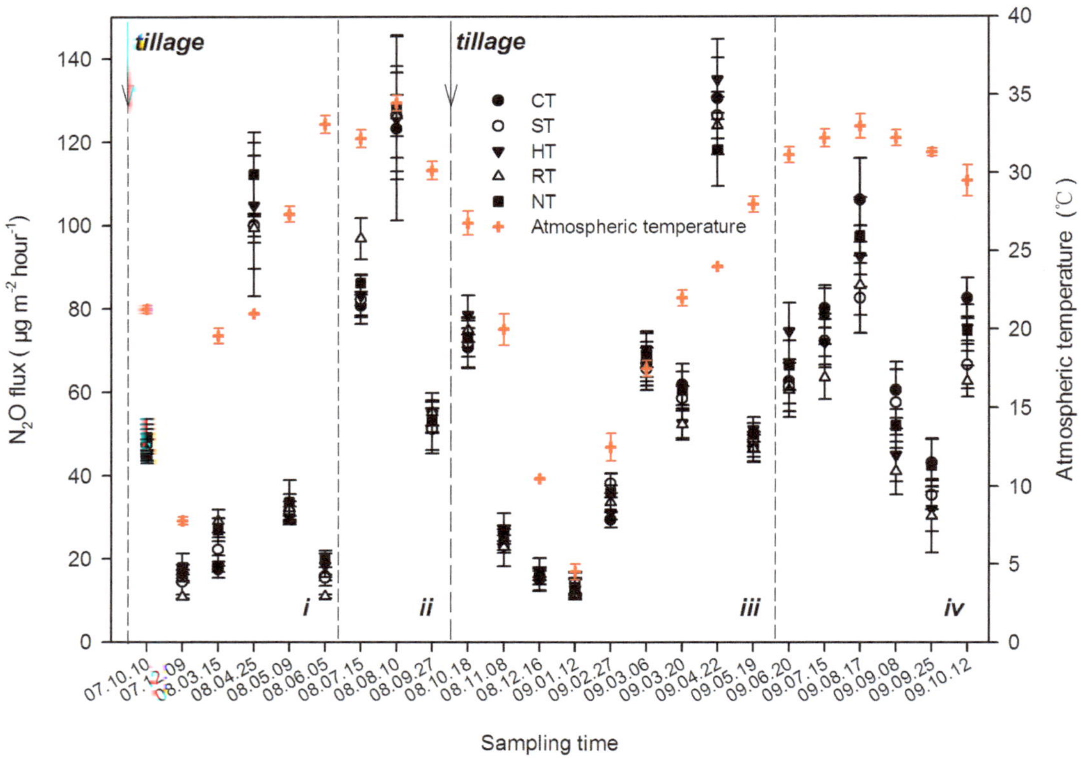

Figure 3. The seasonal characteristics of N_2O flux under the different tillage systems. i and iii were the periods of wheat growth in 2007–2008 and 2008–2009, respectively; ii and iv were the periods of maize growth in 2008 and 2009, respectively. The arrows indicate the time of tilling, and the dotted lines distinguish the different periods of crop growth. The data are means ± SD (n = 3).

were measured under the HT, NT, RT and ST treatments (Figure 5c); the levels in the NT, RT and ST treatments were higher than in the CT treatment by 4.21%, 2.42% and 1.40%, respectively.

Regression Analysis between CH_4, N_2O and Soil Factors

The absorption of CH_4 by the soil in different tillage systems was strongly affected by the soil temperature and soil moisture, and the uptake flux showed a positive correlation with the soil temperature ($R^2 = 0.44$, $P<0.01$; Figure 6a), and a negative correlation was observed with the soil moisture ($R^2 = 0.36$, $P<0.01$; Figure 6b).

The N_2O emission flux and soil temperature were not significantly correlated in this study. However, the N_2O emission flux was significantly related to the soil moisture ($R^2 = 0.63$, $P<0.01$; Figure 7a) and the content of nitrate ($R^2 = 0.50$, $P<0.01$; Figure 7b), which promoted the N_2O emission.

Crop Yields

The crop productivity of the wheat-maize rotation in the two years differed among the tillage systems (Table 3). The highest total productivity in two crop-rotation periods were measured in the ST treatment, with 34.63 t ha^{-1}, which was higher 2%, 6.2%, 0.2% and 25.7% than in the CT, HT, RT and NT treatment. The NT system showed the lowest productivity, only producing 27.54 t ha^{-1} yield in two crop-rotation periods.

Discussion

Effects of Soil Factors on CH_4 Uptake and N_2O Emission

In general, the emissions of CH_4 and N_2O in different seasons are affected by the soil temperature [26,27]. In this study, the CH_4 uptake and N_2O emission were lower in winter and higher in summer (Figures 2 and 3) and significantly related with the change of the soil temperature (Figures 2 and 6a). Similar results have been indicated in previous studies [28,29]. However, the CH_4 uptake flux usually decreased with the irrigation (Figures 3 and 6b). Many studies reported that soil moisture was a limiting factor for CH_4 absorption by the soil [16,30], leading to reduced rates of gas and O_2 diffusion [17].

Sometimes, the seasonal characteristics of N_2O emission exhibit varied trends in different sites, generally affected by temperature, N fertilization or irrigation in a significant linear relationship [27,31], and higher NO_3-N concentrations in wet soils promote the activity of nitrification and denitrification [18]. The higher moisture of the soil in this study promoted N_2O emission (Figures 5b and 7a), and the emission also increased with N fertilizer application (Figures 5c and 7b). In addition, there was no significant correlation between the N_2O emission and soil temperature in the seasonal variation in this study, similar results also found by other studies [32].

Tillage Effect on CH_4 Uptake and N_2O Emission

The fluxes of CH_4 uptake and N_2O emission were related to some soil factors (Figures 6 and 7), and these related soil factors general varied by the different tillage systems (Figure 5), because the changed soil structure and microflora by tillage would further

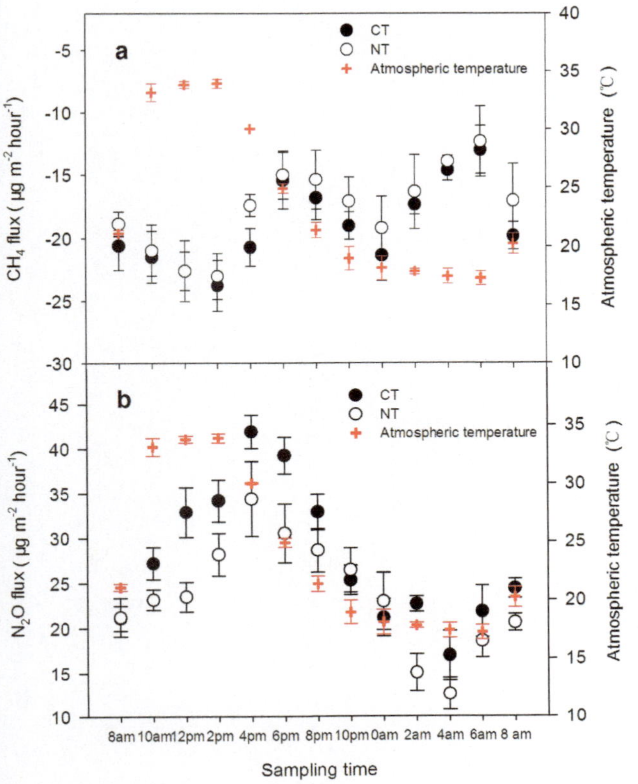

Figure 4. The diurnal flux variations of CH_4 uptake (a) and N_2O emission (b) in the CT and NT treatments. The data were collected from 2nd to 4th of May at 2-hour intervals in 2009. The data are means ± SD (n = 3).

drive emissions of the N_2O and CH_4 from soil [8,15]. For example, a soil with better permeability was a larger absorption sink of CH_4 [5]. In this study, the highest uptake flux of CH_4 was observed in the RT treatment (Table 2), which also contained the highest averaged soil temperature (Figure 5a). The regression analysis showed that there was a significant positive correlation between CH_4 uptake and soil temperature (Figure 6a). In contrast, under the five tillage systems used in this study, the lowest flux of CH_4 uptake was observed under NT (Figure 2), which was consistent with many previous studies [13,14]. The highest moisture and the lowest temperature conditions were also found in the soil of the NT treatment (Figures 5a and 5b). Some previous studies have indicated that the excessive wet condition of soil generally led to compaction of the soil surface without tillage, which blocked CH_4 from entering into soil for oxidation in dry-land farming systems [32–35]. However, others opposite results also indicated that NT could increase CH_4 oxidation because the reduced disturbance of soil could increase the activity of methane-oxidizing bacteria [36,37], while disturbance may negatively affect CH_4 uptake by the soil [38], but sometimes this effect from disturbance is small and can largely be ignored [9,38]. Because the variations of these factors are important driving factors for soil microflora and also responded sensitive by the changed soil structure by tillage.

Similarly, the HT treatment, which had the highest emission of N_2O from the soil because the content of NO_3-N was the highest compared with the other treatments (Figures 5c and 7b). The emission flux of N_2O usually peaked after the stage of N-fertilizer application or irrigation [10], and the highest flux was observed in reduced systems (HT treatment). In some cases, there was a risk of increasing the emission of N_2O under NT or reduced tillage [9]. These methods mostly produced decreased N_2O emission relative to that reported by many previous studies [39,40] because little N_2O was generally produced with a better mixture of soil and residue, and the predominant form of nitrogen was NO_3-N or NH_4-N [14].

Tillage Effect on Crop Yield

Little information was reported on reduced tillage systems effect on crop yields. Sometimes, subsoiling was regarded an effective method to increase wheat production [41–43]. In this study, the highest crop productivity of two rotation periods was shown under the ST and RT systems (Table 3). However, the NT system in this study had the lowest crop total yield, and a similar trends were reported by other studies [5,10]. However, the yields under NT showed dissimilar results in different sites, in which generally reported with the NT increase crop yields compared to conventional tillage [44,45]. Furthermore, relative to other tillage systems, no-tillage grain yields and profits are often dramatically lower during the first few years of adoption. In fact, no tillage have not been widely applied in the world or at least China, because there are actual or perceived problems in crop grown using the

Table 2. The cumulative emissions of CH_4 and N_2O under the different tillage systems.

Item	Cumulative flux (t ha^{-1} year^{-1})				
	CT	ST	HT	RT	NT
The first rotation cycle of wheat-maize system (10/2007–10/2008)					
Cumulative uptake flux of CH_4	1.71b	1.63c	1.51d	1.85a	1.15e
Cumulative emission flux of N_2O	4.18d	4.30c	4.37c	4.45b	4.64a
The second rotation cycle of wheat-maize system (10/2008–10/2009)					
Cumulative uptake flux of CH_4	1.63d	1.81a	1.74b	1.67c	1.20e
Cumulative emission flux of N_2O	4.67a	4.46c	4.57b	4.15d	4.57b
The total cumulative flux in the two years					
CH_4	3.34	3.43	3.25	3.53	2.34
N_2O	8.85	8.76	8.94	8.60	9.22

Different small letters in the same line indicate $P<0.05$. n = 3.

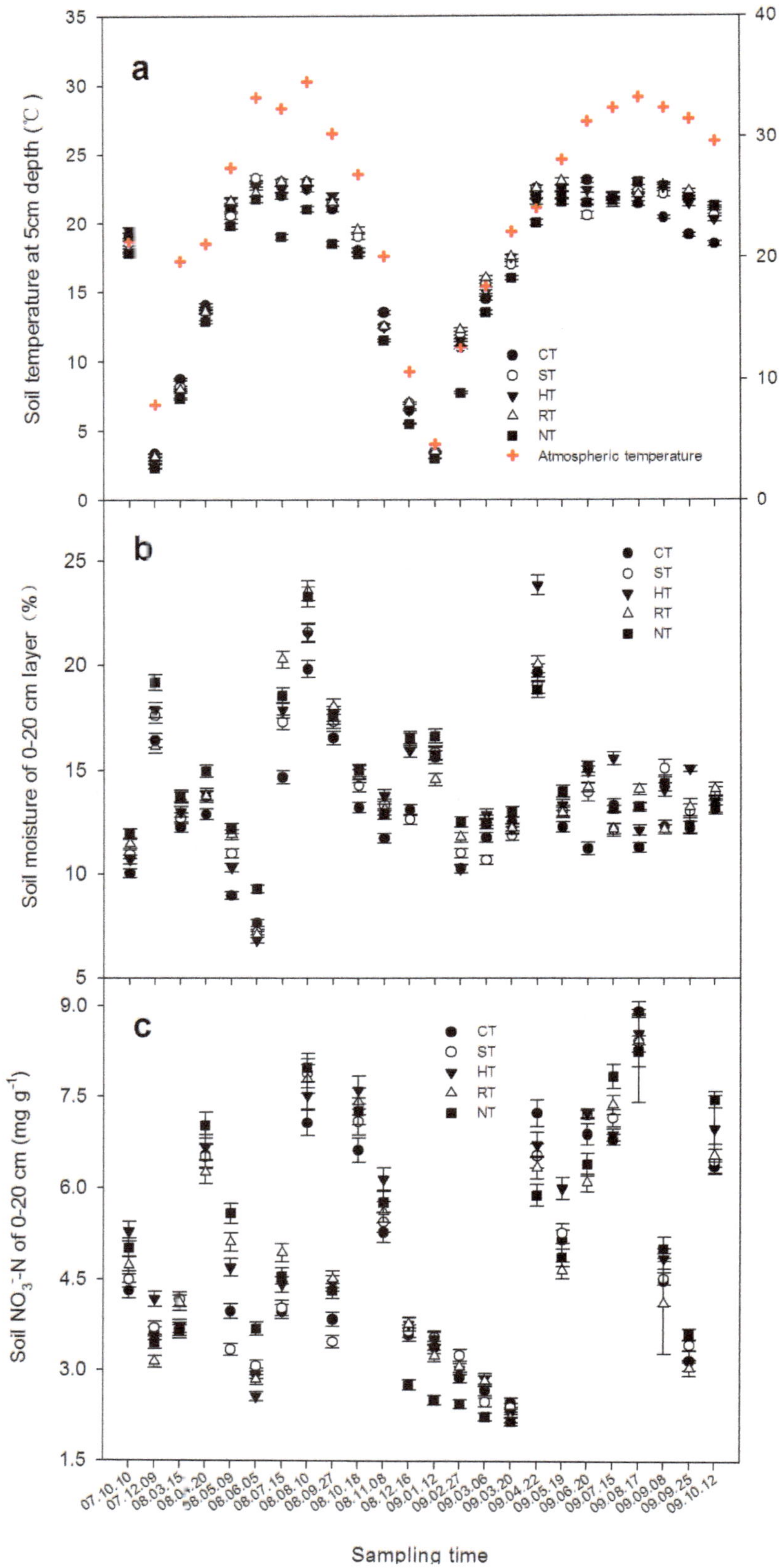

Figure 5. The seasonal variations of the soil temperature at a depth of 5 cm (a), soil moisture at 0–20 cm (b), and soil NO_3^-N content at 0–20 cm (c) under the different tillage systems. The data are means ± SD (n = 3).

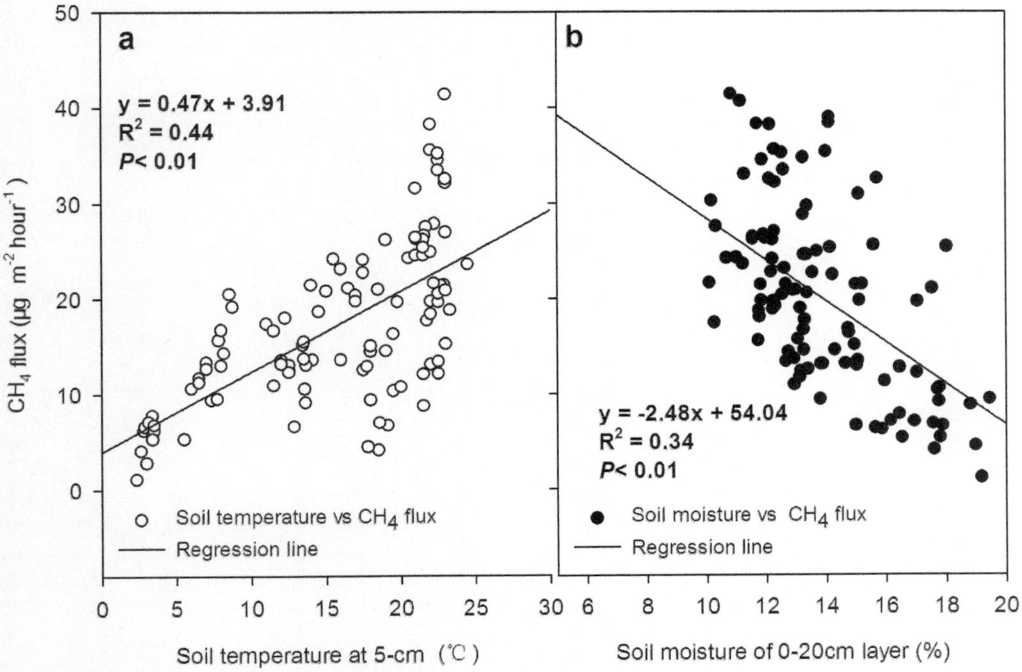

Figure 6. The regression analysis of the CH_4 flux and soil temperature at 5-cm depth (a, n = 105) and soil moisture of the 0–20 cm layer (b, n = 105).

NT, it may have limited adoption by growers. Some specific problems include lower early-season soil temperatures, reduced seed germination and emergence, below-optimal plant populations, poorer weed control, delayed plant development and maturity, increased grain moisture content, and lower grain yield potential [46–51]. Yet no tillage system effects on yield are highly dependent upon soil type, drainage, climate/latitude, and crop rotation [52].

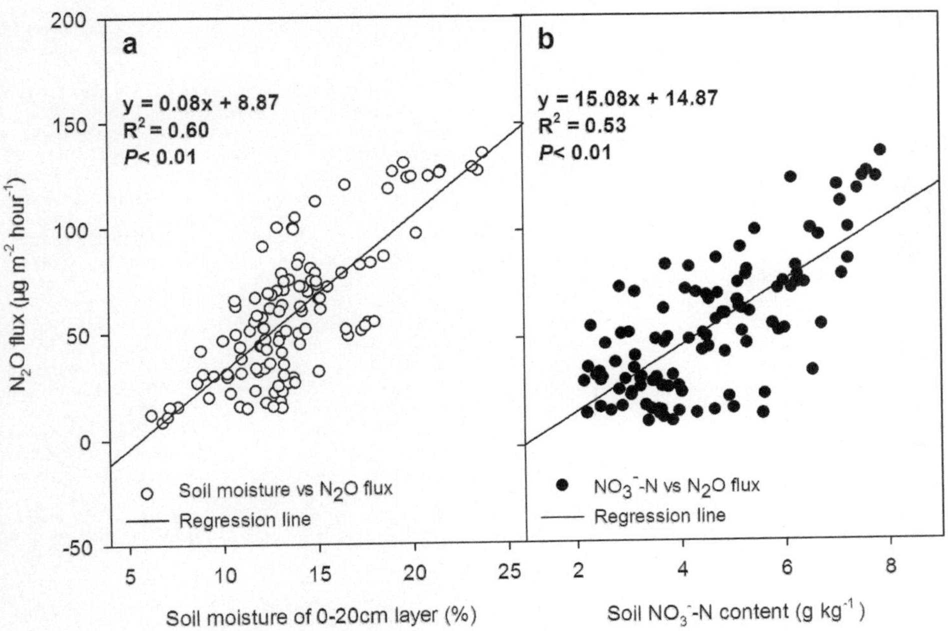

Figure 7. The regression analysis of the N_2O flux and soil moisture of the 0–20 cm layer (a, n = 105) and the soil NO_3^-N of the 0–20 cm layer (b, n = 105).

Table 3. The crop productivity in the different tillage systems.

Wheat-maize cropping-rotation system	Crop yield (t ha^{-1})				
	CT	ST	HT	RT	NT
The first rotation cycle					
Wheat (10/2007~06/2008)	5.90 d	6.18 c	6.58 a	6.35 b	4.65 e
Maize (06/2008~10/2008)	12.19 a	10.98 b	10.10 c	9.92 d	8.83 e
The second rotation cycle					
Wheat (10/2008~06/2009)	5.85 c	6.90 a	6.17 b	6.06 b	4.95 d
Maize yield (06/2009~10/2009)	10.03 c	10.58 b	9.76 d	12.23 a	9.11 e
Total productivity	33.95 b	34.63 a	32.62 c	34.56 a	27.54 d

Different small letters in the same line indicate $P<0.05$. n = 3.
doi:10.1371/journal.pone.0073450.t003

Conclusions

The fluxes of CH_4 and N_2O were significantly affected by no tillage and reduced tillage systems, which also impacted soil factors, such as the soil temperature, moisture and nitrate content. These factors were related with the changes of CH_4 and N_2O fluxes. No-tillage did not reveal a better result for mitigating emissions of CH_4 and N_2O, as well as maintaining a high level of crop yield in this cropping rotation system. Comparison to the conventional tillage and no tillage, under the reduced tillage systems such as subsoiling and rotary tillage, may mitigate emissions of CH_4 and N_2O, and also gain a high level of crop productivity in the wheat-maize cropping rotation system. Sometimes, the both also as an important rotation tillage systems was applied in this region, in particular the subsoiling. The results also provide information on optional tillage in rotation tillage systems for mitigating GHG emissions and improving crop yield.

Author Contributions

Conceived and designed the experiments: ST TN ZL SC. Performed the experiments: ST YW HZ BW NL. Analyzed the data: ST TN. Contributed reagents/materials/analysis tools: ST. Wrote the paper: ST TN.

References

1. Forster P, Ramaswamy V, Artaxo P, Berntsen T, Betts R, et al. (2007) Changes in atmospheric constituents and in radiative forcing. In: Solomon S, Qin D, Manning M, Chen Z, Marquis M, et al., ed. Climate Change 2007: The Physical science basis. Contribution of working group I to the fourth assessment report of the intergovernmental panel on climate change. Cambridge University Press, Cambridge, United Kingdom and New York, USA.
2. Bouwman AF (1990) Exchange of greenhouse gases between terrestrial ecosystems and the atmosphere. In: Bouwman AF, ed. Soils and the Greenhouse Effect. Wiley, Chichester, 61–127.
3. IPCC (2007) Climate change 2007: The physical science basis. Contribution of working group I to the fourth assessment report of the intergovernmental panel on climate change. Cambridge University Press, Cambridge, United Kingdom and New York, USA.
4. Paustian K, Andren O, Janzen HH, Lal R, Smith P, et al. (1997) Agricultural soil as a C sink to offset CO_2 emissions. Soil Use Manage 13: 230–244.
5. Dong YH, Ouyang Z (2005) Effects of organic manures on CO_2 and CH_4 fluxes of farmland. Chin J Appl Ecol 16: 1303–1307.
6. Chatskikh D, Olesen JE (2007) Soil tillage enhanced CO_2 and N_2O emissions from loamy sand soil under spring barley. Soil Till Res 97: 5–18.
7. Willison TW, Webster CP, Goulding KWT, Powlson DS (1995) Methane oxidation in temperate soils: effective of land use and the chemical form of nitrogen fertilizer. Chemosphere 30: 539–546.
8. Lee J, Six J, King AP, Van Kessel C, Rolston DE (2006) Tillage and field scale controls on greenhouse gas emissions. J Environ Qual 35: 714–725.
9. Six J, Ogle SM, Breidt FJ, Conant RT, Mosier AR, et al. (2004) The potential to mitigate global warming with no-tillage management is only realized when practised in the long term. Glob Change Biol 10: 155–160.
10. Bhatia A, Sasmal S, Jain N, Pathak H, Kumar R, et al. (2010) Mitigating nitrous oxide emission from soil under conventional and no-tillage in wheat using nitrification inhibitors. Agr Ecosyst Environ136: 247–253.
11. Zhang HL, Gao WS, Chen F, Zhu WS (2005) Prospects and present situation of conservation tillage. J Chin Agr Uni 10: 16–20.
12. Tian SZ, Ning TY, Zhao HX, Wang BW, Li N, et al. (2012) Response of CH_4 and N_2O Emissions and Wheat Yields to Tillage Method Changes in the North China Plain. PLoS one 7(12): e51206.
13. Elder JW, Lal R (2008) Tillage effects on gaseous emissions from an intensively farmed organic soil in North Central Ohio. Soil Till Res 98: 45–55.
14. Bai XL, Zhang HL, Chen F, Sun GF, Hu Q, et al. (2010) Tillage effects on CH_4 and N_2O emission from double cropping paddy field. Transac CSAE 26: 282–289.
15. Dunfield P, Knowles R, Dumont R, Moore TR (1993) Methane production and consumption in temperate and subarctic peat soils. Soil Biolog Biochem25: 321–326.
16. Zhang XS, Shen SH, Li J (2006) Soil CH_4 uptake in winter wheat field in the North China Plain. J Nanjing Inst Meteorol 29: 181–188.
17. Le Mer J, Roger P (2001) Production, oxidation, emission and consumption of methane by soils: A review. Eur J Soil Biol 37: 25–50.
18. Gregorich EG, Rochette P, Vandenbygart AJ, Angers DA (2005) Greenhouse gas contributions of agricultural soils and potential mitigation practices in Eastern Canada. Soil Till Res 83: 53–72.
19. Robertson G (1993) Fluxes of nitrous oxide and other nitrogen trace gases from intensively managed landscapes: a global perspective. In: Harpwr LA, Mosier AR, Duxbury JM, Rolston DE, ed. Agricultural ecosystem effects on trace gases and global climate change. ASA Special Publication No. 55. ASA, CSSA, SSSA, Madison, wi 95–108.
20. Ma J, Xu H, Cai ZC, Yagi K (2007) Diurnal variation of CH_4 emission from rice field as affected by rice plant. Soils 39(6): 859–862.
21. Zheng XH, Wang MX, Wang YS, Shen RX, Li J, et al. (1998) Comparison of manual and automatic methods for measurement of methane emission from rice paddy fields. Adv Atmos Sci 15(4): 569–579.
22. Wan YF, Lin ED, Li YE, Gao QZ, Qin XB (2005) Studies on closing time in measuring greenhouse gas emission from dry cropland by static chamber method. Chin J Agrometeorol 27(2): 122–124.
23. Tian SZ, Ning TY, Chi SY, Wang Y, Wang BW, et al. (2012) Diurnal variations of the greenhouse gases emission and their optimal observation duration under different tillage systems. Acta Ecologica Sinica 32: 879–888.
24. Liang W, Shi Y, Zhang H, Yue J, Huang GH (2007) Greenhouse gas emissions from northeast China rice fields in fallow season. Pedosphere 17: 630–638.
25. Bao SD (2000) Soil and Agricultural Chemistry Analysis. China Agriculture Press, Beijing.
26. Hassink J (1995) Density fractions of soil macroorganic matter and microbial biomass as predictors of C and C mineralization. Soil Biolog Biocheml 27: 1099–1108.
27. Groffman PM, Hardy JP, Driscoll CT, Fahey TJ (2006) Snow depth, soil freezing, and fluxes of carbon dioxide, nitrous oxide and methane in a northern hardwood forest. Glob Change Biol 12: 1748–1760.
28. Qi YC, Dong YS, Zhang S (2002) Methane fluxes of typical agricultural soil in the north china plain. Rural Ecol Environ 18: 56–60.
29. Dijkstra FA, Morgan JA, von Fischer JC, Follett RF (2011) Elevated CO_2 and warming effects on CH_4 uptake in a semiarid grassland below optimum soil moisture. J Geoph Res 116: 1–9.

30. Von Fischer JC, Butters G, Duchateau PC, Thelwell RJ, Siller R (2009) In situ measures of methanotroph activity in upland soils: Areaction diffusion model and field observation of water stress, J Geoph Res 114:G01015.
31. Cui RY, Chen ZZ, Chen SQ (2001) Progress in research on soil respiration of grasslands. Acta Ecologica Sinica 21: 315–325.
32. Koponen HT, Flojt L, Martikainen PJ (2004) Nitrous oxide emissions from agricultural soils at low temperatures: a laboratory microcosm study. Soil Biolog Biochem 36: 757–766.
33. Omonode RA, Vyn TJ, Smith DR, Hegymegi P, Ga'l A (2007) Soil carbon dioxide and methane fluxes from long-term tillage systems in continuous corn and corn–soybean rotations. Soil Till Res 95: 182–195.
34. Ahmad S, Li C, Dai G, Zhan M, Wang J, et al. (2009) Greenhouse gas emission from direct seeding paddy field under different rice tillage systems in central China. Soil Till Res 106: 54–61.
35. Li DM, Liu MQ, Cheng YH, Wang D, Qin JT, et al. (2011) Methane emissions from double-rice cropping system under conventional and no tillage in southeast China. Soil Till Res 113: 77–81.
36. Smith P, Goulding KWT, Smith KA, Powlson DS, Smith JU, et al. (2001) Enhancing the carbon sink in European agricultural soils: including trace gas fluxes in estimates of carbon mitigation potential. Nut cycle Agroecosyst 60: 237–252.
37. Hütsch BW (1998) Tillage and land use effects on methane oxidation rates and their vertical profiles in soil. Biol Fert Soils 27: 284–292.
38. Robertson GP, Paul EA, Harwood RR (2000) Greenhouse gases in intensive agriculture: Contributions of individual gases to the radiative forcing of the atmosphere. Science 289: 1922–1925.
39. Kessavalou A, Mosise AR, Doran JW (1998) Fluxes of carbon dioxide, nitrous oxide, and methane in grass sod and winter wheat-fallow tillage management. J Environ Qual 27: 1094–1104.
40. Lal R (2004) Soil carbon sequestration impacts on global climate change and food security. Science 304: 1623–1627.
41. He J, Li HW, Gao HW (2006) Subsoiling effect and economic benefit under conservation tillage mode in Northern China. Transact CSAE 22: 62–67.
42. Gong XJ, Qian CR, Yu Y, Zhao Y, Jiang YB, et al. (2009) Effects of Subsoiling and No-tillage on Soil Physical Characters and Corn Yield. J Maize Sci 17: 134–137.
43. Huang M, Wu JZ, Li YJ, Yao YQ, Zhang CJ, et al. (2009) Effects of different tillage management on production and yield of winter wheat in dryland. Transact CSAE 25: 50–54.
44. He J, Li HW, Rasaily RG, Wang QJ, Cai GH, et al. (2011) Soil properties and crop yields after 11 years of no tillage farming in wheat–maize cropping system in North China Plain. Soil Till Res 113: 48–54.
45. He J, Wang QJ, Li HW, Liu LJ, Gao HW (2009b) Effect of alternative tillage and residue cover on yield and water use efficiency in annual double cropping system in North China Plain. Soil Till Res 104: 198–205.
46. Mock JJ, Erbach DC (1977) Influence of conservation-tillage environments on growth and productivity of corn. Agron J 69: 337–340.
47. Kovar JL, Barber SA, Kladivko EJ, Griffith DR (1992) Characterization of soil temperature, water content, and maize root distribution in two tillage systems. Soil Till Res 24: 11–27.
48. Lund MG, Carter PR, Oplinger ES (1993) Tillage and crop rotation affect corn, soybean, and winter wheat yields. J Prod Agr 6: 207–213.
49. Swan JB, Higgs RL, Bailey TB, Wollenhaupt NC, Paulson WH, et al (1994) Surface residue and in-row treatment effects on long-term no-tillage continuous corn. Agron J 86: 711–718.
50. Fortin MC, Pierce FJ (1990) Developmental and growth effects of crop residues on corn. Agron J 82: 710–715.
51. Fortin MC (1993) Soil temperature, soil water, and no-till corn development following in-row residue removal. Agron J 85, 571–576.
52. Boomsma CR, Santini JB, West TD, Brewer JC, McIntyre LM, et al. (2010) Maize grain yield responses to plant height variability resulting from crop rotation and tillage system in a long-term experiment. Soil Till Res 106: 227–240.

Dynamics of Potassium Release and Adsorption on Rice Straw Residue

Jifu Li[1,2], Jianwei Lu[1,2]*, Xiaokun Li[1,2], Tao Ren[1,2], Rihuan Cong[1,2], Li Zhou[1,2]

1 College of Resources and Environment, Huazhong Agricultural University, Wuhan, China, 2 Key Laboratory of Arable Land Conservation (Middle and Lower Reaches of Yangtse River), Ministry of Agriculture, Wuhan, China

Abstract

Straw application can not only increase crop yields, improve soil structure and enrich soil fertility, but can also enhance water and nutrient retention. The aim of this study was to ascertain the relationships between straw decomposition and the release-adsorption processes of K^+. This study increases the understanding of the roles played by agricultural crop residues in the soil environment, informs more effective straw recycling and provides a method for reducing potassium loss. The influence of straw decomposition on the K^+ release rate in paddy soil under flooded condition was studied using incubation experiments, which indicated the decomposition process of rice straw could be divided into two main stages: (a) a rapid decomposition stage from 0 to 60 d and (b) a slow decomposition stage from 60 to 110 d. However, the characteristics of the straw potassium release were different from those of the overall straw decomposition, as 90% of total K was released by the third day of the study. The batches of the K sorption experiments showed that crop residues could adsorb K^+ from the ambient environment, which was subject to decomposition periods and extra K^+ concentration. In addition, a number of materials or binding sites were observed on straw residues using IR analysis, indicating possible coupling sites for K^+ ions. The aqueous solution experiments indicated that raw straw could absorb water at 3.88 g g^{-1}, and this rate rose to its maximum 15 d after incubation. All of the experiments demonstrated that crop residues could absorb large amount of aqueous solution to preserve K^+ indirectly during the initial decomposition period. These crop residues could also directly adsorb K^+ via physical and chemical adsorption in the later period, allowing part of this K^+ to be absorbed by plants for the next growing season.

Editor: Jörg Langowski, German Cancer Research Center, Germany

Funding: This work was supported by Non-profit Research Foundation for Agriculture (201103039), National Natural Science Foundation of China (41301319), Fundamental Research Funds for the Central Universities (2012BQ059) and International Potash Institute Co-operation Program. The funders had no role in study design, data collection and analysis, decision to publish, or preparation of the manuscript.

Competing Interests: The authors have declared that no competing interests exist.

* E-mail: lunm@mail.hzau.edu.cn

Introduction

The current Asian population of 4.3 billion is projected to increase by nearly 0.9 billion people, reaching roughly 5.2 billion, by 2050 [1], which will result in significantly increased regional food demand. Of this, population, 80% will be distributed in China, India and the southeast regions of Asia, posing a challenge to the economic development and social stability of these countries [2]. Rice-based cropping systems are the most productive agroecosystems in these areas and produce the most food for the most people [3]. To meet the food demand of the region, intensification and diversification have been applied as the two main strategies for rice-based cropping systems. In addition to the rise of multiple cropping indexes, fertilization consumption has played a very important role in production increases [4]. Compared with nitrogen and phosphorus fertilizer, potash fertilizer is often ignored by farmers, particularly in Asia [5–7]. Potash resources are comparatively limited [8,9], and in recent years, the higher price of potash on the international market has reduced the demand of potassium, as farmers in the area are unwilling to put more potash into the soil [10]. Soil K deficiency has become a major limiting factor in the modern agricultural process [11]. Therefore, it is of great importance to increase potash supplementation in these regions. K-bearing organic resources such as compost, green manure, farmyard manure and crop straws, particularly abundant crop residues, are again receiving attention from farmers [12].

Annually, the world production of straw is approximately 3.8 billion tons, 74% of which are cereal straws [13]; for rice-based land in Asia, 80% of straw production consists of rice residues [3]. Cereal straws usually have a higher potassium content than other straws (1.2%–1.7%). The results of Kaur and Benipal [14] have shown that returning straw to the field could improve soil available potassium to a significantly great extent than manure. A 30-years field trial conducted by Liao et al. indicated that straw management could increase exchangeable K by 26.4%, nonexchangeable K by 1.8% and SOC 21.0% in comparison to a CK treatment in reddish paddy soil [15]. As straw potassium is primarily present in the form of K^+ ions in the cell fluid [16], the release of K from stubble in field is influenced by rainfall [17]. Duong et al. [18] found that the distance of potassium migration from organic fertilizer is 10 mm. Excepting the K^+ adsorbed or fixed by soil clay particles, 50% of K^+ was retained in the soil solution [14]. Furthermore, farmers prefer to input potash fertilizers one time before sowing. The loss of K in the soil solution from such applications was 1.1- and 14.5- fold that of N and P, respectively

[19]; N and P losses were gradual, while leaching phenomena were observed for K [17]. However, Kozak et al. found that crop residues could intercept a maximum of 29% of the water loss for a given rainfall [20]. Soil water retention is also affected by the organic carbon content, as reported by Rawls et al [21]. These results indicate that crop residues have a positive effect on water absorption. Meanwhile, as a high-quality biological adsorbent, the biochars generated from crop straws can adsorb 0.48–1.40 mol kg^{-1} Cu(II) [22], and even unmodified rice straw can absorb 13.9 mg g^{-1} Cd(II) [23]. However, relatively few data exist regarding the adsorption of potassium by crop residues. Because straw decomposition is a slow and long-term process, a significant quality of plant residues can usually be found on farmland after one season of crop growth.

Since 1980, global warming has received increasing attention [24], bolstered by occasional extreme weather events such as the sustained hot temperature in Europe, North America and Asia in July 2013. Scorching weather causes a water shortage in rice farmland, thereby affecting the absorption of nutrients. The incorporation of residues into the soil may reserve water to slow down seasonal drought and could also adsorb cations. However, the capability of soil residues to preserve water and nutrients during different decomposition periods is unclear, especially, the fixation mechanism of residues for potassium.

Therefore, the aims of this research were as follow: (a) to investigate the characteristics of straw decomposition and K^+ release under flooded conditions; (b) to assess the retention capacity of straw for water and potassium during different decomposition periods; and (c) to ascertain the mechanism of K^+ adsorption on straw residues.

Materials and Methods

Ethics Statement

The authors of this study hereby confirm that no specific permissions were required for our experimental location and activities, as the experimental field belonged to our institute and is employed for scientific research only. The studies had negligible effects on the functioning of the broader ecosystem. The research did not involve measurements on humans or animals, and no endangered or protected species were involved.

Materials

Soil material. Paddy soil was collected from the plow layer (0–20 cm) after the rapeseed harvest in 2012. The sample was mixed, air-dried at room temperature and ground to pass through a 2 mm sieve. Soil physical and chemical properties were measured using conventional methods [25–27]. Soil pH was measured with a glass electrode in a 1:2.5 soil/water solution. Soil organic matter was measured using the dichromate oxidation method, and total nitrogen was measured using the Kjeldahl acid digestion method. Available phosphorus was determined using the Olsen method, and available potassium was measured by flame photometry after NH_4OAc neutral extraction. All parameters were measured three times and are presented as the mean±SD: pH 5.73±0.23, SOM 26.7±1.1 g kg^{-1}, TN 1.09±0.04 g kg^{-1}, Olsen-P 19.4±0.7 mg kg^{-1}, and NH_4OAc-K 107.8±4.3 mg kg^{-1}.

Rice straw material. Rice straw was cut into segments of approximately 2–3 cm and then conserved in a dryer until further use. The initial potassium (K) content of the rice straw was 21.87±0.12 mg g^{-1}.

Methods

Straw decomposition trial. The trial was performed at the experimental base of the College of Resources and Environment beginning on July 7, 2012. Three boxes, constructed of PVC (size 50×30×25 cm), were used for the incubation experiment. A total of 15 kg of dry bulk soil was packed in each box. Before the experiment, the rice straw was dried at 40°C for 3 h, and 10.0 g samples were then accurately weighed into 200 mesh (pore diameter 0.075 mm) nylon bag (size 25×20 cm), and sealed [28]. Each box contained 5 bags of straw, totaling 15 bags in the three boxes. According to the growth period of late rice, we removed nylon bags at 5, 15, 30, 60 and 110 d. During the incubation period, deionized water was added to the boxes to maintain a flooding layer of 1 cm. A schematic diagram of the trial is shown in Figure S1.

On each sampling date, one nylon bag was randomly removed from each box and rinsed with distilled water three times to remove any mud that had adhered to the bag. The residue of the rice straw was then dried at 40°C for 48 h until reaching a constant weight, weighed and ground to pass through a 1 mm sieve. Some of the residue powder was ground again using a mortar, until it passed through a 0.149 mm sieve, to obtain micron particles. The particles were dispersed in an aqueous solution of pH 6.0 to test the zeta potential using zeta potential and nanoparticle size analyzer (ZS90, Melvin British Company UK). Zeta potential is the potential difference between the dispersion medium and the stationary layer of fluid attached to the dispersed particle [29], which indicates the electric potential variation of the residue surface during the decomposition period [22]. All of the residue samples were digested with H_2SO_4-H_2O_2 to determine their potassium content using a flame photometer (M-410, Cole-Parmer USA) and to calculate their potassium release rates. The formulas employed in this study were as follow:

$$\begin{aligned}&\text{Decomposition amount (g)}\\&\quad = \text{Dry matter at 0 d Remaining dry matter at } n \text{ d}\end{aligned} \tag{1}$$

$$\begin{aligned}&\text{Decomposition rate (\%)}\\&\quad = (\text{Decomposition amount/Dry matter at 0 d}) \times 100\%\end{aligned} \tag{2}$$

$$\begin{aligned}&\text{K release amount (K,mg)}\\&\quad = \text{K amount at 0 d} - \text{K remaining at } n \text{ d}\end{aligned} \tag{3}$$

$$\begin{aligned}&\text{K release rate (\%)}\\&\quad = \text{K release amount/K amount at 0 d} \times 100\%\end{aligned} \tag{4}$$

Where n is the day of incubation.

Batches of K sorption experiments. Precise 0.30 g samples of the rice straw residue from different time points were added into 50 mL polythene bottles along with various concentrations of KCl solution (0, 10, 50 and 100 mg K L^{-1}). The bottles were shaken for 4 h at 160 r min^{-1} and filtered to test the K^+ concentration of

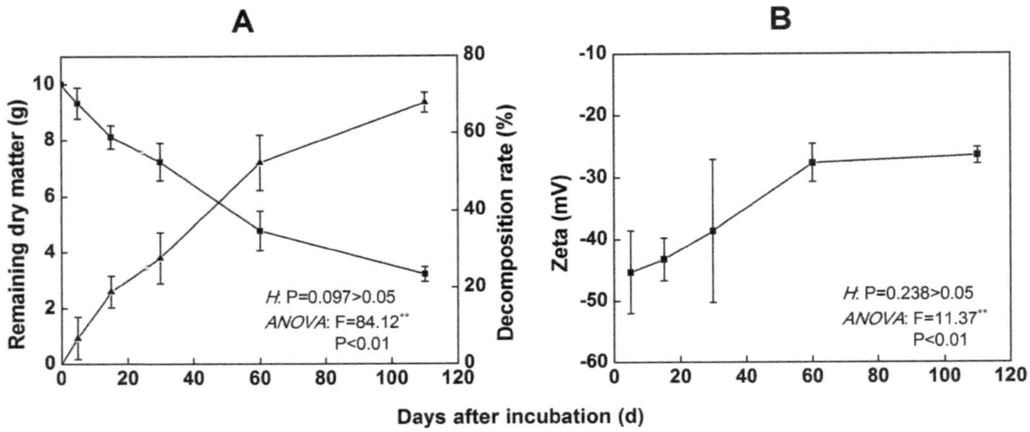

Figure 1. The characteristics of the decomposition and zeta potential of rice straw. The annotations in the panels indicate the homogeneity of variances (H) and ANOVA for the remaining dry matter and zeta potential. The H-test was performed using the Levene test. **indicates significant differences at P<0.01. The values are the means of 3 replicates (±standard deviation).

the equilibrium solution. The experiments were repeated 3 times, and the average values were used for analysis. The K adsorption on straw Q (mg g^{-1}) and the K removing rate R (%) were determined in the following manner [30]:

$$Q = (C_0 - C_t) \times V/m \quad (5)$$

$$R = (C_0 - C_t)/C_0 \times 100\% \quad (6)$$

Where C_0 and C_t are the initial and equilibrium K concentration, respectively, mg L^{-1}; V is the volume of the solution, mL; and m is the mass of the straw residue, g.

According to the K sorption experiment, the maximal Q residue was selected to test the changes of residue structure on K^+ fixation with the help of Fourier transform infrared spectroscopy. (Nexus, Thermo Nicolet USA).

Kinetics of water absorption on rice straw. The water absorption capacity of untreated dried straw [31] was determined by suspending 0.50 g of natural dried straw through a 1.0 mm sieve in a 50 mL beaker with 40 mL deionized water. three replicates were taken at each time interval and cleaned of gravitational water from the straw surface. The material was then weighed on an electronic balance, and the water absorption capacity was calculated.

The water absorption capacity of rice straw residues from different decomposition time points was also determined. A straw sample of 0.30 g was put into a 50 mL beaker with 40 mL of deionized water. After 300 min of immersion, the residues were removed and cleaned of gravitational water before the determination of water absorption capacity. The water absorption capacity of the straw was calculated as follows:

$$M = (M_t - M_0)/M_0 \quad (7)$$

Where M is the water absorption capacity of straw, g g−1; and Mt and M0 are the initial and final mass of straw, respectively, g.

Furthermore, to observe the influence of straw surface pore on the water absorption of rice straw, the surface morphology of residue from different time points was examined using a scanning electron microscope (JSM-6390LV, NTC Japan).

Data Analysis

All analyses were conducted on three replicates. Statistical analyses were performed using OrignPro8.0 software. The means were the average of three replicates, and analysis of variance (ANOVA) was performed according to standard procedures for factorial randomized block designs. Differences at $p<0.05$ level were considered statistically significant, as determined using the least significant difference (LSD) test.

Results

Characteristics of Straw Decomposition and Zeta Potential of Rice Straw Surface

The decomposition dynamics of rice straw during different stages of incubation under flooded conditions are shown in Figure 1A. The average rate of rice straw decomposition was 0.09 g d^{-1} from 0 to 60 d. The decomposition amount accounted for 52.3% (remaining mass 4.77 g) of the total mass after 60 d of incubation. The average rate of decomposition was relatively slow, 0.03 g d^{-1}, from 60 to 110 d. During this period, the decomposition amount accounted for 15.5% of the total mass. The remaining amounts of straw and the accumulation rate were 3.22 g and 67.8%, respectively, after the trial was ended at 110 d.

The zeta potential of the untreated rice straw surface was −56.36 mV in the pH 6.0 aqueous phase (Figure 1B). After 5 d of incubation, the zeta potential increased to −43.2 mV, and it continued to increase to −27.65 mV after 60 d of incubation. On the following days, the zeta potential was not significantly different, ranging from −30 to −20 mV.

Characteristics of Straw K Release

The potassium (K) content of untreated rice straw was 21.87 mg kg^{-1}. After 3 d of immersion, 90% of the total K had been released, and the K content of straw was only 1.47 mg g^{-1} after 5 d of immersion (Figure 2A); this value did not change significantly on the following days. These results indicated that 93.7% of the total potassium could be released into the ambient environment during the early stage of immersion (Figure 2B). After 30 d, the K content of the residues increased slightly and remained

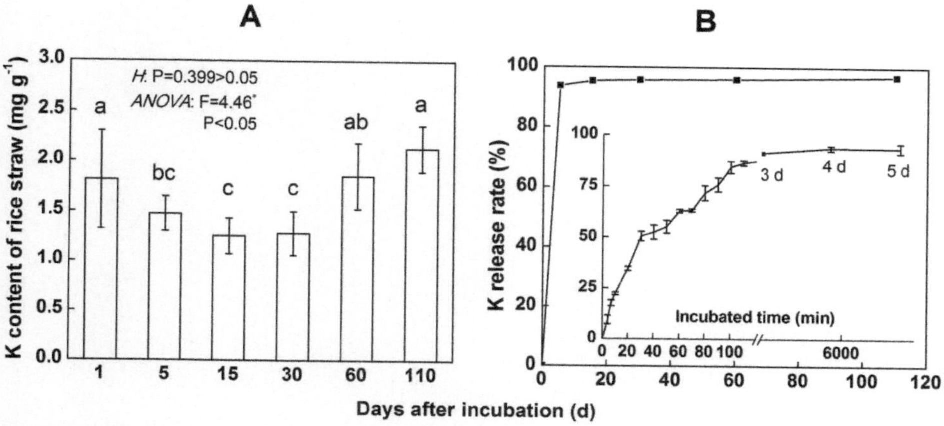

Figure 2. The characteristics of the potassium release of rice straw. The annotation in panel A indicates the homogeneity of variances (H) and ANOVA. The H-test was performed using the Levene test. *indicates significant differences at P<0.05. The values are the means of 3 replicates (±standard deviation). Means with the same letter are not significantly different. The insertion in panel B is the K release rate within 5 d.

at 1–2 mg g^{-1}. At the end of the trial, the K content of the residue and the K release rate were 2.13 mg g^{-1} and 96.9%, respectively.

K Adsorption in Different Decomposition Periods of Straw Residue

Figure 3 shows the adsorption-desorption equilibrium of exogenous potassium over different decomposition periods of straw residue. The results showed that over the entire decomposition period, K^+ ions were, on the whole, released in pure water. When the concentration of K^+ ions was 10 mg L^{-1} in solution, a positive effect of K^+ adsorption was observed after 60 d of decomposition, and the K^+ adsorption capacity of the residue reached 0.13 mg g^{-1} on the 110th d of the experiment. The residue at 15 d showed a positive adsorption effect when the concentration of external potassium increased to 50 mg L^{-1}. Meanwhile, the capacity also increased as decomposition continued. The amount of K^+ adsorption reached a maximum of 0.76 mg g^{-1} after 110 d of incubation, representing a considerable increase of 76.3% compared to its value on the 15th day. When additional increments of K were applied, up to 100 mg L^{-1}, the straw primarily released potassium into solution within 5 d. After 15 d, the adsorption capacity of the residue was higher than that in the trial with 50 mg L^{-1} K supplied over the same period.

Similarly, results of Figure 3B indicated that the K removing rate (R) increased with the decomposition of rice straw. When the concentration of extra K was 10 mg L^{-1}, for example, the value of R was −51.0% at 5 d after incubation, but reached 15.6% after 110 d. At the same time, when the concentration of external K was increased to 50 mg L^{-1}, the removing rate for each period of decomposition except for the first 5 d reached the maximum value. The values of R tended to decrease with increase external K concentration.

FT-IR Analysis for the K Adsorption on Rice Straw Residue

Figure 4 displays the infrared spectra of untreated dried straw both before and after the K adsorption of the rice straw residue on the 110th day after incubation. The results showed that the chemical structure of the straw had undergone a significant change after 110 d of decomposition (Figure 4A–B). New peaks emerged at 3698, 3619 and 779 cm^{-1}, and the existing peaks at 2851, 1086, 798, 695, 528 and 469 cm^{-1} were strengthened to various degrees.

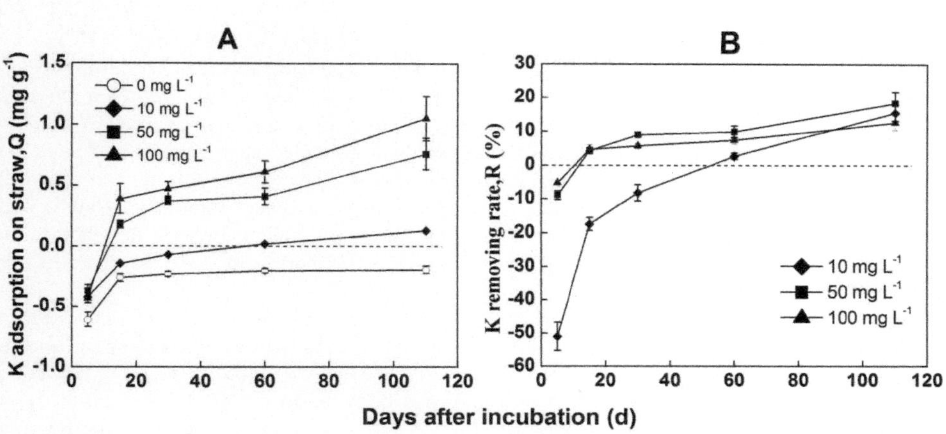

Figure 3. The adsorption of potassium by rice straw residues. The dashed line shows the zero point in the panels. The values are the means of 3 replicates (±standard deviation).

doi:10.1371/journal.pone.0090440.g003

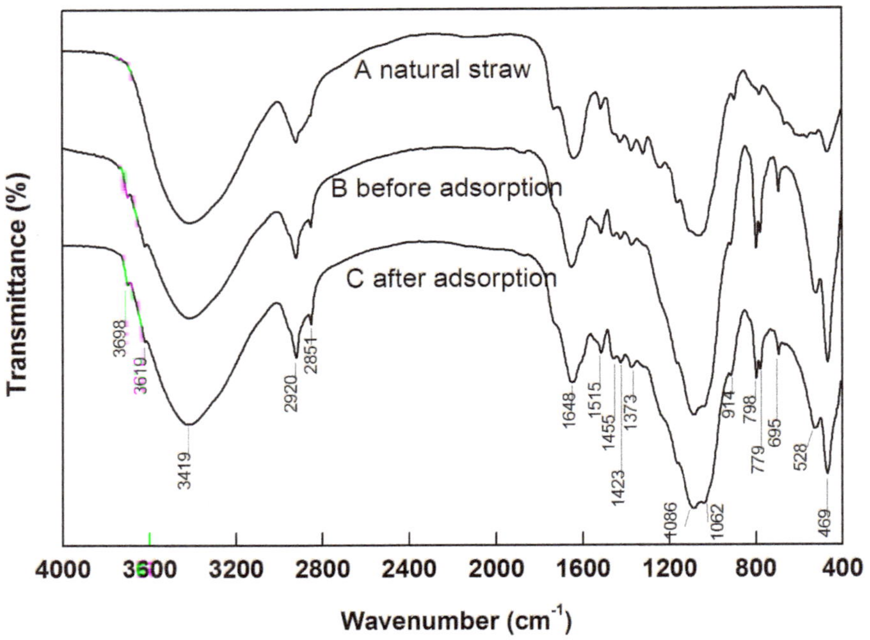

Figure 4. FT-IR spectra of natural dried rice straw and the rice straw residue at 110 d before and after potassium (K) adsorption. Solid line A represents the infrared spectrum of natural dried rice straw. Solid lines B and C represent the infrared spectra of the rice straw at 110 d before and after K adsorption in 50 mg L^{-1} KCl solution, respectively. The data in the panel are the peak wavenumbers.

Figure 4C illustrates the spectra after K adsorption for rice straw after 110 d. The main peaks of the functional groups at 4000–1300 cm^{-1} did not show significant changes. However, in the fingerprint region, the peaks at 798, 695, and 469 cm^{-1} decreased significantly compared to those before K adsorption on the residue.

Kinetics of Water Absorption on Rice Straw

The water absorption kinetic of untreated dried straw are shown in Figure 5A. The results indicated that the water storage capacity of straw reached 2.51 g g^{-1} after 5 min of immersions. From 5 to 90 min, the absorption rate continued to increase at an average rate of 0.01 g (g min) $^{-1}$. After 150 min, the rice straw could not absorb more water at the ambient temperature (23°C), approaching a saturated value of 3.88 g g^{-1}. The water retention also changed significantly during different periods of straw decomposition (Figure 5B). The results showed that the water absorption capacity of the residue continued to rise from 0 to 15 d, and the maximum water content was 5.17 g g^{-1}. Moreover, the water absorption capacity decreased gradually with the extension of straw decomposition. The water content of rice residue was the lowest at 110 d, with a value of 3.00 g g^{-1}.

Discussions

The Differences between Straw Decomposition and K Release

For field straw management, the rate of straw decomposition is influenced by many factors. These variables include internal factors, such as litter quality, material composition and structure [32,33], and external factors, such as temperature, moisture [34], methods of straw utilization (mulching or incorporation), time, length of straw [31] and the concentration of CO_2 in the soil. Litter quality and climate condition are considered to be key factors in the regulation of straw decomposition [35]. The present study was conducted from the beginning of July to the end of October. From 0 to 60 d, the soil temperature reached above 35°C, and this environment did significantly improve the decomposition rate of the straw. However, from 60 to 110 d, the soil temperature decreased to 20–24°C, and the decomposition rate consequently also decreased. After 110 d of degradation, the straw decomposition rate reached 60%. The results presented in Figure 1 illustrate that the decomposition of straw has two main periods: (1) a rapid decomposition stage from 0 to 60 d, during which the structure of the straw surface changed (Figure 6), and the zeta potential increased significantly; and (2) a slow decomposition stage from 60 to 110 d, during which the decomposition rate is significantly lower than before and the zeta potential is basically stable. The characteristic infrared peaks of cellulose were observed at 3412, 2900, 1425, 1370 and 895 cm^{-1} [30]. The characteristic peak of hemicellulose was observed at 1732 cm^{-1} [36], and the absorption peaks characteristic of lignin were seen at 1595 and 1516 cm^{-1}. All of these peaks were not changed dramatically over the study period, as seen in figure 4. Conversely, easily decomposable plant materials such as lipids, pectin, starch and carbohydrates were degraded over one season of crop growth. In addition, negligible amounts of cellulose and hemicellulose were also degraded. Lignin, along with the majority of the cellulose and hemicellulose, may require more time to be degraded by organisms [35].

The characteristics of K release are dramatically different from those of straw decomposition. As K mainly exists in ionic form in plants, it is able to move easily. After 5 d of immersion, more than 90% of potassium had been released from the straw. This result coincides with that of Rodríguez-Lizana et al. [17], who reported that the K of sunflower residue decreased by 98% over the study period while only 43% of the residue itself degraded over the same

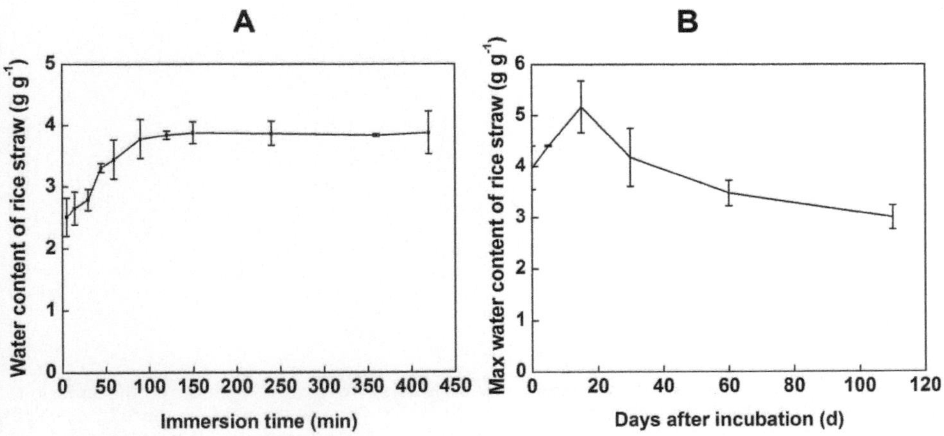

Figure 5. The kinetics of water absorption for rice straw. Panel A shows the changes of water absorption in dried rice straw. Panel B shows the changes of water absorption of straw residue for different decomposition periods. The values are the means of 3 replicates (±standard deviation).

time. Thus, the maintenance of residues under conservation tillage requires the application of large amount of K to the soil.

Mechanism of K Adsorption on Straw Residues

Currently, the ability of crop straw to serve as a bioadsorbent material has been widely studied and has received increasing attention, especially regarding the possibility of reducing the pollution of heavy metals such as Hg(II), Cd(II), and U(VI) [23,37]. The adsorption of metal ions mainly relies on physical adsorption and chemical adsorption. Usually, the latter process is dominated, as reported by Ke et al., who observed that modified rice straw coupled with metal ions to form organic compounds [38]. However, few report exist regarding the coupling interaction between heavy metals and residues during different decomposition period. In particular, alkali metal atoms (Li, Na, K) rarely form complexes with organic ligands because of their unique structures (which lack an empty orbital in the valence shell). When the active functional groups in rice straw bound Cd, nearly all the infrared bands from 3423 to 1000 cm^{-1} decreased in intensity [23]. In the residues after K adsorption, the functional group signatures at 4000 to 1300 cm^{-1} did not change (Figure 4). These results indicate that potassium chelated with C = C, C-O or carboxylic acids on straw is impossible. Moreover, greater focus has been placed in previous studies on the changes in the physical or chemical properties of soil after straw management [39], or on the conversion among diverse forms of nutrients [11]; the straw

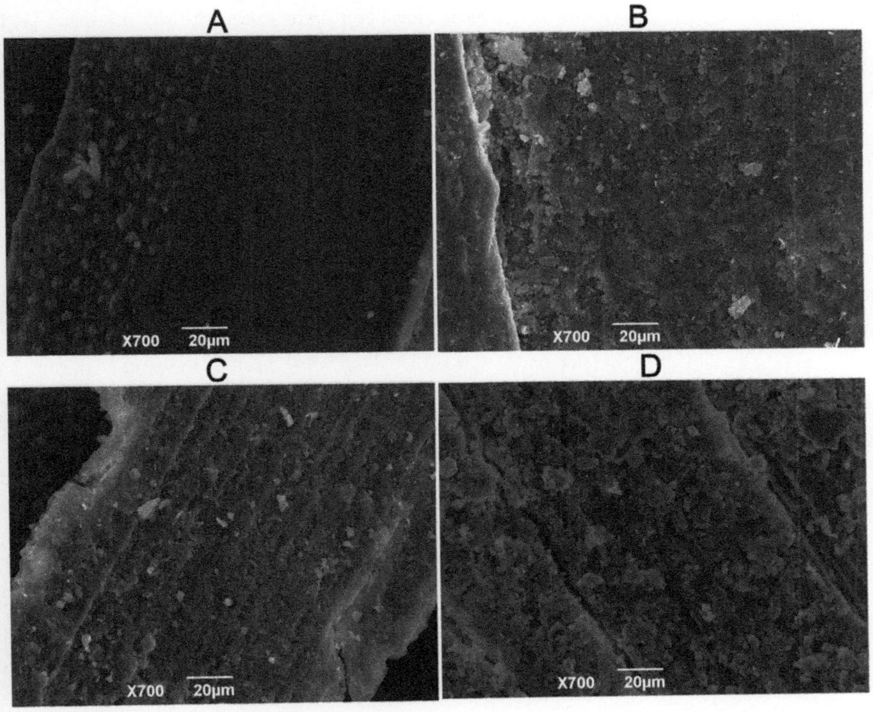

Figure 6. SEM micrographs of the rice straw residue surface on different days after incubation: A 0 d, B 15 d, C 60 d and D 110 d.

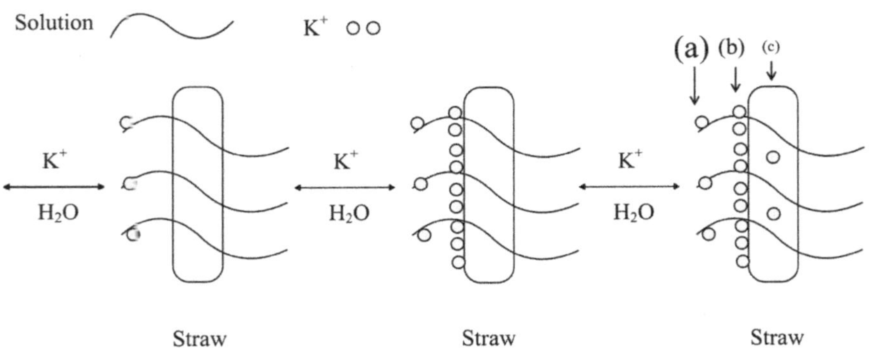

Figure 7. Schematic of the adsorption process of potassium (K^+) on the rice straw residue surface. (a) K^+ retained on the straw in the form of an aqueous solution absorbed by residue; (b) K^+ adsorbed via electrostatic interaction; and (c) K^+ fixed on the residue surface by chemically interfacial adsorption. The size of the three letters indicates the amount of K^+ reserved on the residue by the corresponding process.

residue itself has therefore often been ignored. The adequate prediction of K dynamics and bioavailability is further complicated by the limited knowledge of the reversibility of K adsorption in soil. However, it was noted in the present study that rice straw residue could adsorb potassium despite releasing most of its potassium at the beginning of the experiment. At the same time, rape (canola) and wheat straw at different points in decomposition could also adsorb a portion of the K^+ ions, indicating that the role of crop residues in potassium adsorption is widespread (Figure S2). The present study provides both direct and indirect observations demonstrating that the adsorption mechanism can be described by the following three aspects:

(1) Physically interfacial adsorption: electrostatic adsorption. Due to higher content of potassium in natural straw, the rate of K release is significantly greater than the rate of K adsorption capacity during the earlier stage of decomposition. As decomposition continues, the zeta potential of the straw surface increases. Straw residues could adsorb K^+ ions from the surrounding environment (Figure 3). The straw K adsorption capacity is a function of the decomposition period and the extra K concentration, indicating an equilibrium between the K pool adsorbed on the residue and that in the bulk solution. Residues had an especially strong adsorption capacity after long-term decomposition. Therefore, as vast majority of K fertilizer inputs to the soil are one-time events, a portion of these inputs could be replaced by accumulated debris (or organic matter). Considering the K uptake by plants when the K concentration in the soil solution is reduced, crop residues could release K to replenish the soil solution [18].

(2) Chemically interfacial adsorption. The infrared spectra of straw residues after K adsorption on the 110th day of decomposition showed decreased intensity of the peaks at 798, 695 and 458 cm^{-1} in the fingerprint region. This results suggested that binding sites existed on the residue surface that could couple with K^+ ions. For peak at 798 cm^{-1} indicates the presence of C-H stretching or Si-O stretching, and Lu et al has demonstrated that the peaks at 795 and 466 cm^{-1} represent symmetric Si-O stretching vibration and Si-O deformation vibration, respectively [40]. The peak at 695 cm^{-1} likely reflects the vibration of nitrogenous or boric compounds [41,42]. These results confirmed that the compounds containing Si, B and N on the residue surface were involved in coupling with K^+.

(3) Solution fixed on residue. A study of the water retention of the 11 main varieties of crop straw, as conducted by Iqbal et al., showed that the water retention of straw was not necessarily linked with content of its components (cellulose, hemicellulose, lignin) but was significant related with its porosity and the variation of this porosity during decomposition. These authors found that the water content of rice straw reached its maximum after 6 h of immersion [31]. According to the present research, the capacity of water absorbed on the straw reached saturation after 150 min of immersion, which was consistent with the pervious study. The water content of the decomposed residue reached its maximum of 5.17 g g^{-1} after 15 d of incubation. Due to the differences of porosity, the change of the water content of the decomposed rice residue differed from that of maize residue, which showed higher water retention as incubation proceeded [31]. On the fifth day, the rice straw displayed a regular and compact surface structure, showing no difference with the raw material [43], which likely relied on its inner surface to absorb water (Figure 6). After 15 d of incubation, pectin, starch and other biodegradable materials on the surface of straw were depredated and fell off; the resulting surface was rougher and retained more water. After 60 d, the easily decomposable materials on the straw surface were completely decomposed, and the lignin and cellulose were exposed to the surface. Due to the reduction of porosity, the water retention decreased.

Briefly, the K adsorption process can be described with the following steps (Figure 7): (1) K^+ is transferred from the solution to the rice straw surface; (2) fractional K^+ ions are transferred from the rice straw surface to active sites; and (3) the water content on the surface is maximized, and the K^+ on the residue surface reaches equilibrium with the bulk solution. Therefore, under the same conditions, the amount of K retention via water absorption in straw is greater than that from physical adsorption and chemical adsorption. The latter two methods may, however, play an additional roles in the potassium retention.

The loss of soil nutrients (especially N and K) occurs primarily through runoff and leakage [44,45]. In China and South Asia, rice-based cropping systems such as two- or three-cropping rotation modes are nearly universally deployed [3]. Because of the large population, food demand and soil nutrient depletion in these regions, potassium deficit has become a serious problem [46]. For these regions in particular, the combination of rain, hot and intensive farming increases the risk of nutrient losses and subsequent loss of soil fertility. The present results indicate that straw residues at different points in the decomposition process have different capacities for adsorbing potassium and retaining water. The accumulation of residues in the soil not only reduces water loss, mitigates soil erosion and provides nutrients [47,48], but can also maintain soil nutrients, enhance soil fertility and promote the development of modern agriculture.

Supporting Information

Figure S1 Schematic diagram of the experimental setup of the straw decomposition trial. The depth of the bulk soil in boxes was approximately 22 cm, and that of the flooding layer was 1 cm. The nylon bags were buried into the soil according to the orientation in the diagram.

Figure S2 The adsorption of potassium (K) on wheat and rape straws for different decomposition periods. The added K concentration is 50 mg L^{-1}. The annotations in the panels are the homogeneity of variances (H) and ANOVA. The H-test was performed using the Levene test. **indicates significant differences at $P<0.01$. The values are the means of 3 replicates (±standard deviation). The means with the same letter are not significantly different.

Acknowledgments

We gratefully acknowledge the reviewer and the editor for their constructive comments and suggestions. We also thank Wenjun Zhang for her comments on revising the manuscript.

Author Contributions

Conceived and designed the experiments: JWL JFL TR. Performed the experiments: JFL RHC. Analyzed the data: JFL JWL XKL. Contributed reagents/materials/analysis tools: XKL RHC LZ. Wrote the paper: JFL JWL LZ.

References

1. United Nations DoEaSA, Population Division (2013) World Population Prospects: The 2012 Revision, Highlights and Advance Tables. Working Paper No. ESA/P/WP.228.
2. Alexandratos N, Bruinsma J (2012) World agriculture towards 2030/2050: the 2012 revision. Rome: FAO. pp. ESA Working paper No.12–03.
3. Singh B, Shan YH, Johnson-Beebout SE, Singh Y, Buresh RJ (2008) Crop Residue Management for Lowland Rice-Based Cropping Systems in Asia. Advances in Agronomy 98: 117–199.
4. Stewart WM, Dibb DW, Johnston AE, Smyth TJ (2005) The contribution of commercia l fertilizer nutrients to food production. Agronomy Journal 97: 1–6.
5. Regmi AP, Ladha JK, Pasuquin E, Pathak H, Hobbs PR, et al. (2002) The role of potassium in sustaining yields in a long-term rice-wheat experiment in the Indo-Gangetic Plains of Nepal. Biology and Fertility of Soils 36: 240–247.
6. Panaullah GM, Timsina J, Saleque MA, Ishaque M, Pathan ABMBU, et al. (2006) Nutrient Uptake and Apparent Balances for Rice-Wheat Sequences. III. Potassium. Journal of Plant Nutrition 29: 173–187.
7. Wang XB, Hoogmoed WB, Cai DX, Perdok UD, Oenema O (2007) Crop residue, manure and fertilizer in dryland maize under reduced tillage in northern China: II nutrient balances and soil fertility. Nutrient Cycling in Agroecosystems 79: 17–34.
8. Sheldrick WF, Syers JK, Lingard J (2003) Soil nutrient audits for China to estimate nutrient balances and output input relationships. Agriculture Ecosystem & Environment 94: 341–354.
9. Pathak H, Mohanty S, Jain N, Bhatia A (2009) Nitrogen, phosphorus, and potassium budgets in Indian agriculture. Nutrient Cycling in Agroecosystems 86: 287–299.
10. Evenett SJ, Frédéric J (2012) Trade, competition, and the pricing of commodities. London: Centre for Economic Policy Research.
11. Li XK, Lu JW, Wu LS, Chen F (2009) The difference of potassium dynamics between yellowish red soil and yellow cinnamon soil under rapeseed (Brassica napus L.)-rice (Oryza sativa L.) rotation. Plant and Soil 320: 141–151.
12. Tejada M, Hernandez M, Garcia C (2009) Soil restoration using composted plant residues: Effects on soil properties. Soil and Tillage Research 102: 109–117.
13. Lal R (2005) World crop residues production and implications of its use as a biofuel. Environ Int 31: 575–584.
14. Kaur N, Benipal DS (2006) Effect of crop residue and farmyard manure on K forms on soils of long term fertility experiment. Indian Journal Crop Science 1: 161–164.
15. Liao YL, Zheng SX, Nie J, Xie J, Lu YH, et al. (2013) Long-Term Effect of Fertilizer and Rice Straw on Mineral Composition and Potassium Adsorption in a Reddish Paddy Soil. Journal of Integrative Agriculture 12: 694–710.
16. Jordan CF (1985) Nutrient cycling in tropical forest ecosystems. Principles and their application in management and conservation. New York: Wiley.
17. Rodríguez-Lizana A, Carbonell R, González P, Ordóñez R (2010) N, P and K released by the field decomposition of residues of a pea-wheat-sunflower rotation. Nutr Cycl Agroecosyst 87: 199–208.
18. Duong TTT, Verma SL, Penfold C, Marschner P (2013) Nutrient release from composts into the surrounding soil. Geoderma 195–196: 42–47.
19. Lin CW, Luo CY, Pang LY, Huang JJ, Tu SH (2011) Effect of Different Fertilization Methods and Rain Intensities on Soil Nutrient Loss from a Purple Soil. Scientia Agricultura Sinica 44: 1847–1854.
20. Kozak JA, Ahuja LR, Green TR, Ma LW (2007) Modelling crop canopy and residue rainfall interception effects on soil hydrological components for semi-arid agriculture. Hydrological Processes 21: 229–241.
21. Rawls WJ, Pachepsky YA, Ritchie JC, Sobecki TM, Bloodworth H (2003) Effect of soil organic carbon on soil water retention. Geoderma 116: 61–76.
22. Tong XJ, Li JY, Yuan JH, Xu RK (2011) Adsorption of Cu(II) by biochars generated from three crop straws. Chemical Engineering Journal 172: 828–834.
23. Ding Y, Jing DB, Gong HL, Zhou LB, Yang XS (2012) Biosorption of aquatic cadmium(II) by unmodified rice straw. Bioresource Technology 114: 20–25.
24. Vitousek PM (1994) Beyond Global Warming: Ecology and Global Change. Ecology 75: 1861–1876.
25. Walkley A (1947) A critical examination of a rapid method for determining organic carbon in soils: Effect of variations in digestion conditions and of inorganic soil constituents. Soil Science 63: 251–264.
26. Olsen SR, Cole CV, Watanabe FS, Dean LA (1954) Estimation of available phosphorus in soils by extraction with sodium bicarbonate. Washington, DC: USDA Circ. 939. U.S. Gov. Print. Office. 1–19.
27. van Reeuwijk LP (1992) Procedures for soil analysis 3rd edn. Wageningen, the Netherlands: ISRIC.
28. Daudu CK, Muchaonyerwa P, Mnkeni PNS (2009) Litterbag decomposition of genetically modified maize residues and their constituent Bacillus thuringiensis protein (Cry1Ab) under field conditions in the central region of the Eastern Cape, South Africa. Agriculture, Ecosystems & Environment 134: 153–158.
29. Honary S, Zahir F (2013) Effect of Zeta Potential on the Properties of Nano-Drug Delivery Systems -A Review (Part 1). Tropical Journal of Pharmaceutical Research 12: 255–264.
30. Cao W, Dang Z, Yi XY, Yang C, Lu GN, et al. (2013) Removal of chromium (VI) from electroplating wastewater using an anion exchanger derived from rice straw. Environmental Technology 34: 7–14.
31. Iqbal A, Beaugrand J, Garnier P, Recous S (2013) Tissue density determines the water storage characteristics of crop residues. Plant and Soil 367: 285–299.
32. Thomas MB, Spurway MI, Stewart DPC (1998) A review of factors influencing organic matter decomposition and nitrogen immobilisation in container media. In The International Plant Propagators' Society Combined Proceedings 48: 66–71.
33. Zhu JX, Yang WQ, He XH (2013) Temporal dynamics of abiotic and biotic factors on leaf litter of three plant species in relation to decomposition rate along a subalpine elevation gradient. PLoS ONE 8(4): e62073.
34. Pal D, Broadbent FE (1975) Influence of moisture on rice straw decomposition in soils. Soil Science Society of America Journal 39: 59–63.
35. Wang X, Sun B, Mao J, Sui Y, Cao X (2012) Structural Convergence of Maize and Wheat Straw during Two-Year Decomposition under Different Climate Conditions. Environmental Science & Technology 46: 7159–7165.
36. Chen XL, Yu J, Zhang ZB, Lu CH (2011) Study on structure and thermal stability properties of cellulose fibers from rice straw. Carbohydrate Polymers 85: 245–250.
37. Rocha CG, Zaia DAM, Alfaya RVS, Alfaya AAS (2009) Use of rice straw as biosorbent for removal of Cu(II), Zn(II), Cd(II) and Hg(II) ions in industrial effluents. J Hazard Mater 166: 383–388.
38. Ke X, Zhang Y, Li PJ, Li RD (2009) Study on mechanism of chestnut inner shells removal of heavy metals from acidic solutions. Journal of Shenzhen University (Science & Engineering) 26: 72–76 (In Chinese).
39. Karami A, Homaee M, Afzalinia S, Ruhipour H, Basirat S (2012) Organic resource management: Impacts on soil aggregate stability and other soil physico-chemical properties. Agriculture, Ecosystems & Environment 148: 22–28.
40. Lu P, Hsieh YL (2012) Highly pure amorphous silica nano-disks from rice straw. Powder Technology 225: 149–155.
41. Liao LB, Wang LJ, J.W Y, Fang QF (2010) Mineral material of modern testing technology. Beijing: Chemical industry press (In Chinese).
42. Peter L (2011) Infrared and raman spectroscopy: principles and spectral interpretation. Elsevier.

43. Yu G, Yano S, Inoue H, Inoue S, Endo T, et al. (2010) Pretreatment of Rice Straw by a Hot-Compressed Water Process for Enzymatic Hydrolysis. Applied Biochemistry and Biotechnology 160: 539–551.
44. Sims JT, Goggin N, McDermott J (1999) Nutrient management for water quality protection: integrating research into environmental policy. Water Science and Technology 39: 291–298.
45. Lai FY, Yu G, Gui F (2006) Preliminary study on assessment of nutrient transport in the Taihu basin based on SWAT modeling. Science in China: Series D Earth Sciences 49: 135–145.
46. Cakmak I (2002) Plant nutrition research: priorities to meet human needs for food in sustainable ways. Plant Soil 247: 3–24.
47. Yang SM, Malhi SS, Li FM, Suo DR, Xu MG, et al. (2007) Long-term effects of manure and fertilization on soil organic matter and quality parameters of a calcareous soil in NW China. Journal of Plant Nutrition and Soil Science 170: 234–243.
48. Liu Y, Gao MS, Wu W, Tanveer SK, Wen XX, et al. (2013) The effects of conservation tillage practices on the soil water-holding capacity of a non-irrigated apple orchard in the Loess Plateau, China. Soil and Tillage Research 130: 7–12.

13

Representation of Dormant and Active Microbial Dynamics for Ecosystem Modeling

Gangsheng Wang[1,2]*, Melanie A. Mayes[1,2], Lianhong Gu[1,2], Christopher W. Schadt[1,3]

1 Climate Change Science Institute, Oak Ridge National Laboratory, Oak Ridge, Tennessee, United States of America, 2 Environmental Sciences Division, Oak Ridge National Laboratory, Oak Ridge, Tennessee, United States of America, 3 Biosciences Division, Oak Ridge National Laboratory, Oak Ridge, Tennessee, United States of America

Abstract

Dormancy is an essential strategy for microorganisms to cope with environmental stress. However, global ecosystem models typically ignore microbial dormancy, resulting in notable model uncertainties. To facilitate the consideration of dormancy in these large-scale models, we propose a new microbial physiology component that works for a wide range of substrate availabilities. This new model is based on microbial physiological states and the major parameters are the maximum specific growth and maintenance rates of active microbes and the ratio of dormant to active maintenance rates. A major improvement of our model over extant models is that it can explain the low active microbial fractions commonly observed in undisturbed soils. Our new model shows that the exponentially-increasing respiration from substrate-induced respiration experiments can only be used to determine the maximum specific growth rate and initial active microbial biomass, while the respiration data representing both exponentially-increasing and non-exponentially-increasing phases can robustly determine a range of key parameters including the initial total live biomass, initial active fraction, the maximum specific growth and maintenance rates, and the half-saturation constant. Our new model can be incorporated into existing ecosystem models to account for dormancy in microbially-driven processes and to provide improved estimates of microbial activities.

Editor: Jonathan H. Badger, J. Craig Venter Institute, United States of America

Funding: This research was funded by the Laboratory Directed Research and Development (LDRD) Program of the Oak Ridge National Laboratory (ORNL) and by the U.S. Department of Energy Biological and Environmental Research (BER) program. ORNL is managed by UT-Battelle, LLC, for the U.S. Department of Energy under contract DE-AC05-00OR22725. The funders had no role in study design, data collection and analysis, decision to publish, or preparation of the manuscript.

Competing Interests: The authors have declared that no competing interests exist.

* E-mail: wangg@ornl.gov

Introduction

Ecologically-important processes such as soil organic carbon and nutrient cycling largely depend on the active fraction of microbial communities [1]. At any given time in a given environment, microorganisms can be in active, dormant, or dead states [2]. Dormancy is considered an evolutionary strategy designed to maintain the genetic code until conditions improve to allow replication [3]. When environmental conditions are unfavorable for growth, e.g., resource limitation, microbes may enter a reversible state of low to zero metabolic activity to alleviate the loss of biomass and metabolic functions [4,5]. The maintenance coefficient (i.e., maintenance cost of C per unit microbial biomass C per unit time) can be two to three orders of magnitude lower in dormant microbes than in metabolically active microbes [6,7]. Many soils have slow organic matter turnover rates with seasonal changes in substrate supply, temperature, and moisture. The complexity of soils in space and time may result in uneven distributions of multiple potentially limiting resources, leading to significant rates of dormancy even when some resources are abundant. When spatial and temporal complexity is combined with differential resource partitioning among species in a community, high rates of dormancy could be a prominent feature in soil systems. Thus it is essential to understand dormancy in order to more accurately predict how active microorganisms contribute to ecosystem processes such as decomposition and nutrient turnover [1].

A complicating factor in studying microbial dormancy is that no single approach can be easily employe to simultaneously measure individual microbial states (active, dormant or dead), and a combination of different techniques is required. Differential staining is often used to segregate physiological states with direct microscopic counting of bacteria and fungi. 'Life-indicating' stains that require the presence of 'standard' physiological abilities, such as the esterase activity needed for fluorescein diacetate cleavage, may distinguish active from dormant+dead cells [8]. When combined with general-purpose stains, these strains can distinguish dormant cells by difference [9]. Combining membrane-permeant with membrane-impermeant nucleophilic stains (e.g., SYTO-9 and propidium iodide respectively) may distinguish live from dead, but not active from dormant [10,11]. Active microbes may or may not be 'viable' with common culture-based techniques, which complicates classification and measurement of dormancy phenomena [5]. Methods such as direct plating, serial dilution and most probable number (MPN) techniques will not distinguish between active and dormant organisms [12]. Substrate Induced Respiration (SIR) or Substrate Induced Growth Response (SIGR) method [13,14] can distinguish active and dormant communities if growth respiration curves are modeled (using initial exponentially-increasing respiration); however, the technique often needs to be combined with microscopy or chloroform fumigation/extraction in order to obtain total live microbial biomass [15,16].

Despite limitations in distinguishing active, dormant and dead microbial biomass, abundant evidence indicates that the majority of environmental microorganisms in a given community may be dormant under natural conditions [1,17]. Alvarez *et al.* [18] reported that only 3.8–9.7% of the total biomass is active in a Typic Argiudoll soil from the Argentinean Pampa. Khomutova *et al.* [19] showed that the fraction of active microbial biomass ranged from 0.02% to 19.1% in the subkurgan paleosoils of different age and 9.2–24.2% in modern background soils. Microbial biomass measured through SIR or SIGR is thought to reflect only the active portion because the maintenance respiration of dormancy biomass is negligible in the initial exponentially-increasing phase [13,16,20]. Through a mathematical analysis of respiration curves, Van de Werf & Verstraete [21] examined 16 soils and found that 4–49% of the total biomass was in an active state; and the active component in undisturbed natural ecosystems (18.8±8.8%, mean±standard deviation) was about 70% of that in arable agricultural soils (25.7±14.8%). Stenström *et al.* [22] showed that the fraction of active biomass typically varied from 5% to 20% in soils without recent addition of substrates. Lennon & Jones [5] found much lower active fractions in soils (18±15%) than in marine (65±19%) and fresh (54±11%) water environments. From the above studies it seems conservative to extrapolate that the active fraction is very likely below 50% of live microbes under most natural soil conditions.

Microbially-mediated processes have been incorporated into ecosystem models [23–28] although continued development is still required to bring microbial processes into global climate models [29–31]. However, these recent models do not consider physiological state changes and assume that measures of microbial biomass constitute the active biomass. The exclusion of dormancy from the microbially-driven ecosystem processes could result in incorrect estimates of total live microbial biomass, which further leads to deficiencies in model parameterization and predictions of soil organic carbon and nutrients.

Generally, there are two strategies to represent physiological states in microbial-ecology models: one is to explicitly separate the total live biomass into two pools, i.e., active and dormant [4,32]; the other is to directly regard the active fraction (i.e., ratio of active biomass to total live biomass) as a state variable [33,34]. Both of these two approaches predict the total live biomass, active and dormant biomass, and the flux between the active and dormant components. Apparently the introduction of the 'active fraction' as a state variable in the latter approach simplifies the model structure since the adaptive variation of microbial composition might be represented by one single variable (active fraction) [34,35]. However, another state variable indicating the microbial biomass pool size (e.g., total live biomass, active biomass or dormant biomass) is still essential for ecosystem modeling since the carbon and nutrient fluxes are pool-size dependent. For example, if we define active fraction and total microbial biomass as state variables, the active and dormant biomass could be determined by them, and the net flux between active and dormant fractions and other related fluxes could also be computed according to the active and dormant biomass constrained by mass balance. The above-mentioned modeling efforts have shown that adequate representation of dormancy and the transitions between the dormant and active states is crucial for modeling important microbially-mediated ecosystem processes.

Here, we review state-of-the-art microbial dormancy modeling approaches and discuss the rationales of these models with a focus on transformation processes between active and dormant states. We propose an improved synthetic microbial physiology model based on accepted assumptions and examine the model behavior with theoretical and experimental analyses. In this paper, the 'total microbial biomass' refers to the 'total live microbial biomass' unless otherwise stated. Our objective is to clarify the applicability of existing microbial dormancy models and provide a new theoretical basis for representing microbial activity and dormancy in ecosystem models.

Dormancy In Microbial Models

Transformation between active and dormant states

Although Buerger *et al.* [36] argued that dormant microbial cells could reactivate stochastically and might be independent of environmental cues, environmental factors such as substrate availability are often thought to control the transformation between active and dormant states [5]. Most models (see Appendix S1) distinguish the active biomass pool from the dormant pool and define them as two state variables (B_a and B_d) (Fig. 1). Only active microbes (B_a) can uptake substrate and produce new cells. The connection between the active and dormant states is a reversible process including two directional sub-processes, i.e., dormancy (from active to dormant) and reactivation (or resuscitation, from dormant to active). Losses from active biomass include growth respiration and maintenance (maintenance respiration, mortality, enzyme synthesis, etc.) [23]. Dormant microbes still require energy for maintenance and survival although at a lower metabolic rate [5].

The net transformation rate ($B^N_{a\to d}$) from active to dormant state is the difference between the flux from active to dormant ($B_{a\to d}$) and the flux from dormant to active state ($B_{d\to a}$), i.e., $B^N_{a\to d}=B_{a\to d}-B_{d\to a}$. The models of Hunt [37] (Equation S1-1 in Appendix S1) and Gignoux *et al.* [38] (Equation S1-2 in Appendix S1) directly formulate the net flux ($B^N_{a\to d}$) without explicit components for $B_{a\to d}$ and $B_{d\to a}$. The direction of the net flux depends on the maintenance requirement relative to the substrate availability. If the available substrate is less than the maintenance requirement, there is a positive net flux from active to dormant pool, and vice versa. In addition, Hunt [37] assumed a 'buffer zone' for the change of states: when the maintenance requirement surpasses the substrate supply but the deficit is within a small fraction (1% d^{-1}) of B_a, there is no flux between the two states.

Some models define rates for both dormancy and reactivation. In the model of Ayati [39] (Equation S1-3 in Appendix S1), the

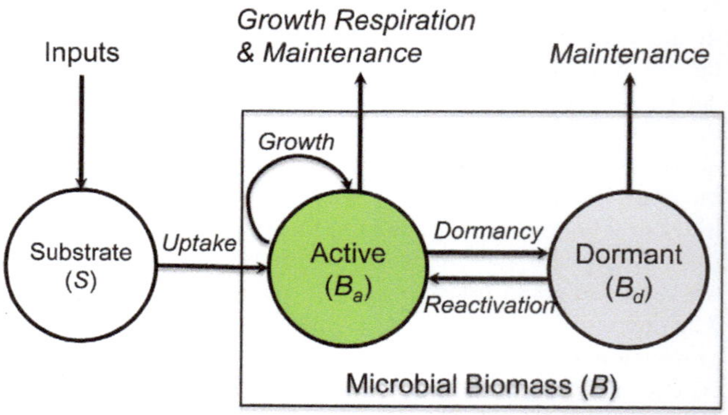

Figure 1. Active and dormant microbial biomass pools in microbial physiology models (modified from Fig. 2 in Lennon & Jones, 2011).

dormant rate ($\gamma_{a\to d}$) increases with declining substrate concentration, and the reactivation ($\gamma_{d\to a}$) only occurs when substrate concentration is higher than the half-saturation constant (K_s). Konopka [32] modified the potential rates for dormancy and reactivation by the relative growth rate ($\mu/\mu_{\max}$, ratio of true specific growth rate to maximum specific growth rate), i.e, the two rates are multiplied by (1-$\mu/\mu_{\max}$) and $\mu/\mu_{\max}$, respectively (Equation S1-4 in Appendix S1). Similarly, Jones & Lennon [40] postulated two complementary rates (1-*R* and *R*) for dormancy and resuscitation (Equation S1-5 in Appendix S1).

Two other models also explicitly formulate the two conversion rates between states but do so using concepts of probability. Bär *et al.* [41] used two complementary factors (1-*J* and *J*) to represent the probability for the transition between active and dormant state in addition to an identical potential rate constant for the two processes (Equation S1-6 in Appendix S1). The conceptual model of Locey [42] applies a deterministic dormant rate and a stochastic resuscitation rate (Equation S1-7 in Appendix S1). The potential resuscitation rate is modified by (1-*p*), where *p* is the probability that a disturbance in the active pool will result in the immigration of one individual from the metacommunity. The probability (*J*) in Bär *et al.* [41] is explicitly calculated from the environmental cues (e.g., soil moisture), while the cause of the probability (*p*) in Locey [42] is not elucidated.

Switch function model

In addition to the dormancy and reactivation processes, a key concept in the model developed by Stolpovsky *et al.* [4] is 'switch function' (Equation S1-8 in Appendix S1). The switch function (θ) determines the fraction of active cells taking up dissolved organic carbon (DOC). This function refers to the growth fraction in active biomass (B_a) that consumes substrate and thus is not the same as the active fraction (r) in total biomass (B). Furthermore, the dormancy and reactivation fluxes are set to be proportional to (1-θ) and θ, respectively. θ is formulated by the Fermi-Dirac statistics [4]. Another feature of this model is the consideration of 'depth' of dormancy in reactivation, where the reactivation rate is negatively dependent on the duration of dormancy. The switch function model includes at least 15 model parameters and it is difficult to compute the Gibbs energy change of the oxidation of DOC [4].

The switch function (θ) sets it apart from the conventional Michaelis-Menten (M-M) or Monod kinetics because of its new perspective of thermodynamics. According to the M-M kinetics [43], the substrate saturation level represents the fraction of enzyme-substrate complex (*ES*) in active enzyme (E_0), where the substrate saturation level is formulated by $S/(K_s + S)$ with S and K_s being the substrate concentration and the half-saturation constant [43]. When the M-M kinetics is applied to describe microbial uptake of substrate, the substrate (or combined with TEA) saturation level (i.e., $\mu(S, TEA)$ in Equation S1-8a of Appendix S1) is a measure of the actively growing fraction in the active microbial community. The switch function is also determined by the saturation levels of substrate and terminal electron acceptor (TEA), i.e., $\mu(S, TEA)$ (Equation S1-8e in Appendix S1). Mathematically the inclusion of both the switch function (θ) and the M-M kinetics (i.e., $\mu(S, TEA)$) might result in double counting of the impact of substrate and TEA. We would recommend using the switch function (θ) to modify the microbial uptake rate if the Gibbs energy change of the oxidation of substrate (ΔG) is tractable and the thermodynamic threshold (G_0) and the steepness of the step function (*st*) are identifiable.

Physiological state index models

As an alternative to models with two microbial biomass pools (i.e., active and dormant), a further state variable indicating the dormant or active fraction in total biomass has been proposed. Wirtz [44] developed a simple index ($r_d = 0.5$–1.0) representing the dormant microbial biomass as a fraction of the steady-state total biomass (B_{stat}) under the condition of $B_d << B_a$. In case of a net loss of total biomass ($dB/dt<0$), the dormant biomass $B_d = B_{stat}\cdot r_d$; otherwise ($dB/dt>0$), $B_d = B_{stat}\cdot(1-r_d)$. This model has a sudden change of dormant biomass at the transition point (i.e., $dB/dt = 0$) since $r_d>0.5$.

Different from the dormant index of Wirtz [44], the concept of an active index (i.e., index of physiological state) of soil microbial community has been employed in soil carbon and nutrient cycling models [33,34]. The index of physiological state (r), referring to the activity state, is often defined as the ratio of metabolically active microbial biomass to the total soil microbial biomass [22,33,34].

In the Synthetic Chemostat Model (SCM), the rate of change of the state variable r is described as follows [33,45]:

$$\frac{dr}{dt} = \frac{1}{B}\frac{dB}{dt}\cdot(\phi - r) = \mu\cdot(\phi - r) \tag{1}$$

with

$$\phi = \phi(S) = S^n/(K_r + S^n), \text{or} \phi = S/(K_r + S) \tag{2}$$

where $r = B_a/B$, representing the fraction (hereinafter referred to as 'active fraction') of active biomass in total biomass; μ is the specific growth rate of total biomass; ϕ denotes the saturation level of substrate (S); the simple power ($n = 1$) has been widely used [35] and, in this case ($n = 1$), K_r is called the half-saturation constant.

Blagodatsky & Richter [34] used the expression $\mu(S) = \mu_{\max}\cdot\phi(S)$ in their model development. This expression was not derived in the original definition of the specific growth rate (see Equation 3) by Panikov [45] and because its validity cannot be inferred, the concepts will not be addressed here.

According to Panikov's derivation [45], the specific growth rate (μ) follows the general definition [46,47]:

$$\mu = \frac{1}{B}\frac{dB}{dt} \tag{3}$$

Based on Equations 1 and 3, we can derive (see Equation S2-1 in Appendix S2):

$$dB_a/dt = \phi\cdot(dB/dt) \tag{4}$$

$$dB_d/dt = (1-\phi)\cdot(dB/dt) \tag{5}$$

We find that the model described by Equation 1 is not applicable under low substrate availability, as described below. Generally, the rates of change in biomass pools (B, B_a, and B_d) can be expressed as

$$dB/dt = g^{\pm}(S,B_a) - f^{+}(S,B_d) \tag{6}$$

$$dB_a/dt = g^{\pm}(S,B_a) - B^N_{a\to d} \quad (7)$$

$$dB_d/dt = -f^{+}(S,B_d) + B^N_{a\to d} \quad (8)$$

where $B^N_{a\to d}$ denotes the net dormancy flux; $g^{\pm}(S, B_a)$ is a function that represents the difference between the substrate uptake and the maintenance requirements of B_a, i.e., the net growth of B_a; and $f^+(S, B_a)$ is a function denoting the maintenance and survival energy costs of B_d. The superscript '±' in $g^{\pm}$ indicates the function value of g could be positive at high S or non-positive when the substrate uptake cannot satisfy the maintenance requirements of B_a at low S. The superscript '+' in f^+ implies $f \geq 0$. Note that the function $f^+(S, B_a)$ is not necessarily dependent on S [4].

From Equations 4, 6 and 7, we can obtain

$$B^N_{a\to d} = (1-\phi)\cdot g^{\pm}(S,B_a) - \phi\cdot f^{+}(S,B_d) \quad (9)$$

The two terms in the right side of Equation 9 may be regarded as the conversion of B_a to B_d (i.e., $B_{a\to d}$) and the transformation of B_d to B_a (i.e., $B_{d\to a}$), respectively. At high S resulting in $g \geq 0$, Equation 9 may be one of the possible expressions for $B_{a\to d}$ and $B_{d\to a}$. However, at low S leading to $g<0$ and $B_{a\to d}<0$, i.e., no active cells become dormant under insufficient substrate, which is inconsistent with the strategy of dormancy for microorganisms when faced with unfavorable environmental conditions [5].

Based on the above analysis, we conclude that the physiological state index model (Equation 1) needs to be improved. In other words, the empirical assumption that the steady state active fraction (r^{ss}) approaches the substrate saturation level (ϕ^{ss}) may not be necessary because this assumption could lead to impractical flux (Equation 9) between dormant and active states under low substrate availability.

A Synthetic Microbial Physiology Model

Based on the aforementioned review and analysis, we have developed a synthetic microbial physiology model component relating to substrate availability. As indicated by Fig. 1, the growth and maintenance functions of active microbes (B_a) are characterized by the maximum specific growth rate (μ_G) and maintenance rate (m_R); whereas the dormant microbes (B_d) cost energy to maintain their basic cellular functions at a much lower specific maintenance rate (denoted by $\beta\cdot m_R$, where $\beta<1$) [48].

General assumptions

First we define the substrate saturation level (ϕ) as

$$\phi = S/(K_S + S) \quad (10)$$

where the parameter K_s is the half saturation constant for substrate uptake as indicated by the M-M kinetics [43].

Based on the above review of existing dormancy models, the following assumptions are accepted in our new model: (1) the dormancy rate is proportional to the active biomass and the reactivation rate is proportional to the dormant biomass, i.e., $B_{a\to d} \propto B_a$ and $B_{d\to a} \propto B_d$; (2) under very high substrate concentration ($S>>K_s$), $\phi\to 1$, $B_{a\to d}\to 0$ and $B_{d\to a} \geq 0$; (3) under very low substrate ($S<<K_s$), $\phi\to 0$, $B_{a\to d} \geq 0$ and $B_{d\to a}\to 0$; (4) based on the assumptions (1–3), we derive that $B_{a\to d} \propto (1-\phi)\cdot B_a$ and $B_{d\to a} \propto \phi\cdot B_a$; (5) further we assume that the maximum specific maintenance rate for active microbes (m_R with units of h^{-1}) controls both transformation processes since the maintenance energy cost is the key factor regulating the dormancy strategy [5,37,38]. As a result we postulate that

$$B_{a\to d} = (1-\phi)\cdot m_R\cdot B_a \quad (11a)$$

$$B_{d\to a} = \phi\cdot m_R\cdot B_d \quad (11b)$$

Model Description

Equations 11a and 11b only describe the transformation between the active and dormant states. They need to be linked to a microbial growth and maintenance model for depicting microbial dynamics. Our recent work to develop the Microbial-ENzyme-mediated Decomposition (MEND) model [23] suggested that it might be adapted to serve this purpose due to its focus on microbial processes for which we have developed a firm theoretical basis [47,49]. Combining Equations 11a and 11b with the MEND model [23,47], we express the microbial physiology component (see Fig. 1) as a group of differential equations

$$dS/dt = I_s - \frac{1}{Y_G}\cdot\frac{\phi}{\alpha} m_R\cdot B_a \quad (12a)$$

$$dB/dt = d(B_a + B_d)/dt = (\phi/\alpha - 1)\cdot m_R\cdot B_a - (\beta\cdot m_R)\cdot B_d \quad (12b)$$

$$dB_a/dt = (\phi/\alpha - 1)\cdot m_R\cdot B_a - (1-\phi)\cdot m_R\cdot B_a + \phi\cdot m_R\cdot B_d \quad (12c)$$

$$dB_d/dt = -(\beta\cdot m_R)\cdot B_d + (1-\phi)\cdot m_R\cdot B_a - \phi\cdot m_R\cdot B_d \quad (12d)$$

where t is the time scale; ϕ is defined by Equation 10; I_s is the input to substrate pool; Y_G is the true growth yield; m_R denotes the specific maintenance rate at active state (h^{-1}); $\alpha = m_R/(\mu_G + m_R)$ is the ratio of m_R to the sum of maximum specific growth rate (μ_G) and m_R, $\alpha \in (0,0.5)$ since usually $m_R \leq \mu_G$; and β (0–1) is the ratio of dormant maintenance rate to active maintenance rate, i.e., (β m_R) denotes the maximum specific maintenance rate at dormant state.

In summary, there are five parameters (α, β, m_R or μ_G, Y_G, K_s) in the proposed model (hereinafter referred to as the MEND model). From Equations 12b and 12c, we can derive the change rate of active fraction (r) (see Equation S2-2 in Appendix S2)

$$dr/dt = m_R\cdot[(\phi - r) + (\phi/\alpha + \beta - 1)\cdot r\cdot(1-r)] \quad (12e)$$

This equation for r is more complicated than Equation 1 but still practical, given currently available data. Additionally, it implies that r needs not approach ϕ at steady state in our model, whereas $r \equiv \phi$ at steady state is required by the model of Panikov [45].

Steady state analysis

Assuming the input (I_s) is time-invariant, we can obtain the steady state solution to the above new MEND model (see Equations S2-3(a–e) in Appendix S2). Fig. 2 shows the steady

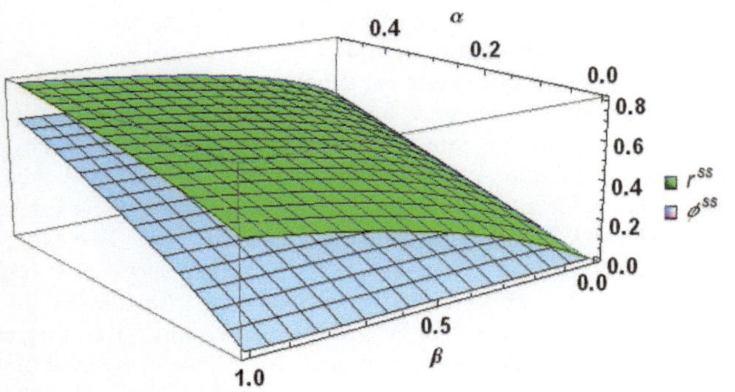

Figure 2. Steady state active fraction (r^{ss}) and substrate saturation level (ϕ^{ss}) as a function of α and β; $\alpha = m_R/(\mu_G+m_R)$, μ_G and m_R (h^{-1}) are maximum specific growth rate and specific maintenance rate for active microbial biomass, respectivly; β denotes the ratio of dormant specific maintenance rate to m_R.

state active fraction (r^{ss}) and substrate saturation level (ϕ^{ss}) as a function of the two physiological indices, i.e., α (0–0.5) and β (0–1). Both r^{ss} and ϕ^{ss} positively depend on α and β and $r^{ss} \geq \phi^{ss}$ for any combinations of α and β. If we consider two extreme values of $\beta \to 0$ or $\beta \to 1$, the r^{ss} and ϕ^{ss} (see Equations S2-4 and S2-5 in Appendix S2) can be simplified to

$$r^{ss}_{\beta\to 0} = \phi^{ss}_{\beta\to 0} = \alpha \tag{13a}$$

$$\begin{cases} r^{ss}_{\beta\to 1} = \frac{1+\sqrt{1+8\alpha}}{4} \\ \phi^{ss}_{\beta\to 1} = \frac{4\alpha-1+\sqrt{1+8\alpha}}{3+\sqrt{1+8\alpha}} \end{cases} \tag{13b}$$

Equation 13 and Fig. 2 indicate that: (1) the steady state active fraction (r^{ss}) is equal to ϕ^{ss} and they are identical to $\alpha = m_R/(\mu_G+m_R)$ only under the condition of $\beta \to 0$; (2) the upper bound of r^{ss} is approximately 0.8 at $\alpha \to 0.5$ and $\beta \to 1$; and (3) with $\alpha \leq 0.5$, the maximum r^{ss} is ca. 0.5 if the magnitude of β is around 0.001–0.01 [6]. This threshold value (0.5) of r^{ss} is a reasonable estimate that can explain how the measured active fraction of microbes in undisturbed soils is usually considerably less than the total biomass [5,21,22].

Model simplification under sufficient substrate condition

As mentioned in the Introduction, SIR or SIGR method can distinguish active from dormant composition and the data from these experiments have been widely used to estimate the active microbial biomass and the maximum specific growth rate [13,14]. The simplification of the microbial model under excess substrate has also been employed to estimate maximum specific growth rate (μ_G), active microbial biomass (B_a), and/or total microbial biomass (B) using the SIR or SIGR data [1,13,35]. Here we show the simplification of our model (Equation 12) for conditions appropriate to SIGR or SIR experiments, e.g., the short-term period of exponentially-increasing respiration of active biomass following substrate addition. We will test our reduced and full model with the SIGR data of Colores *et al.* [13] in the next section.

Under sufficient substrate (i.e., $S >> K_s$ in Equation 10 thus $\phi \to 1$), Equations 12(a–e) can be simplified and integrated for initial conditions, i.e., $S = S_0$, $B = B_0$ and $r = r_0$ at $t = 0$ (see Equations S2-6 and S2-7 in Appendix S2):

$$S(t) = S_0 - \frac{B(t) - B_0}{Y_G(1-\alpha)} \tag{14a}$$

$$B(t) = B_0 r_0 \cdot e^{(1/\alpha - 1)m_R t} + B_0(1-r_0)\cdot[\alpha \cdot e^{(1/\alpha-1)m_R t} + (1-\alpha)\cdot e^{-m_R t}] \tag{14b}$$

$$r(t) = \frac{[r_0 + \alpha(1-r_0)]\cdot e^{(m_R/\alpha)\cdot t} - \alpha(1-r_0)}{[r_0 + \alpha(1-r_0)]\cdot e^{(m_R/\alpha)\cdot t} + (1-\alpha)(1-r_0)} \tag{14c}$$

The CO_2 production rate, $v(t)$, during the exponential growth stage is derived as an explicit function of t (see Equation S2-7d in Appendix S2):

$$v(t) = \frac{dCO_2}{dt} = \frac{B_0(1-Y_G)}{Y_G}\left\{[(m_R/\alpha)\cdot r_0 + m_R\cdot(1-r_0)]\cdot e^{(1/\alpha-1)m_R t} - [m_R\cdot(1-r_0)]\cdot e^{-m_R t}\right\} \tag{14d}$$

The respiration rate, $v(t)$, is associated with two exponential items, i.e., $e^{(1/\alpha-1)m_R t} = e^{\mu_G t}$and $e^{-m_R t}$. Considering an extreme case that $m_R << \mu_G$ (i.e., $\alpha \to 0$), Equations 14(b–d) can be further simplified to Equations S2-8(b–d) (see Appendix S2). Equations S2-8b and S2-8c (denoting $B(t)$ and $r(t)$, respectively) are similar to Equations 11 and 10 in Panikov & Sizova [35], respectively. However, Equation S2-8d (for $v(t)$) is different from Equation 13 of Panikov & Sizova [35], where a constant '*A*' was added to the exponential. Equation S2-8d is identical to Equation 7 derived for SIGR experiments in Colores *et al.* [13].

Panikov & Sizova [35] used their Equation 13 to fit respiration rates during the exponentially-increasing (i.e., no substrate limitation) phase (see Fig. 2 in Panikov & Sizova [35] for data and curve fittings). However, these data are based on glucose-induced respiration that includes both basal respiration of native SOC and respiration due to the addition of glucose [13]. The basal respiration rate may be regarded as a constant in certain cases (see Colores *et al.* [13] and data in Fig. 1 of Blagodatsky *et al.* [50]). The constant '*A*' representing the basal respiration rate was included in Equation 13 of Panikov & Sizova [35] in order to fit the combined respiration from the addition of glucose and basal respiration. However, this constant '*A*' cannot be derived from such governing equations as Equations S2-6(a–c) (see Appendix S2) that assume respiration is the sole result of substrate addition. In other words, the equations do not include basal respiration. Certainly, our predicted respiration could include basal respiration as long as (i) a basal respiration rate is added to Eq. 14d *ad hoc* or (ii) Equations S2-6(a–c), or more commonly Equations 12(a–e), are linked to a soil organic matter (SOM) decomposition model, which can produce decomposed native soil C in addition to the respiration of substrate addition. Because Equation 13 of Panikov & Sizova [35] is not linked to a native C decomposition model, fitting the model to combined native C and substrate respiration data is not appropriate.

Model test I: substrate-induced respiration

In this section, we used the respiration data from ^{14}C-labeled glucose SIGR experiments by Colores et al. [13] to calibrate our

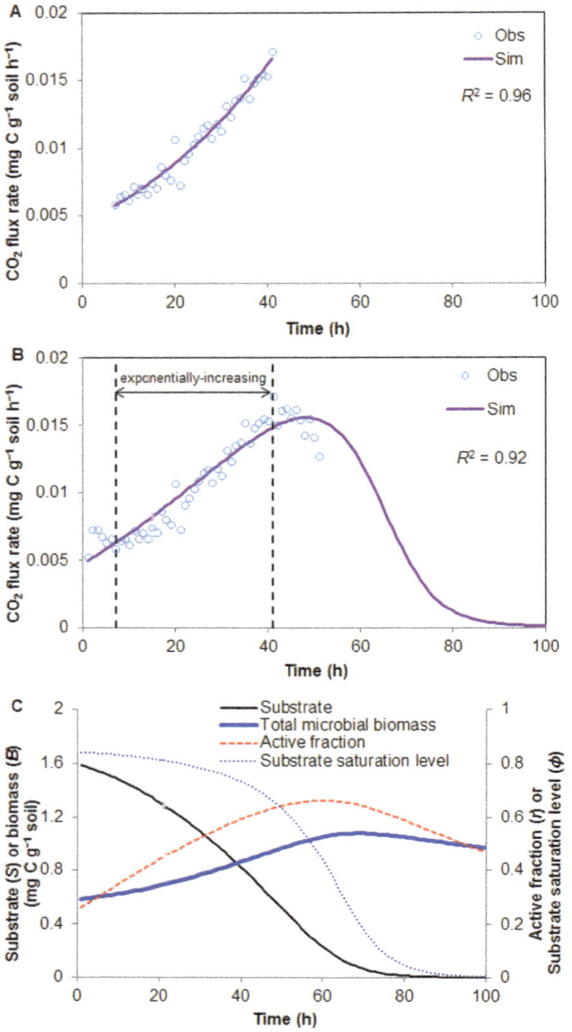

Figure 3. MEND model simulations against the respiration rates due to added ^{14}C-labeled glucose in Colores et al. [13]. (a) Fitting of the respiration rates in the exponentially-increasing phase using Equation 14, 'Obs' and 'Sim' denote observed and simulated data, respectively. (b) Fitting of the respiration rates in both exponentially-increasing and non-exponentially-increasing phases using Equation 12. (c) Simulated substrate (*S*), total live microbial biomass (*B*), active fraction (*r*) and substrate saturation level (ϑ) based on Equation 12.

MEND model. The respiration data only represented the CO_2 production from the added substrate and did not include basal respiration from the native C.

First we employed Equation 14d to fit the respiration rates during the exponentially-increasing stage and the result is shown in Fig. 3a (see original data in Fig. 3 of Colores et al. [13]). The true growth yield (Y_G) was set to 0.5 according to Colores *et al.* [13]. There are four undetermined parameters (B_0, r_0, μ_G, α) in Equation 14d (with $m_R = \mu_G \cdot \alpha/(1-\alpha)$). We found that only the maximum specific growth rate (μ_G) could be determined with high confidence (coefficient of variation (CV) = 5%) from the exponentially-increasing respiration rates. The CVs of the other three optimized parameters (B_0, r_0, α) were as high as 55–77% (Table 1). However, the initial active microbial biomass ($B_{a0} = B_0 \times r_0$) had a lower uncertainty (CV = 20%) compared to B_0 and r_0. The above results indicate that the exponentially-increasing respiration rates can only be used to obtain μ_G and B_{a0}.

We then conducted numerical simulations in terms of all data including both exponentially-increasing and non-exponentially-increasing respiration rates (Fig. 3b). The non-exponentially-increasing respiration rates include the lag period before the exponentially-increasing phase and the respiration at longer times after the rates cease to increase exponentially [13]. The latter phase is likely because of the substrate saturation levels (ϕ) become limiting to respiration. We used Equations 12a, 12b, 12e and the corresponding expression for CO_2 flux rate, to allow the substrate saturation level (ϕ) to change with time. Additionally, we used the ranges of μ_G determined above. We used the SCEUA (Shuffled Complex Evolution at University of Arizona) algorithm [51,52] to determine model parameters. The SCEUA is a widely used stochastic optimization algorithm for calibrating hydrological and environmental models [51].

When exponentially-increasing and non-exponentially-increasing data are included together, the CVs of all parameters (B_0, r_0, μ_G, α, K_s, β) are within 25% except β with a high CV of 76% (Table 1). The optimized μ_G values (0.030 ± 0.001 h^{-1}) are almost the same as obtained by Colores *et al.* [13]. Model estimates of $\alpha = 0.228 \pm 0.031$ indicate that the maximum specific maintenance rate of active microbes (m_R) is about 30% of μ_G and thus cannot be ignored. The initial active biomass (B_{a0}) is 0.145 ± 0.004 mg C g^{-1} soil (Table 1), which is lower than the values (0.194 ± 0.004 mg C g^{-1} soil) using the SIGR method [13]. This is likely due to the inclusion of maintenance respiration (characterized by m_R, see Equation 14d) in our model even for the exponentially-increasing stage; thus a lower B_{a0} could produce similar CO_2 flux to the case with higher B_{a0} that does not include the contributions from maintenance respiration. Our results also show that the initial active fraction (r_0) is 28.5 ± 6.4% and β is 0.025 ± 0.019. The magnitude of β is comparable to the estimation by Anderson & Domsch [6,7]. In addition, the half-saturation constant (K_s) was estimated as 0.275 ± 0.038 mg C g^{-1} soil, which is very close to the values derived from 16 soils by Van de Werf & Verstraete [21]. This K_s value indicates the substrate saturation level (ϕ) is higher than 0.7 before the transition from exponentially-increasing to non-exponentially-increasing phase (Fig. 3c). The changes of substrate (*S*), total microbial biomass (*B*) and active fraction (*r*) with time are also shown in Fig. 3c. In conclusion, the five parameters (B_0, r_0, μ_G, α, K_s) can be effectively determined using both exponentially-increasing and non-exponentially-increasing respiration rates, whereas β may also be determined but with a relatively high uncertainty (CV = 76%) than the other parameters.

Through this experimental analysis, we identified the need for isotopic data to discriminate between basal and substrate-induced respiration. We also discovered that the exponentially-increasing period due to substrate addition can be used to identify only a select set of model parameters (i.e., μ_G and B_{a0}) as also demonstrated by the method of Colores *et al.* [13]. These parameters, however, can be further applied to longer-term respiration experiments to enable fitting to obtain the remainder of model parameters by using our MEND model. Thus, we have found a new and unique solution to identify different parameters as a function of time, and to effectively use isotopic labeling to yield a specific set of model parameters.

Model test II: intermittent substrate supply

In order to further validate this additional physiological component in the MEND model, we also tested it against a laboratory experimental dataset with intermittent substrate supply [4]. In addition to the substrate, another limiting factor (i.e.,

Table 1. MEND model parameters values used for simulation of respiration rates due to added ^{14}C-labeled glucose in Colores et al. [13].

Parameter	Exponentially-increasing respiration*			All data†			Description
	Mean	SD‡	CV§	Mean	SD	CV	
B_0	0.504	0.279	55%	0.525	0.080	15%	Initial microbial biomass, (mg C g^{-1} soil)
r_0	0.394	0.263	67%	0.285	0.064	23%	Initial active fraction
μ_G	0.027	0.001	5%	0.030	0.001	3%	Maximum specific growth rate (h^{-1})
α	0.185	0.142	77%	0.228	0.031	13%	$m_R/(\mu_G+m_R)$, m_R is maximum specific maintenance rate for active microbes (h^{-1})
K_s	—	—	—	0.275	0.038	14%	Half-saturation constant for substrate (mg C g^{-1} soil)
β	—	—	—	0.025	0.019	76%	Ratio of dormant maintenance rate to m_R
Y_G	0.5	—	—	0.5	—	—	True growth yield, constant
B_{a0}	0.135	0.027	20%	0.145	0.004	3%	Initial active biomass (mg C g^{-1} soil), calculated by $B_0 \times r_0$

*Only the respiration rates during exponentially-increasing phase are used.
†All data including both exponentially-increasing and non-exponentially-increasing respiration.
‡SD: standard deviation.
§CV: Coefficient of variation.

oxygen, O_2) was included in this study. For this reason, we also introduced one more parameter (K_o: half saturation constant for O_2) to represent the limitation of O_2 on the microbial processes sketched in Fig. 1. Similar to substrates, the saturation level of O_2 is computed as $O_2/(O_2+K_o)$, where O_2 denotes the concentration of oxygen. The simulated oxygen concentrations by Stolpovsky et al. [4] were used as an input to our model. We used the SCEUA algorithm to determine the six model parameters in addition to the initial value for active fraction (r_0).

A summary of the seven parameters (one of them is r_0) and their fitted values is presented in Table 2. The initial active fraction (r_0) has a median of 0.925 with the 95% confidence interval (CI) of [0.628–1.000]. It means that a high r_0 is required for this experiment, but not necessary to be 1.0 set by Stolpovsky et al. [4]. The model and data are not sensitive to β since its 95% CI covers a wide range from 0.001 to 1. The reason is that the experiment only lasts for a very short time (33 h) so the influence of low metabolic rate at dormant state is insignificant.

Fig. 4 shows that the simulated total biomass (B) and substrate (S) concentrations agree very well with the observations (the coefficients of determination are 0.98 and 0.78 for biomass in Fig. 4a and substrate in Fig. 4b, respectively). Our simulation results indicate that, under limited O_2 between 12h and 24 h of the experiment, the active biomass decreases and the dormant biomass increases. As a result, the active fraction (r) declines from ca. 0.9 to 0.7 (Fig. 4a). For the same period Stolpovsky et al. [4] predicted a decrease of r from 1.0 to ca. 0, which means that all active biomass becomes dormant. Although there were not adequate measurements to confirm either prediction, our predicted changes in the active fraction (r) appear to be more reasonable during such a short experimental time period. This demonstration also shows that our model is capable of producing reasonable change in total, active, and dormant microbial biomass in response to substrate supply as well as an important forcing function (O_2).

Conclusions

We show that the physiological state index model (Equation 1) of Panikov [33] can be improved by eliminating the assumption that the steady state active fraction (r^{ss}) approaches the substrate saturation level (ϕ^{ss}). In particular, the model of Panikov [33] indicates that no active cells become dormant under insufficient substrate, which disregards the general nature of the strategy of dormancy in microorganisms when faced with unfavorable environmental conditions [5]. Our analysis also implies that the estimate of respiration rates under sufficient substrate by the

Table 2. MEND model parameter values used for simulation of the experiment described in Fig. 3 of Stolpovsky et al. (2011).

Parameter	Fitted Value*	Initial Range	Description
m_R	0.032, [0.011–0.048]	0.001–0.1	Specific maintenance rate for active biomass (h^{-1})
α	0.099, [0.045–0.181]	0.001–0.50	$m_R/(\mu_G+m_R)$, μ_G is specific growth rate (h^{-1})
K_s	3.110, [1.387–5.652]	0.1–9.0	Half-saturation constant for substrate (mg L^{-1})
Y_G	0.573, [0.463–0.600]	0.2–0.6	Growth yield factor (–)
K_o	0.0008, [0.0007–0.001]	0.005–0.1	Half-saturation constant for dissolved oxygen (mM)
β	0.351, [0.001–1.000]	0.001–1	Ratio of dormant maintenance rate to m_R
r_0	0.925, [0.628–1.000]	0–1	Initial fraction of active biomass (–)

*Medians and 95% confidence intervals of the fitted values from 100 optimization runs, i.e., 100 different random seeds are used for the stochastic optimization algorithm.

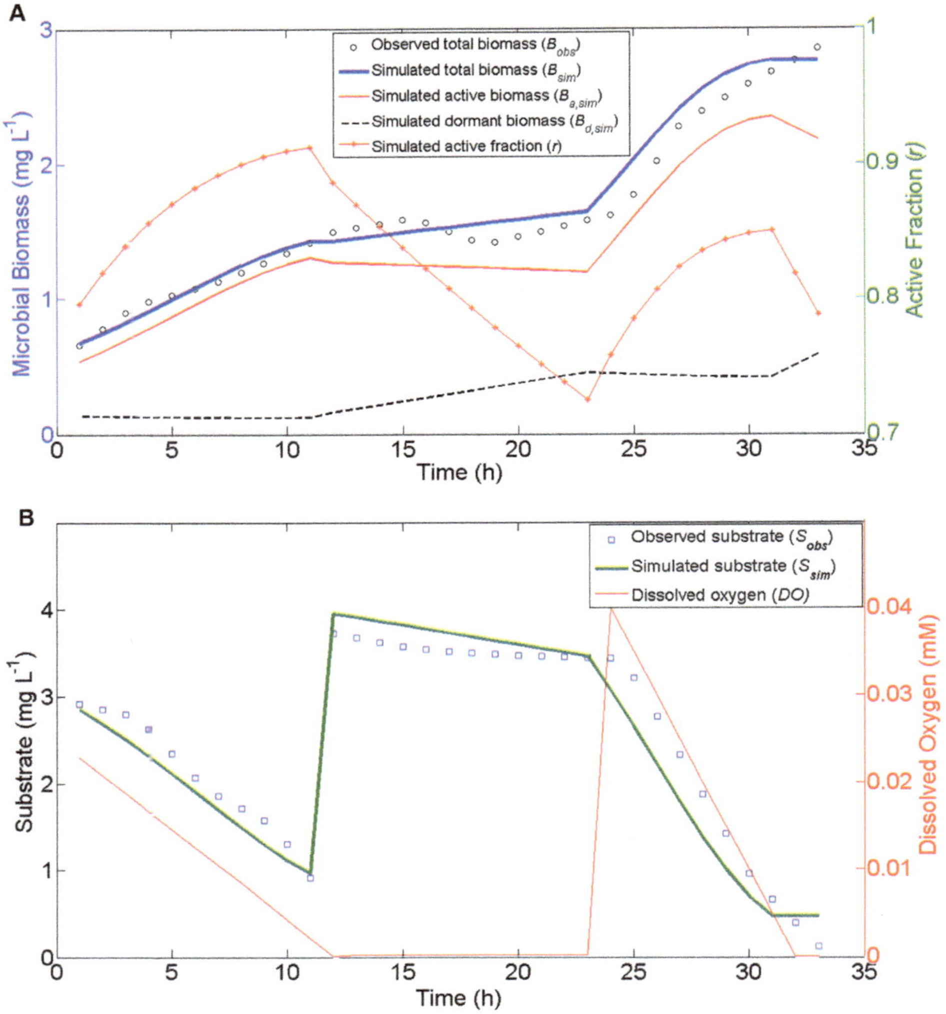

Figure 4. MEND model simulations against the experimental dataset used by Stolpovsky et al. (2011). (a) total live biomass, active and dormant biomass, and active fraction; (b) observed and simulated substrate concentration and prescribed O_2 concentration. There are three manipulations on the substrate and oxygen: (1) at time 0, the substrate (3 mg/L) and O_2 (0.025 mM) are added to the system; (2) after 12 h, the same amount of substrate is injected; (3) at 24 h, additional O_2 (0.04 mM) is injected to the system. The observed concentrations of substrate and total biomass are hourly data interpolated from the original observations in Stolpovsky et al. (2011). We scaled the substrate concentrations (with units of mM in original data) to match the magnitude of biomass concentration in units of mg/L.

physiological state index model is deficient. Pertaining to the switch function model, we argue that the switch function (θ) is also determined by the substrate (or combined with other impact factors) saturation level thus we would recommend using the switch function to modify the microbial uptake rate if the Gibbs energy change of the oxidation of substrate (ΔG) is tractable and the thermodynamic threshold (G_0) and the steepness of the step function (st) are identifiable. Based on the generally accepted assumptions summarized from existing dormancy models, we postulate a synthetic microbial physiology component to account for dormancy. Both the steady state active fraction (r^{ss}) and substrate saturation level (ϕ^{ss}) can be expressed as functions of two physiological indices: α and β. The index $\alpha = m_R/(\mu_G+m_R)$ is composed of μ_G and m_R denoting the maximum specific growth and maintenance rates, respectively, for active microbes. The index β represents the ratio of dormant to active maintenance rates. The value of r^{ss} is no less than ϕ^{ss}, and is equal only under the condition of $\beta \rightarrow 0$, where they are both identical to α. The upper bound of r^{ss} is ca. 0.8 at $\alpha \rightarrow 0.5$ and $\beta \rightarrow 1$. The maximum r^{ss} is ca. 0.5 if β (≤ 0.01) following the estimation of Anderson & Domsch [6]. It is evident that r^{ss} could be attenuated further by other limiting factors. The application of the MEND microbial physiology model to an experimental dataset with intermittent substrate supply shows satisfactory model performance (the determination coefficients are 0.98 and 0.78 for microbial biomass and substrate, respectively). The case study on the SIGR dataset indicate that the exponentially-increasing respiration rates can only be used to determine μ_G and B_{a0} (initial active biomass), while the major parameters (B_0, r_0, μ_G, α, K_s) can be effectively determined using both exponentially-increasing and non-exponentially-increasing respiration rates.

In conclusion, the microbial physiology model presented here can be incorporated into existing ecosystem models to account for dormancy in microbially-mediated processes. We have illustrated the impacts of substrate and oxygen availabilities on the physiological states through this study. Other environmental factors, such as soil temperature and soil water potential, could also be introduced into this framework to affect the transformation processes between the two microbial compositions. The changes in the physiological states of microbes could further alter the microbially-driven carbon and nutrient dynamics in ecosystems. Traditional measures of microbial biomass include the entire microbial population, even though dormancy is an important evolutionary strategy for preservation of microbial genetics and function until conditions for growth and replication improve. Parameterizing microbial ecosystem models assuming the entire population is active could therefore lead to significant errors. The approach described here provides a tractable and testable method to include dormancy as a response to external forcing.

Acknowledgments

The authors thank Dr. Sindhu Jagadamma for her helpful comments. Thanks also go to the two anonymous reviewers for their constructive comments.

Author Contributions

Conceived and designed the experiments: GW MAM. Analyzed the data: GW MAM CWS. Wrote the paper: GW MAM LG CWS. Mathematical analysis: GW LG.

References

1. Blagodatsky SA, Heinemeyer O, Richter J (2000) Estimating the active and total soil microbial biomass by kinetic respiration analysis. Biology and Fertility of Soils 32: 73–81.
2. Mason C, Hamer G, Bryers J (1986) The death and lysis of microorganisms in environmental processes. FEMS microbiology letters 39: 373–401.
3. Price PB, Sowers T (2004) Temperature dependence of metabolic rates for microbial growth, maintenance, and survival. Proceedings of the National Academy of Sciences of the United States of America 101: 4631–4636.
4. Stolpovsky K, Martinez-Lavanchy P, Heipieper HJ, Van Cappellen P, Thullner M (2011) Incorporating dormancy in dynamic microbial community models. Ecological Modelling 222: 3092–3102.
5. Lennon JT, Jones SE (2011) Microbial seed banks: the ecological and evolutionary implications of dormancy. Nature Reviews Microbiology 9: 119–130.
6. Anderson TH, Domsch KH (1985) Determination of ecophysiological maintenance carbon requirements of soil microorganisms in a dormant state. Biology and Fertility of Soils 1: 81–89.
7. Anderson TH, Domsch KH (1985) Maintenance carbon requirements of actively-metabolizing microbial populations under *in situ* conditions. Soil Biology and Biochemistry 17: 197–203.
8. Adam G, Duncan H (2001) Development of a sensitive and rapid method for the measurement of total microbial activity using fluorescein diacetate (FDA) in a range of soils. Soil Biology and Biochemistry 33: 943–951.
9. Jones KH, Senft JA (1985) An improved method to determine cell viability by simultaneous staining with fluorescein diacetate-propidium iodide. Journal of Histochemistry & Cytochemistry 33: 77–79.
10. Boulos L, Prévost M, Barbeau B, Coallier J, Desjardins R (1999) LIVE/DEAD® BacLight™: application of a new rapid staining method for direct enumeration of viable and total bacteria in drinking water. Journal of Microbiological Methods 37: 77–86.
11. Stocks S (2004) Mechanism and use of the commercially available viability stain, BacLight. Cytometry Part A 61: 189–195.
12. Schulz S, Peréz-de-Mora A, Engel M, Munch JC, Schloter M (2010) A comparative study of most probable number (MPN)-PCR vs. real-time-PCR for the measurement of abundance and assessment of diversity of alkB homologous genes in soil. Journal of Microbiological Methods 80: 295–298.
13. Colores GM, Schmidt SK, Fisk MC (1996) Estimating the biomass of microbial functional groups using rates of growth-related soil respiration. Soil Biology and Biochemistry 28: 1569–1577.
14. Anderson JPE, Domsch KH (1978) A physiological method for quantitative measurement of microbial biomass in soils. Soil Biology & Biochemistry 10: 215–221.
15. Jenkinson DS, Powlson DS (1976) The effects of biocidal treatments on metabolism in soil-V: A method for measuring soil biomass. Soil Biology and Biochemistry 8: 209–213.
16. Lodge DJ (1993) Nutrient cycling by fungi in wet tropical forests. In: Isaac S, editor. Aspects of Tropical Mycology: Symposium of the British Mycological Society Cambridge, UK: Cambridge University Press. pp. 37–57.
17. Yarwood S, Brewer E, Yarwood R, Lajtha K, Myrold D (2013) Soil microbe active community composition and capability of responding to litter addition after 12 years of no inputs. Applied and Environmental Microbiology 79: 1385–1392.
18. Alvarez C, Alvarez R, Grigera M, Lavado R (1998) Associations between organic matter fractions and the active soil microbial biomass. Soil Biology and Biochemistry 30: 767–773.
19. Khomutova T, Demkina T, Demkin V (2004) Estimation of the total and active microbial biomasses in buried subkurgan paleosoils of different age. Microbiology 73: 196–201.
20. Orwin KH, Wardle DA, Greenfield LG (2006) Context-dependent changes in the resistance and resilience of soil microbes to an experimental disturbance for three primary plant chronosequences. Oikos 112: 196–208.
21. Van de Werf H, Verstraete W (1987) Estimation of active soil microbial biomass by mathematical analysis of respiration curves: calibration of the test procedure. Soil Biology and Biochemistry 19: 261–265.
22. Stenström J, Svensson K, Johansson M (2001) Reversible transition between active and dormant microbial states in soil. FEMS Microbiology Ecology 36: 93–104.
23. Wang G, Post WM, Mayes MA (2013) Development of microbial-enzyme-mediated decomposition model parameters through steady-state and dynamic analyses. Ecological Applications 23: 255–272.
24. Moorhead DL, Lashermes G, Sinsabaugh RL (2012) A theoretical model of C- and N-acquiring exoenzyme activities, which balances microbial demands during decomposition. Soil Biology and Biochemistry 53: 133–141.
25. Sinsabaugh RL, Manzoni S, Moorhead DL, Richter A (2013) Carbon use efficiency of microbial communities: stoichiometry, methodology and modelling. Ecology Letters 16: 930–939.
26. Schimel JP, Weintraub MN (2003) The implications of exoenzyme activity on microbial carbon and nitrogen limitation in soil: a theoretical model. Soil Biology & Biochemistry 35: 549–563.
27. Lawrence CR, Neff JC, Schimel JP (2009) Does adding microbial mechanisms of decomposition improve soil organic matter models? A comparison of four models using data from a pulsed rewetting experiment. Soil Biology & Biochemistry 41: 1923–1934.
28. Wang G, Post WM, Mayes MA, Frerichs JT, Jagadamma S (2012) Parameter estimation for models of ligninolytic and cellulolytic enzyme kinetics. Soil Biology and Biochemistry 48: 28–38.
29. Todd-Brown KE, Hopkins FM, Kivlin SN, Talbot JM, Allison SD (2012) A framework for representing microbial decomposition in coupled climate models. Biogeochemistry 109: 19–33.
30. Treseder KK, Balser TC, Bradford MA, Brodie EL, Dubinsky EA, et al. (2012) Integrating microbial ecology into ecosystem models: challenges and priorities. Biogeochemistry 109: 7–18.
31. Wieder WR, Bonan GB, Allison SD (2013) Global soil carbon projections are improved by modelling microbial processes. Nature Clim Change 3: 909–912.
32. Konopka A (1999) Theoretical analysis of the starvation response under substrate pulses. Microbial Ecology 38: 321–329.
33. Panikov NS (1996) Mechanistic mathematical models of microbial growth in bioreactors and in natural soils: explanation of complex phenomena. Mathematics and Computers in Simulation 42: 179–186.
34. Blagodatsky S, Richter O (1998) Microbial growth in soil and nitrogen turnover: a theoretical model considering the activity state of microorganisms. Soil Biology and Biochemistry 30: 1743–1755.
35. Panikov NS, Sizova MV (1996) A kinetic method for estimating the biomass of microbial functional groups in soil. Journal of Microbiological Methods 24: 219–230.
36. Buerger S, Spoering A, Gavrish E, Leslin C, Ling L, et al. (2012) Microbial scout hypothesis, stochastic exit from dormancy, and the nature of slow growers. Applied and Environmental Microbiology 78: 3221–3228.
37. Hunt HW (1977) A simulation model for decomposition in grasslands. Ecology 58: 469–484.
38. Gignoux J, House J, Hall D, Masse D, Nacro HB, et al. (2001) Design and test of a generic cohort model of soil organic matter decomposition: the SOMKO model. Global Ecology and Biogeography 10: 639–660.
39. Ayati BP (2012) Microbial dormancy in batch cultures as a function of substrate-dependent mortality. Journal of Theoretical Biology 293: 34–40.

40. Jones SE, Lennon JT (2010) Dormancy contributes to the maintenance of microbial diversity. Proceedings of the National Academy of Sciences 107: 5881–5886.
41. Bär M, Hardenberg J, Meron E, Provenzale A (2002) Modelling the survival of bacteria in drylands: the advantage of being dormant. Proceedings of the Royal Society of London Series B: Biological Sciences 269: 937–942.
42. Locey KJ (2010) Synthesizing traditional biogeography with microbial ecology: the importance of dormancy. Journal of Biogeography 37: 1835–1841.
43. Wang G, Post WM (2013) A note on the reverse Michaelis–Menten kinetics. Soil Biology and Biochemistry 57: 946–949.
44. Wirtz KW (2003) Control of biogeochemical cycling by mobility and metabolic strategies of microbes in the sediments: an integrated model study. FEMS Microbiology Ecology 46: 295–306.
45. Panikov NS (1995) Microbial growth kinetics. London, UK: Chapman & Hall. 378 p.
46. Pirt SJ (1965) Maintenance energy of bateria in growing cultures. Proceedings of the Royal Society of London Series B-Biological Sciences 163: 224–231.
47. Wang G, Post WM (2012) A theoretical reassessment of microbial maintenance and implications for microbial ecology modeling. FEMS Microbiology Ecology 81: 610–617.
48. Hoehler TM, Jorgensen BB (2013) Microbial life under extreme energy limitation. Nat Rev Micro 11: 83–94.
49. Beeftink HH, Van der Heijden RTJM, Heijnen JJ (1990) Maintenance requirements: energy supply from simultaneous endogenous respiration and substrate consumption. FEMS microbiology letters 73: 203–209.
50. Blagodatsky S, Yevdokimov I, Larionova A, Richter J (1998) Microbial growth in soil and nitrogen turnover: model calibration with laboratory data. Soil Biology and Biochemistry 30: 1757–1764.
51. Duan QY, Sorooshian S, Gupta V (1992) Effective and efficient global optimization for conceptual rainfall-runoff models. Water Resources Research 28: 1015–1031.
52. Wang G, Xia J, Chen J (2009) Quantification of effects of climate variations and human activities on runoff by a monthly water balance model: A case study of the Chaobai River basin in northern China. Water Resources Research 45: W00A11.

14

Long-Term Monitoring of Rainfed Wheat Yield and Soil Water at the Loess Plateau Reveals Low Water Use Efficiency

Wei Qin[1,2], Baoliang Chi[1]*, Oene Oenema[2]*

1 Dryland Agriculture Research Centre, Shanxi Academy of Agricultural Sciences, Taiyuan, Shanxi, China, 2 Environmental Sciences, Wageningen University, Wageningen, The Netherlands

Abstract

Increasing crop yield and water use efficiency (WUE) in dryland farming requires a quantitative understanding of relationships between crop yield and the water balance over many years. Here, we report on a long-term dryland monitoring site at the Loess Plateau, Shanxi, China, where winter wheat was grown for 30 consecutive years and soil water content (0–200 cm) was measured every 10 days. The monitoring data were used to calibrate the AquaCrop model and then to analyse the components of the water balance. There was a strong positive relationship between total available water and mean cereal yield. However, only one-third of the available water was actually used by the winter wheat for crop transpiration. The remaining two-thirds were lost by soil evaporation, of which 40 and 60% was lost during the growing and fallow seasons, respectively. Wheat yields ranged from 0.6 to 3.9 ton/ha and WUE from 0.3 to 0.9 kg/m^3. Results of model experiments suggest that minimizing soil evaporation via straw mulch or plastic film covers could potentially double wheat yields and WUE. We conclude that the relatively low wheat yields and low WUE were mainly related to (i) limited rainfall, (ii) low soil water storage during fallow season due to large soil evaporation, and (iii) poor synchronisation of the wheat growing season to the rain season. The model experiments suggest significant potential for increased yields and WUE.

Editor: Eric Gordon Lamb, University of Saskatchewan, Canada

Funding: The authors have no support or funding to report.

Competing Interests: The authors have declared that no competing interests exist.

* E-mail: Baoliangchi@hotmail.com (BC); Oene.Oenema@wur.nl (OO)

Introduction

Water scarcity is a growing global concern [1–4]. For rainfed agriculture, this pressure may become more severe under climate change due to the expected more erratic rainfall and longer dry spells [5–8]. Currently, rainfed agriculture covers 80% of the world's cultivated land and accounts for 60% of crop production [9]. However, crop yield and water use efficiency (WUE) are often low in rainfed agriculture, especially in arid and semi-arid areas due to, for example, degraded soils, erratic rainfall and poor water management [10]. Many of these areas in Africa and Asia face also rapid population growth. Hence, there is a pressing need to increase crop yields and WUE in rainfed agriculture [11–15].

Water use efficiency is commonly defined as crop yield over evapotranspiration (ET), where ET is the sum of soil evaporation (E) and crop transpiration (T) [16–19]. The latter (T) is a direct consequence of crop production, while E is 'unproductive' water loss. The central question for rainfed agriculture in arid and semi-arid regions is 'how to transform unproductive water loss (E) into productive water use (T)'. Unfortunately, the partitioning between E and T is often not well-known, due to difficulties and high cost in distinguishing E and T in the field. As a result, water use of crops is commonly reported as evapotranspiration (ET) [16,17,20]. This lack of information makes it difficult to assess how much of the evaporative water loss can be used for increasing yields by appropriate measures.

Crop growth simulation models have the potential of providing more comprehensive insights into the functioning of soil-crop systems, and can be helpful to explore options for increasing yield and WUE [21–23]. These models though are simplified representations of parts of reality and therefore require testing in the real world. Fortunately, numerous field studies have examined the effect of water availability, with or without irrigation treatments, on crop yield and WUE [16–20]. In principle, these results can be used to calibrate and test the crop growth simulation models. However, most of these field studies are short-term (2–5 years) and focus on the growing season only, largely ignoring the water balance during the fallow season. Furthermore, crop yields and WUE show large variations due to differences in soils, climate conditions and crop husbandry practices. For example, in the North China Plain, annual precipitation ranges from 400 to 650 mm and wheat grain yields roughly range from 1 to 3 ton/ha/year under rainfed conditions. With 200 to 300 mm of irrigation, grain yield can be increased by 60 to 100% and WUE can be increased by 20 to 40% [16]. Globally, WUE of wheat shows even larger variation ranging from approximately 0.2 to 1.8 kg/m^3 [24]. Evidently, this wide range suggests considerable scope for improvement, but the underlying causes of WUE is not always well-known and irrigation water is not available on most places.

The study reported here has the objectives (i) to calibrate a water-driven crop growth model on the basis of monitoring data from a long-term field experiment with rainfed winter wheat, (ii) to analyse the components of the water balance of this field, and (iii) to explore the potential for increasing crop yield and WUE through model experiments. The field experiment was situated in the dryland of the Loess Plateau, Shanxi, China. Winter wheat was grown for 30 consecutive years and soil water content (0–200 cm) was measured every 10 days. The FAO AquaCrop model was chosen as simulation model because it is a water-driven crop growth model that can separately calculate E and T, and simulates the final crop yield as function of water use [23,25]. The AquaCrop model uses canopy ground cover as the basis to calculate T and to separate E and T. Crop yield is then calculated as the product of biomass and harvest index (HI). The principles and modules of the AquaCrop model are well-documented in a series of AquaCrop publications [23,25–28]. Compared to some other crop growth models, AquaCrop requires relatively few input parameters [25,28]. Such a limited number of input parameters facilitates model calibration and utilization for different crops and under different management strategies [23,26,27,29–36].

Materials and Methods

Site Information

The long-term monitoring site is located in Beizhang, Linyi county in Shanxi province on the Loess Plateau (35° 9′ 3.83″ N, 110° 34′ 25.40″ E, Altitude: 491 m) in China. The authority is Dryland Agriculture Research Centre, Shanxi Academy of Agricultural Sciences. We have the permission to conduct the study on this site. We confirm that the field studies did not involve endangered or protected species.

The site has a semi-arid climate with extensive monsoonal influence, which is dry and cold in winter, rainy and hot in summer. Rainfall in June to September accounts for more than 70% of annual rainfall. Average annual rainfall was 517 mm in the period of 1980–2010 with large annual variations, from a minimum of 331 mm to a maximum of 832 mm. Mean annual sunshine duration is 2270 h, annual average temperature 13.5°C and mean annual potential evaporation is 1340 mm. The soil is a typical Loessial soils (Calcic Luvisols) of the Loess Plateau [37–40]. Soil slope is <1% and soil texture is silt loam, with a small proportion of clay. The soil was rather homogeneous in texture and key physical properties (Table 1). Maximum soil water holding capacity was a significant fraction of the total annual rainfall (Table 2).

The top soil (30 cm) contained 5.8 g/kg of organic C and 0.55 g/kg of total N. Mean available N, P and K were 63 mg/kg, 14 mg/kg and 142 mg/kg, respectively, in 2007. Since 1983, mean available P and K have been increased by 7 mg/kg and 10 mg/kg, respectively [41]. Total soil organic C was measured by dry combustion combined with elemental C analysis. Total N was measured by Kjeldahl method. Soil mineral N was extracted with 1 M KCl and analysed by the cadmium reduction method. Available P was extracted with 0.5 M $NaHCO_3$. Available K was extracted with 1.0 M NH4OAc.

Experimental Design and Measurements

The long-term crop yield and soil moisture monitoring experiment started in 1980. There were no experimental treatments. The size of the field is 0.5 ha. Winter wheat was planted each year between 25 September and 5 October, depending on the actual climate and soil water conditions. Sowing rate was 150 kg seed/ha. The growing period of the winter wheat is approximately 245 days (from 1 October to 1 June of the next year). The fallow season is about 120 days (from 2nd June to 30th September) (Figure 1). Local bred cultivars (Jinmai) were used for all years. Information about crop parameters is listed in Table 3. Fertilizers were applied at planting at a rate of 127.5 kg/ha of N as urea and 90 kg/ha of P_2O_5 as superphosphate, for all years. At harvest, grain yields were measured in 5 random plots (2 m^2 per plot) selected from two diagonal lines of the field. Dry matter content of grain and straw was measured at the laboratory after drying at 70°C. Soil water contents were measured in 16 different layers up to 2 m depth every 10 days by using the gravity method. The top 10 cm was sampled in 5 cm intervals. From 0.1 to 1 m, samples were taken at 10 cm intervals, and from 1 to 2 m at 20 cm intervals (Figure 2). Each sample consisted of 5 subsamples, taken randomly on two diagonal lines across the field. Observations during soil sampling over years revealed that spatial variations in soil profile were small.

Water Balance and WUE Estimations

Mean monthly rainfall data were collected at a near-by meteorological station (Linyi station, 35° 10′ 7.02″ N, 110° 46′ 44.59″ E, Altitude: 441 m), which is 25 km away from the experimental field. The water balance for both fallow and crop growing seasons reads as follows:

$$R + I \pm \Delta S = E + T + R_r + P_e \quad (1)$$

where R is rainfall, I is irrigation, ΔS is the change of soil water content, E is soil evaporation, T is crop transpiration, R_r is runoff and P_e is percolation. All units are presented in mm or in m^3/ha.

Irrigation was not applied in this study. Furthermore, runoff and percolation (leaching) were small and disregarded. Hence, the water balance of the fallow seasons was simplified to:

$$R \pm \Delta S = E \quad (2)$$

The water balance of the growing season was simplified to:

$$R \pm \Delta S = E + T \quad (3)$$

We used formula (2) and the recorded rainfall and soil water content data to calculate evaporation during the fallow season and formula (3) to calculate ET during the growing season. The partitioning of E and T was done by the AquaCrop model.

Water use efficiency (WUE, kg/m^3) was calculated as:

$$WUE = grain\,yield\,/\,ET \quad (4)$$

where ET is evapotranspiration (mm), the sum of E and T during the growing season.

Calibration and Validation of the AquaCrop Model

The FAO AquaCrop model is a water-driven crop growth model for the simulation of crop biomass and yield as function of water availability [23,25]. AquaCrop requires 4 main sets of input data, i.e. climate data (rainfall, minimum and maximum temperature and reference evapotranspiration (ETo)), crop parameters, soil data and field management data. Climate data were collected from a near-by meteorological station (Linyi station, 35° 10′ 7.02″ N, 110° 46′ 44.59″ E, Altitude: 441 m), which is 25 km away from the experimental field. ETo was calculated by FAO Penman-Monteith equation as described in Allen et al. [42].

Table 1. Soil physical properties for different layers up to 200 cm.

Soil layer	Bulk density	Field Capacity	Wilting point	Texture, (particle size, μm in %)			
cm	g/cm³	v/v in %	v/v in %	>63 μm	63–20 μm	20–2 μm	<2 μm
5	1.34	31.4	6.8	7.5	41.0	32.4	19.2
10	1.34	31.4	6.8	7.5	41.0	32.4	19.2
20	1.39	32.9	8.5	8.3	41.8	31.2	18.8
30	1.43	34.5	8.4	6.9	44.3	30.1	18.7
40	1.39	33.6	7.0	6.9	44.3	30.1	18.7
50	1.43	33.6	7.3	6.9	44.3	30.1	18.7
60	1.39	32.8	7.0	4.5	40.5	31.4	23.6
70	1.36	30.9	7.2	4.5	40.5	31.4	23.6
80	1.25	27.6	6.9	4.5	40.5	31.4	23.6
90	1.28	27.9	8.3	4.5	40.5	31.4	23.6
100	1.32	28.3	8.2	4.5	40.5	31.4	23.6
120	1.30	29.3	7.8	3.0	43.4	34.4	19.2
140	1.32	29.7	7.9	3.0	43.4	34.4	19.2
160	1.31	29.5	7.9	2.4	40.4	38.7	18.5
180	1.32	29.7	7.9	2.4	40.4	38.7	18.5
200	1.32	29.7	7.9	2.4	40.4	38.7	18.5

Soil data (Tables 1 and 2) and field management data were derived from measurements and from recordings at the monitoring site. Crop data and parameters were derived from measurements, literature [23,25–27] and by calibration (Table 3).

The objective of calibration and validation was to achieve the best match between simulated outputs and monitoring data (soil moisture and crop yield) for all 30 years, using a common procedure [27,29–36]. We randomly selected 15 years' data from the monitoring field for calibration and used the other 15 years' data for validation of the calibrated model. In the calibration step, we used the year-specific climate, soil and initial soil water data as fixed input data. Then we adjusted some of the crop parameters (see Table 3), based on our understanding of crop growth, development and crop responses to water deficits, until the differences between simulated output and monitoring data were minimal. We repeated this process for all selected 15 years and ultimately obtained a satisfactorily index of agreement ($d = 0.92$). Then, the calibrated model was validated with the other 15 years' data, and again obtained an acceptable index of agreement ($d = 0.93$).

Data Analysis

The root mean square error (RMSE) [43] has been widely used to evaluate the performance of a model [30,44–47]. However, Willmott and Matsuura [44] pointed out that, compared to RMSE, the mean absolute error (MAE) is a better indicator and therefore the evaluation of model performance should be based on the MAE. In this study, we used the MAE, the mean bias error (MBE) and the Willmott index of agreement (d) to evaluate the model performance.

The MAE measures the weighted average magnitude of the absolute errors and was calculated as follows:

$$MAE = \frac{1}{n}\sum_{i=1}^{n}|\mathrm{Mi} - \mathrm{Oi}| \qquad (5)$$

where n is the number of observations, Mi is the modelled yield or soil water content and Oi is the observed yield or soil water content.

The MBE indicates whether the model is under or over predicting the observed values and also indicates the uniformity of error distribution. Positive MBE values indicate over prediction, negative values indicate under prediction and a value of zero

Table 2. Soil moisture holding capacity of the soil profile (0–200 cm).

	Total water, mm(mm)	Available water, mm(mm)	In volume, % (v/v)
Saturation	860	711	43
Field capacity	596	447	30
70% FC	462	312	23
50% FC	373	223	19
Wilting point	149	0	8

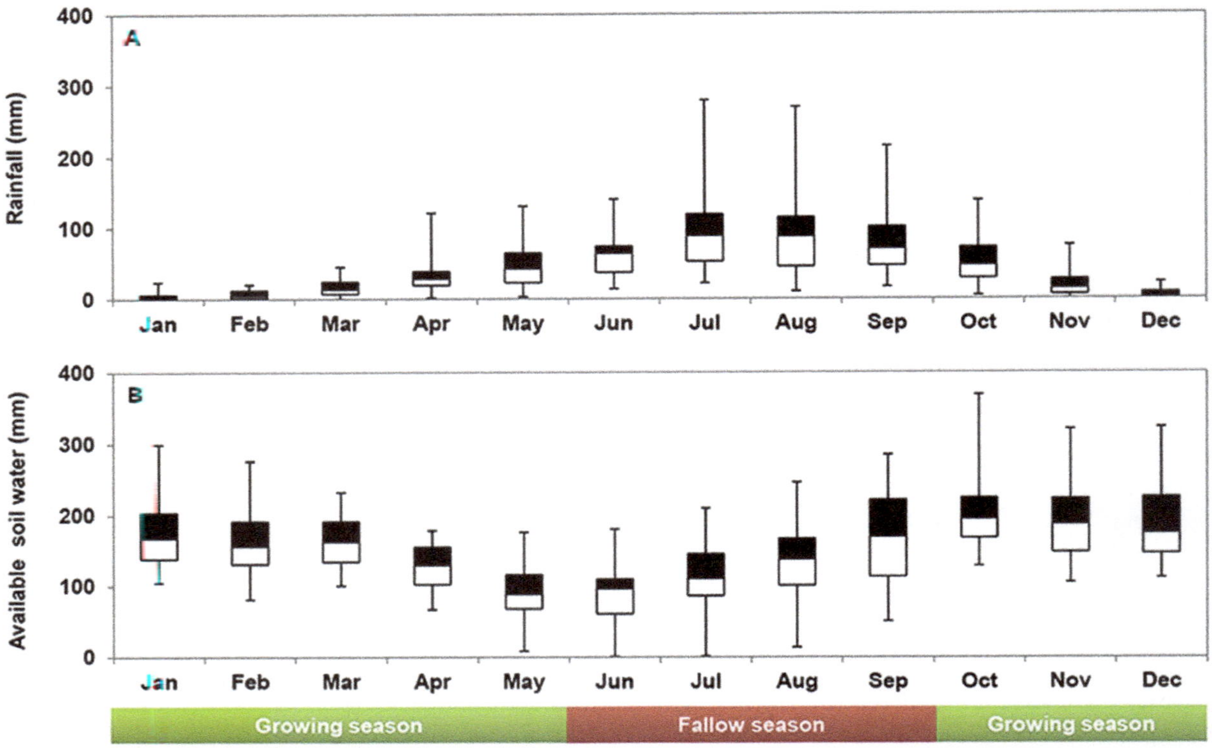

Figure 1. Monthly rainfall and available soil water during the period 1980–2010. A: monthly rainfall distribution; B: monthly available water content. Black and white boxes show 75 and 25% percentile values. Whiskers show maximum and minimum values (the same applies to other figures). The growing season of winter wheat was 245 days from 1 October to 1 June. of the next year, highlighted in green bar. Fallow season was 120 days from 2nd Jun. to 30th Sep., highlighted in red bar.

indicates equal distribution between negative and positive values. The MBE was calculated as follows:

$$MBE = \frac{1}{n}\sum_{i=1}^{n}(\mathrm{Mi} - \mathrm{Oi}) \tag{6}$$

The Willmott index of agreement (d) [43] has values ranging from 0 to 1. A value close to 1 suggests a good model performance. The agreement index d was calculated as follows:

$$d = 1 - \frac{\sum_{i=1}^{n}(\mathrm{Mi} - \mathrm{Oi})^2}{\sum_{i=1}^{n}(|\mathrm{Mi} - \bar{\mathrm{O}}| + |\mathrm{Oi} - \bar{\mathrm{O}}|)^2} \tag{7}$$

where $\bar{\mathrm{O}}$ is the average value of the observed yield or soil water content.

Model Experiments

To investigate the potentials of minimizing soil evaporation so as to increase crop yield and WUE, we set up six model experiments as follows:

Experiment 1 (E1): Reference A; winter wheat was planted at 50% of field capacity (FC) (low soil water content)

Experiment 2 (E2): E1+organic mulch (straw)

Experiment 3 (E3): E1+plastic film cover

Experiment 4 (E4): Reference B; winter wheat was planted at 70% of FC (high soil water content)

Experiment 5 (E5): E4+organic mulch (straw)

Experiment 6 (E6): E4+plastic film cover

In the model experiments, we used E1 and E4 as references to simulate crop growth under relatively low (with 50% FC) and high (with 70% FC) soil water content at winter wheat seeding (Table 2). The range from 50 to 70% FC largely represented the initial soil water content for the planting period. The low value is representative for no water harvesting, and the high values is representative for water harvesting during the fallow period via mulching and/or covers. Model experiments E2 and E5 aimed at testing the effects of organic mulch during the growing season, and experiments E3 and E6 aimed at testing the effects of plastic film during the growing season on crop yield and water balance. The effectiveness of the soil evaporation reduction by organic mulch and plastic film cover during the growing season were estimated at 50 and 90%, respectively, which are default values of the AquaCrop model [25]. However, since plastic film will cover only about 80% of the wheat planted field, the overall soil evaporation reduction by plastic film was set at 72%. In practice, the plastic film covers are used for the growing season only and destroyed or removed after harvesting the crops. We ran these six experiments for the period of 1980–2010.

Table 3. Full set of crop parameters used in this study.

Parameters	Contents and values	Source
Crop development	in calendar days	
From sowing to emergence	7 days	Field observation
From sowing to max. canopy	200 days	Field observation
From sowing to flowering	200 days	Field observation
From sowing to senescence	210 days	Field observation
From sowing to maturity	245 days	Field observation
Length for building up Harvest Index (HI)	44 days	Field observation
Duration of flowering	5 days	Field observation
Plant density	300 plants/m^2	Field observation
Sowing rate	150 kg seed/ha	Field observation
1000 seed mass	40 g	Field observation
Germination rate	80%	Field observation
Max. root depth	2 m	Field observation
Reference HI	42%	Field observation
Max HI	48%	Field observation
Crop water productivity	15 g/m^2	Calibrated
Canopy development		
Initial canopy cover	4.5%	Field observation
Canopy expansion	2.9%/day	Calculated by AquaCrop
Max. canopy cover	90%	Field observation
Canopy decline	7.2%/day (39 days)	Calculated by AquaCrop
Thresholds temperatures		
Base temperature for biomass production	0°C	Calibrated
Upper temperature for biomass production	26°C	Calibrated
Range of cold stress for biomass production	0–14°C	Calibrated
Range of cold stress for pollination	4–9°C	Calibrated
Range of heat stress for pollination	32–37°C	Calibrated
Water extraction pattern throughout the root zone		
Upper 1/4 (0–0.5 m)	40%	Calibrated
Second 1/4 (0.5–1 m)	30%	Calibrated
Third 1/4 (1–1.5 m)	20%	Calibrated
bottom 1/4 (1.5–2 m)	10%	Calibrated
Water stresses		
Canopy expansion	Moderately tolerant (upper = 0.25, lower = 0.65, shape factor = 5)	Calibrated
Stomatal closure	Extremely sensitive (upper = 0.25, shape factor = 2.5)	Calibrated
Early canopy senescence	Tolerant (upper = 0.75, shape factor = 2.5)	Calibrated
Aeration stress	Moderately tolerant (5 vol%)	Calibrated
Evapotranspiration		
Soil evaporation coefficient	Effect of canopy shelter in late season = 50%	AquaCrop default value
Crop transpiration coefficient	1.1 (reduction with age = 0.15%/day)	AquaCrop default value
Fertilities stresses	Not considered	AquaCrop default value

Results

Patterns of Monthly Rainfall and Soil Water Content

Most of the rainfall occurs in the summer from June to September, accounting for more than 70% of the total annual rainfall while winter wheat was seeded in the end of September or beginning of October, and harvested in the end of May or beginning of June. Hence, most of the rain fell when wheat had matured already, i.e., during the fallow period. Therefore, the growing season of winter wheat was poorly synchronized to the rain season. On the other hand, due to concentrated rainfall in June to September, mean soil water content in the upper two metre was highest in October when winter wheat was seeded. Thereafter, the soil water content gradually decreased, due to soil

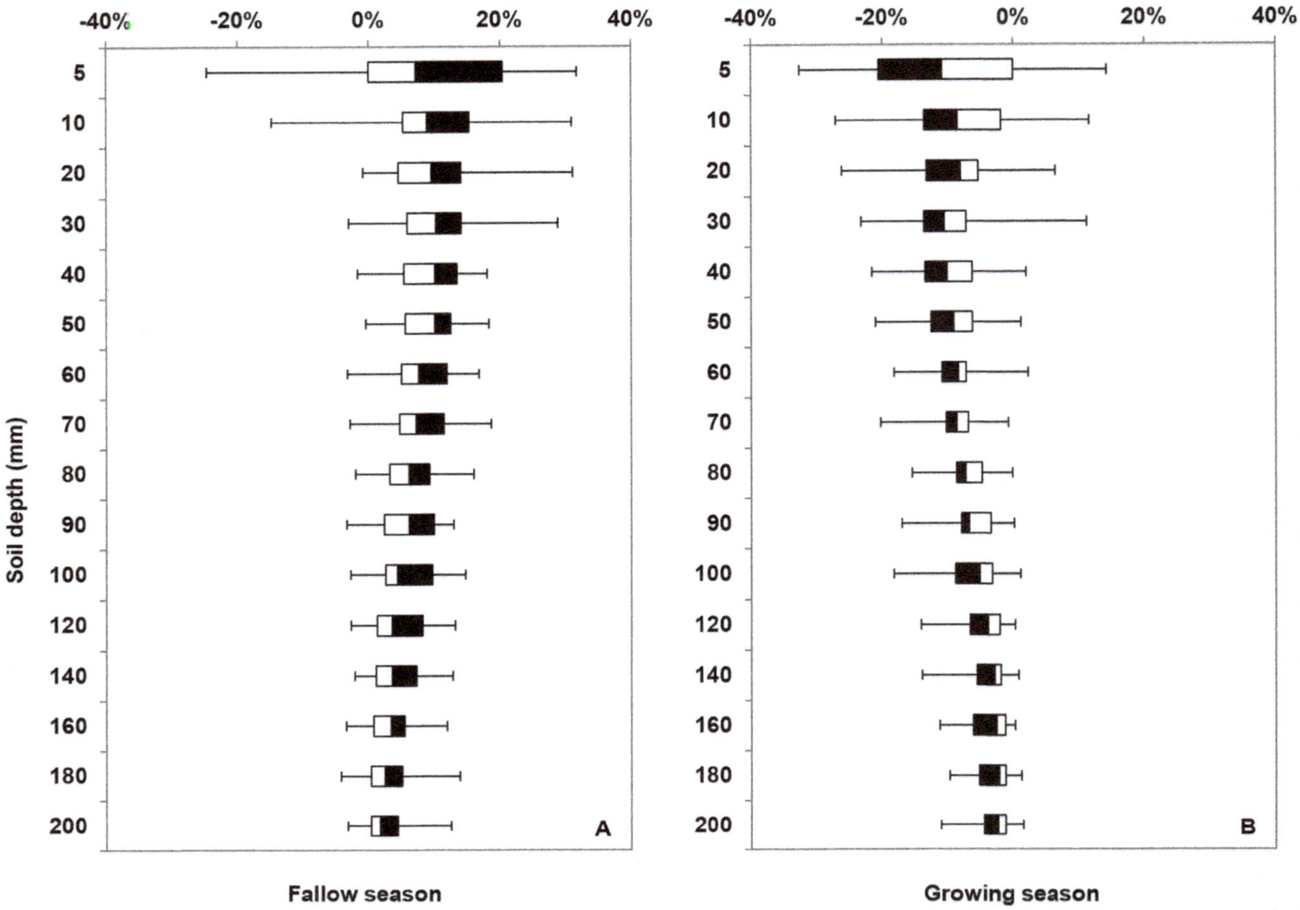

Figure 2. Changes in soil water content in the upper 2 meter of the soil during the period of 1980–2010. A: water stored in the upper 2 meter during the fallow season, presented in positive percentage (in volume, v/v). B: water depletion in the upper 2 m during the growing season, presented in negative percentage (in volume, v/v). The soil water changes for fallow season were calculated as the soil water content of October deducted by that of June. Similarly, the soil water changes for growing season were calculated as the soil water content of June deducted by that of October in the previous year.

evaporation and crop transpiration. Mean soil water content was lowest in June, at the beginning of the rain season when the winter wheat was harvested (Figure 1).

The changes of the soil water content (for 16 layers in the upper 2 meter of the soil) during the fallow and growing seasons are shown in Figure 2. The changes in water content in the fallow season mirror the changes during the growing season, i.e. water stored in the soil during the fallow season largely equalled to soil water depletion during the growing season. The soil water changes for fallow season were calculated as the soil water content of October deducted by that of June. Similarly, the soil water changes for growing season were calculated as the soil water content of June deducted by that of October in the previous year. Soil water content mostly changed in the top 1 m, accounting for 70% of total soil water change (±120 mm) (Figure 2). We cannot exclude that some of the seasonal variations are caused by slight spatial variations in soil characteristics.

Model Calibration and Performance

The AquaCrop model was calibrated using the climate data, soil data, field management data and monitoring data. The full set of crop parameters is listed in Table 3. The performance of the model on the simulated crop yield and soil water balance is shown in Figure 3. The relationship between observed and modelled grain yield of the calibrated model for the second set of 15-years data (validation step) was almost as good as for the whole set of data (not shown). For the 30 years' data, the relationship between observed and modelled grain yield had a correlation coefficient of (R^2) of 0.77, and the index of agreement (d) was 0.93 (Figure 3A). Similarly, the relationship between observed and modelled soil water content showed a R^2 of 0.78 and an index of agreement (d) of 0.93 (Figure 3B). Mean absolute errors (MAE) were 311 kg/ha and 25 mm, respectively. Mean bias errors (MBE) were −168 kg/ha and −12 mm in simulating yields and soil water balance, respectively, suggesting that the model slightly underestimated grain yields and soil water contents by 8 and 9%, respectively. We conclude that the performance of the AquaCrop model was acceptable for doing further simulations.

Water Balance and Partitioning of E and T

AquaCrop was used to estimate soil evaporation and crop transpiration, and the water balance for both fallow and growing seasons during the period 1980–2010 (Figure 4). There was only a very marginal change in soil water content when considering the water balance of the total season (fallow+growing season) over the 30 years' period; total rainfall was nearly equal to the sum of soil

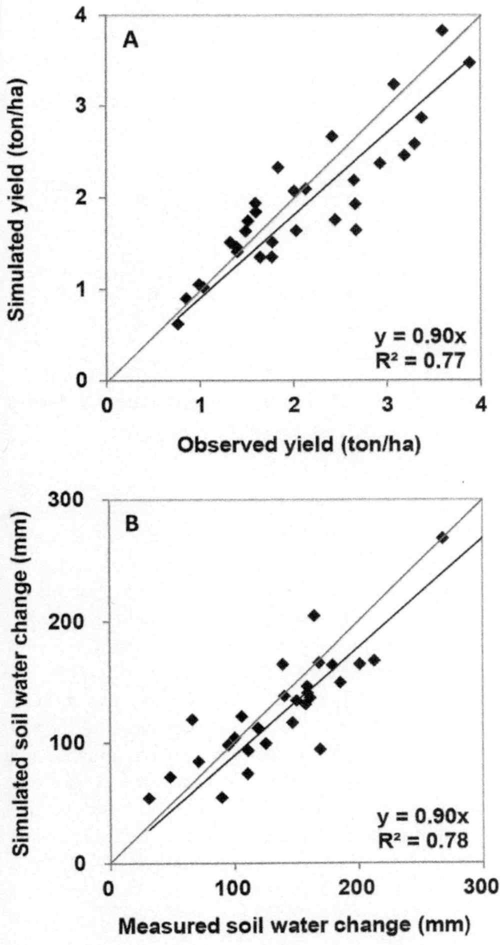

Figure 3. The AquaCrop Model simulations on yield (A) and soil water change (B). Diagonal lines represent 1:1 lines.

evaporation and crop transpiration. Crop transpiration accounted for approximately one-third of total seasonal rainfall. The remaining two-thirds were assumed to be lost by soil evaporation (Figure 4A), but we cannot exclude that a small fraction of this evaporative loss was actually lost by leaching and/or runoff. Mean rainfall in the fallow season was 323 mm, of which 207 mm (64%) was lost by soil evaporation and 116 mm (36%) was stored in the soil (Figure 4B). During the growing season, mean crop transpiration was 185 mm (57%) and soil evaporation was 137 mm (43%), of which rainfall and soil moisture contributed 194 (60%) and 128 mm (40%), respectively (Figure 4C). Mean rainfall of the growing season was slightly larger than mean crop transpiration.

Wheat Yield and WUE

Due to limited amounts of available water and the irregular rainfall pattern, wheat yield and WUE were rather low, ranging from 0.6 to 3.9 ton/ha/year and from 0.3 to 0.9 kg/m^3, respectively. Relationships between yield and annual rainfall ($p = 0.028$), and between yield and growing season rainfall ($p = 0.027$) were highly significant (Figures 5). The relationship between yield and rainfall during the fallow season was not significant ($p = 0.167$). Furthermore, there were significant linear relationships between wheat grain yield and ET, T, and WUE, with coefficients of determination (R^2) of 0.61, 0.68 and 0.72, respectively (Figure 5). Given the slope (0.01) between yield and T, an increase of 1 mm available water can produce 10 kg grain per hectare.

Model Experiments

In model experiment E1 (winter wheat planted at 50% of FC), mean wheat yield and WUE were 2 ton/ha and 0.6 kg/m^3, respectively (Figure 6). In model experiment E2, with organic mulch during the growing season, mean yield increased to 2.3 ton/ha and WUE increased to 0.8 kg/m^3. In model experiment E3, with plastic film during the growing season, mean yield increased further to 2.5 ton/ha and WUE to 1.0 kg/m^3.

Similarly, in model experiment E4 (winter wheat planted at 70% of FC), mean yield and WUE were 2.9 ton/ha and 0.8 kg/m^3, respectively. With organic mulch (E5), mean yield and WUE increased to 3.2 ton/ha and 0.9 kg/m^3, and with plastic film cover (E6), mean yield and WUE increased further to 3.5 ton/ha and 1.1 kg/m^3. Our model experiments show that both organic mulch and plastic film cover could significantly improve yield and WUE, but the impact of plastic cover was bigger than the organic mulch mainly due to high effectiveness in reducing soil evaporation. Moreover, increasing water storage in the soil during the fallow season, so that available soil water content increases from 50 to 70% of FC in autumn at the time of winter wheat seeding, is at least as effective as a plastic cover during the growing season.

Discussion

We successfully calibrated and validated the AquaCrop model on the basis of the long-term monitoring data (30 years) of rainfed winter wheat on the Loess Plateau of northern China. The full set of crop parameters (Table 3) may provide also guidance to future studies and further model calibrations and validations. We also quantified four key components of the water balance, i.e., rainfall, changes in soil water, soil evaporation and crop transpiration, for both fallow and growing seasons by combining empirical data from a long-term wheat monitoring site with calculated results using the AquaCrop model. In the end, we also explored the potential for increasing wheat yield and WUE through model experiments.

The relationship between observed and modelled wheat yield had a R^2 of 0.77, slope of 0.9 and an index of agreement (d) of 0.93 (Figure 3). Mkhabela and Bullock [47] reported a R^2 of 0.66, slope of 0.96, index of agreement (d) of 0.99 between observed and modelled wheat yields. Araya et al. [48] reported a R^2>0.80 when simulating barley biomass and grain yield. Stricevic et al. [49] reported a R^2>0.84 when simulating yields of maize, sunflower and sugar beet. Similarly for simulating soil water content, we found a R^2 of 0.78, slope of 0.9 and an index of agreement (d) of 0.93 (Figure 3). Mkhabela and Bullock [47] reported a R^2 of 0.9 and a slope of 0.9 for simulating soil water content. Hence, the performance of AquaCrop for our dryland wheat field is largely comparable with that of other modelling studies.

Water stress limited the crop yield at this site. According to our model simulations, water stress has led to suboptimal yields in essentially all years, for both low and relatively high wheat yields. For example, a very low grain yield (0.6 ton/ha) was recorded in the year 2000, when water stress for leaf expansion and stomatal closure started already at the 54th day after planting. In contrast, water stress for leaf expansion and stomatal closure started to occur only from day 156 after planting in 2003, when grain yield was 3.8 ton/ha. In both cases, water stress occurred before flowering stage (~200 days after planting). Water stress leads to

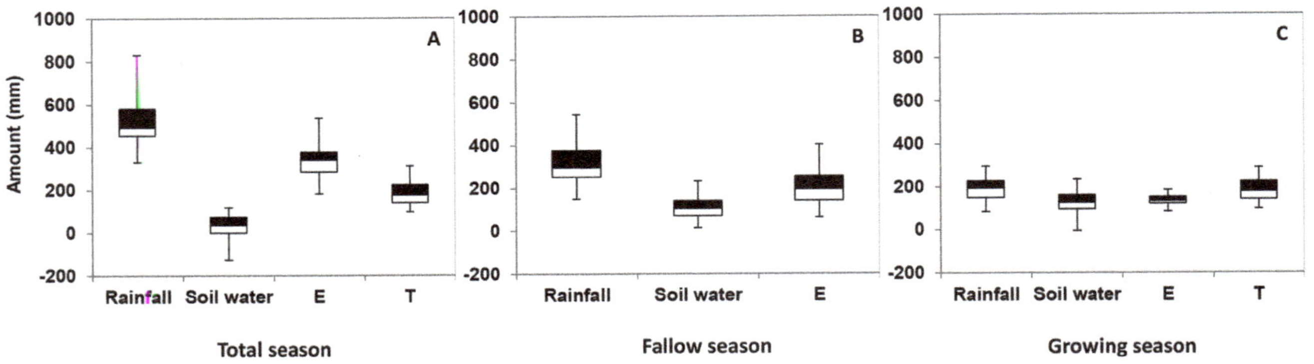

Figure 4. Water balance (rainfall, change in soil water, E, T) of the total season (A), fallow season (B) and growing season (C) during the period of 1980–2010. Total season means the sum of fallow season and growing season.

low grain yield and low WUE, but depending on the stage and duration of the water stress [17,50].

We found significant linear relationships between wheat yield and ET, and between wheat yield and T (Figure 5), in line with some other studies [17,51,52]. Mean T/ET ratio in our study was only 58% during the growing season, which was 8–12% lower than that reported by Liu et al. [53] and Kang et al. [54] but highly in line with that reported by Wang et al. [55]. The higher T/ET ratio in the studies of Liu et al. [53] and Kang et al. [54] was probably due to the irrigation treatments where the crop had more water for transpiration. It is well-known that crop yield and WUE are often lower in rainfed agriculture than in irrigated agriculture [17,20,50], but depending also on possible nutrient, weed, and disease stresses and irrigation management.

Advanced technologies, such as precision irrigation, are for a long time available but unfortunately not affordable and applicable to the farmers of the Loess Plateau, mainly because of the high cost relative to the low value of cereals [56]. Therefore, we focused on low-cost options, such as straw mulch and plastic film cover because those are the most accessible and low cost materials for farmers to implement in the field. Minimizing soil evaporation could save water for crop transpiration, and thereby increase wheat yield and WUE. Our model experiments suggest that wheat yield can be improved significantly by minimizing soil evaporation via organic mulch and plastic film cover, especially also during the fallow period. Mulching with crop residues can decrease soil evaporation and increase soil water retention. Plastic film cover can significantly increase crop yield and WUE, and promote crop growth during early growth when temperature is

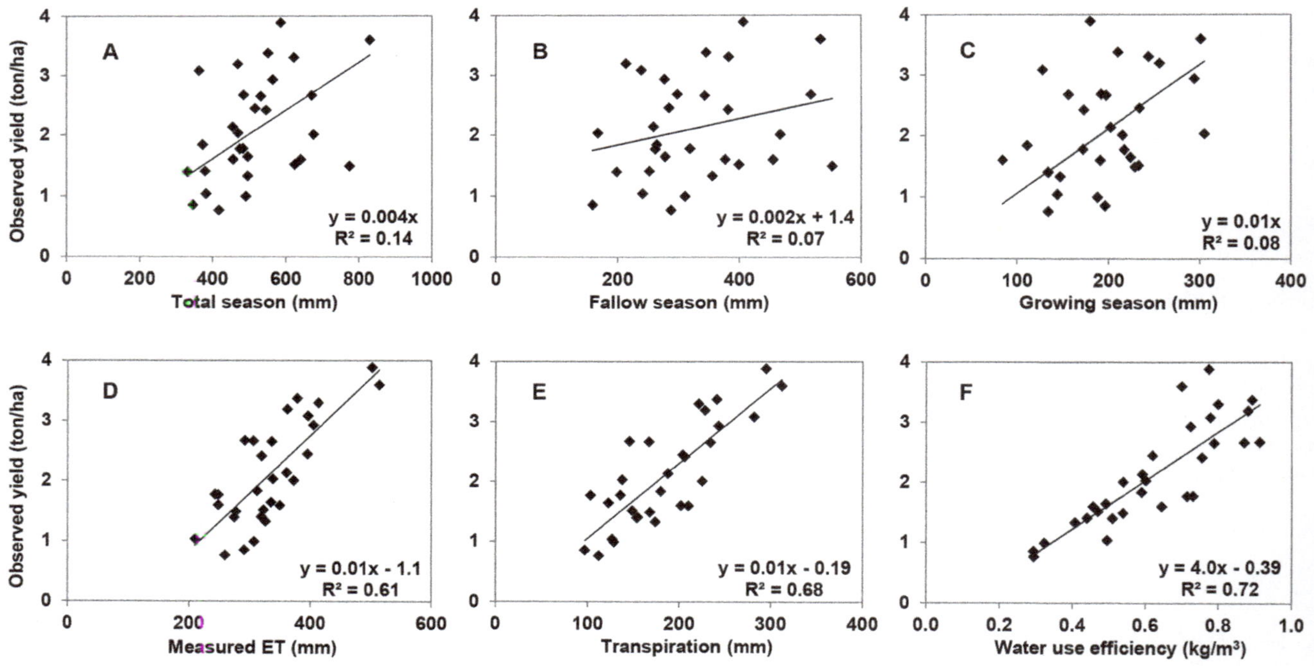

Figure 5. Relationships between observed yield and total rainfall (A), rainfall in fallow season (B), rainfall in growing season (C), measured ET (D), Transpiration (E) and WUE (F). The significant level is 0.028 for (A), 0.167 for (B), 0.027 for (C). For D, E and F, the significant level is all smaller than 0.01.

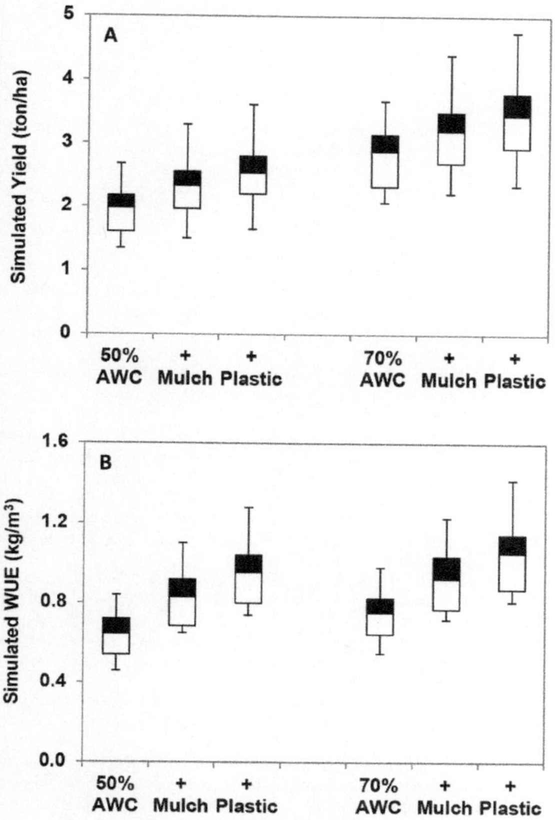

Figure 6. Results of six model experiments; effects of straw mulch and plastic film on wheat yield (A) and water use efficiency (B). Experiment E1: Planting at 50% of FC; Experiment E2: E1+organic mulch; Experiment E3: E1+plastic cover; Experiment E4: Planting at 70% of FC; Experiment E5: E4+organic mulch; Experiment E6: E4+plastic cover.

low. Our results show that crop yields can be increased by ~0.9 ton/ha through increasing soil water storage during the fallow period. Crop yields can be increased further by on average ~0.3 ton/ha through straw mulch and by ~0.5 ton/ha on through plastic film covers during the growing season. At the same time, WUE increases on average by 0.2 to 0.6 kg/m^3. These results are in line with results reported by Deng et al. [57] and others [58–60].

Evidently, increased rainwater harvesting during the fallow season is an effective option. Straw mulch significantly reduces the evaporative water losses during the fallow season and is conducive to the infiltration of rain water in the soil. Plastic film covers are less applicable during the fallow season, because they may limit the infiltration of rain water and thereby increase runoff. Reduced tillage can also improve soil water storage. According to a recent study of Hou et al. [61], rotational tillage (rotation of no-tillage and subsoiling) could significantly increase soil water storage during the summer fallow and wheat growing season compared with conventional tillage. They found that rotational tillage increased wheat yields by 10%, and WUE by 7.5%, respectively.

Conclusions

Low wheat yield at the monitoring site was largely due to (i) limited rainfall, (ii) low soil water storage during fallow season because of high water loss via soil evaporation, and (iii) the poor synchronisation of the wheat growing season to the rainfall distribution season. Although water was limited, on average only one-third of the total available water was actually used by the crop for transpiration. The remaining two-thirds was lost by soil evaporation, 60% during the fallow season and 40% during the growing season. Our model experiments suggest that minimizing soil evaporation via organic mulch or plastic film covers can significantly increase wheat yield and WUE. More importantly, these increases can be realized by the application of relatively low cost measures. Further studies are needed to test the effectiveness of these measures in the field.

Acknowledgments

We thank Xiaoguang Yang, from College of Resources and Environmental Sciences, China Agricultural University and Junqi Zhu, from Wageningen University and Research Centre, for providing the climate data for this study. We also thank Xinxin Wang, from Wageningen University and Research Centre, for helping us with statistics.

Author Contributions

Conceived and designed the experiments: WQ BC OO. Performed the experiments: WQ BC. Analyzed the data: WQ BC. Contributed reagents/materials/analysis tools: WQ BC OO. Wrote the paper: WQ BC OO.

References

1. de Wit CT (1958) Transpiration and crop yields. Verslag Landbouwkundig Onderzoek, Wageningen, Netherlands: Wageningen University.
2. Marschner H (1995) Mineral Nutrition of Higher Plants. Academic Press, San Diego: 889 p.
3. Piao SL, Ciais P, Huang Y, Shen ZH, Peng SS, et al. (2010) The impacts of climate change on water resources and agriculture in China. Nature 467: 43–51.
4. Hoekstra AY, Mekonnen MM (2012) The water footprint of humanity. Proceedings of the National Academy of Sciences of the United States of America 109: 3232–3237.
5. IPCC (2007) Climate Change 2007: Impacts, Adaptation and Vulnerability. Contribution of Working Group II to the Fourth Assessment Report of the International Panel of Climate Change. Cambridge Univ Press, UK.
6. Marris E (2008) Water: More crop per drop. Nature 452: 273–277.
7. Milly PCD, Betancourt J, Falkenmark M, Hirsch RM, Kundzewicz ZW, et al. (2008) Climate change - Stationarity is dead: Whither water management? Science 319: 573–574.
8. Hanjra MA, Qureshi ME (2010) Global water crisis and future food security in an era of climate change. Food Policy 35: 365–377.
9. World Water Assesment Programme (2009) The 3rd United Nations World Water Development Report: Water in a Changing World. Paris: UNESCO, and London: Earthscan.
10. Rockstrom J, Karlberg L, Wani SP, Barron J, Hatibu N, et al. (2010) Managing water in rainfed agriculture-The need for a paradigm shift. Agricultural Water Management 97: 543–550.
11. Vitousek PM, Mooney HA, Lubchenco J, Melillo JM (1997) Human domination of Earth's ecosystems. Science 277: 494–499.
12. Smil V (2000) Feeding the world. A challenge for the twenty-first century. MIT Press, Cambridge, Massachusetts, USA.
13. Tilman D, Cassman KG, Matson PA, Naylor R, Polasky S (2002) Agricultural sustainability and intensive production practices. Nature 418: 671–677.
14. Sachs JD (2008) Common Wealth. Economics for a Crowded Planet. The Penguin Press, New York: 386 p.
15. Godfray HC, Beddington JR, Crute IR, Haddad L, Lawrence D, et al. (2010) Food security: the challenge of feeding 9 billion people. Science 327: 812–818.
16. Zhang H, Wang X, You M, Liu C (1999) Water-yield relations and water-use efficiency of winter wheat in the North China Plain. Irrigation Science 19: 37–45.
17. Kang SZ, Zhang L, Liang YL, Hu XT, Cai HJ, et al. (2002) Effects of limited irrigation on yield and water use efficiency of winter wheat in the Loess Plateau of China. Agricultural Water Management 55: 203–216.
18. Zhang XY, Pei D, Chen SY, Sun HY, Yang YH (2006) Performance of double-cropped winter wheat-summer maize under minimum irrigation in the North China Plain. Agronomy Journal 98: 1620–1626.

19. Siahpoosh MR, Dehghanian E (2012) Water Use Efficiency, Transpiration Efficiency, and Uptake Efficiency of Wheat during Drought. Agronomy Journal 104: 1238–1243.
20. Zhang XY, Chen SY, Sun HY, Wang YM, Shao LW (2010) Water use efficiency and associated traits in winter wheat cultivars in the North China Plain. Agricultural Water Management 97: 1117–1125.
21. Whisler FD, Acock B, Baker DN, Fye RE, Hodges HF, et al. (1986) Crop Simulation-Models in Agronomic Systems. Advances in Agronomy 40: 141–208.
22. de Wit CT, Vankeulen H (1987) Modeling Production of Field Crops and Its Requirements. Geoderma 40: 253–265.
23. Steduto P, Hsiao TC, Raes D, Fereres E (2009) AquaCrop-The FAO Crop Model to Simulate Yield Response to Water: I. Concepts and Underlying Principles. Agronomy Journal 101: 426–437.
24. Zwart SJ, Bastiaanssen WGM, de Fraiture C, Molden DJ (2010) A global benchmark map of water productivity for rainfed and irrigated wheat. Agricultural Water Management 97: 1617–1627.
25. Steduto P, Hsiao TC, Fereres E, Raes D (2012) FAO Irrigation and drainage paper 66, Crop yield response to water, Food and Agriculture Organization of the United Nations.
26. Raes D, Steduto P, Hsiao TC, Fereres E (2009) AquaCrop-The FAO Crop Model to Simulate Yield Response to Water: II. Main Algorithms and Software Description. Agronomy Journal 101: 438–447.
27. Hsiao TC, Heng L, Steduto P, Rojas-Lara B, Raes D, et al. (2009) AquaCrop-The FAO Crop Model to Simulate Yield Response to Water: III. Parameterization and Testing for Maize. Agronomy Journal 101: 448–459.
28. Todorovic M, Albrizio R, Zivotic L, Saab MTA, Stockle C, et al. (2009) Assessment of AquaCrop, CropSyst, and WOFOST Models in the Simulation of Sunflower Growth under Different Water Regimes. Agronomy Journal 101: 509–521.
29. Farahani HJ, Izzi G, Oweis TY (2009) Parameterization and Evaluation of the AquaCrop Model for Full and Deficit Irrigated Cotton. Agronomy Journal 101: 469–476.
30. Dominguez A, Tarjuelo JM, de Juan IA, Lopez-Mata E, Breidy J, et al. (2011) Deficit irrigation under water stress and salinity conditions: The MOPECO-Salt Model. Agricultural Water Management 98: 1451–1461.
31. Geerts S, Raes D, Garcia M, Taboada C, Miranda R, et al. (2009) Modeling the potential for closing quinoa yield gaps under varying water availability in the Bolivian Altiplano. Agricultural Water Management 96: 1652–1658.
32. Geerts S, Raes D, Garcia M, Miranda R, Cusicanqui JA, et al. (2009) Simulating Yield Response of Quinoa to Water Availability with AquaCrop. Agronomy Journal 101: 499–508.
33. Andarzian B, Bannayan M, Steduto P, Mazraeh H, Barati ME, et al. (2011) Validation and testing of the AquaCrop model under full and deficit irrigated wheat production in Iran. Agricultural Water Management 100: 1–8.
34. Salemi H, Soom MAM, Lee TS, Mousavi SF, Ganji A, et al. (2011) Application of AquaCrop model in deficit irrigation management of Winter wheat in arid region. African Journal of Agricultural Research 6: 2204–2215.
35. Araya A, Keesstra SD, Stroosnijder L (2010) Simulating yield response to water of Teff (Eragrostis tef) with FAO's AquaCrop model. Field Crops Research 116: 196–204.
36. Abedinpour M, Sarangi A, Rajput TBS, Singh M, Pathak H, et al. (2012) Performance evaluation of AquaCrop model for maize crop in a semi-arid environment. Agricultural Water Management 110: 55–66.
37. ISS-CAS (2012) China Soil Scientific Database Website. Institute of Soil Science, Chinese Academy of Science (ISS-CAS), China. Available: http://www.soil.csdb.cn/. Accessed 2012 Oct 1.
38. IUSS (2006) World reference base for soil resources 2006. A framework for international classification, correlation and communication, 2nd edition. World Soil Resources Reports No 103 FAO, Rome.
39. Wang XB, Cai DX, Hoogmoed WB, Oenema O, Perdok UD (2007) Developments in conservation tillage in rainfed regions of North China. Soil & Tillage Research 93: 239–250.
40. Wang XB, Wu HJ, Dai K, Zhang DC, Feng ZH, et al. (2012) Tillage and crop residue effects on rainfed wheat and maize production in northern China. Field Crops Research 132: 106–116.
41. Xie WY, Zhou HP, Guan CL, Zhang JJ, Yan XY, et al. (2011) Spatial Distribution of Soil Active Nutrients in Farmland in Shanxi Province. Journal of Shanxi Agricultural Sciences.
42. Allen RG, Pereira LS, Raes D, Smith M (1998) Crop evapotranspiration. Guidelines for computing crop water requirements. FAO irrigation and drainage paper no. 56, 300 p.
43. Willmott CJ (1982) Some Comments on the Evaluation of Model Performance. Bulletin of the American Meteorological Society 63: 1309–1313.
44. Willmott CJ, Matsuura K (2005) Advantages of the mean absolute error (MAE) over the root mean square error (RMSE) in assessing average model performance. Climate Research 30: 79–82.
45. Hu K, Li B, Chen D, Zhang Y, Edis R (2008) Simulation of nitrate leaching under irrigated maize on sandy soil in desert oasis in Inner Mongolia, China. Agricultural Water Management 95: 1180–1188.
46. Hu KL, Li Y, Chen WP, Chen DL, Wei YP, et al. (2010) Modeling Nitrate Leaching and Optimizing Water and Nitrogen Management under Irrigated Maize in Desert Oases in Northwestern China. Journal of Environmental Quality 39: 667–677.
47. Mkhabela MS, Bullock PR (2012) Performance of the FAO AquaCrop model for wheat grain yield and soil moisture simulation in Western Canada. Agricultural Water Management 110: 16–24.
48. Araya A, Habtu S, Hadgu KM, Kebede A, Dejene T (2010) Test of AquaCrop model in simulating biomass and yield of water deficient and irrigated barley (Hordeum vulgare). Agricultural Water Management. 1838–1846.
49. Stricevic R, Cosic M, Djurovic N, Pejic B, Maksimovic L (2011) Assessment of the FAO AquaCrop model in the simulation of rainfed and supplementally irrigated maize, sugar beet and sunflower. Agricultural Water Management. 1615–1621.
50. Zhang HP, Oweis T (1999) Water-yield relations and optimal irrigation scheduling of wheat in the Mediterranean region. Agricultural Water Management 38: 195–211.
51. Huang MB, Gallichand J, Zhong LP (2004) Water-yield relationships and optimal water management for winter wheat in the Loess Plateau of China. Irrigation Science 23: 47–54.
52. Kang SZ, Zhang L, Liang YL, Dawes W (2003) Simulation of winter wheat yield and water use efficiency in the Loess Plateau of China using WAVES. Agricultural Systems 78: 355–367.
53. Liu CM, Zhang XY, Zhang YQ (2002) Determination of daily evaporation and evapotranspiration of winter wheat and maize by large-scale weighing lysimeter and micro-lysimeter. Agricultural and Forest Meteorology 111: 109–120.
54. Kang SZ, Gu BJ, Du TS, Zhang JH (2003) Crop coefficient and ratio of transpiration to evapotranspiration of winter wheat and maize in a semi-humid region. Agricultural Water Management 59: 239–254.
55. Wang P, Song XF, Han DM, Zhang YH, Zhang B (2012) Determination of evaporation, transpiration and deep percolation of summer corn and winter wheat after irrigation. Agricultural Water Management 105: 32–37.
56. Robert PC (2002) Precision agriculture: a challenge for crop nutrition management. Plant and Soil 247: 143–149.
57. Deng XP, Shan L, Zhang HP, Turner NC (2006) Improving agricultural water use efficiency in and and semiarid areas of China. Agricultural Water Management 80: 23–40.
58. Chakraborty D, Garg RN, Tomar RK, Singh R, Sharma SK, et al. (2010) Synthetic and organic mulching and nitrogen effect on winter wheat (Triticum aestivum L.) in a semi-arid environment. Agricultural Water Management 97: 738–748.
59. Gao YJ, Li Y, Zhang JC, Liu WG, Dang ZP, et al. (2009) Effects of mulch, N fertilizer, and plant density on wheat yield, wheat nitrogen uptake, and residual soil nitrate in a dryland area of China. Nutrient Cycling in Agroecosystems 85: 109–121.
60. Chakraboyty D, Nagarajan S, Aggarwal P, Gupta VK, Tomar RK, et al. (2008) Effect of mulching on soil and plant water status, and the growth and yield of wheat (Triticum aestivum L.) in a semi-arid environment. Agricultural Water Management 95: 1323–1334.
61. Hou XQ, Li R, Jia ZK, Han QF, Wang W, et al. (2012) Effects of rotational tillage practices on soil properties, winter wheat yields and water-use efficiency in semi-arid areas of north-west China. Field Crops Research 129: 7–13.

15

Identification of Factors Influencing the Restoration of Cyanobacteria-Dominated Biological Soil Crusts

Chongfeng Bu[1,2*], Shufang Wu[1,3], Yongsheng Yang[2], Mingguo Zheng[4]

1 Institute of Soil and Water Conservation, Northwest A & F University, Yangling, Shaanxi, China, **2** Institute of Soil and Water Conservation, Chinese Academy of Sciences and Ministry of Water Resources, Yangling, Shaanxi, China, **3** College of Water Resources and Architectural Engineering, Northwest A & F University, Yangling, Shaanxi, China, **4** Institute of Geographic Sciences and Natural Resources Research, Chinese Academy of Sciences, Beijing, China

Abstract

Biological soil crusts (BSCs) cover >35% of the Earth's land area and contribute to important ecological functions in arid and semiarid ecosystems, including erosion reduction, hydrological cycling, and nutrient cycling. Artificial rapid cultivation of BSCs can provide a novel alternative to traditional biological methods for controlling soil and water loss such as the planting of trees, shrubs, and grasses. At present, little is known regarding the cultivation of BSCs in the field due to lack of knowledge regarding the influencing factors that control BSCs growth. Thus, we determined the effects of various environmental factors (shade; watering; N, P, K, and Ca concentrations) on the growth of cyanobacteria-dominated BSCs from the Sonoran Desert in the southwestern United States. The soil surface changes and chlorophyll *a* concentrations were used as proxies of BSC growth and development. After 4 months, five factors were found to impact BSC growth with the following order of importance: $NH_4NO_3 \approx$ watering frequency>shading>$CaCO_3 \approx KH_2PO_4$. The soil water content was the primary positive factor affecting BSC growth, and BSCs that were watered every 5 days harbored greater biomass than those watered every 10 days. Groups that received NH_4NO_3 consistently exhibited poor growth, suggesting that fixed N amendment may suppress BSC growth. The effect of shading on the BSC biomass was inconsistent and depended on many factors including the soil water content and availability of nutrients. KH_2PO_4 and $CaCO_3$ had nonsignificant effects on BSC growth. Collectively, our results indicate that the rapid restoration of BSCs can be controlled and realized by artificial "broadcasting" cultivation through the optimization of environmental factors.

Editor: Brett Neilan, University of New South Wales, Australia

Funding: The National Natural Science Foundation of China (Grant Nos 41071192 & 40701096), the West Light Foundation of the Chinese Academy of Sciences (No. B2008132), and the 111 project (No. B12007) provided funding for this research. The funders had no role in study design, data collection and analysis, decision to publish, or preparation of the manuscript.

Competing Interests: The authors have declared that no competing interests exist.

* E-mail: buchongfeng@163.com

Introduction

Biological soil crusts (BSCs) are highly complex communities composed of mosses, cyanobacteria, lichens, bacteria, and fungi with soil particles [1,2]. BSCs, which are widely distributed in arid and semiarid areas of hot and cold zones, impact biogeochemical cycles and hydrological processes and function to prevent erosion [3,4,5,6]. BSCs are known to enrich soil nutrients, enhance soil stability, and reduce soil erosion by wind and water [7,8,9]. For example, in the Loess Plateau, moss-dominated BSC were shown to reduce water runoff and sediment loss from soils by >30% and >80%, respectively [10]. In a separate wind tunnel experiment, moss- and cyanobacteria-dominated BSCs reduced the wind erosion rate of soils by more than 90% as compared to uncovered desert soils [11]. Therefore, attempts have been made to establish artificial BSCs to promote soil stability and to reduce losses due to wind and water erosion. Wei [12] proposed the concept of BSC carpet engineering, i.e., the creation of artificial BSCs to control sand movement. Subsequent research on BSC inoculation has progressed rapidly, especially regarding the use of artificial cyanobacteria-dominated BSC cultures. For instance, in a 200-ha experimental soil plot established in the Hobq Desert [13,14,15], an artificial cyanobacterial crust enhanced the activity of soil communities and accelerated soil development [16]. Moreover, the average thickness of this artificial BSC ranged from 2.23 to 5.36 mm, with coverage of 70%, after 3 years of growth, suggesting that this approach may have long-term positive effects on soil stability [17].

BSCs are generally classified into moss, lichen, and cyanobacterial crusts based on their predominant compositions. Mosses with stems, leaves, and rhizoids are considered the higher plant and have relatively strong photosynthetic capacities. Cyanobacteria, which usually act as the pioneer organisms for desert soils, are considered the lower plant characterized by low photosynthesis. Lichens are complexes of algal cells enveloped within epiphyte mycelia and can endure extremely arid environments [18]. These three organisms are characterized by different physiological properties and require multifarious environments and conditions for rapid growth, which results in the complexity of artificial restoration for each type of BSC. The influences affecting the development of BSCs can be divided into internal and external factors. Internal factors include the physiological characteristics of the species present in the BSCs and the interactions between these species, whereas external factors involve soil moisture, light intensity, temperature, and the availability of nutrients. A previous study [19] showed that the production of biomass and exopolysaccharides

(EPS) by *Microcoleus vaginatus* Gom. was not linearly related to temperature, light, or the availability of nutrients. Li et al. [20] found that the colonization of cyanobacteria and algae in the earliest successional stages of BSCs was facilitated by a higher soil pH and total potassium content in the topsoil. Recent research [21] showed that BSCs amended with composted sewage sludge promoted N and C fixation and increased chlorophyll *a*. However, the frequent watering of these BSCs led to a lower diversity of cyanobacteria populations. In addition, the organic matter content, soil water content, total N, total P, bioavailable P, bioavailable K, pH, electrical conductivity, and total salt content also significantly affected the abundance of microorganisms in BSCs [22]. External factors, such as those described above, are likely to significantly influence key ecological processes such as nitrogen fixation. *Collema tenax* is the dominant N-fixing lichen in many BSCs, and the gross photosynthesis of this species was surprisingly found to be suppressed by the addition of nutrients such as P, K, and Zn [8]. It remains unclear, however, how nutrient amendment affects biogeochemical processes in the long term and how this may influence BSC stability [23].

For cyanobacterial BSCs, knowledge on the rapid cultivation through the direct broadcasting of crust samples remains scarce. The experimental study of multiple factors will be beneficial to realizing the rapid restoration of cyanobacterial crusts in the field. Based on crust sampling in the field, lab cultivation, and the investigation of biomass and surface changes, the present study examined the effects of light intensity, watering frequency, and nutrient amendment of N, P, K, and Ca on the development of cyanobacteria-dominated BSCs located in the Sonoran Desert in the southwestern United States. The results of this work improve our understanding of the factors that affect BSC development, and this information may narrow gaps between application and research to restore and engineer BSCs in arid and semiarid regions.

Materials and Methods

Soil Sampling and Preparation

Samples were collected from a plot regulated by Arizona State University and located in the Sonoran Desert near Phoenix, Arizona, USA (32°59′27″N, 111°45′38″W). Cyanobacterial crusts were collected from the soil surface (<10 mm depth), and the underlying soil was sampled from a depth >10 mm as the substrate. The crust sample and soil substrate were transported to the laboratory and sifted using a 5-mm sieve; the materials were then air dried and kept for use in subsequent experiments. This study was approved by the Department of Life Sciences, Arizona State University.

The crust sample was dominated by cyanobacteria, and no moss or lichen was observed. The dominant cyanobacteria at the end of the experiment was identified via microscopy as *Microcoleus vaginatus* Gom.

Properties of the Crust and Soil Substrate

The initial chlorophyll *a* concentration of the BSC sample, determined using an acetone extraction method [24], was 1.28±0.08 μg/cm^2, a value much lower than that of other well-developed cyanobacterial crusts (typically 10 to 50 μg/cm^2, [25,26]). The physical and chemical soil properties of both the crust and substrate samples were determined using standard methods [27], and the results are shown in Table 1.

Table 1. Selected physical and chemical properties of the soils of the study.

Sample	Particle size distribution			Total N (g kg^{-1})	Total P (g kg^{-1})	Total K (g kg^{-1})	$CaCO_3$ (g kg^{-1})	TOC (%)	pH
	Sand (g kg^{-1})	Silt (g kg^{-1})	Clay (g kg^{-1})						
Crust	60.1	34.4	5.5	0.60	0.61	0.37	1.2	0.44	7.44
Substrate	72.5	23.3	4.2	0.20	0.44	0.30	1.2	0.16	7.72

Table 2. Treatments used in the cyanobacteria-dominated crust cultivation experiments.

Treatment																
	W10															
	S								NS							
	N				NN				N				NN			
	KP		NKP		KP		NKP		KP		NKP		KP		NKP	
	Ca	NCa	Ca	NCa	Ca	NCa	Ca	NCa	Ca	NCa	Ca	NCa	Ca	NCa	Ca	NCa
Treatment No.	1	2	3	4	5	6	7	8	9	10	11	12	13	14	15	16
Combination	W10+S+N+KP+Ca	W10+S+N+KP	W10+S+N+Ca	W10+S+N	W10+S+KP+Ca	W10+S+KP	W10+S+Ca	W10+S	W10+N+KP+Ca	W10+ N+KP	1W10+N+Ca	W10+N	W10+KP+Ca	W10+KP	W10+Ca	W10

Treatment																
	W5															
	S								NS							
	N				NN				N				NN			
	KP		NKP		KP		NKP		KP		NKP		KP		NKP	
	Ca	NCa	Ca	NCa	Ca	NCa	Ca	NCa	Ca	NCa	Ca	NCa	Ca	NCa	Ca	NCa
Treatment No.	17	18	19	20	21	22	23	24	25	26	27	28	29	30	31	32
Combination	W5+S−N+KP+Ca	W5+S+N+KP	W5+S+N+Ca	W5+S+N	W5+S+KP+Ca	W5+S+KP	W5+S+Ca	W5+S	W5+N+KP+Ca	W5+N+KP	W5+N+Ca	W5+N	W5+KP+Ca	W5+KP	W5+Ca	W5

Note: W10:300 ml deionized water was added to each sample every 10 days. W5:300 ml deionized water was added to each sample every 5 days. S: each sample was covered by a 60% shade cloth at 20 cm above the pot. NS: no shading. N: 2.10 g NH_4NO_3 added to each pot. NN: no NH_4NO_3 added to any pots. KP: 1.05 g KH_2PO_4 added to each pot. NKP: no H_2PO_4 added to any pots. Ca: 2.10 g $CaCO_3$ added to each pot. NCa: no $CaCO_3$ added to any pots.

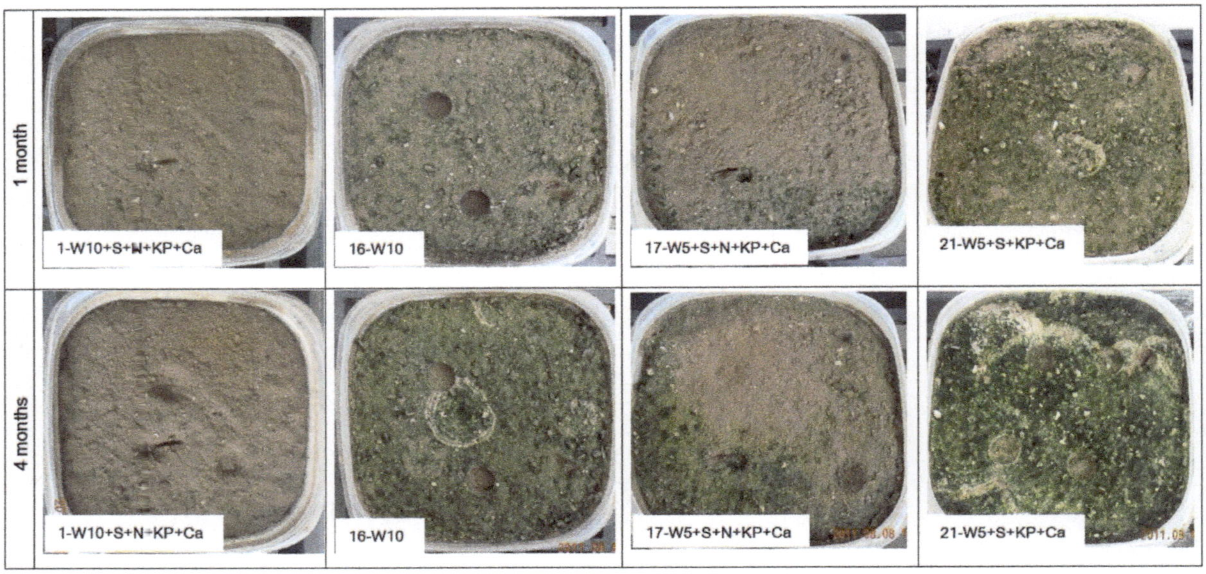

Figure 1. Visible changes in the soil surfaces after 1 and 4 months of cultivation in selected treatments.

Experimental Design

Substrate soil (approximately 1,000 g) was encased in 15 cm×15 cm pots at a depth of 4.5 cm. Nutrient amendments (described below) were stirred into the crust sample, and the crust sample (approximately 200 g) was then homogenized and spread on top of the soil substrate to a depth of 0.5 cm in each pot.

Five external factors were evaluated: frequency of watering [5 or 10 day interval (W5 and W10, respectively)]; shade (S) or no shade (NS); NH_4NO_3 amendment (N) or no NH_4NO_3 amendment (NN); KH_2PO_4 amendment (KP) or no KH_2PO_4 amendment (NKP); and $CaCO_3$ amendment (Ca) or no $CaCO_3$ amendment (NCa). Full factorial combinations of these conditions were tested as described in Table 2. Three replicate growth experiments were performed for each treatment.

The treatment groups were watered every 5 or 10 days using a dripping system until each soil pot was saturated by 300 ml water. Double-deionized water was used to avoid the introduction of additional nutrients. A knitted black shade cloth, which excluded 60% of all incident light, was placed 20 cm above each pot in the shaded treatments. Nutrient amendments were supplied at the following levels: 2.1 g of NH_4NO_3 (300 kg/ha), 1.05 g of KH_2PO_4 (150 kg/ha), and 2.1 g of $CaCO_3$ (300 kg/ha). Pots were incubated in a greenhouse with natural light for a total of 4 months, spanning April to August. The greenhouse was ventilated continuously and did not allow the incubating pots to be exposed to rain. The greenhouse temperatures fluctuated with the changing seasons.

Indices and Data Analysis

The chlorophyll *a* content of the cyanobacteria-dominated BSCs was used as a proxy for their growth as biomass. At the end of the experiment, only the upper layer (≈ 4 mm) of the crust was used for the measurement of chlorophyll *a*. Exopolysaccharides (EPS, indicating the metabolic capacity of cyanobacterial crust) and soil organic matter (SOM) were measured using a combustion method and the phenol-sulfuric acid (PSA) test, respectively [27,28]. Changes at the soil surface were recorded using a digital camera, and the crust coverage was measured using wire meshes with 100 grid squares (5×5 mm). Using the program SPSS 14.0, a paired-sample *t*-test (16 samples) was used to determine the significance of each factor with respect to the chlorophyll *a* concentration. The differences between each set of two factors were analyzed using a one-way ANOVA and LSD analysis (16 samples). The corresponding index is presented as the mean ± standard error.

Results

Changes in the Crust Surface Area

Green surfaces indicative of photosynthetic pigments (e.g., chlorophyll *a*) were not present in any of the pots at the start of the experiment. However, the cyanobacterial crusts developed rapidly, as indicated by the development of green pigment after only 1 month. The coverage increased continually, and the surface color became darker over time, as shown in the photographs of the four treatment groups (treatments 1, 16, 17, and 21; Figure 1) which are the most representative of the surface differences. Figure 1 showed as well that, the treatment groups without NH_4NO_3 generally grew much faster than the other treatment groups. The treatment groups exposed to NH_4NO_3 developed patchy crusts, although additional watering produced better growth in these treatments [e.g., treatments 1 (W10+S+N+KP+Ca) and 17 (W5+S+N+KP+Ca)]. Watering every 5 days in combination with shade produced the best growth. The effects of $CaCO_3$ and KH_2PO_4 on the cyanobacterial crusts were not significant based on the image analysis alone.

In addition to the visual illustration described above, the corresponding coverage of the cyanobacterial crust for those 4 treatments after 1 and 4 months was consistent with the photographic evidence. For example, in treatment groups 16 (less watering) and 21 (more watering), both without NH_4NO_3, the crust coverage reached 82.2±2.19% and 93.3±6.43%, respectively, within 1 month and increased to 95.7±1.86% and 98.3±0.33%, respectively, after 4 months. For treatment groups 1 (less watering) and 17 (more watering), both with NH_4NO_3, the crust coverage was 2.8±2.84% and 12.3±9.61%, respectively, after 1 month and increased to only 4.0±1.15% and 26.7±10.7%,

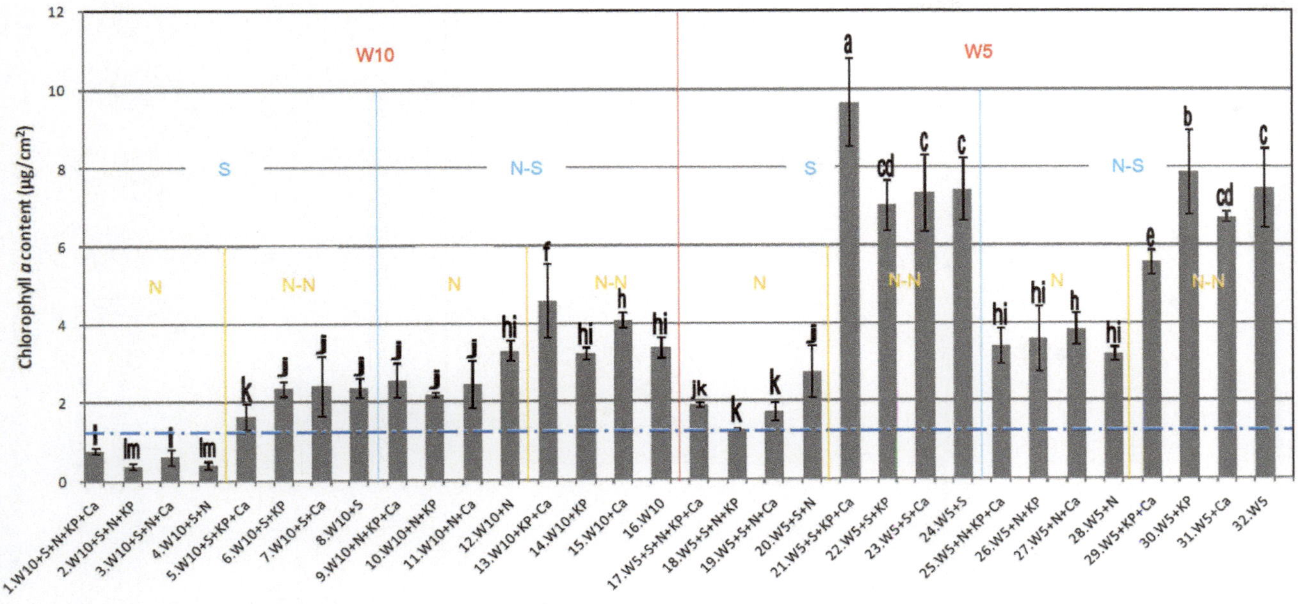

Figure 2. Chlorophyll *a* content in each of the 32 treatments after 4 months of cultivation. Note: W10:300 ml deionized water was added to each sample every 10 days. W5:300 ml deionized water was added to each sample every 5 days. S: each sample was covered by a 60% shade cloth positioned 20 cm above the pot. NS: no shading. N: 2.10 g NH_4NO_3 added to each pot. NN: no NH_4NO_3 added to any pots. KP: 1.05 g KH_2PO_4 added to each pot. NKP: no H_2PO_4 added to any pots. Ca: 2.10 g $CaCO_3$ added to each pot. NCa: no $CaCO_3$ added to any pots. The capital letters indicate the different treatments, while different lower case letters above the bars indicate significant differences between any two treatments at $P<0.01$.

respectively, after 4 months. These results indicated that the crust coverage of the treatment without NH_4NO_3 can reach high levels within a relatively short time, and additional watering enhanced this trend.

Differences of Crust Biomass

Figure 2 shows the chlorophyll *a* levels of each treatment group after 4 months. The watering frequency, light intensity, and different nutrient amendments significantly influenced the crust development. The frequency of watering significantly increased the growth of crust biomass. The comparison of the treatment

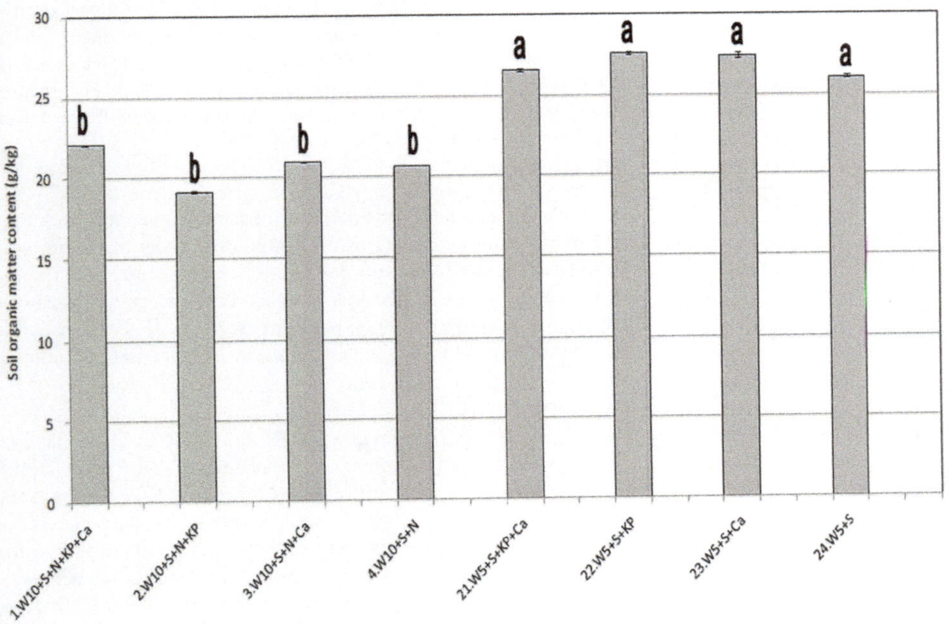

Figure 3. Soil organic matter content of the cyanobacterial crust layer in the treatments with the highest (21–24) and lowest (1–4) measured biomass. Note: Different letters above the bars indicate significant differences ($P<0.01$) between treatments.

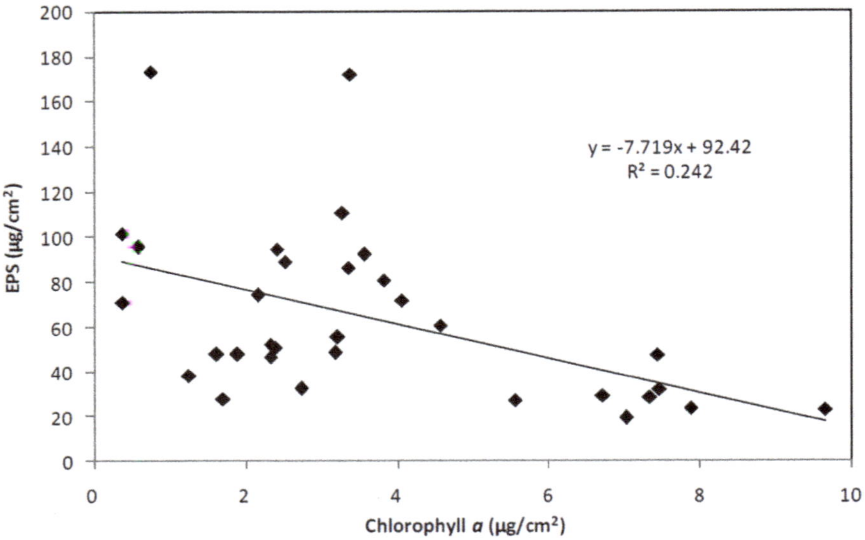

Figure 4. Relationship between chlorophyll *a* and exopolysaccharides (EPS).

groups watered every 10 days (1–16) and the same treatments watered every 5 days (17–32) clearly indicated that a greater frequency of watering produced higher biomass. The highest biomass and best crust growth were produced under sufficient watering (treatments 21–24). However, the effects of watering on cyanobacterial growth varied with different the nutrient levels. The amendment of NH_4NO_3 inhibited the development of the cyanobacterial crust. In particular, the biomass levels of the crusts treated with NH_4NO_3 were consistently lower than those of the untreated crusts in the treatment groups that were watered every 5 days. The effects of KH_2PO_4 and $CACO_3$ on the development of crusts were inconsistent; i.e., higher/lower biomass levels were observed in the presence/absence of both KH_2PO_4 and $CACO_3$.

For the 5-day watering frequency treatments, the highest biomass was observed in the two shaded treatment groups [21 (W5+S+KP+Ca) and 23 (W5+S+Ca)]. This result indicated that shade could positively affect the development of crust biomass when the soil moisture content was high. However, the effects of shading were not consistent and often inhibited biomass accumulation. For example (Figure 2), shading led to the development of less crust biomass in the treatment groups watered every 10 days (treatment groups 1–16); this effect was more pronounced in the groups treated with NH_4NO_3 (1–4, 9–12). Shade inhibited crust growth in the treatment groups that received NH_4NO_3 (17–20, 25–28) but led to a slight enhancement of growth in the groups that were not treated with NH_4NO_3 (21–24, 29–32). Thus, the effect of shade may vary according to other factors, including growth stage, season, and availability of nutrients.

Further statistical analyses were performed to quantify the effects of the five factors on crust development. The paired-sample *t*-test showed that NH_4NO_3 and watering frequency had highly significant effects ($P<0.01$), whereas the degree of shading had a lower, although significant, effect on crust biomass development ($P<0.05$). $CaCO_3$ and KH_2PO_4 had no significant effects on crust biomass development. The one-way ANOVA and LSD analysis indicated there was no significant difference between treatments that received NH_4NO_3 and the frequency of watering. Likewise, no significant difference was observed between treatments that received $CaCO_3$ and KH_2PO_4. The effects of the five factors on cyanobacteria growth (in descending order of importance) were as follows: NH_4NO_3 ≈ watering frequency>shade>$CaCO_3$ ≈ KH_2PO_4. The significant differences between all treatments (Figure 2) showed a similar trend as in the statistical analysis presented above.

Crust Organic Matter and EPS

To further examine the changes in crust development, the treatments with the lowest (treatments 1–4, watered every 10 days, not treated with NH_4NO_3) and highest (treatments 21–24, watered every 5 days, treated with NH_4NO_3) biomass levels were selected for the measurement of the soil organic matter in the cyanobacterial crust layer after 4 months of cultivation (Figure 3). This analysis indicated that the treatments with higher biomass levels contained more soil organic matter. We also measured the EPS contents as an indication of the metabolic capacity of the cyanobacterial crusts. The EPS contents of the crust layers that received the same treatments (Figure 4) were inversely proportional to their chlorophyll *a* contents and soil organic matter levels.

Discussion

For a certain cyanobacterial crust, the external factors including the soil water content, availability of nutrients, light, and temperature are the main factors for rapid crust growth and development. Our results were both consistent and inconsistent with the findings of the previous studies. The reasons or interpretations behind these conclusions are discussed as follows.

Nutrients Factor

In general, BSCs catalyze the fixation of CO_2 and N_2 and function to enrich the nutrient composition of soil [29,30,31,32]. Thus, we assumed that nutrient amendment would facilitate the growth of BSCs, however, our results surprisingly contradicted this hypothesis. Specifically, the application of NH_4NO_3 inhibited the development of cyanobacterial crusts in our study, regardless of the soil moisture content or shade level, although BSCs are known to possess a high capacity for N_2 fixation [33]. It was postulated that the disruption of the N_2 fixation component of the niche inhibited the cyanobacteria growth. This result was in contrast to previous reports that most BSC types require significant amounts of C and N nutrients to maintain vigorous development [21,34]. Furthermore, we found that higher K and Ca levels did not benefit the development of cyanobacterial crusts, which also contrasts with the results of a previous report [20]. This result may be due to the secondary roles of these nutrients in controlling crust development in comparison to the primary variables, such as light intensity and watering frequency. Clearly, these different results or abnormal phenomenon cannot be explained solely based on the current experiments. Further studies are warranted to determine how and why cyanobacterial crusts from different regions differentially respond to nutrient amendment, with the goal of developing a holistic understanding of the environmental conditions that promote optimal crust development.

Soil Moisture Factor

The soil water content was found to represent a critical and positive factor for the development of BSCs in this study, a finding that is consistent with the results of previous studies. For example, the optimum moisture levels for net and gross photosynthesis in the cyanobacteria-dominated BSCs of China's Loess Plateau were 40%–80% of the field water capacity [10]. Likewise, a study conducted in the Negev Desert indicated a strong positive relationship between the daylight wetness duration and the chlorophyll content of the crust, whereas the moisture content played an important role in stabilizing the BSC habitats [35]. As expected, our results were consistent with those of previous reports indicating that a higher soil water content or longer wetness duration significantly promotes the growth and development of cyanobacterial crust. Therefore, increasing the surface soil water and extending the wetness duration should be the most important mechanisms for rapidly restoring crusts in the field.

Light and Temperature Factor

Previously, it was shown that the production of biomass was not linearly correlated with the temperature, light, or the availability of nutrients in the culture. Instead, the optimal biomass levels of *Microcoleus vaginatus* Gom. were produced when the temperature was 30°C ±2°C, the light was 600–700 $\mu E/m^2 \cdot s$, and the renewal rate was 35% [19]. Zhao et al. [10] showed that the light compensation point and saturation point of a cyanobacteria-dominated BSC in the hilly Loess Plateau of China were < 10 $\mu mol/m^2 \cdot s$ and 800 $\mu mol/m^2 \cdot s$, respectively. In our study, light intensity also impacted the development of crust biomass. However, compared with previous reports focused on single factor, our experiment was conducted under variable temperature, light intensity, and soil water content conditions similar to the natural environment. For example, the ranges of light intensity with and without shading were 1.1–401.6 $\mu mol/m^2 \cdot s$ and 8.4–1164.3 $\mu mol/m^2 \cdot s$, respectively on July 28th; meanwhile, the temperature fluctuated at 27.2–36.4°C and 27.8–39.4°C with and without shading, respectively, on July 28th. The total reduction of light associated with the shade cloth was 53.6%–86.9%, whereas the effect of shading on the temperature was minimal (0.6–3.0°C). Clearly, shading simultaneously changed the light intensity and temperature, which most likely resulted in actual temperature or light values that were lower than the optimal level on some days but higher on other days. For this reason, both inhibitory and promoting effects of shading were observed in the study, indicating that shading or the shading level in the field must be considered to obtain the optimal conditions.

Relationship between Biomass and EPS

Unexpectedly, the EPS content was not positively correlated with the biomass level, as reported by a previous study [19]. We hypothesized that EPS might be consumed by heterotrophic organisms in the short term (during this period between the lasting watering event and EPS measurement, EPS was measured on the 5th and 10th day after the last watering event for the W5 and W10 treatments, respectively) or to produce filamentous cells of cyanobacteria in the soil crusts. These explanations would account for the higher biomass and lower EPS levels in these crusts. In addition, compared with studies conducted in the field [36,37], our crust was cultivated in lab conditions and maintained for only a relatively short period (4 months), which may explain the negative relationship between the biomass and EPS. Further experimental studies are necessary to clearly address this question.

Feasibility of Enhanced Restoration of Cyanobacterial Crusts

Wang et al. [16] and Rao et al. [38] both independently assessed the feasibility of using cyanobacterial culture inoculation to promote BSC formation in desert areas in Inner Mongolia. They found that crust coverage could reach 48.5% with an accompanying increase in the chlorophyll *a* content during the second year. However, they used a "culture inoculation" method experiencing isolation, purification, propagation and inoculation, which has a high cost and technical complexity. In this study, we attempted to use a "broadcasting" method through directly spreading the cyanobacterial crusts 'diluted' on the soil substrate and created the appropriate conditions for realizing rapid crust development, which is characterized by convenience and easy operation. The results showed that the spreading of cyanobacterial crust organisms onto a soil substrate with the appropriate regulation of the environment is also effective and feasible for the development of cyanobacteria-dominated BSCs. If effective techniques for condition regulation are developed, the current easy method of crust restoration could be widely applied and engineered in the field.

Conclusions

Our study demonstrated that the cultivation and development of cyanobacteria-dominated crusts through a "broadcasting" method is feasible. Soil moisture is a primary factor influencing the growth of cyanobacteria and significantly promoted the development of cyanobacteria-dominated BSCs. In contrast, amendment with NH_4NO_3 had significant negative effects on the growth of cyanobacterial crusts in the current experiment, whereas KH_2PO_4 and $CaCO_3$ had nonsignificant effects on growth. Shading may positively or negatively impact the crust growth depending on whether the temperature and light intensity under shading are lower or higher than the required optimal condition. Five factors impact BSC growth with the following order of importance: $NH_4NO_3 \approx$ watering frequency > shading > $CaCO_3 \approx KH_2PO_4$. The specific reason for the negative correlation between EPS and biomass should be studied in further research. In the field, we suggested that maintaining high soil moisture content by reducing runoff and increasing infiltration facilitates the growth and restoration of cyanobacterial crusts, meanwhile fertilization with NH_4NO_3 should be avoided. Our understanding of the factors that influence BSC development will facilitate studies regarding the restoration and cultivation of BSCs.

Acknowledgments

We deeply appreciate the kind assistance of Dr. Natalie Myers, Dr. Ferran Garcia-Pichel, Ruth M. Potrafka, Dr. Yevgeniy Marusenko, and Martha Camacho at Arizona State University during the experimental process. We also wish to thank Dr. Gale W. Jeff for assistance with the language and revision.

Author Contributions

Conceived and designed the experiments: CB SW. Performed the experiments: CB SW. Analyzed the data: CB MZ. Contributed reagents/materials/analysis tools: CB. Wrote the paper: CB SW YY.

References

1. Belnap J (2003) The world at your feet: desert biological soil crusts. Front Ecol Environ 1: 181–189.
2. Meeting B (1991) Biological surface features of semiarid lands and deserts. In: Skujins J, editor. Semiarid and sand deserts: soil resource and Reclamation. New York: Marcel Dekker. 257–293.
3. Tighe M, Haling RE, Flavel RJ, Young IM (2012) Ecological succession, hydrology and carbon acquisition of biological soil crusts measured at the micro-scale. PLoS ONE 7(10): e48565. doi:10.1371/journal.pone.0048565.
4. Angel R, Matthies D, Conrad R (2011) Activation of methanogenesis in arid biological soil crusts despite the presence of oxygen. PLoS ONE 6(5): e20453. doi:10.1371/journal.pone.0020453.
5. Bowker MA, Maestre FT, Escolar C (2009) Biological crusts as a model system for examining the biodiversity ecosystem function relationship in soils. Soil Biol Biochem 42: 1–13.
6. Bu CF, Cai GQ, Zhang XC, Ma L (2008) Review on developmental characteristics and ecological functions of soil crust. Prog Geogr 27: 26–31 (in Chinese with English abstract).
7. Bowker MA, Belnap J, Chaudhary VB, Johnson NC (2008) Revisiting classic water erosion models in drylands: The strong impact of biological soil crusts. Soil Biol Biochem 40: 2309–2316.
8. Zhang YM, Wang HL, Wang XQ, Yang WK, Zhang DY (2006) The microstructure of microbiotic crust and its influence on wind erosion for a sandy soil surface in the Gurbantunggut Desert of Northwestern China. Geoderma 132: 441–449.
9. Zhao YG, Xu MX, Belnap J (2010) Response of biocrusts' photosynthesis to environmental factors: a possible explanation of the spatial distribution of biocrusts in the Hilly Loess Plateau region of China. Acta Ecol Sin 30: 4668–4675 (in Chinese with English abstract).
10. Bu CF, Wu SF, Yang YS, Han FP, Meng J (2013) Interactive effects of moss-dominated biological soil crusts and different types of vegetation on erosion and soil water. Soil Use Manage Under review.
11. Wang XQ, Zhang YM, Zhang WM, Han ZW (2004) Wind tunnel experiment of biological crust effect on wind erodibility of sand surface in gurbantunggut desert, Xinjiang. J Glaciol Geocryol 26: 632–638.
12. Wei JC (2005) Biocarpet engineering using microbiotic crust for controlling sand. Arid Zone Res 22: 287–288 (in Chinese with English abstract).
13. Hu CX, Liu YD, Song LR, Zhang DL (2002) Effect of desert soil algae on the stabilization of fine sands. J Appl Phycol 14: 281–292 (in Chinese with English abstract).
14. Rao BQ, Liu YD, Hu CX, Li DH, Shen YW (2009) The technology of man-made algal crust and its applications in control of desertification. Acta Hydrobiol Sin 33: 756–761 (in Chinese with English abstract).
15. Wang WB, Liu YD, Li DH, Hu CX, Rao BQ (2009) Feasibility of cyanobacteria inoculation for biological soil crusts formation in desert area. Soil Biol Biochem 41: 926–929.
16. Tang DS, Wang WB, Li DH, Hu CX, Liu YD (2009) Effects of artificial algal crust on soil enzyme activities of Hopq desert, China. Acta Hydrobiol Sin 31: 339–344 (in Chinese with English abstract).
17. Xie ZM, Liu YD, Hu CX, Chen LZ, Li DH (2007) Relationships between the biomass of algal crusts in fields and their compressive strength. Soil Biol Biochem 39: 567–572.
18. Bu CF, Wu SF, Zhang KK, Yang YS, Gao GX (2013). Biological Soil Crusts: an Eco-adaptive Biological Conservative Mechanism and Implications for Ecological Restoration. Plant Biosyst DOI:10.1080/11263504.2013.819820.
19. Xie ZM, Liu YD, Chen LZ, Hu CX, Li DH, et al. (2008) The effects of different cultivation conditions on the biomass and exopolysaccharide production by *Microcoleus Vaginatus* Gom. Acta Hydrobiol Sin 32: 272–275 (in Chinese with English abstract).
20. Li XR, He MZ, Zerbe S, Li XJ, Liu LC (2010) Micro-geomorphology determines community structure of biological soil crusts at small scales. Earth Surf Proc Land 35: 932–940.
21. Maestre FT, Martín N, Díez B, Lopez-Poma R, Santos F, et al. (2006) Watering, fertilization, and slurry inoculation promote recovery of biological crust function in degraded doils. Microbial Ecol 52: 365–377.
22. Wu H, Wang HL, Liang SM, Nie HL, Zhang YM (2006) Temporal-spatial dynamics of distribution patterns of microorganism relating to biological soil crusts in the Gurbantunggut Desert. Chinese Sci Bull 51: 124–131.
23. Bu CF, Wu SF, Xie YS, Zhang XC (2013) The Study of Biological Soil Crusts: Hotspots and Prospects. CLEAN: Soil Air Water 41(9): 899–906.
24. Tsujimura S, Nakahara H, Ishida N (2000) Estimation of soil algal biomass in salinized irrigation land: a comparison of culture dilution and chlorophyll a extraction methods. J App Phycol 12: 1–8.
25. Bowker MA, Reed SC, Belnap J, Phillips SL (2002) Temporal variation in community composition, pigmentation, and Fv/Fm of desert cyanobacteria soil crusts. Microbial Ecol 43: 13–25.
26. Lan SB, Wu L, Zhang DL, Hu CX, Liu YD (2011) Ethanol outperforms multiple solvents in the extraction of chlorophyll-*a* from biological soil crusts. Soil Biol Biochem 43: 857–861.
27. Institute of Soil Science, Chinese Academic Sciences (1978) Analysis of physical and chemical properties of soil. Shanghai: Shanghai Science and Technology Press. (in Chinese).
28. Dubois M, Gilles KA, Hamilton JK, Rebers PT, Smith F (1956) Colorimetric method for determination of sugars and related substances. Anal Chem 28: 350–356.
29. Belnap J, Harper KT, Warren SD (1994) Surface disturbance of cryptobiotic soil crusts: Nitrogenase activity, chlorophyll content, and chlorophyll degradation. Arid Soil Res Rehab 8: 1–8.
30. Evans RD, Johansen JR (1999) Microbiotic crusts and ecosystem processes. Crit Rev Plant Sci 18: 183–225.
31. Guo YR, Zhao HL, Zuo XA, Sam D, Zhao XY (2008) Biological soil crust development and its topsoil properties in the process of dune stabilization, Inner Mongolia, China. Environ Geol 54: 653–662.
32. Wu N, Zhang YM, Downing A (2009) Comparative study of nitrogenase activity in different types of biological soil crusts in the Gurbantunggut Desert, Northwestern China. J Arid Environ 73: 828–833.
33. Housman DC, Powers HH, Collins AD, Belnap J (2006). Carbon and nitrogen fixation differ between successional stages of biological soil crusts in the Colorado Plateau and Chihuahuan Desert. J Arid Environ 66: 620–634.
34. Chen YQ, Zhao YG, Ran MY (2011) Influence of 4 nutrients on the development of moss crust. J Northwest A&F Univ (Nat. Sci. Ed.) 39(5): 44–50 (in Chinese with English abstract).
35. Kidron GJ, Vonshak A, Abeliovich A (2009) Microbiotic crusts as biomarkers for surface stability and wetness duration in the Negev Desert. Earth Surf Process Land 34: 1594–1604.
36. Chen LZ, Liu YD, Song LR (2002) The function of exopolysaccharides of *Microcoleus* in the formation of desert soil. Acta Hydrobiol Sin 26(2): 155–159.
37. De Philippis R, Vincenzini M (2003) Outermost polysaccharidic investments of cyanobacteria: Nature, significance and possible applications. Recent Res Devel Microbiol 7: 13–22.
38. Rao BQ, Wang WB, Lan SB, Li DH, Hu CX. (2009) Development characteristics and distribution of microorganisms within 3-year-old artificial algal crusts in Hopq desert. Acta Hydrobiol Sin 33: 937–944 (in Chinese with English abstract).

16

Integrating Stand and Soil Properties to Understand Foliar Nutrient Dynamics during Forest Succession Following Slash-and-Burn Agriculture in the Bolivian Amazon

Eben N. Broadbent[1,2,3]*, Angélica M. Almeyda Zambrano[1,3,4], Gregory P. Asner[1], Marlene Soriano[5], Christopher B. Field[1,2], Harrison Ramos de Souza[6], Marielos Peña-Claros[5,7], Rachel I. Adams[8], Rodolfo Dirzo[2], Larry Giles[1]

1 Department of Global Ecology, Carnegie Institution for Science, Stanford, California, United States of America, **2** Department of Biology, Stanford University, Stanford, California, United States of America, **3** Sustainability Science Program, Kennedy School of Government, Harvard University, Cambridge, Massachusetts, United States of America, **4** Department of Anthropology, Stanford University, Stanford, California, United States of America, **5** Instituto Boliviano de Investigación Forestal, Santa Cruz de la Sierra, Bolivia, **6** Centro de Investigación y Preservación de la Amazonia, Universidad Amazónica de Pando, Cobija, Bolivia, **7** Forest Ecology and Forest Management Group, Wageningen University, Wageningen, the Netherlands, **8** Plant and Microbial Biology, University of California, Berkeley, California, United States of America

Abstract

Secondary forests cover large areas of the tropics and play an important role in the global carbon cycle. During secondary forest succession, simultaneous changes occur among stand structural attributes, soil properties, and species composition. Most studies classify tree species into categories based on their regeneration requirements. We use a high-resolution secondary forest chronosequence to assign trees to a continuous gradient in species successional status assigned according to their distribution across the chronosequence. Species successional status, not stand age or differences in stand structure or soil properties, was found to be the best predictor of leaf trait variation. Foliar $\delta^{13}C$ had a significant positive relationship with species successional status, indicating changes in foliar physiology related to growth and competitive strategy, but was not correlated with stand age, whereas soil $\delta^{13}C$ dynamics were largely constrained by plant species composition. Foliar $\delta^{15}N$ had a significant negative correlation with both stand age and species successional status, – most likely resulting from a large initial biomass-burning enrichment in soil ^{15}N and ^{13}C and not closure of the nitrogen cycle. Foliar %C was neither correlated with stand age nor species successional status but was found to display significant phylogenetic signal. Results from this study are relevant to understanding the dynamics of tree species growth and competition during forest succession and highlight possibilities of, and potentially confounding signals affecting, the utility of leaf traits to understand community and species dynamics during secondary forest succession.

Editor: Carl J. Bernacchi, University of Illinois, United States of America

Funding: This work was supported by the Carnegie Institution for Science, Stanford University, a National Science Foundation (NSF) doctoral dissertation improvement grant and a Global Change Education Program Graduate Research Education Fellowship (GCEP GREF) from the U.S. Department of Energy. This work was partially conducted while E. Broadbent was a doctoral fellow and A. Almeyda Zambrano was a Giorgio Ruffolo Fellow in the Sustainability Science Program at Harvard University. Support from Italy's Ministry for Environment, Land and Sea is gratefully acknowledged. The funders had no role in study design, data collection and analysis, decision to publish, or preparation of the manuscript.

Competing Interests: The authors have declared that no competing interests exist.

* E-mail: eben@amazonico.org

Introduction

Secondary forests cover a large and expanding portion of tropical forests worldwide [1]. These forests provide valuable ecosystem services [2], including biodiversity corridors and refugia [3], wildlife habitat, water filtration, and forest products [4]. In addition, carbon uptake by secondary forests is an important factor in greenhouse gas emissions [5], with 30% of deforested areas in the Brazilian Amazon having been abandoned and now in some stage of regrowth [6]. Forest regeneration following slash-burn agriculture is of particular importance, as this activity has resulted in 50% of annual deforestation and 25% of estimated net greenhouse gas emissions in Asia [7]. Due to their fast growth rates, these forests may help alleviate deforestation and degradation pressure on existing old-growth forests [8]. Given the importance of secondary forests, a detailed understanding of the successional processes governing the development of their structure, soil properties and species composition is critical. However, despite numerous studies on successional processes, substantial uncertainty persists regarding their growth rates [5], nutrient dynamics [9], and the interactions between land use history and successional trajectories [8]. This is, in part, a result of the numerous factors influencing regeneration, including soil type [3] and nutrient availability [10], previous land use intensity [11], fire history [12], topography [13], and distance to seed trees [14].

During secondary forest succession, changes in stand structural attributes occur in sync with changes in soil properties and species composition. Succession is typically divided into distinct structural phases characterized by a unique suite of species [15] traditionally divided into successional/functional guilds [16–19]. The first phase of succession lasts only 1–5 years and is dominated by herbs, shrubs and climbers. During the second phase, from approximately 3–30 years following abandonment, pioneer species, with low wood density, rapid growth rates and high light requirements, develop a short stature closed canopy resulting in phase one species being shaded out. A transition then occurs within the species regeneration guild from short-lived 'pioneers of initiation' to longer-lived 'pioneers of exclusion' [20], which while still having high light requirements and rapid growth rates, are able to gain taller statures more typical of a mature forest. In the final phase slow growing shade tolerant species with high wood density replace the pioneers, as most pioneer species seedlings are incapable of growing in the increasingly shaded understory [21,22], resulting in a composition approaching that of mature forest [17,23]. Waring and Running (2007) [24], and see Oliver and Larson (1990) [25] as applied to tropical successional systems in Chazdon (2008) [26], refer to the structural phases of forest succession as stand initiation, stem exclusion and understory reinitiation phases, with the final old growth stage, in both stand structure and composition, being reached 100 to 400 years post-abandonment.

At the species scale, forest succession theory has typically grouped species according to regeneration requirements into species successional categories or guilds [27]. Although common, this approach is limited and studies are beginning to investigate the dynamics and ecological implications of approaches incorporating more detailed species successional classifications [28,29]. Such studies have the potential to elucidate gradients of change in successional species not easily seen when using categorical classifications [23]. New approaches have been undertaken by increasing the number of successional categories [28], referred to as plant functional types, and by developing continuous gradients of successional status using multi-variate methods [23,30]. An improved understanding of how successional status affects leaf traits in secondary forests is necessary given: (a) an increased interest in linkages between plant functional traits and species assembly processes [31,32]; and (b) the increasing use of leaf traits as indicators of ecosystem nutrient cycling and limitation [33].

Foliar properties, including nutrients and isotopes, are being increasingly used to describe the dynamics of plant communities [34,35] and may provide new insights into community successional dynamics [31]. Within most terrestrial ecosystems, nitrogen (N) and phosphorus (P) availability are the primary limiters of plant growth [34] and the foliar N:P ratio has been of particular focus [36]. Its use has highlighted changes from a conservative N cycle in early secondary sites to a conservative P cycle later in succession [37], with an increase in the foliar N:P ratio being used to indicate a shift to P limitation of ecosystem processes [38]. Co-limitation by N and P is also possible [39], similar to that which can occur during primary succession [40]. Further insights into ecosystem dynamics have been revealed by combining foliar nutrient concentrations with carbon and nitrogen stable isotopes. The carbon isotope ratio ($\delta^{13}C$) is representative of leaf intercellular processes and water use efficiency, integrating photosynthetic activity throughout the leaf's lifespan (41). Foliar $\delta^{13}C$ is correlated with a broad range of plant functional characteristics, including leaf size and thickness, stomatal density, and gas exchange metabolism [41] and leaf mass per area (LMA; [42]). The nitrogen stable isotope ratio functions as more of an ecosystem scale integrator determined by internal processes and varying input-output balances, with decreasing foliar $\delta^{15}N$ generally representing a tightening or closure of the N cycle [43].

Although leaf traits have potential to improve our understanding of forest succession dynamics, few studies have been conducted on the factors, including stand or soil properties, constraining leaf trait variation [28,41,43]. In particular, few studies have investigated the drivers - including soil isotope variation [44,45] - of foliar isotope variation within different successional tree species during succession [46,47]. We use a high-resolution forest succession chronosequence following slash-burn agriculture to evaluate the biotic and abiotic predictors of leaf trait variation in 20 tropical tree species encompassing a continuous gradient from early to late successional status. Biotic predictors include stand structural characteristics, taxonomic and phylogenetic analyses, and species successional position, calculated as the stand age at which each species becomes most abundant. Abiotic predictors include a suite of soil properties, including fertility and structure measurements. Our overarching interest is whether leaf trait variation during forest succession is explained principally by changes in: (a) stand age, (b) forest structure, or (c) soil properties, or, alternatively, by (d) shifts along a continuous gradient of species varying in life strategy. In addition, we consider phylogenetic signal as a predictor of leaf trait variation. Results from this study are relevant to better understanding forest regeneration following disturbance, a carbon sink of global importance, as well as forest community and species dynamics in general.

Materials and Methods

Study Sites

This study was carried out in the community of Molienda (municipality of Bolpebra and Department of Pando) in the Bolivian Amazon (11°26′28.189″ S, 69°09′30.06″ W), with the assistance of community members who worked with us, allowed us access to their properties, and shared their extensive knowledge of the area. The forest is considered lowland tropical moist forest with mildly undulating topography, has a mean annual rainfall of 1800 mm, and has a pronounced dry season extending from May to September [48]. In this community, most households rely on slash-and-burn agriculture as their principal food source, with preference to opening agricultural areas within primary forest. Patches of current slash-and-burn agriculture, usually less than 3 ha in extent (hereafter referred to as agriculture), and successional forests growing on abandoned agricultural fields are dispersed throughout the primary forest, which dominates the landscape. Forest stands used in this study were identified through interviews with long-term residents. Information of each stand was cross-validated using important historical events and through triangulation via interviews with multiple community members. Only stands with similar topography, hydrology, and land use history were included. Stands were distributed widely throughout the landscape (>500 m apart) to minimize spatial auto-correlation. In total, 15 successional stands $\leq$3 ha in extent and surrounded by primary forest, with ages ranging from 4–47 years, and two primary forest stands were identified, giving this chronosequence among the highest temporal resolutions and range identified in our literature review. All stands were initially primary forest, which were cleared, burned and then used for growing beans, corn, rice and yucca for 2–3 years prior to abandonment. No stands underwent wildfires, logging, or had been reentered for agricultural use post-abandonment. These stands, being first-cycle, are therefore representative of the lowest intensity of proceeding land use in the Amazon [11].

Forest Structure and Composition

Forest inventories were conducted using one 10×80 m belt transect randomly selected within each stand with the requirement that all transect area was at least 20 m from the stand edge. All trees, both living and dead, $\geq$2 m in height were measured. Botanical samples were collected and brought to the Centro de Investigación y Preservación de la Amazonia (CIPA) herbarium in Cobija, Bolivia for identification. All trees were mapped to Cartesian coordinates within each transect and, for each tree, we quantified diameter at breast height (DBH; 1.3 m; cm), height (m), and crown exposure (CE), defined using a five-point scale (Clark and Clark 1992) in which 1 = no direct light or low amount of lateral light, 2 = intermediate or high amount of lateral light, 3 = vertical light in part of the crown, 4 = vertical light in the whole crown, and 5 = exposed crown with direct light coming from all directions (i.e., emergent). Liana infestation was defined for each tree using a scale of 1 = none, to 4 = completely covered. The approximate percentage of each tree's crown volume was estimated for the following four categories: new, senescent, and mature leaves or no leaves present. Wood density was estimated for all tree species at the highest taxonomic resolution possible using the web-based wood density database [49], and information derived from Nogueira et al. (2007) [50] and Fearnside (1997) [51]. In cases where species identification was not available (29% of living stems), individuals were given the mean calculated wood density of the forest stand. Biomass (kg) was calculated using the equation for tree biomass described in Chave et al. (2005) [52]:

$$\text{Biomass} = 0.0509 \times \text{Wood Density} \times DBH^2 \times \text{Tree Height}$$

For each stand we also calculated the average and maximum tree height and DBH which we refer to as height_{avg}, height_{max}, DBH_{avg}, and DBH_{max}, respectively, and used the mean value of all trees within each stand to describe the other stand structural variables.

We used the Shannon Weiner index to compare compositional diversity among forest stands. The Chao-Jaccard (CJ) dissimilarity index (zero = no dissimilarity, one = complete dissimilarity) was run between all stand ages (excluding dead stems) [53] using the *vegdist* and *mantel* tests in the 'vegan' package in R. Significance of compositional differences among all stand ages was tested using the Pearson method of the Mantel test with 1000 permutations. We developed a continuous metric, termed species successional status, calculated separately for each tree species as the median stand age in which the each species occurred weighted according to the number of stems of each species occurring within each study plot. Species occurrences were also used by Chazdon et al. (2011) [54] to classify generalist and specialist tree species in tropical habitats. Species successional status had a significant positive relationship with stand age (Adj-R^2 = 0.48, P<0.0001, N= 1479; Figure 1), while the lower R^2 value indicated that tree species occurred across a wide range of stand ages enabling the subsequent comparative analyses of the effects of stand age and species successional status on leaf trait variation. This approach differs from that of Peña-Claros (2003) [23], and used by Poorter (2004) [30], which used correspondence analysis to assign a value of 0 (earliest) to 100 (latest) for successional status, as it directly provides a successional status age for each species enabling direct comparison with measurements of stand structure and soil properties. Species data from primary forest stands were not used in the calculation of successional age as stand age was unknown.

Soil Properties

Soil cores (2×4″; AMS, Inc., American Falls, ID) were collected for three randomly chosen locations at depths of 0–10, 10–20, and 20–30 cm. Soil samples were then aggregated for each depth in the field, oven dried at 50°C for 72 hours, lightly ground and sieved to 2 mm to remove coarse particles, including roots and stones. Mineral fractions weighing ~150 grams were placed into 50 ml polypropylene centrifuge tubes for transportation to the Department of Global Ecology for further analysis. pH was measured on fresh soil samples in solution immediately following collection using a hand-held pH meter (Hanna Instruments, Inc.,Woonsocket, RI). Bulk density was determined using the approach described in Elmore and Asner (2006) [180] using the soil corer to obtain a soil from a known volume for each depth. Bulk density samples were oven dried at 70°C for 96 hours, and sieved to 2 mm to remove roots and stones. The mineral fraction of the soil sample was weighed using a portable electronic scale (Ohaus, Inc., Pine Brook, NJ). The volume of fractions larger than 2 mm was recorded using the displacement method. The bulk density per depth was calculated as the mineral fraction sample mass divided by the volume (170–206 ml), adjusted for the volume of the large fraction (mean±standard deviation was 2.6±1.4% of volume). Soil mass was calculated as Mg ha^{-1} to 30 cm depth using the average of the three bulk density measurements and adjusted for the mass >2 mm.

The soil samples were ground to a fine powder using a Wiley Mill and elemental content (%C; %N) for carbon and nitrogen and $\delta^{13}C$ and $\delta^{15}N$ isotope ratios were quantified using a Carlo Erba EA 1110 C:N combustion (NC2500, CE Instruments, Milan, Italy) coupled with an isotope ratio mass spectrometer (IRMS Delta Plus, Finnigan Mat, San Jose, CA) operating in a continuous flow mode. Standards used for carbon and nitrogen isotopes are PDB and AIR, respectively. Data are expressed in δ (‰) notation [177,178] where:

$$\Delta = \left(R_{sample}/R_{standard} - 1\right) * 1000$$

with R equal to the ratio of ^{13}C:^{12}C or ^{15}N:^{14}N for the sample and standard. In this context, a positive δ value means the sample has more of the heavier isotope than the standard and vice-versa. We use the terminology of enrichment (i.e., positive value or less negative trend; becoming heavier) or depletion (i.e., negative value or more negative trend; becoming lighter) of the heavier isotope versus the standard as described by Dawson et al. (2002) [41]. Individual soil depths were then aggregated and homogenized to get average 0–30 cm depth soil samples. These samples were measured for extractable phosphorus using the weak Bray and Sodium Bicarbonate methods (ppm; P1 and P2, respectively), soil pH (saturated paste method), extractable cations (ppm; K, Mg, Ca, Na and H) using 1.0 ammonium acetate @ pH 7.0, and soil texture using NaHexametaphosphate+hydrometer (%) at A&L laboratories (Modesto, CA). Cation exchange capacity (CEC) was calculated as the sum of K, MG, Ca, Na and H (meq/100 g). All variables were converted from meq/100 g, % or ppm to a 0–30 cm per hectare scale using the soil mass (kg ha^{-1}).

Foliar Properties

Tree species selected for foliar analyses were identified to encompass a wide taxonomic range and to be present in as many age stands as possible. Two to three top of canopy trees of each study species were randomly chosen from each stand provided the particular species was present. In total, 149 tree individuals encompassing 20 species, or approximately 10% of all tree species,

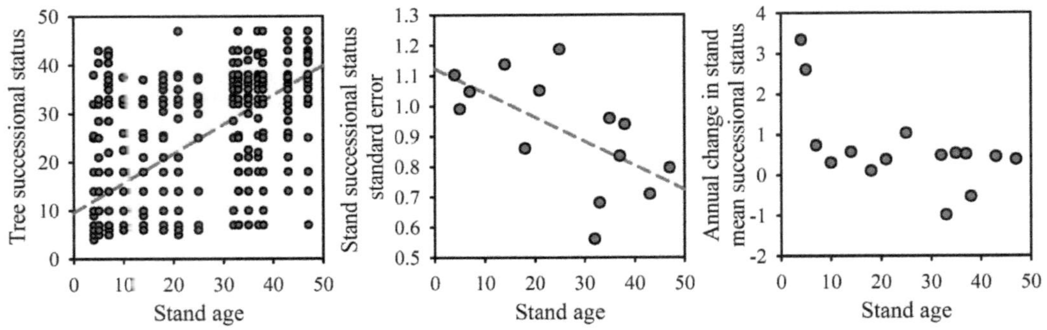

Figure 1. Linear relationship between stand age (years) and species successional status mean (years; left), standard error (SE; middle) and annual change (years; right). Stand values are calculated using all tree stems within each secondary forest stand.

were selected. From these individuals we collected 10–15 fully expanded mature leaves from two separate locations within the full sunlight portion of each individual's crown either by hand or using a shotgun. Leaf samples were oven dried at 60°C for 72 h, sealed in plastic bags and stored in an air-conditioned room prior to transportation to the Carnegie Institution's Department of Global Ecology at Stanford University. Foliar samples were aggregated to the scale of sample tree and ground to a fine powder using a Wiley Mill (Thomas Scientific, Swedesboro, NJ). Foliar N (TKN; mg g^{-1}) and P (TKP; mg g^{-1}) were extracted using a sulfuric acid/hydrogen peroxide digest and quantified using simultaneous colorimetric N and P analyses on a rapid flow autoanalyzer (OI Analytical, College Station, TX), using the ammonium molybdate ascorbic acid method [179]. Elemental content (%C; %N) for carbon and nitrogen and $\delta^{13}C$ and $\delta^{15}N$ isotope ratios were quantified on leaf samples aggregated from each tree using a C:N combustion analyzer coupled with an isotope ratio mass spectrometer as described in the soil methods. Foliar C:N values were calculated using % values and N:P ratios were calculated using Alpkem values.

Statistical Analysis

Statistical analyses for this and all following sections were carried out using JMP v.7.0.1 (SAS Institute, Inc.) and in R v.2.9.2 (http://www.R-project.org). The raw stand, tree, soil and foliar data used in this study is provided as File S1. Summary statistics in the following sections refer to mean ± standard deviation. We use an alpha-level for significance testing of 0.05 throughout, but provide data on less significant findings for general interpretation. First, we measured overall changes in stand structural and soil properties through regressions versus stand age. Soil analyses are aggregated soil depth samples, with the exception - when indicated - of a multiple regression analysis for soil $\delta^{13}C$ and $\delta^{15}N$ including stand age, soil depth (5, 15 and 25 cm) and their interaction as predictor variables. While we compared secondary forest values with those from the primary forest plots using a two-sided t-test, results from these analyses were used for descriptive purposes only given the small primary forest sample size ($N = 2$ for soils and stand values). Second, we investigated relationships between leaf traits and a suite of predictor variables. Leaf traits were treated as response variables and included: (a) %C, %N, and their ratio; (b) N (mg g^{-1}), P (mg g^{-1}), and their ratio; and (c) the stable isotope ratios $\delta^{13}C$ and $\delta^{15}N$. Potential predictors of leaf trait variation were: (a) stand age, (b) stand structure, (c) soil properties, and (d) species successional status. Third, to test for changes in plant isotope fractionation during succession – potentially indicative of changes in the importance of mycorrhizal fungi [43] – linear regressions were run between foliar-minus-soil isotope values, referred to as $\Delta\delta^{13}C_{plant\text{-}soil}$ or $\Delta\delta^{15}N_{plant\text{-}soil}$ [55], and stand age and species successional status.

Whereas stand age and species successional status were individual values, the stand structure and soil properties groups were each composed of 18 unique, but often correlated, variables. To summarize each of these groups we used the first two axes of separate Principal Components Analyses (PCAs) [56]. To identify the most important predictor variables for each leaf trait we used the *bestglm* command (best subsets approach) based on the Akaike Information Criteria (AIC) in R and then used changes in the adjusted R^2 value to select the best number and combination of predictor variables for each leaf trait. Pearson correlations were used to assess relationships among all predictor variables. We further explored relationships between leaf traits, individual stand, and soil variables through regression analysis. To directly test the importance of stand age versus species successional status on leaf trait variation we used the ANOVA command in R to compare linear regression models including only stand age or species successional status to a model including both variables. To test if individual species followed the same relationships as those found across the species community we used linear regressions between leaf traits and stand age for four species (*Cecropia polystachya*, *Miconia* sp., *Jacaranda cuspidifolia* and *Inga* sp.) representing early to later successional statuses and occurring across a wide range of stand ages. In the case of foliar $\delta^{13}C$, in which intra-species patterns opposed those at the community scale, we ran an additional multiple linear regression model using stand age, species successional status and their interaction as predictor variables. Data for leaf trait analyses were transformed, when significantly different from normal as indicated using the *shapiro.test* command in R, using either the Box Cox, logarithmic, square root, or exponential transformation in R.

Third, we tested separately for taxonomic and phylogenetic sources of leaf trait variation, similar to the approach used by Swenson and Enquist (2007) [57]. Taxonomic differences in each leaf trait were tested using a Kruskal-Wallis test in R. A phylogram of phylogenetic relationships among our study species, based on molecular data compiled into a mega-tree, was constructed using Phylomatic (www.phylodiversity.net/phylomatic/), which is a standard approach used in over 46 peer-reviewed articles (www.citeulike.org/group/4921/library/). Further relationships among the Fabaceae were resolved following the Tree of Life Web Project (www.tolweb.org/Fabaceae) and Wojciechowski et al. (2004) [58], and among Moraceae following Zerega et al. (2005) [59]. Angiosperm node ages were calculated from Wikstrom et al. (2001) [60] and Hedges et al. (2006) [61] (www.timetree.org). The

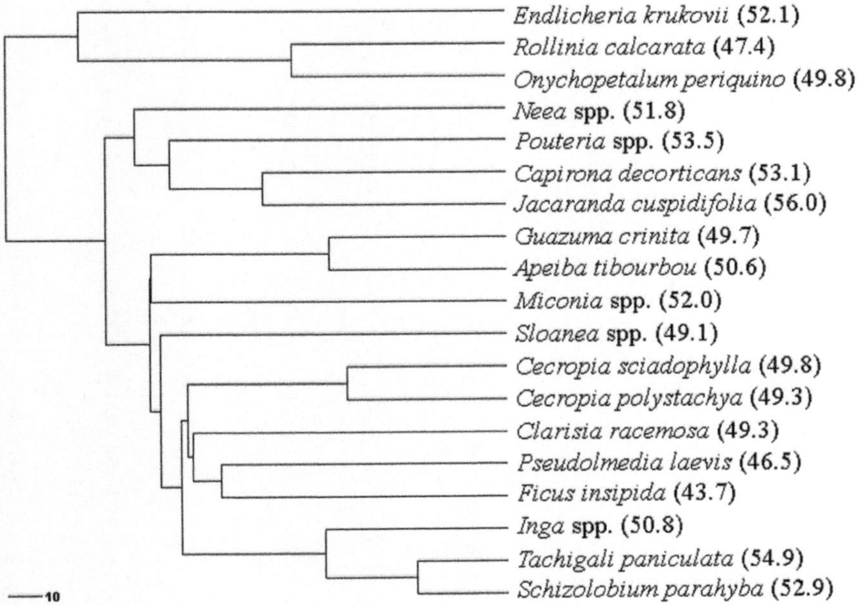

Figure 2. Ultrametric tree of species used for leaf trait phylogenetic signal analyses. Foliar %C is provided after the species name as it was the only trait identified as having significant phylogenetic signal. The scale bar (lower left corner) represents 10 million years.

resultant phylogeny had five (of 18) soft polytomies the ages of which were estimated using Phylocom BLADJ (Branch Length ADJuster; www.phylodiversity.net/bladj; Webb et al. 2008 [62]), following which the phylogeny was converted to an ultrametric tree (i.e., branch lengths consistent with estimated relative time of divergence; Figure 2).

Tests for phylogenetic signal (i.e., do related taxa have more similar leaf traits) were run separately for each leaf trait and for species successional status using the *phylosignal* module of the *picante* package in R [63]. The *phylosignal* module tests for the presence of phylogenetic signal by comparing observed patterns of a leaf trait to a null model produced by randomly shuffling taxa labels across the tips of the phylogenetic tree [64], providing a *P*-value of signal significance, which does not provide information about trait evolution, and a *K*-statistic which tests for evolutionary processes by comparing trait data to an evolutionarily null model in which a *K*-statistic of one is equal to a Brownian motion model of evolution [64]. Although *K*-statistic values greater than one indicate conservatism of traits versus random or convergent evolution (~0), we do not attempt to make any evolutionary arguments in this study [65], given the incomplete status of our phylogenetic tree [66].

Results

Forest Structure and Composition

Forest structure and species richness and diversity information for all study sites are summarized in Table 1. Data were collected from 17 stands encompassing secondary forest ages 4 to 47 and two primary forest stands. We identified 205 species from 1,892 individual trees. The most abundant species was *Jacaranda cuspidifolia* (Bignoniaceae) which had 157 individuals and dominated the developing stand stage. One hundred and thirty eight of our species were represented by less than five individuals, while 78 were represented by only one. The average tree crown contained 65%, 8% and 4% of mature, senescent and new leaves, respectively, while 23% of the crown volume remained vacant. Three distinct phases were identified within our chronosequence (Figure 3), corresponding to stand initiation, stem exclusion and understory reinitiation stages [24,25]. During phase one the percentage of dead trees decreased rapidly from greater than 60% to less than 10% in the 4 and 7 year old stands, respectively. Phase two demonstrated a stem exclusion peak as shown through an increase, then decrease, in understory stem density. Phase three showed a steady increase in understory stem density with little stem mortality. Biomass increased according to a positive Michaelis-Menten asymptotic relationship with stand age during forest succession, while both tree height and DBH for the entire community and for emergent trees only, exhibited significant linear and quadratic relationships (Figure 4). Peaks in tree height and DBH were found between 30–40 years post-abandonment. The Pearson mantel statistic showed significant changes in community composition among the study stands for the entire community ($R = 0.37$, $P = 0.001$, $N = 1892$) and within emergent trees only ($R = 0.43$, $P = 0.001$, N = 1892). While the entire tree species community – dominated by the understory - became more similar to primary forest species composition ($R^2 = 0.75$, $P < 0.0001$, $N = 15$) during succession, no such pattern was shown in the emergent trees, which remained completely dissimilar (i.e., Chao-Jaccard = 1) at all stand ages. Stand age had significant positive correlations with both tree community species richness (Adj-$R^2 = 0.62$, $P < 0.0001$, $N = 15$) and species diversity (Adj-$R^2 = 0.64$, $P < 0.0001$, $N = 15$), with positive trends between stand age and emergent tree species richness (Adj-$R^2 = 0.15$, $P = 0.0819$, $N = 15$) and diversity (Adj-$R^2 = 0.15$, $P = 0.0824$, $N = 15$). Community species richness was also significantly positively correlated with community species diversity (Adj-$R^2 = 0.93$, $P < 0.0001$, $N = 14$). The standard error of species successional statuses, calculated across all stems in each stand, had a significant negative relationship with stand age ($R^2 = 0.38$, $P < 0.0001$, $N = 14$; Figure 1). While rate of taxonomic compositional change was greatest following 20 years, as indicated through species similarity

Table 1. Stand characteristics of all successional and primary (Pr) forest study sites (mean±std. dev.).

	Stand age (years)																
	4	**5**	**7**	**10**	**14**	**18**	**21**	**25**	**32**	**33**	**35**	**37**	**38**	**43**	**47**	**Pr-1**	**Pr-2**
Tree density (# ha^{-1})	1537.5	1375.0	1637.5	1912.5	950.0	1450.0	1237.5	775.0	1350.0	1837.5	1237.5	1025.0	1375.0	1487.5	1700.0	1325.0	1387.5
Dead trees (# ha^{-1})	937.5	237.5	125.0	137.5	137.5	162.5	100.0	87.5	75.0	50.0	75.0	62.5	50.0	62.5	100.0	100.0	62.5
Basal area (m^2 ha^{-1})	10.6	13.2	32.1	24.8	33.4	32.1	31.6	28.2	43.7	36.1	27.3	17.3	30.5	43.3	54.0	34.5	36.3
Biomass (Mg ha^{-1})	27.3	36.1	158.3	101.8	152.0	118.2	196.6	133.4	304.9	243.7	158.4	114.3	204.9	269.8	460.5	282.3	325.6
CE (avg± std. dev.; 1–5)*	4.5±0.7	3.5±1.5	2.3±1.3	3.2±1.4	3.1±1.5	2.9±1.4	2.6±1.4	3.6±1.4	2.1±1.3	1.9±1.2	2.6±1.3	2.8±1.6	2.7±1.3	2.5±1.3	2.6±1.4	1.9±1.2	1.7±1.0
Tree height (avg±std. dev.; m)	8.1±2.3	8.7±3.0	12.6±4.9	10.2±3.9	13.3±7.2	10.9±5.7	13.4±6.9	13.0±6.0	10.3±5.0	10.7±6.7	10.6±6.6	8.2±4.3	10.7±6.9	11.1±5.9	11.4±6.5	11.8±6.1	11.1±6.3
Tree height (max; m)	14	16	25	19	28	27	30	26	28	30	27	30	28	27	38	30	36
DBH (avg± std. dev.; cm)**	8.9±3.0	10.3±4.0	13.7±7.9	11.7±5.3	17.9±11.4	14.0±9.3	15.6±9.1	19.2±9.8	15.5±13.2	12.5±9.8	13.6±9.8	11.6±9.0	13.2±10.4	15.3±11.8	14.6±13.9	13.5±12.3	13.3±12.6
DBH (max; cm)**	17.7	20.0	59.3	30.0	69.0	67.0	45.8	52.0	78.5	49.5	55.2	73.0	50.8	52.8	102.0	82.0	78.3
Lianas (avg ±std. dev.; 1–4)	2.1±1.0	1.5±1.1	1.3±0.8	1.6±1.0	1.8±1.0	1.8±1.2	1.7±1.1	1.5±1.0	1.5±0.9	1.7±1.1	1.8±1.1	2.5±1.3	1.7±1.0	1.3±0.7	1.8±1.1	1.4±0.6	1.5±0.9
Mature leaves (avg±std. dev.; %)	65.0±23.3	56.2±23.5	66.0±22.3	67.0±25.5	52.7±31.1	65.2±31.9	55.1±31.1	41.1±32.7	72.6±25.5	75.0±23.3	65.8±31.4	75.6±22.7	72.0±25.5	62.4±28.6	76.3±27.8	70.2±23.4	72.8±22.0
New leaves (avg±std. dev.; %)	5.4±9.7	4.1±8.7	3.4±9.0	3.7±12.6	4.7±18.3	5.7±14.3	5.7±12.0	4.4±13.4	5.0±17.7	3.4±11.4	6.6±17.7	0.9±5.7	2.4±7.0	2.4±11.6	2.7±11.1	6.7±13.8	4.4±10.9
Senescent leaves (avg±std. dev.; %)	13.5±10.8	14.0±12.1	7.0±8.6	10.3±12.9	10.5±11.5	6.2±9.1	7.6±9.6	8.6±9.1	6.5±10.0	6.7±9.8	6.6±9.0	6.2±10.3	7.5±9.1	9.7±10.1	5.1±8.8	6.0±12.7	4.5±7.7
No leaves (avg±std. dev.; %)	16.0±19.4	26.0±22.8	23.6±20.4	19.0±21.6	32.2±30.9	22.8±27.1	31.7±29.6	46.0±32.1	15.9±17.7	15.0±18.7	21.1±26.6	18.1±22.6	18.2±22.1	25.6±27.2	16.0±21.6	17.0±15.2	18.2±16.7
Wood density (avg± std. dev.; kg m^{-3})	456±112.3	402.7±130.4	496.3±124.2	521.3±95.2	459.2±128.3	464.8±132.9	485.6±107.1	466.9±97.9	611.9±91.3	654.0±177.3	526.1±91.0	620.8±96.9	578.1±88.3	579.2±91.1	601.2±83.5	601.3±59.2	606.4±83.1

Table 1. Cont.

	Stand age (years)																
	4	5	7	10	14	18	21	25	32	33	35	37	38	43	47	Pr-1	Pr-2
Richness (# sp transect-1)	16	18	37	17	25	27	23	15	51	55	44	40	50	57	49	38	43
Diversity (Shannon-Weiner)	2.3	2.3	3.0	1.8	2.9	2.7	2.6	2.3	3.7	3.4	3.5	3.4	3.7	3.8	3.5	3.3	3.5
CJ vs. primary***	0.99	0.98	0.87	0.97	0.94	0.89	0.93	0.96	0.75	0.45	0.68	0.65	0.65	0.31	0.31		

*Crown Exposure (CE) index, **Diameter at Breast Height (1.3 m), ***Chao-Jaccard community composition similarity versus primary forest sites.

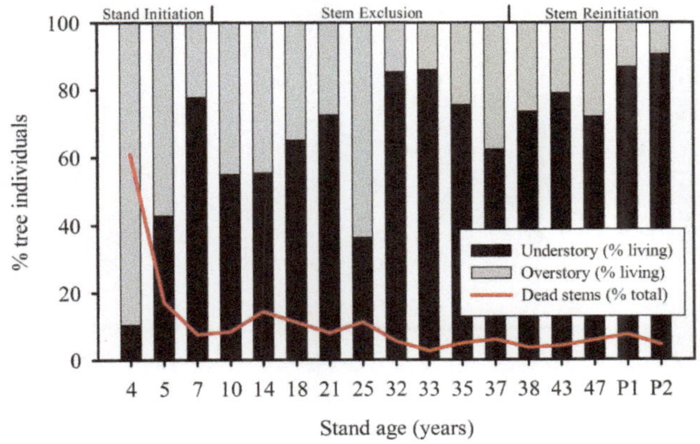

Figure 3. Percentage total living trees in understory and overstory crown exposure (CE) positions; 1–3, and 4–5, respectively. Stand development phases (top) correspond to those described by Waring and Running (2007) [24] and Oliver and Larson (1990) [25].

to primary forest (Figure 4), the rate of change in stand mean successional status was greatest prior to 10 years after which it reduced greatly (Figure 1).

Stand PCA axes 1–3 explained 47%, 24% and 9% of variation across all stand structural variables, respectively (Table 2 stand axes). Stand PCA axis one had strong positive correlations with biomass, basal area, species diversity and $height_{max}$ and negative correlations with number of dead trees and the % crown with senescent leaves. As the strongest correlations were with biomass, we refer to this as the biomass axis. PCA axis two had strong positive correlations with % crown with no leaves, DBH_{avg} and $height_{avg}$ and negative correlations with tree density, liana infestation, and % crown with mature leaves. As the strongest correlations were with $height_{avg}$ and DBH_{avg}, we refer to this axis as the structure axis. While the biomass axis had a significant positive linear relationship with stand age (PCA1 = −4.20 + 0.17 * Stand age, $R^2 = 0.71$, $P<0.0001$, $N = 15$) the structure axis did not. Secondary forests, as compared to primary forests, had significantly lower biomass (178±110 vs. 304±31), wood density (528±75 vs. 604±4), DBH_{max} (54.8±22.1 vs. 80.2±2.6), % crown in mature leaves (64.5±9.8 vs. 71.5±1.84) and greater mean crown exposure class (2.86±0.65 vs. 1.8±0.14), liana infestation (1.71±0.30 vs. 1.45±0.07) and % crown in senescent leaves (8.4±2.7 vs. 5.3±1.1).

Soil Properties

Soil properties for all study sites are summarized in Table 3 and Figure 5. Significant relationships were found between stand age and soil $\delta^{13}C$ (negative, $R^2 = 0.45$, $P = 0.0171$, $N = 12$), CEC (positive, $R^2 = 0.31$, $P = 0.0592$, $N = 12$) and soil N (positive, $R^2 = 0.34$, $P = 0.0461$, $N = 12$). While significant positive relationships were found between stand age and total 0–30 cm depth soil mass (Mg ha^{-1}) ($R^2 = 0.55$, $P = 0.0058$, $N = 12$), and individually for 10–20 cm ($R^2 = 0.55$, $P = 0.0058$, $N = 12$) and 20–30 cm ($R^2 = 0.55$, $P = 0.0058$, $N = 12$), no significant relationship was found for 0–10 cm. The percentage of soil mass >2 mm (mean±STD of 2.47±0.74%) did not show significant changes with stand age, although a trend of an increasing fraction in the 0–10 cm depth and decreasing in the 20–30 cm depth was seen.

Strong positive trends were found, significant following exclusion of the youngest successional stand due to extensive charcoal

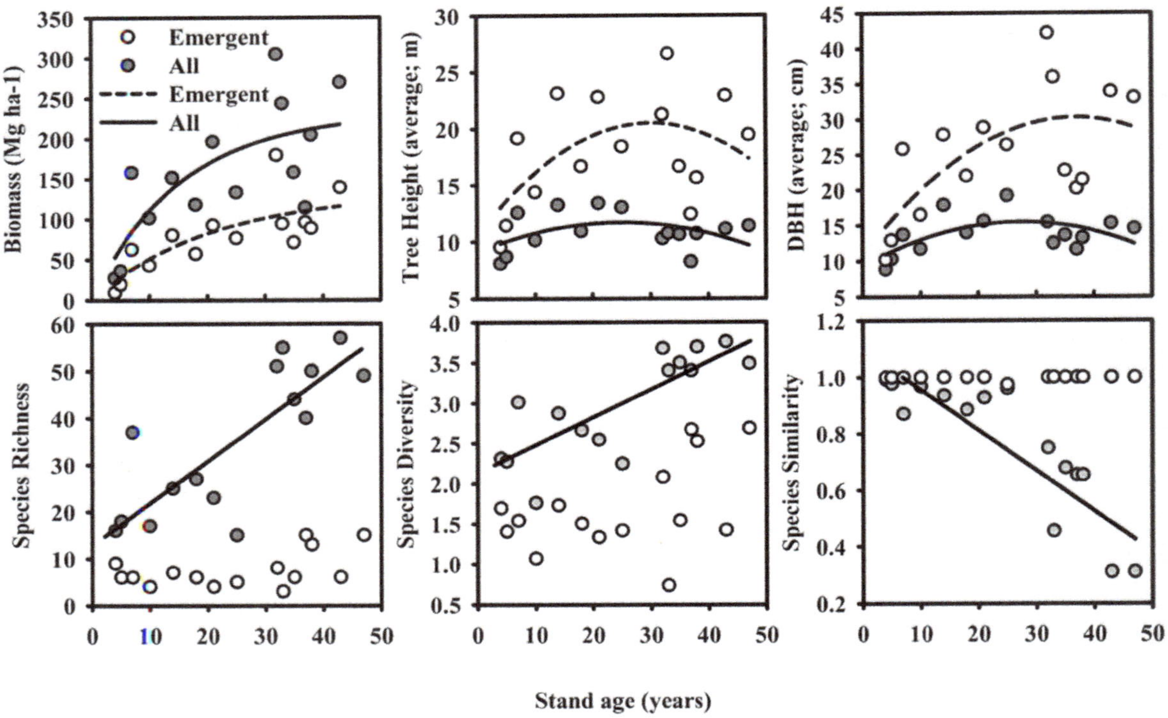

Figure 4. Stand structural and species characteristics for successional forest study sites. Species diversity is calculated using the Shannon-Weiner metric and species similarity is calculated using a Chao-Jaccard index versus primary forest composition. Michaelis-Menten relationship is shown between biomass and stand age, while quadratic are shown between tree height and diameter at breast height (DBH) and stand age.

Table 2. Eigenvectors of stand and soil Principal Components Analysis (PCA) axes.

Stand biomass axis			**Stand structure axis**			**Soil texture axis**			**Soil fertility axis**		
Stand properties		**PCA 1**	**Stand properties**		**PCA 2**	**Soil properties**		**PCA 1**	**Soil properties**		**PCA 2**
Richness	+	0.30711	Mean DBH	+	0.43933	Clay	−	0.37631	Nitrogen	+	0.44133
Biomass	+	0.30329	No leaves	+	0.42061	Sand	+	0.33806	K	+	0.37087
Max height	+	0.2997	Mean height	+	0.4199	C	+	0.33283	Carbon	+	0.35208
CE	−	0.29305	Mature leaves	−	0.36626	P1	+	0.31279	H	+	0.32176
Diversity	+	0.29001	Tree density	−	0.2818	Silt	−	0.30596	$\delta^{15}N$	−	0.31279
Chao-Jaccard	−	0.28963	Liana	−	0.21519	CEC	+	0.2773	Mg	+	0.28983
Wood density	+	0.28943	Dead trees	−	0.2048	P2	+	0.2607	CEC	+	0.2672
Basal area	+	0.28547	Wood density	−	0.17084	Base Saturation	+	0.25635	$\delta^{13}C$	+	0.25618
Senescent leaves	−	0.28039	Basal area	+	0.15914	Mg	+	0.23177	Na	+	0.1867
Max DBH	+	0.27384	New leaves	+	0.15187	$\delta^{13}C$	−	0.21944	Sand	−	0.18151
Dead trees	−	0.2258	Max height	+	0.14433	$\delta^{15}N$	−	0.20182	Clay	+	0.17175
Mature leaves	+	0.20452	Max DBH	+	0.11745	Na	−	0.17779	Silt	+	0.12216
New leaves	−	0.14245	Richness	−	0.11551	Soil mass	−	0.14391	P2	−	0.04887
No leaves	−	0.12287	Chao-Jaccard	+	0.11432	Bulk density	−	0.14209	P1	+	0.04834
Mean DBH	+	0.10854	Senescent leaves	−	0.06739	Carbon	+	0.0838	Soil mass	−	0.03816
Mean height	+	0.07784	Biomass	+	0.06094	H	+	0.06165	Bulk density	+	0.03518
Tree density	+	0.0664	Diversity	−	0.05918	K	−	0.05983	Base Saturation	−	0.03317
Liana	−	0.02621	CE	−	0.03973	Nitrogen	+	0.04247	C	+	0.02439

Stand and soil properties and units are provided in Tables 1 and 3, respectively.

Table 3. Soil properties for all successional and primary (Pr) forest study sites.

	Stand age (years)														
	5	**7**	**10**	**14**	**18**	**21**	**25**	**32**	**35**	**37**	**38**	**43**	**47**	**Pr-1**	**Pr-2**
Carbon (C; Mg ha^{-1})	25.6	15.1	18.6	23.4	20.6	22.9	23.3	19.4	34.7	17.8	28.6	24.4	27.1	25.7	20.4
Nitrogen (N; kg ha^{-1})	2971.6	1714.9	2222.3	2670.9	2532.5	2600.1	2189.5	2538.8	3422.9	2322.9	2883.6	2521.3	2920.4	3210.8	2365.5
$\delta^{15}N$	9.7	9.8	10.8	9.2	9.0	8.0	7.1	8.3	7.6	10.0	8.8	8.9	9.1	9.2	9.1
$\delta^{13}C$	−27.4	−29.3	−27.6	−28.1	−28.2	−28.1	−27.9	−28.2	−28.2	−28.0	−28.2	−29.2	−28.2	−27.1	−27.4
Phosphorus (P1; kg ha^{-1})*	9.3	7.2	8.4	8.6	2.9	7.3		7.3	10.1	1.9	7.8	11.6	7.9	9.9	2.8
Phosphorus (P2; kg ha^{-1})**	10.7	10.9	12.0	11.2	10.9	10.9		7.3	11.2	3.9	7.8	15.4	15.7	7.9	7.0
Calcium (Ca; kg ha^{-1})	253.0	138.7	256.4	305.4	279.5	348.9		690.0	418.7	180.6	223.1	1465.1	253.4	269.4	1366.4
Potassium (K; kg ha^{-1})	284.6	93.7	87.6	148.3	128.4	235.2		194.9	196.2	213.1	108.5	98.2	209.0	264.8	141.6
Magnesium (Mg; kg ha^{-1})	201.9	53.2	82.5	171.4	152.4	150.8		254.7	169.4	121.4	80.9	196.0	107.9	134.3	357.2
Sodium (Na; kg ha^{-1})	27.8	23.2	34.3	34.7	31.1	53.4		29.5	31.7	44.4	35.7	28.1	42.9	45.8	41.2
Hydrogen (H; kg ha^{-1})	32.3	21.9	32.2	26.4	36.5	40.3		73.4	48.9	31.4	31.3	27.2	67.4	47.8	17.6
CEC (kmol ha^{-1})***	69.8	36.4	55.3	60.8	67.3	78.1		134.5	89.8	57.5	53.2	120.0	95.6	80.6	120.5
Base saturation (%)	54.0	40.3	42.2	57.0	46.2	48.9		45.9	45.9	45.9	41.6	77.5	30.1	41.2	85.5
Sand (Mg ha^{-1})	1797.2	2478.1	2331.4	2256.3	1823.4	2338.6	1682.1	2344.1	1956.7	1261.4	1800.0	2559.8	2218.4	1358.0	2251.3
Silt (Mg ha^{-1})	1155.4	666.6	894.2	836.8	1099.8	668.2	1195.5	669.7	1060.5	1572.8	1334.5	863.6	881.0	1515.9	713.1
Clay (Mg ha^{-1})	613.3	478.2	766.5	642.5	694.6	624.6	597.8	626.1	716.9	1058.9	744.8	431.8	833.8	1073.8	531.4
Soil mass (Mg ha^{-1})€	3565.9	3623.0	3992.2	3735.6	3617.8	3631.4	3475.4	3639.9	3734.1	3893.1	3879.3	3855.1	3933.2	3947.8	3495.8
Bulk density (Avg. 0–30 cm; g cm^{-3})	1.233	1.235	1.357	1.300	1.235	1.237	1.180	1.269	1.303	1.336	1.323	1.309	1.338	1.399	1.183

*Weak Bray Extraction,
**Strong Bray Extraction,
€Adjusted for mass >2 mm,
***Cation Exchange Capacity calculated as the sum of Ca, K, Mg, Na and H (meq/100 g).

from the recent burn, between stand age and soil carbon (Mg ha^{-1}), ($R^2 = 0.31$, $P = 0.0589$, $N = 12$) - but not soil organic matter (%) or %C. A negative trend was identified between soil base saturation and stand age ($R^2 = 0.29$, $P = 0.0854$, $N = 11$). No significant relationships were found between stand age and soil phosphorus [as measured through P1 and P2 methods], or sand, silt and clay content (Mg ha^{-1}), although soil P1 had a significantly negative relationship (with a trend in P2) with soil clay content ($R^2 = 0.38$, $P = 0.0321$, $N = 12$) which was not correlated with stand age. While soil %C had no correlation with bulk density, a strong negative trend was found between soil organic matter (%) and bulk density ($R^2 = 0.26$, $P = 0.0921$, $N = 12$).

The multiple regression model of soil $\delta^{15}N$ which included soil depth was highly significant ($R^2 = 0.23$, $P = 0.0285$, $N = 38$), and showed a negative and positive relationship between soil $\delta^{15}N$ and stand age (t-ratio $= -2.07$, F-ratio $= 4.29$, P-value $= 0.0459$) and depth (t-ratio $= 2.25$, F-ratio $= 5.08$, P-value $= 0.0307$), respectively, and no significant interaction (although see Figure 6 for trend). While the model of soil $\delta^{13}C$ was also significant ($R^2 = 0.28$, $P = 0.0111$, $N = 37$), only a positive relationship between $\delta^{13}C$ (which became less negative) and soil depth was found (t-ratio $= 2.81$, F-ratio $= 7.90$, P-value $= 0.0083$) - although a negative trend was found between soil $\delta^{13}C$ (which became more negative) and stand age (t-ratio $= -1.70$, F-ratio $= 2.90$, P-value $= 0.0983$) which was significant when aggregated as described above. Soil $\delta^{15}N$ within the secondary forest stands at mean 7.5, 15 and 25 cm depths were 8.25 ± 1.26, 9.24 ± 1.20, and 9.32 ± 0.97 (‰), respectively, and soil $\delta^{13}C$ values at these depths were -28.43 ± 0.46, -28.14 ± 0.96, and -27.50 ± 0.99 (‰), respectively. However, recognizing the sample size constraints ($N = 6$), no pattern of increasing soil $\delta^{15}N$ was found within the two primary forest stands ($P = 0.3594$) with soil $\delta^{15}N$ of 9.15 ± 0.07, 9.65 ± 0.49, and 8.60 ± 0.28 (‰) for the three depths, respectively, while soil $\delta^{13}C$ did become significantly less negative with depth ($R^2 = 0.83$, $P = 0.0079$, $N = 6$; varying from -28.3 to -26.4 from 7.5 to 25 cm depth, respectively).

Soil PCA axes 1–3 explained 30%, 22%, and 15% of variation across all soil property variables, respectively (Table 2 soil axes). Soil PCA axis one was positively correlated with P1, P2, C, Mg, CEC, base cations, and sand content and negatively correlated with stable isotopes, and silt and clay content. As the strongest correlations were with the soil texture measures we refer to this axis as the soil texture axis. Soil PCA axis two was positively correlated with C, N, $\delta^{13}C$, K, MG, H, and negatively correlated with $\delta^{15}N$. As strong correlations were identified throughout this group, and in particular with soil C, we refer to this axis as the soil fertility axis [67]. While the texture axis had no significant relationship with stand age, the soil fertility axis had a significant positive linear relationship (PCA-2 $= -2.73 + 0.10$ * Stand age, $R^2 = 0.42$, $P < 0.0313$, $N = 11$). Secondary forests, as compared to primary forests, had significantly lower soil $\delta^{13}C$ (-28.20 ± 0.53 vs. -27.25 ± 0.21 (‰), respectively) and higher P2 (10.66 ± 3.24 vs. 7.45 ± 0.64, respectively).

Leaf Traits

The tree species, and their descriptive statistics, selected for inclusion in foliar analyses are provided in Table 4. While most leaf traits were significantly inter-correlated, foliar $\delta^{15}N$ was only correlated, positively, with foliar %C and $\delta^{13}C$ (Table 5). Prior to assessing leaf trait relationships with our six predictor variables, we used Pearson correlations to understand their relationships

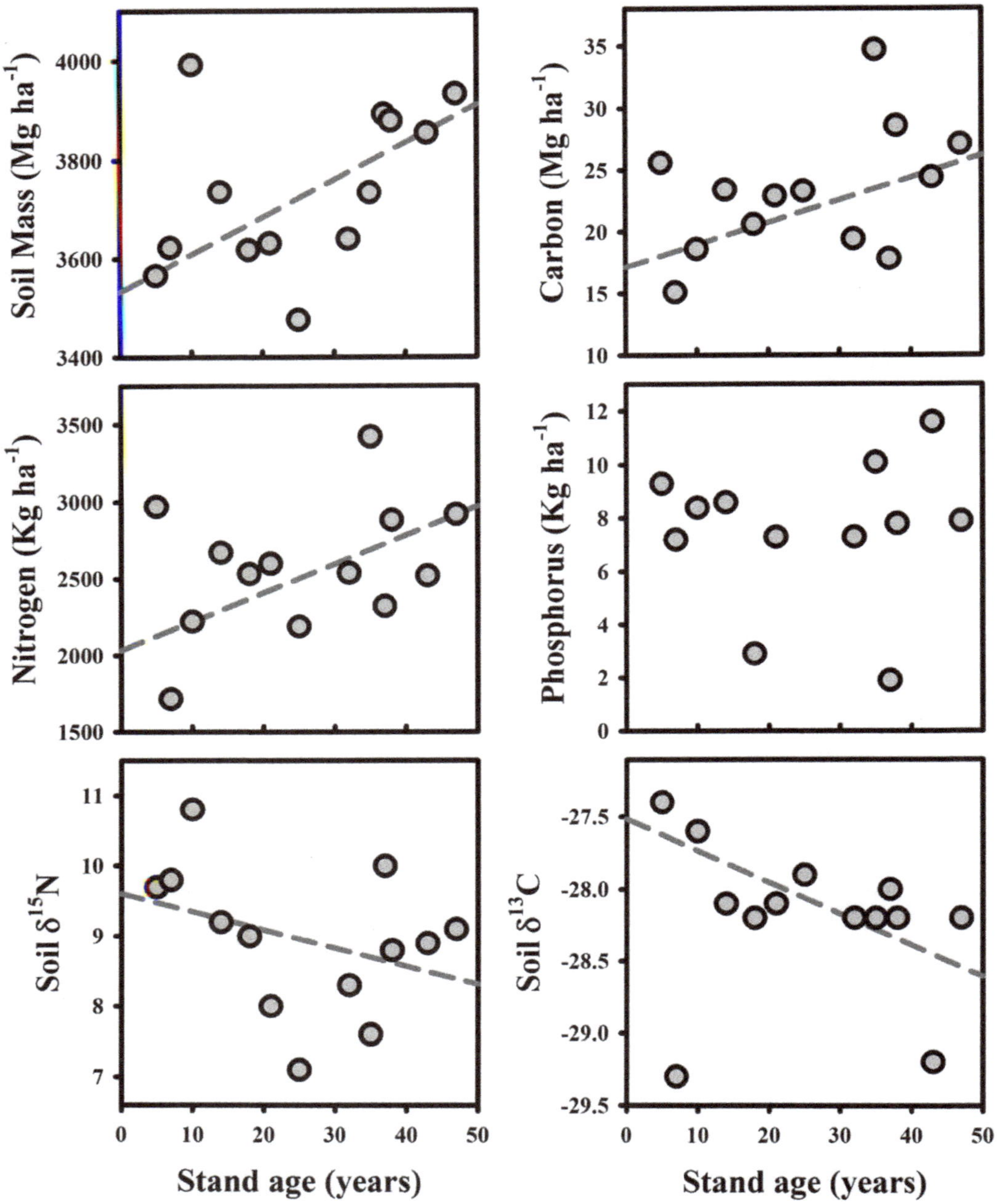

Figure 5. Selected soil properties for successional forest study sites. Significant linear regressions are shown in grey ($N = 13$). Linear regressions for soil $\delta^{15}N$ were significant using separate depth points ($N = 39$) and for soil $\delta^{13}C$ following exclusion of the 7 year old stand outlier ($N = 12$).

(Table 6). Stand age was significantly correlated with all of the predictor variables except for soil texture, which was only correlated with the stand structure axis. Species successional status was correlated with both the stand biomass and soil texture axes. The best subsets regression using these predictor variables showed that while stand age had two significant relationships with leaf traits, specifically positive with foliar $\delta^{13}C$ (i.e., less negative) and negative with $\delta^{15}N$, species successional status had five significant relationships, with foliar N (+), %N (+), C:N (−), N:P (+) and $\delta^{13}C$ (more negative) (Table 7 and Figure 7). Foliar $\delta^{15}N$ had a significant positive relationship with soil $\delta^{15}N$ ($R^2 = 0.246$, $P <$ 0.0001, $N = 232$) whereas none was found between foliar and soil $\delta^{13}C$. A significant negative correlation, although weak, was found between $\Delta\delta^{15}N_{plant\text{-}soil}$ and stand age ($R^2 = 0.029$, $P = 0.0089$, $N = 232$, but after exclusion of the youngest stand $R^2 = 0.119$, $P <$ 0.0001, $N = 204$) and species successional status ($R^2 = 0.118$, $P <$ 0.0001, $N = 232$), whereas a significant negative correlation was found only between $\Delta\delta^{13}C_{plant\text{-}soil}$ and species successional status ($R^2 = 0.149$, $P < 0.0001$, $N = 227$) (Figure 8).

The stand biomass and structure axes had significant correlations with foliar $\delta^{15}N$ - positive and negative, respectively - and soil texture had a significant positive correlation with foliar P and

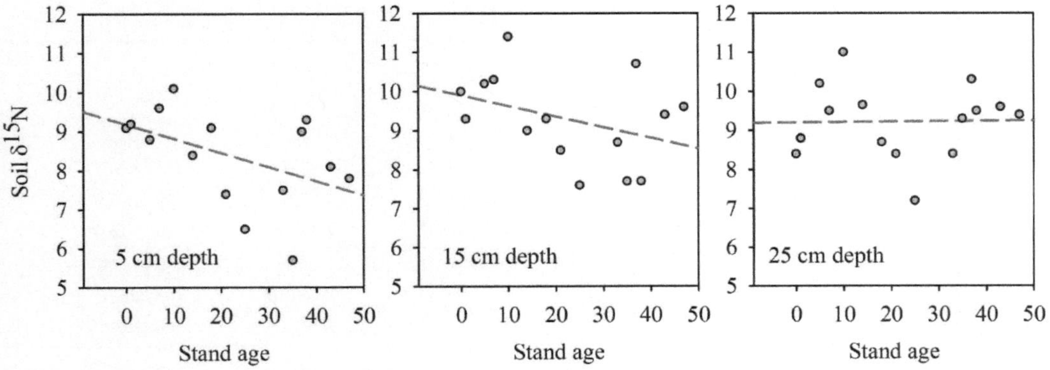

Figure 6. Soil $\delta^{15}N$ values versus stand age by soil depth. A significant relationship exists between stand age and $\delta^{15}N$ for all samples ($R^2=0.26$, $P=0.0881$, $N=12$) and a trend at 5 cm depth ($R^2=0.26$, $P=0.0881$, $N=12$).

negative correlation with foliar N:P. Of all leaf traits only foliar %C was not significantly correlated with at least one predictor variable. For all leaf traits, species successional status had significantly higher explanatory power than stand age (Table 8). We then tested if correlations in leaf traits with stand age found across the species community also occurred within four individual species. No species had a significant relationship between foliar N, P, or N:P and stand age. Three species of four (*Inga* sp. being the exception) had significant negative relationships between foliar $\delta^{15}N$ and stand age (R^2 ranged from 0.19–0.26) and two species (*Miconia* sp. and *J. cuspidifolia*) had significant positive relationships between foliar $\delta^{13}C$ (less negative) and stand age (R^2 ranged from 0.09–0.13). Last, we tested for taxonomic and phylogenetic leaf trait signal. Although significant differences existed among species for all foliar variables ($R^2=0.38$ to 0.54, $P<0.0001$), significant phylogenetic signal (i.e., more than by chance alone) was found for only foliar %C ($K=1.2272$, $P=0.0073$). We do not make a case for phylogenetic trait conservatism or any evolutionary arguments, especially as our significant *K*-statistic value was only marginally larger than one.

Table 4. Characteristics of tree species selected for analysis of foliar properties.

	Common name		Scientific name						
Species ID	Portuguese	Spanish	Genus	Species	Family	Wood density (kg m^{-3})	Successional age (years)	Total (sampled) stems**	Foliar n α
1	Ambaibo	Ambaibillo	Cecropia	polystachya	Cecropiaceae	280	6	76(14)	28 α
2	Favera	Serebo	Schizolobium	parahyba	Fabaceae	360	7	7(3)	6
3	Gameleira	Gomellera	Ficus	insipida	Moraceae	482	10	7(2)	4
4	Vela blanca		Miconia	sp.	Melastomataceae	587	10	128(22)	44 α
5	Ambaibo	Ambaibo	Cecropia	sciadophylla	Cecropiaceae	280	14	57(9)	18
6	Pente de Macaco	Peine de Mono	Apeiba	tibourbou	Tiliaceae	292	14	39(2)	4
7	Marupa	Chepereque	Jacaranda	cuspidifolia	Bignoniaceae	518	18	157(28)	56 α
8	Mutanba	Mutanba	Guazuma	crinita	Sterculiaceae	540	25	14(5)	10
9	Inga	Pacai	Inga	sp.	Fabaceae	603	32	86(20)	40 α
10	Urucuseco	Urucurana	Sloanea	sp.	Elaeocarpaceae	513	33	16(4)	8
11	Joao Mole	Joao Mole	Neea	sp.	Nyctaginaceae	454	33	36(11)	22
12	Louro preto	Laurel	Endlicheria	krukovii	Lauraceae	588	33	40(5)	10
13	Biorana	Trompillo	Pouteria	sp.	Sapotaceae	769	35	21(4)	8
14	Pama	Nui	Pseudolmedia	laevis	Moraceae	593	35	31(2)	4
15	Enviara caju	Piraquina	Onychopetalum	periquino	Annonaceae	619	37	27(8)	16
16	Guayabochi	Guayabochi	Capirona	decorticans	Rubiaceae	840	37	16(2)	4
17	Guariuba	Murure	Clarisia	racemosa	Moraceae	550	37	9(2)	4
18	Jachi	Palo santo	Tachigali	paniculata	Fabaceae	600	41	16(2)	4
19	Breu branco		Rollinia	calcarata	Annonaceae	520	43	16(2)	4
20	Balsamo	Balsamo	Myroxylon	balsamum	Fabaceae	796	43	3(2)	4

*Across all study sites, α = selected for individual species response analyses.

Table 5. Pearson correlations among leaf trait variables.

	P	N:P	C%	N%	C:N	$\delta^{15}N$	$\delta^{13}C$
N	0.66(264)***	0.24(263)***	0.12(260)*	0.85(262)***	−0.85(259)***	0.04(263)	−0.04(258)
P		−0.55(263)***	−0.28(259)***	0.61(261)***	−0.66(258)***	0.08(262)	0.33(258)***
N:P			0.49(258)***	0.14(260)*	−0.07(257)	−0.07(261)	−0.44(257)***
C%				0.15(259)*	−0.03(259)	0.25(259)***	−0.27(254)***
N%					−0.99(259)***	0.07(260)	−0.07(255)
C:N						−0.04(258)	0.03(253)
$\delta^{15}N$							0.20(258)**

Data is the correlation value (*N*) and *P*-value expressed as *<0.1, **<0.05, ***<0.01, and ****<0.001.

Discussion

We used a high-resolution forest succession chronosequence to evaluate potential predictors of leaf trait variation. We analyzed correlations among leaf traits and stand age, species successional status, and PCA-derived axes of stand biomass and structure and soil texture and fertility. While all leaf traits, with the exception of foliar %C, had significant correlations with at least one predictor variable, the strongest relationships were found between foliar stable isotopes and stand age and species successional status. Although our discussion focuses on leaf trait variation during secondary forest succession, we also discuss patterns identified in stand structure and composition and soil properties, focusing on variation in soil $\delta^{13}C$ and $\delta^{15}N$.

Stand Structure and Composition

The rapid decline in dead stems in the early stages of succession were likely a result of both remnant dead stems from the slash and burn cycle and rapid turnover in species composition during the first 10 years of succession. Areas having undergone multiple slash and burn cycles would be less likely to have remnant dead trees, however our plots had undergone only 2–3 years of agriculture prior to abandonment. Prach et al. (1993) [68] documented rapid turnover in species composition during the early stages of succession (<10 years) which, subsequently, slowed dramatically in a successional sere ecosystem. Similar changes in the rates of change in species composition and forest structure are highlighted by Chazdon et al. (2007) [141] for tropical forests. Our results indicated a similar progression with rapid species turnover, as shown through decreases in composition differences as compared to primary forest, beginning 15 years post-abandonment with the fastest rate of change (i.e., species turnover) during the first 10 years of forest succession (Figure 1), simultaneous with a reduction in the standard deviation of successional status. Such changes are in line with those predicted by the "Initial Floristic Composition" (IFC) hypotheses, which states that early successional stands will contain a large proportion of the species dominating in later stages of succession [69]. In our case, the rapid reduction in stand mean successional status in the early stages is due to high mortality of early pioneer species while the high variance in the early succession is due to the prevalence of later succession species (Figure 1). In older stands, pioneer species have died off due to differences in growth rates, longevity and shade-tolerance [70], reducing the variance of successional status in those stands.

Further insights are possible by separating overstory species from the community in general. The increase in community species richness was not seen in the overstory, similar to results found by Guariguata et al. (1997) [71] and Norden et al. (2009) [72], whereas slight increases in diversity were seen. This represents continual shifts in overstory composition, versus addition through succession. However, the increasing species richness differs from that postulated by the IFC hypotheses, as new species are continually added throughout succession and earliest stands having low richness and high dominance by a few short-lived pioneers. The species similarity changes are on par with those hypothesized by Oliver and Larson (1990) [25] and Waring and Running (2007) [24]. Although the understory approximated primary forest composition after only 50 years, the overstory remained completely different (Figure 4), similar to that found by Peña-Claros (2003) [23]. Waring and Running (2007) [24] state that primary forest composition will be attained between 100–400 years post-abandonment. However, it is not possible to simply extrapolate vertical growth rates of shade tolerant trees now dominant in the understory to predict composition changes in the overstory as many have negligible growth rates [73], with changes

Table 6. Pearson correlations among the predictor variables stand age (years), species successional status (status), and the PCA derived axes of stand biomass and structure and soil texture and fertility.

	Status	Biomass	Structure	Texture	Fertility
Stand age	0.63(265)****	0.86(265)****	0.1(265)*	−0.01(223)	0.67(195)****
Status		0.54(265)****	0.02(265)	0.05(223)	0.46(195)****
Biomass			0.24(265)****	0.08(223)	0.39(195)****
Structure				0.64(223)****	0.01(195)
Texture					0.03(195)

Data is the correlation value (*N*) and *P*-value expressed as *<0.1, **<0.05, ***<0.01, and ****<0.001.

Table 7. Best subsets models of leaf traits versus stand age, species successional status, and Principal Component Analysis (PCA) derived gradients in stand biomass and structure, and soil texture and fertility.

			PCA Axes						
Leaf traits	**Stand age**	**Species Successional Status**	**Biomass**	**Structure**	**Texture**	**Fertility**	**Adj-R^2**	***P*-value**	**N**
N	–	2.54 (6.46)**	–	–	–	–	0.021	0.0116	265
P	–	–	–	–	2.61 (6.83)***	–	0.026	0.0096	222
N:P	–	4.09 (16.76)$^{\phi}$	–	–	−3.79 (14.40)$^{\phi}$	–	0.115	<0.0001	221
C %	–	–	–	–	–	–	–	–	260
N %	–	2.40 (5.74)***	–	–	–	–	0.018	0.0173	262
C:N	–	−2.44 (5.97)***	–	–	–	–	0.019	0.0152	259
$\delta^{13}C$	3.58 (12.85)$^{\phi}$	−7.93 (62.86)$^{\phi}$	–	–	–	–	0.202	<0.0001	258
$\delta^{15}N$	−7.54 (56.92)$^{\phi}$	–	4.30 (18.50)$^{\phi}$	−3.17 (10.02)***	–	–	0.234	<0.0001	263

Data represents the *t*-ratio (*F*-ratio) and *P*-value significance, with increasing * representing *P*-values of 0.1, 0.05, 0.01, and 0.001, respectively, and ϕ <0.0001.

occurring in pulse events following disturbance related gap openings [74].

By 50 years, our secondary forests were similar in structure to our primary forest plots. Biomass was nearing the asymptote and average tree height and DBH had both begun to decrease. Similar dynamics have been seen during succession in the Amazon [23] and Panama [22], with the earlier stages of succession having the greatest rates of biomass accrual [75]. This relationship has been described as resulting from die-off of pioneer species simultaneous to establishment of long lived shade tolerant species, with more than 200 years being estimated to be required to attain primary forest biomass [76]. We found a quadratic relationship between stand age and the height of our emergent trees in which the mean height of emergent trees attained maximum values in intermediate stands. We postulated this to result from early successional species, with no recruitment in the understory, prior to species turnover to dominance by shade tolerant species [76,77]. We found indications towards this occurring with the trees contributing the most to

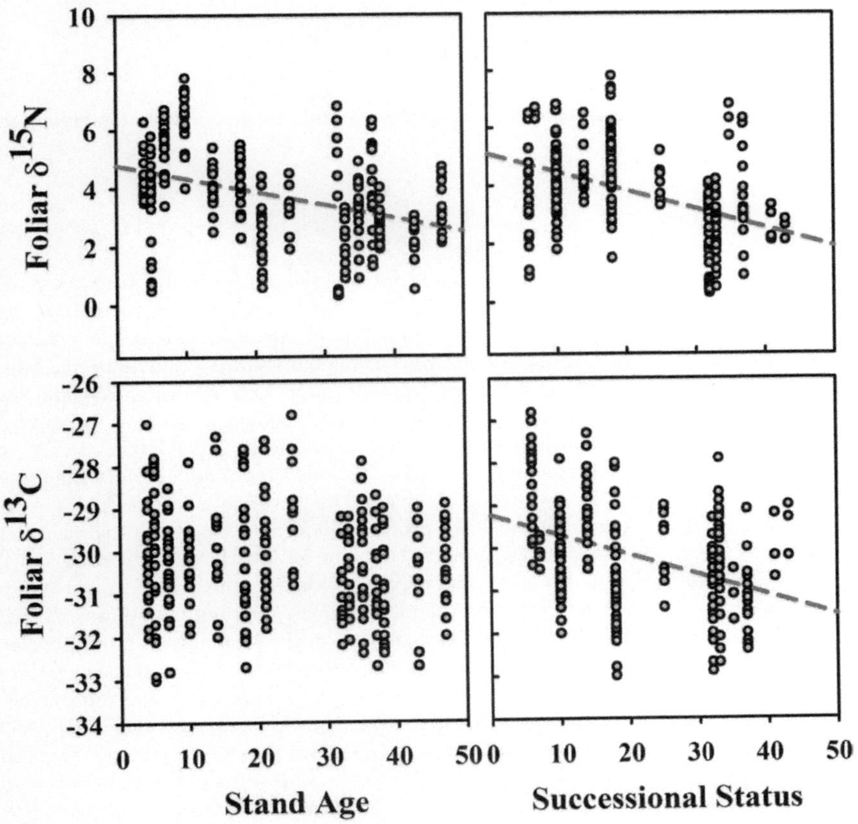

Figure 7. Foliar $\delta^{15}N$ and $\delta^{13}C$ values versus stand age and species successional status. Significant linear regressions are shown in grey.

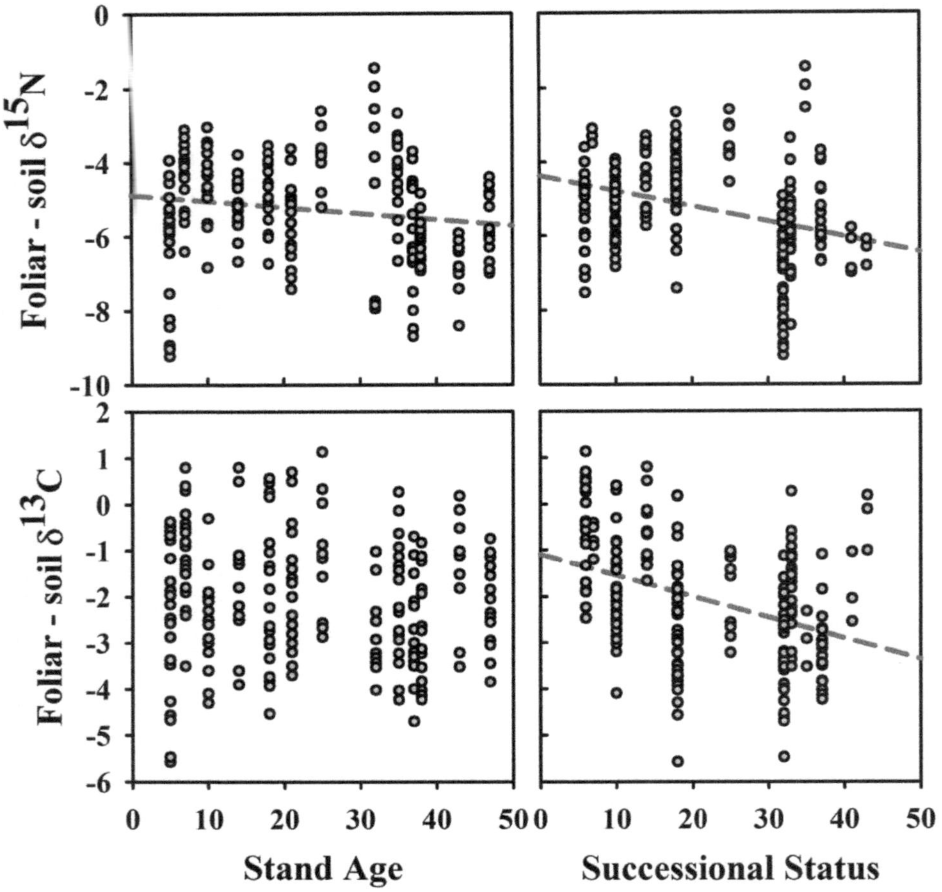

Figure 8. Foliar-minus-soil $\delta^{15}N$ and $\delta^{13}C$ values versus stand age and species successional status. Significant linear regressions are shown in grey.

overstory composition in the intermediate aged stands to be early successional species growing past their identified species successional statuses ($R^2 = 0.42$, $P = 0.0658$, $N = 13$).

Soil Properties

Changes in soil properties during secondary forest succession typically include increases in soil carbon [78]. Although the opposite was found by Schedlbauer and Kavanagh (2008) [45], in former pastures in Costa Rica, who explained that active non-crystalline clays and aluminum-humus linkages may have resulted in higher carbon stability during land use than that found in other studies. Other studies in the tropics have found pasture soils tending to have less carbon storage than cultivated soils [79], increasing extractable soil N and decreasing P stocks [80], and decreases in bulk density [81] (but see [82]), among other possibilities [9]. Generally, rapid decreases in total soil organic carbon (SOC) occur within the first 50 years following forest conversion, followed by slower losses until reaching a new lower equilibrium after 100 years. However, forest soils have significant memory and forest derived SOC represents >80% of the total pool if cultivation occurs for less than 5 years [83]. Global patterns of soil nitrogen are primarily controlled by mean annual precipitation and temperature that determine input from atmospheric deposition and N fixation, but during the slash-and-burn process, significant ecosystem nitrogen is lost through biomass removal, volatilization from combustion [84], and denitrification and leaching [85].

During stand development, we found significant increases in soil mass that, in part, resulted in increasing pools of carbon and nitrogen, but not phosphorus. In our case, increases in soil mass partly resulted from compaction following the slash-and-burn agricultural activity which reduced bulk density during the soil preparation, crop planting, and plant growth and harvesting, similar to that found by Brown and Lugo (1990) [187], contrary to that found following more intensive and more extended land uses such as pasture, such as found by Davidson et al. (2007) [37] in the top 0–10 cm depth profile, although when only considering this depth results from our study are similar. Feldpausch et al. (2004) [80] identified similar patterns during young successional stands (<14 years), with soil nitrogen increasing but phosphorus moving from below- to above-ground. Such dynamics are typical during succession as available soil phosphorus declines during nutrient redistribution from vegetation growth due to virtually no primary minerals remaining in the highly weathered soils [86]. Soil nitrogen, however, can increase through symbiotic nitrogen fixation, especially during early stages of succession [87,88] and atmospheric deposition [89]. These changes were highlighted by the soil fertility axis, driven largely through N, increasing with stand age. Reductions in soil $\delta^{15}N$, although a non-significant

Table 8. Regressions of foliar variables versus stand age (years)/species successional status (years).

Foliar variables	Combined	Stand age	Successional status
Linear regression models and model comparisons			
N:P	0.05 (8) 260***	0.01 (2) 261 NS	0.05 (15) 261***
		14***	1 NS
C:N	0.01 (2) 256 NS	0.00 (1) 257 NS	0.01 (4) 257*
		4	0 NS
$\delta^{15}N$	0.24 (43) 260****	0.18 (57) 261****	0.22 (74) 261****
		23****	9**
$\delta^{13}C$	0.20 (33) 258****	0.01 (3) 258 NS	0.17 (52) 258****
		31.3****	5.39*
Standardized slope values			
N:P	−0.57 NS/1.90***	NS	0.62
C:N	NS	NS	−1.65
$\delta^{15}N$	−0.33**/−0.52****	−0.66	−0.73
$\delta^{13}C$	0.02****/−0.07****	NS	−0.05

Statistics provided are Adj-R^2 (F) *df* and *P*-value. Model comparisons are conducted using the F-test in R and *P*-values are presented as: blank <0.1, *<0.05, **<0.01, ***<0.001, ****<0.0001.

negative trend with stand age, also formed an important contributor of the soil fertility axis.

Soil $\delta^{15}N$ is determined by the equilibrium of $\delta^{15}N$ inputs, outputs and internally through redistribution via plant uptake [55] and most soils have positive $\delta^{15}N$ values due to accumulated losses [90]. In general, ecosystems with high rates of nitrogen fractionation resulting in soil ^{15}N enrichment are viewed as having a leaky or open N cycle, typically with abundant N, versus ecosystems with a conservative or closed N cycle, and therefore reduced N loss, having reduced ^{15}N enrichment [39,91]. Input processes include: (a) soil ^{15}N depletion through atmospheric inputs, including from precipitation [92], combustion [93], and nitrogen deposition [94]; and (b) soil ^{15}N enrichment via symbiotic nitrogen fixation related fractionation during uptake [95–98]. Output processes include: (a) soil ^{15}N enrichment through selective loss of ^{14}N during decomposition related nitrification and denitrification [94,99], including gaseous ^{14}N losses to the atmosphere [100]; and (b) soil ^{15}N enrichment through hydrologic leaching loss of ^{15}N depleted nitrogen (relative to the soil) produced during nitrification, denitrification and ammonia volatilization [101]. Redistribution processes include: (a) soil ^{15}N enrichment through discrimination against ^{15}N during biological nitrogen fixation, or creation of ^{15}N depleted compounds during decomposition and stabilization, and subsequent loss and/or assimilation of leachate from such activities at depth resulting in soil ^{15}N depletion [98]; and (b) discrimination against ^{15}N during plant uptake resulting in redistribution of depleted ^{15}N from mineral soil at depth to plant biomass and eventually to the surface horizon [43].

Patterns of soil $\delta^{15}N$ in depth profiles, now well established [99,102], result from multiple factors, including: (a) the soil surface can become ^{15}N depleted as litterfall accumulates at the soil surface while deeper soils can become ^{15}N enriched due to increased mycorrhizal fungal activity [98]; (b) N loss from nitrification and denitrification can result in either: (i) a steady soil ^{15}N enrichment with depth given abundant N availability [98]; or (ii) enriched soil ^{15}N at intermediate depth in arbuscular mycorrhizal (AM) systems and/or sites of higher available nitrogen [103]; and/or (c) increases in soil fungi with depth resulting in soil ^{15}N enrichment [104]. Differences in soil texture may also play a role in defining soil $\delta^{15}N$, with increasing clay % generally accompanying soil ^{15}N enrichment [105], as well as land use intensity altering soil $\delta^{15}N$ [106]. Of particular importance, burning alters the ^{15}N pattern in soil profiles by eliminating the most ^{15}N depleted organic layer [98] resulting in significant soil $\delta^{15}N$ enrichment in the upper 20 cm [107].

In our secondary forest stands, soil $\delta^{15}N$ both decreased with stand age and increased with soil depth. In the primary forest however, we found that soil $\delta^{15}N$ was greatest at intermediate depth followed by decreases as would be expected for AM systems or those with higher N availability [103]. Differences along the stand age gradient were unlikely a result of soil clay content as no significant changes were found [105]. The pattern of soil ^{15}N enrichment with depth may develop rapidly due to the large soil ^{15}N enrichment in the upper 20 cm from combustion fractionation during the burning process [107], preferentially moving ^{14}N downwards. Such dynamics would occur most rapidly during the agricultural phase and in early successional stands with less well developed root mats and reduced bulk density but continue throughout all stands. Although no significant interaction effect was found showing reduced profile significance with stand age, as would have been expected if our secondary forest ^{15}N profiles were developing towards those found in the primary forests, this could take longer than our secondary stands studied [98].

The subsequent significant soil ^{15}N depletion with stand age, which occurred most rapidly in the shallower depths most affected by burning (see trend in Figure 6), would then be due to simultaneous influence of: (a) atmospheric deposition of isotopically lighter ^{14}N [94], of especial importance in this area due to the intense annual fire season [108]; (b) preferential leaching of soil ^{14}N, and ^{15}N depleted products, during the intense annual wet season [101,109]; (c) changes in plant root distribution and increased foraging depth causing changes in the utilization of soil inorganic N pools [94] or acting as a biological pump to the surface via litterfall [183], of depleted ^{15}N - although the soil ^{15}N values found in deeper soils (>30 cm) were not measured in this study; (d) possible mycorrhizal fungal activity [98] – although such activities typically enrich soil ^{15}N, and/or (e) mobilization of recalcitrant soil N to active pools [37], with recalcitrant pools being depleted in ^{15}N if formed prior to the combustion event. Compton et al. (2007) [43] studied soil $\delta^{15}N$ in secondary forests up to 115 years post agricultural abandonment in Rhode Island, USA and, with the exception of the organic horizon in which $\delta^{15}N$ decreased, found ^{15}N enrichment with stand age across all depths – although they found the same pattern of ^{15}N enrichment with depth in the soil as identified in our study. These differences are likely due to very different land use intensities between the two studies, with the stands studied by Compton et al. (2007) [43] having undergone far greater land use intensity, including both mechanized agricultural and/or high intensity cattle grazing, prior to abandonment. Similar results – to Compton et al. (2007) [43] – obtained by Billings and Richter (2006) [44] may be also explained through the far greater mechanized agriculture prior to abandonment.

Variation in soil $\delta^{13}C$ is generally thought to be a result of the dominant plant species composition, as decomposition related fractionation is small relative to that during carbon fixation [102], and is therefore driven by changes in the relative abundance of C_3 (i.e., most plants) to C_4 (i.e., corn, sugar cane, most tropical pasture grasses) plants [45]. Differences in foliar carbon isotope composition between C_3 and C_4 species reflect differences in their photosynthetic pathways, with C_3 species have a $\delta^{13}C$ range of −32 to −20% whereas C_4 species range from −17% to −9% [110]. However, soil ^{13}C enrichment also occurs during auto- and heterotrophic soil CO_2 efflux - which is ^{13}C depleted - from decomposition of surface litter [111], and during oxidation of soil organic matter and soil humification processes [112–114], or possibly depletion from root respiration [115]. The isotopic composition of autotrophic respiration (i.e., leaves, twigs, roots) is most likely derived from young newly fixed carbon, with the related $\delta^{13}C$ value, whereas heterotrophic respiration (i.e., decomposition) will likely have a different isotopic signature depending on the carbon turnover rate of the labile available carbon pools [114,116]. Such differences have been identified in soils throughout the tropics [117,118], with a $\delta^{13}C$ of −25% commonly used to represent the value of the stable pool [119]. Profiles showing increasing $\delta^{13}C$ with depth, which occur independent of soil type [120], are commonly found under conditions of stable vegetation cover and low soil disturbance (i.e., tilling) [121] and result from reduced decomposition related isotope fractionation with depth [112] and decreases in the size of soil organic matter fractions [122]. In these situations, surface carbon is generally of young origin and labile with increasingly old and recalcitrant SOM pools with depth [119].

Changes in soil $\delta^{13}C$ during forest succession reflect: (a) variation in the turnover rates of soil organic matter (SOM), including decreased stability of some previously stable – potentially ^{13}C enriched - C pools following burning [119] – including changes in relative contribution of microbial vs. plant soil organic matter [114]; (b) changes in vegetation, including shifts between C_3 and C_4 species, as soil ^{13}C composition greatly reflects that of the dominant vegetation [102], due to low rates of fractionation during decomposition relative to fixation [123] – although Billings and Richter (2006) [44] found very slow incorporation of new plant carbon into soil horizons beyond the uppermost (<10 cm) layers; and (c) a large flux of depleted ^{13}C plant lignin and root organic matter, or remnant charcoal, from combustion during the deforestation process [119,181] – which can result in confusion as pasture carbon isotope composition could appear similar to intact forests. Decreased soil $\delta^{13}C$ in abandoned pastures and agricultural areas, as compared to forest, has been well established [45,83] with increasing historical years of cultivation of C^4 species having a positive (less negative) relationship with soil $\delta^{13}C$ values (forest soil $\delta^{13}C = -24$, vs. 17 and 60 years of cultivation = −23 and −16, respectively; [83]).

Although the soil ^{13}C depletion with stand age in our study contrasted that found by Billings and Richter (2006) [44] whose stands were regrowing on sites subjected to intensive agriculture of cotton (a C_3 species), it was similar to that found by Lopez-Ulloa et al. (2005) [124] who studied forest stands regrowing on sites opened through slash-burn and used as pasture. In our study, changes in soil $\delta^{13}C$ were unlikely dominated by a C_4 species signal related to the agricultural (or pasture) phase as identified in studies following more intense land use [44]. However, although subsistence agricultural species grown in our study area were primarily C_3, including bananas, beans, rice and yucca, the dominant crops included corn, a C_4 species, and rice. More likely, the decrease in soil $\delta^{13}C$ seen with increasing stand age represents a gradual return to stable primary forest values following: (a) a large signal from burning related ^{13}C enrichment during the slash-and-burn process [119], (b) a smaller ^{13}C enrichment signal from the corn, a C_4 species, cultivated during the 1–2 year no till agriculture period [110], (c) a transition to C_3 species dominated forest composition with more stable disturbance dynamics, and (d) continual input of ^{13}C depleted carbon during the dry season from wildfires [125]. Soil $\delta^{13}C$ in our study did increase with depth as expected given reduced rates of decomposition; with the development of a linear relationship between ^{13}C enrichment and increasing depth occurring in the primary stands as expected given their more stable forest structure and plant composition [121]. Our primary forest soil $\delta^{13}C$ were considerably more enriched than those in our oldest secondary forest stands, possibly indicating long term ^{13}C enrichment via leaching.

Leaf Traits

At a global scale leaf traits have been found to represent a continuous gradient dubbed the 'leaf economic spectrum', ranging from short-lived high photosynthetic capacity leaves to long-lived, thick, low photosynthetic capacity leaves [126]. However, further analyses have revealed that the range within groups along this gradient is often larger than the differences among them [127] – in part related to broad changes in mean annual temperature and precipitation – and shifts from long-lived low nutrient leaves in low fertility soils to high nutrient content leaves on more fertile sites [128]. Leaf trait variation within a primary forest (i.e., a relatively stable environment) is further constrained by phylogenetic, taxonomic or functional group differences [28,129–132], growth environment [133–135] and propagation strategy [27], among many factors [8]. Understanding variation in leaf traits through forest succession (i.e., a highly dynamic environment) integrates across these groups as it moves though variation in inter- and intra-species competition, differences in plant growth and reproductive strategy, and plant-soil feedbacks [23,136,137].

During secondary forest succession, as well as over longer term ecosystem succession processes [175,176], foliar trait dynamics may provide information unique from that shown in the stand or soil properties, or largely result from overall stand changes or differences in soil properties. Understanding these factors is critical to enable the use of leaf traits, including nutrient concentrations and stable isotope ratios, to understand successional dynamics. We start by investigating, for each measured leaf trait, possible phylogenetic control over leaf trait variation, then focus on how changes in stand structure and soil properties may influence their expression. In addition, we focus on situations where leaf traits may follow different trends than those occurring belowground and thereby provide additional information.

First, we investigate whether differences in leaf traits are a result of phylogenetic control (Figure 2). Powers and Tifflin (2010) [130] found that inter-specific variation accounted for 57–83% of leaf trait variance across 87 tropical dry forest tree species whereas He et al. (2010) [129] found that 27% of leaf trait variation was due to phylogenetic differences in a grassland ecosystem. These contrast with our results, which showed phylogenetic signal only in leaf %C. However, our results strongly agree with those of Fyllas et al. (2009) [131] and Asner and Martin (2011) [138], indicating that that phylogeny accounts for 50–80% of the variation in foliar %C, depending upon site characteristics. As such our results can be explained by considering our study system which represents a large gradient of stand structure and species composition, in which the stand age or species successional status gradients outweigh differences among species. In any individual stand however it is entirely plausible that inter-species differences are of primary importance to explain leaf trait variation, although Letcher (2010) [139] found significant over-dispersion of species - across the phylogenetic community - at stand scale during forest succession as a result of rapid transition through species 'functional' groups (i.e., along the species successional status gradient). In particular, Letcher (2010) [139] found that no phylogenetic structure existed in the youngest stands potentially resulting from the rapid rate of change in species, as indicated in Figure 1 for mean stand species successional status. Fyllas et al. (2009) [131] analyzed leaf traits from 508 species distributed across a range of soil types and precipitation regimes and found that foliar %C (as identified in our study), %N and Mg concentration were highly taxonomical constrained. However, foliar P, K, Ca and $\delta^{13}C$ were influenced by site growing conditions, with soil fertility being the most important predictor for all variables. Mean annual temperature (MAT) was negatively related to foliar N, P and K, and MAP was positively related to foliar %C and $\delta^{13}C$. Townsend et al. (2007) [36] used foliar N:P ratios to show that differences in growth latitude or mean annual precipitation (MAP) had a non-significant influence on foliar N:P ratios within the tropics, while large significant differences were found among species, between the dry and wet season, and with soil order. Pringle et al. (2010) [140] found no phylogenetic signal among the leaf traits of trees growing in a seasonally dry tropical forest in Mexico, which they explained as indicating that selective pressures (i.e., functional convergence) constrained leaf traits.

In the context of forest succession, phylogenetic signal over all successional stands may be less relevant to understanding temporal dynamics than how species functional qualities change through time [8,28]. In particular, given successional theory, it is more likely that changes in a suite of functional characteristics are occurring through succession [142], in part due to environmental niche partitioning and strategy differentiation [182]. The importance in functional differentiation during succession was supported by the lack of phylogenetic signal across the species successional status gradient of our studied trees, although further analyses are required to unravel phylogenetic – successional status interactions during succession. Huc et al. (1994) [143] studied differences in leaf traits among categorical divisions of species successional status in the French Guiana. Wood cellulose $\delta^{13}C$, leaf gas exchange and leaf water potential were shown to differ significantly between pioneer and late stage successional guild tree species growing in a common garden [143], and the authors highlighted the need for additional research on successional guild control over ecophysiological function [29].

Given significant changes in soil N and P pools, as expected given successional theory predicting changes from N to P limitation during forest succession [144], we anticipated a corresponding positive shift in the foliar N:P ratio. Our results show, that in our study area, changes in foliar N:P are related to changes in species successional status (Table 7). Although stands younger than 10 years of age do have stand average foliar N:P value less than 14–16, the N (lower) to P (higher) limitation threshold described by Townsend et al. (2007) [36] in a literature review, most are above this threshold indicating that significant N limitation is not occurring in our study area. This was expected given the large N pulse seen in the soil $\delta^{15}N$ of younger stands, while N in older stands would have accumulated due to biological N fixation and atmospheric deposition. Although foliar N varied with successional status, foliar P varied only with the soil texture gradient, with foliar P tracking changes in soil P. Soil P however did not vary with stand age, indicating these changes are unrelated to successional process, but instead had a significant negative correlation with soil clay. These results are similar to those identified by Silver et al. (2000) [75] who found that when soil clay content increased, labile P decreased while total P increased, due to P being easily complexed with exchangeable Al and Fe. Our soil P analyses (weak Bray) were designed to assess the labile P pools [145,146], which would have the greatest effect on foliar P concentrations. Therefore, in this study, our results indicate that soil P represents variation in soil clay content resulting from differences in our chronosequence not related to successional processes, which was then mirrored in the foliar P content. Changes in foliar N however resulted from, to a much lesser extent, increasing N availability during stand development, and primarily from shifts in species successional growth strategies, indicative of species growing on nutrient rich sites [128], and similar to results from Yan et al. (2006) [142] showing increasing foliar N among three tree species increasing in successional status which was attributed to changes from conservative to resource spending in later successional species due to increases in soil fertility. Our data shows this pattern to occur across both the entire species community as well as within individual species.

While foliar $\delta^{15}N$ reflects in large part the isotopic composition of the available soil nitrogen pool [55] – which in turn results from species composition [190] and N cycle openness [91], various sources of fractionation exist which can provide additional information over soil $\delta^{15}N$ alone, e.g., biological nitrogen fixation [147] discriminates against ^{15}N during soil N to plant transfer decreasing foliar $\delta^{15}N$. During forest succession, theory predicts decreasing N limitation and therefore greater openness of the N cycle, resulting in increased soil ^{15}N fractionation and therefore ^{15}N enrichment. Contrary to this however, simultaneous reductions in foliar and soil $\delta^{15}N$ during succession have been found in several studies. Wang et al. (2007) [148] attributed decreases in soil and foliar $\delta^{15}N$ on an old field in northern Virginia following intensive agriculture because of increases in woody ectomycorrhizal (EM) and herbaceous vesicular-arbuscular mycorrhizal (VAM) – also referred to as arbuscular mycorrhizal (AM) due to vesicles not being produced by all fungi - hosted fungi causing discrimination against soil ^{15}N during uptake. They highlight that VAM abundance increases into early successional stands and then reduces in intermediate aged stands when EM hosted fungi (which discriminate more against ^{15}N than VAM) increase [149,150]. In addition to finding community level dynamics of decreasing foliar $\delta^{15}N$ they also identified decreasing patterns among individual species, similar to our findings. They conclude that soil $\delta^{15}N$ is the major factor driving foliar $\delta^{15}N$ due to a highly significant positive relationship (R^2>0.90). In our study, we likewise found a highly significant relationship, although the R^2 value of 0.25 indicated that other processes might be constraining foliar $\delta^{15}N$ dynamics, such as root development, which likely provides access to increasingly depleted N at depth [43].

Compton et al. (2007) [43] use the divergence between foliar and soil $\delta^{15}N$ values ($\Delta\delta^{15}N_{plant\text{-}soil}$) to address drivers of foliar $\delta^{15}N$ variation, including rooting depth, as it allows for direct comparison of N fractionation during plant uptake, with increasing values indicating increased fractionation. Although we perform a similar study of divergence, we recognize that our soil analyses were limited to <30 cm whereas rooting depths are well distributed up to 2 meters depth [151], and sometimes extend much deeper (>8 meters in some primary forests in the Amazon; [152]). Divergence results, in part, from: (a) differences in the rate and abundance of EM and AM activity [184]; (b) changes in plant efficiency of mineral N cycling during succession, as has been found for increasing water availability [101]; and (c) changes in plant uptake between organic/NH_4^+ (cool temperate) and $N0_3^-$ (tropical) [55], although climatic differences would be minimal among our co-located study stands.

As mycorrhizal activity results in depletion of foliar ^{15}N we expected an increasing $\Delta\delta^{15}N_{plant\text{-}soil}$ divergence during forest succession. This trend would be most pronounced if soil $\delta^{15}N$ were being measured at the same depths that mycorrhizal activity was greatest, and diluted if soil from shallower depths having substantial litter input and therefore soil ^{15}N depletion was used. During forest succession, simultaneous increases in rooting depth – as we collected leaves from top-of-canopy positions - would influence this relationship as we found soil ^{15}N enrichment with depth, but further detailed analysis of soil mycorrhizal activity and soil $\delta^{15}N$ to depths of at least 3–5 meters would be required to interpret possible interactions. While stand age and species successional status both exhibited the significant increases in divergence as indicative of increased mycorrhizal activity, species successional status was a much stronger predictor. Although we do not have data regarding specific mycorrhizal relationships among our study species, an increasing probability of mycorrhizal associations, and in particular AM – versus EM - associations which the vast majority of tropical tree species have [185,186], occurring in later successional status species could explain this correlation – thus changes in species successional status during stand age would drive $\Delta\delta^{15}N_{plant\text{-}soil}$ divergence depletion via an iterative feedback loop, simultaneous to a general trend of ^{15}N depletion occurring across the species community as well as within individual species through stand age which indicates that ecological processes other than species successional status transitions are driving overall ecosystem $\delta^{15}N$ depletion patterns – as discussed previously above – but such transitions may be accelerating ongoing ^{15}N depletion patterns and in particular at shallower soil depths incorporating increasingly more depleted ^{15}N via organic matter input.

Although patterns of decreasing $\delta^{15}N$ were found by our study, the driving factors likely differ in our study from those of Compton et al. (2007) [43] and Wang et al. (2007) [148]. The successional chronosequence used in these studies did not undergo a burning cycle directly prior to abandonment, which significantly enriches soil ^{15}N, rather they propose that their decreases resulted dominantly from: (a) shifts in rooting depth; (b) fractionation during plant uptake, (c) increased mycorrhizal activity (although no N-fixers were present in the Compton et al. (2007) [43] study site), and (d) changes in soil N cycling rates. Processes dominating our study N dynamics were likely more similar to those Davidson et al. (2007) [12] who found that foliar N, N:P, and $\delta^{15}N$ all increased during forest succession. Likewise, we found increases in foliar N:P, driven by increases in N but no change in P, but conversely found decreasing foliar $\delta^{15}N$. We propose that although similar dynamics of N cycle recuperation (i.e., increasing leakiness and therefore N fractionation with stand development) are occurring within our study area, changes in foliar $\delta^{15}N$ are more representative of a decrease in soil $\delta^{15}N$ following high initial enrichment from combustion during the slash-and-burn process. The study plots selected by Davidson et al. (2007) [37], although likely burned at various points, had existed long enough in an agricultural or pasture state to stabilize at more depleted $\delta^{15}N$ values prior to abandonment, with increasing time in pasture shown to result in increasingly depleted soil $\delta^{15}N$ values [99].

Variation in foliar $\delta^{13}C$ results primarily from differences in the ratio of internal (c_i) to atmospheric (c_a) CO_2 concentrations, in which a decreased internal CO_2 concentration relative to atmospheric (i.e., via closed stomata or rapid photosynthesis) results in decreased ^{13}C discrimination and enrichment (less negative) foliar $\delta^{13}C$ values [153]. In C_3 plants, foliar ^{13}C discrimination occurs primarily during photosynthetic CO_2 uptake by ribulose1,5-bisphosphate (R_uP_2) carboxylase [154] and during gaseous diffusion through the boundary layer and leaf stomata [155], and is moderated by the CO_2 diffusion rate into the leaf and rate of carboxylation [155]. Therefore photosynthesis has a depleted ^{13}C uptake and tends to enrich the surrounding atmosphere while respiration has a depleted release and tends to deplete the atmosphere in ^{13}C [156]. Factors affecting the c_i/c_a ratio include: (a) variation in the concentration of source CO_2 [153]. However, differences in atmospheric CO_2 are likely minor and not relevant in our study, and respiration derived CO_2, which is ^{13}C depleted ($\delta^{13}C\sim-28\%$ [157]) relative to atmospheric (and respiration itself causes little fractionation [158]), is not significantly different when leaves are sampled in locations of unimpeded air circulation [42] or greater than 7 m above ground in an intact primary tropical forest [157,159]; (b) C_3 or C_4 photosynthesis [110], which result in typical foliar $\delta^{13}C$ values of approximately -25% and -12%, respectively [110]; (c) a wide variety of environmental stresses which may result in alter the ratio of conductance to photosynthesis [160], such as soil drought which influences conductance more than photosynthesis [161,162], with increasing drought resulting in enriched (less negative) foliar $\delta^{13}C$ values; (d) differences in internal gas exchange in thick leaves, such as found by Vitousek et al. (1990) [42] where enriched (less negative) foliar $\delta^{13}C$ (ranging from approximately -29.5 to -24.5 from low to higher elevations, respectively) was found within *Metrosideros polymorpha* at higher elevations as a result of increasing internal resistance to CO_2 diffusion to sites of carboxylation within the leaf, associated with increased leaf mass per area (LMA), as well as thicker mesophylls and increased %N [163]; (e) differences in time-integrated water use efficiency (WUE) – assessed as the ratio of leaf assimilation rate of CO_2 (A) to leaf water vapor conductance (g) – with increasing WUE (i.e., decreased water vapor conductance relative to photosynthetic capacity) resulting in enriched (less negative) foliar $\delta^{13}C$ [153]; and (f) differences in growth location, such as understory environments with decreased irradiance (low A) and higher relative humidity (open stomata) having greater c_i/c_a ratio resulting in depleted (more negative) foliar $\delta^{13}C$ values [155].

Differences in foliar $\delta^{13}C$, resulting from combinations of the above mentioned factors, may vary in relation to: (a) species differences; (b) spatial location; and/or (c) disturbance history. Bonal et al. (2007) [47] investigated foliar physiology of seedlings of species belonging to different successional groups, in a greenhouse experiment on seedlings, and found that although pioneer (fast growing early successional) species had higher assimilation rates, reduced water use efficiency resulted in depleted (more negative) foliar $\delta^{13}C$ values than fast-growing late successional species. Interestingly, they found no linear pattern of foliar $\delta^{13}C$ change, but rather a V pattern, suggesting a transition among factors driving discrimination, and highlighted the need to study foliar $\delta^{13}C$ dynamics along more complete successional gradients. Such non-linear variation might result from differences in seedling growth strategies, with: (a) low WUE resulting in ^{13}C depletion (greater discrimination) in pioneer species; (b) high A but higher WUE resulting in relative ^{13}C enrichment in intermediate successional species; and (c) very low rates of $C0_2$ assimilation offsetting high WUE causing ^{13}C depletion in late successional species. Huc et al. (1994) [143] had similar findings using $\delta^{13}C$ of cellulose in wood cores from adult trees growing in artificial stands of pioneer and late stage forest species. They found that pioneer species had the greatest ^{13}C discrimination, resulting from: (a) lower WUE; (b) high maximum conductance values; and (c) high specific hydraulic conductance, indicating an increased competitive ability for water and nutrient uptake in pioneer species.

Variation in foliar LMA and $\delta^{13}C$ often follow predictable spatial patterns within forests. Domingues et al. (2005) [164] found LMA to increase and foliar $\delta^{13}C$ to become enriched (less negative) with increasing height in an Amazonia old growth forest as a result of increased A values (due to increased irradiance) resulting in reduced c_i – increased c_i/c_a ratio. Ometto et al. (2002)

[159] had similar findings in the Brazilian Amazon, which they attributed to gradients in light and humidity – and specifically not to variation in source CO_2 isotope composition, which varied by only 3% throughout the full canopy profile - affecting the ratio of leaf photosynthetic capacity and stomatal conductance. However, light gradients do not necessarily result in variation in carbon isotope discrimination as A and g typically change simultaneously keeping c_i/c_a, and therefore foliar ^{13}C fractionation, relatively constant [188] – at least within an individual species with low leaf plasticity, as leaves growing under increased light availability can have higher LMA values [42]. Additional variation may also result from increasing tree height with succession, as height increases have been found to result in ^{13}C enrichment due to reduced conductance and increases in LMA [165] – although this pattern is non-linear and most pronounced above 30 meters which is greater than the maximum tree height in all our secondary forest stands. Although not the subject of their analyses, variation in height within a forest may occur simultaneous with changes in species composition resulting in taxonomically constrained variation in foliar $\delta^{13}C$.

Foliar $\delta^{13}C$ variation also occurs within individual tree crowns. Waring and Silvester (1994) [166] found that the aspect of exposure and branch length accounted for 6% of the foliar $\delta^{13}C$ variation within individual crowns of *Pinus radiata*, with shorter branches or shaded aspect of exposure being related to enriched (less negative) foliar $\delta^{13}C$ values. These differences resulted from decreased stomatal conductance on the longer branches. Differences in foliar $\delta^{13}C$ among study sites typically results from a combination of the factors identified above. For example, Ehleringer et al. (1986) [167] found a clear decrease in foliar $\delta^{13}C$ from disturbed to undisturbed sites, indicating decreased CO_2 concentrations and increased water use efficiency (WUE). They explain this difference however through large changes in irradiance as altered through leaf canopy position and overstory density – likely causing changes in LMA.

We identified a highly significant decrease in foliar $\delta^{13}C$ with increasing species successional status, but not with increasing stand age. Many of the factors identified above are: (a) related to either stand structural or soil properties, which would change with stand age and therefore result in no linear pattern; or (b) were normalized for across our study stands, such as topographic differences, elevation or climatic conditions. As all our leaves were collected from top of canopy full sunlight and atmospherically well circulated positions, differences in source CO_2 or illumination are unlikely to have had a significant impact. Increases in internal CO_2 diffusion resistance, as identified by Vitousek et al. (1990) [42], are also unlikely play a major role, as early successional species (and most others as well) have, relative to high elevation *Metrosideros polymorpha*, thin leaves. In addition, internal diffuse resistance would result in reduced discrimination (enriched foliar $\delta^{13}C$) in the later successional species having greater LMA – the opposite to the positive relationship between increasing ^{13}C depletion with species successional status we found in our analyses.

Given the significant relationship with species successional status, but not stand age, described above, variation in foliar $\delta^{13}C$ appears to be almost entirely related to species differences in leaf physiology and growth strategy correlated to their species successional status. While we did not collect data on LMA for our study species, Poorter et al. (2004) [32] quantified leaf traits of <2 m tall saplings across a successional gradient in the Bolivia Amazon and found significant decreases in specific leaf area (converse of LMA), nitrogen, water content and increases in C:N and lignin in later stage successional species. If similar patterns in LMA exist across adult individuals, as was found by Reich et al. (1995) [29] in the Venezuelan Amazon where later stage successional species had increasingly lower rates of photosynthesis, conductance and SLA, and increased leaf life span and toughness, then decreasing conductance relative to A could result in depleted (more negative) foliar $\delta^{13}C$ values. The strong negative relationship between foliar $\delta^{13}C$ and species successional status, but not stand age, further emphasizes that this trend is occurring simultaneous to – but independent of - the negative relationship shown between soil $\delta^{13}C$ and stand age, which is in large part driven through changes in the foliar $\delta^{13}C$ input during transition through species succession status. Given the non-significant relationship between stand age and foliar $\delta^{13}C$, the significant increase in $\Delta\delta^{13}C_{\text{plant-soil}}$ we identified with increasing stand age indicates that either: (a) the incorporation of the foliar $\delta^{13}C$ into the soil slows with increasing stand age; (b) that a significant source of soil $\delta^{13}C$ enrichment develops with increasing stand age; or (c) a significant source of soil $\delta^{13}C$ diminishes with stand age. Changes in foliar $\delta^{13}C$ incorporation into the soil are directly related to the rapid increases in the quantity of decomposers during early stand development which can cause an exponential increase in the rate of overall decomposition, as the decomposers have access to abundant high quality litterfall from fast growing pioneer species, followed by reduced but stable rates of decomposition later in succession of the lower quality litterfall from slower growing tree species [31,168]. However, increasing soil ^{13}C enrichment via auto- and hetero-trophic respiration and microbial fractionation during decomposition could serve to offset soil ^{13}C depletion resulting from: (a) the loss of the initial large ^{13}C enrichment pulse following the burning; (b) incorporation of increasingly depleted ^{13}C organic matter into the soil; and (c) transition to all C^3 species, and thereby causing the increasingly negative $\Delta\delta^{13}C_{\text{plant-soil}}$ relationship.

Integration and Conclusion

During forest succession competition among species for resources, including light, water and soil nutrients, play an important role in species composition [169]. Considering variable resources, alterations in composition during forest succession are likely partially explained through resource niche partitioning [170]. Tree species from differing successional guilds may express different competitive capacities or nutrient use strategies during stand development [171], including shifts in nutrient use strategies from 'conservative consumption' to 'nutrient spending' – although this is most related to shifts in plant functional type rather than community-scale dynamics [142]. The interactions between these factors, including differences among taxonomic and physiological and functional diversity, has been described as being one of the least understood themes in tropical forest ecology [28,172]. It is difficult to separate soil and foliar processes as they develop through complex feedbacks [136]. For example, changing successional vegetation plays a crucial role in making nutrients from the total soil pool available to plants [82] which feeds back to influence which species are best adapted to succeed and their spatial distribution [189]. Processes, such as the introduction of exotic species [173], can alter dynamics - similar to the continuous dynamic of feedback restructuring that occurs during secondary succession. Successional dynamics and biomass accumulation following abandonment depend in large part on the initial disturbance [174].

In this study, we found soil N to increase during forest succession while P remained stable. This was expected given the increasing influence of biological nitrogen fixation during succession and progressive depletion or incorporation of P into aboveground biomass. Increasing foliar %N values were shown with increasing species successional status indicating changes in growth and competitive strategy under conditions of varying N availability. Foliar P however was related to soil texture, not soil P concentration, and therefore represented differences among the plots not related to forest succession. The significant soil $\delta^{15}N$ depletion with stand age likely does not represent a progressive tightening of the nitrogen cycle; rather it represents the balance of other fractionation related nitrogen cycling processes in a post-fire induced ^{15}N enrichment context. While foliar $\delta^{15}N$ is largely determined by the soil isotope composition, increasing divergence with species successional status indicates an increase in nitrogen isotope fractionation in older stands by later succession species, possibly as result of increased mycorrhizal activity and changes in root foraging depth. Soil $\delta^{13}C$ is however in large part determined by the plant composition, with few plant-soil carbon isotope feedbacks, that is again partly determined by the initial burning related soil ^{13}C enrichment.

Foliar $\delta^{13}C$ depletion with increasing succession status, but not stand age, highlights the changes in tree growth and competition strategy during succession from thin fast growing leaves with rapid A and low WUE to slow growing species with thick well protected long lasting leaves having reduced rates of A but increased WUE. Increasing divergence between soil and foliar $\delta^{13}C$ may be a result of increases in the quantity - but less likely the rate – of carbon cycling processes, such as decomposition, as stands age. Foliar %C was unrelated to stand age or species successional status but had significant phylogenetic signal, which was not found across the species successional status gradient. Phylogenetic constraints on foliar %C have been identified in other studies in less dynamic study systems, and likely other leaf traits in our study would exhibit phylogenetic signal under more stable (i.e., primary forest) conditions. The interplay among soil and foliar factors, with more specific quantification of the competitive processes during forest succession, requires further investigation and will be the focus of future work.

Acknowledgments

We thank T. Tobeck, N. Gurwick, C. Andreassi, and K. Mertes for help with laboratory analyses; M. Gilbert, J. Johnson, K. Cahill, K. Gerow and A. Storer for help with statistical analyses; A. Rominger, E. Kurten and J. Savage for help with phylogenetic analyses; E. Soriano and the many members of our field crew from Brazil, Peru and Bolivia for help with field data collection and A. Duchelle and F. Brown for sharing their expertise in this region; R. Chazdon, E. Davidson, P. Vitousek and B. Houlton for valuable comments throughout; D. Julio and the community of Molienda in Bolivia for working with us, allowing access to their properties, and for sharing their extensive knowledge of the area. We thank W. Clark, N. Dickson, N.M. Holbrook and the fellows of the Sustainability Science Program, especially S. Rantala, F. Mello, R. Fishman, B. Dillon and M. Gilbert, for help and guidance during the writing and analysis process. We thank our anonymous reviewers for their time and many useful insights.

Author Contributions

Conceived and designed the experiments: ENB AMAZ MS. Performed the experiments: ENB AMAZ MS HRDS. Analyzed the data: ENB AMAZ MS HRDS GPA CBF MPC RA RD. Contributed reagents/materials/analysis tools: LG. Wrote the paper: ENB.

References

1. Asner GP, Rudel TK, Aide TM, Defries R, Emerson R (2009) A contemporary assessment of change in humid tropical forests. Conservation Biology 23: 1386–1395.
2. Chazdon RL (2008) Beyond deforestation: Restoring forests and ecosystem services on degraded lands. Science 320: 1458–1460.
3. Moran EF, Brondizio ES, Tucker JM, Silva-Forsberg MC, McCracken S, et al. (2000) Effects of soil fertility and land-use on forest succession in Amazonia. Forest Ecology and Management 139: 93–108.
4. Wadsworth FH (1997) Forest production for tropical America. Agriculture Handbook 710. United States Department of Agriculture Forest Service, Washington, DC.
5. Fearnside PM, Guimaraes WM (1996) Carbon uptake by secondary forests in Brazilian Amazonia. Forest Ecology and Management 80: 35–46.
6. Houghton RA, Skole D, Nobre CA, Hackler J, Lawrence K, et al. (2000) Annual fluxes of carbon from deforestation and regrowth in the Brazilian Amazon. Nature 403: 301–304.
7. Lawrence D (2005) Biomass accumulation after 10–200 years of shifting cultivation in Bornean rain forest. Ecology 86: 26–33.
8. Guariguata M, Ostertag R (2001) Neotropical secondary forest succession: changes in structural and functional characteristics. Forest Ecology and Management 148: 185–206.
9. Ostertag R, Marin-Spiotta E, Silver WL, Schulten J (2008) Litterfall and decomposition in relation to soil carbon pools along a secondary forest chronosequence in Puerto Rico. Ecosystems 11: 701–714.
10. Quesada CA, Lloyd J, Schwarz M, Baker TR, Phillips OL, et al. (2009) Regional and large-scale patterns in Amazon forest structure and function are mediated by variations in soil physical and chemical properties. Biogeosciences Discuss 6: 3993–4057.
11. Gehring C, Denich M, Vlek PL (2005) Resilience of secondary forest regrowth after slash-and-burn agriculture in central Amazonia. Journal of Tropical Ecology 21: 519–527.
12. Davidson EA, Zarin DJ, Brondizio E, Vieira ICG, Sa T, et al. (2005) Legacy of Fire Slows Carbon Accumulation in Amazonian Forest Regrowth. Frontiers in Ecology and the Environment 3: 365–369.
13. Castilho CV, Magnusson WE, Nazare R, Lima AP, Higuchi N, et al. (2006) Variation in aboveground tree live biomass in a central Amazonian Forest: Effects of soil and topography. Forest Ecology and Management 234: 85–96.
14. Guevara S, Purata SE, Van der Maarel E (1986) The role of remnant forest trees in tropical secondary succession. Vegetatio 66: 77–84.
15. Budowski G (1965) Distribution of tropical American rain forest species in the light of successional processes. Turrialba 15: 40–42.
16. Finegan B (1984) Forest succession. Nature 312: 109–115.
17. Finegan B (1996) Pattern and process in neotropical secondary rain forests: the first 100 years of succession. Trends in Ecology and Evolution - Cell 11: 793–796.
18. Kennard DK (2002) Secondary forest succession in a tropical dry forest: patterns of development across a 50-year chronosequence in lowland Bolivia. Journal of Tropical Ecology 18: 53–66.
19. Pinard M, Putz F, Rumiz D, Guzmán R, Jardim A (1999) Ecological characterization of tree species for guiding forest management decisions in seasonally dry forests in Lomerio, Bolivia. Forest Ecology and Management 113: 201–213.
20. Denslow JS (1996) Functional group diversity and responses to disturbance. Biodiversity and ecosystem processes in tropical forests. (eds G.H. Orians, R. Dirzo & J.H. Cushman), 127–151. Springer-Verlag, Berlin.
21. Denslow JS (1987) Tropical rainforest gaps and tree species diversity. Annual Review of Ecology and Systematics. 30: 761–766.
22. Denslow JS, Guzman SG (2000) Variation in stand structure, light and seedling abundance across a tropical moist forest chronosequence, Panama. Journal of Vegetation Science 11: 201–212.
23. Peña-Claros M (2003) Changes in forest structure and species composition during secondary forest succession in the Bolivian Amazon. Biotropica 35: 450–461.
24. Waring RH, Running SW (2007) Temporal changes in forest structure and function. Forest Ecosystems: Analysis at Multiple Scales. 149–182. Elsevier, Burlington, MA.
25. Oliver C, Larson B (1990) Forest stand dynamics, 467 pp. MaGraw-Hill, New York.
26. Chazdon RL (2008) Chance and determinism in tropical forest succession. Pages 384–408 in W. Carson and S. A. Schnitzer, editors. Tropical forest community ecology. Wiley-Blackwell Publishing.
27. Swaine MD, Whitmore TC (1988) On the definition of ecological species groups in tropical rain forests. Plant Ecology 75: 81–86.
28. Chazdon RL, Finegan B, Capers RS, Salgado-Negret B, Casanoves F, et al. (2010) Composition and dynamics of functional groups of trees during tropical forest succession in northeastern Costa Rica. Biotropica 42: 31–40.

29. Reich P, Ellsworth D, Uhl C (1995) Leaf carbon and nutrient assimilation and conservation in species of differing successional status in an oligotrophic Amazonian forest. Functional Ecology 9: 65–76.
30. Poorter L, van de Plassche S, Boot RG (2004) Leaf traits and herbivory rates of tropical tree species differing in successional status. Plant Biology 6: 746–754.
31. Garnier E, Cortez J, Billès G, Navas M-L, Roumet C, et al. (2004) Plant functional markers capture ecosystem properties during secondary succession. Ecology 85: 2630–2637.
32. Poorter L (2007) Are species adapted to their regeneration niche, adult niche, or both? The America Naturalist 169: 433–442.
33. Ordoñez JC, van Bodegom PM, Witte JPM, Wright IJ, Reich PB, et al. (2009) A global study of relationships between leaf traits, climate and soil measures of nutrient fertility. Global Ecology and Biogeography 18: 137–149.
34. Gusewell S (2004) N:P ratios in terrestrial plants: variation and functional significance. New Phytologist 164: 243–266.
35. Koerselman W, Meuleman AF (1996) The vegetation N:P ratio: a new tool to detect the nature of nutrient limitation. The Journal of Applied Ecology 33: 1441–1450.
36. Townsend AR, Cleveland CC, Asner GP, Bustamante MMC (2007) Controls over foliar N:P ratios in tropical rain forests. Ecology 88: 107–118.
37. Davidson EA, de Carvalho CJR, Figueira AM, Ishida FY, Ometto JPHB, et al. (2007) Recuperation of nitrogen cycling in Amazonian forests following agricultural abandonment. Nature 447: 995–998.
38. Tessier JT, Raynal DJ (2003) Use of nitrogen to phosphorus ratios in plant tissue as an indicator of nutrient limitation and nitrogen saturation. Journal of Applied Ecology 40: 523–534.
39. Davidson EA, Howarth RW (2007) Nutrients in synergy. Nature 449: 1000–1001.
40. Hedin L, Vitousek P, Matson P (2003) Nutrient losses over four million years of tropical forest development. Ecology 84: 2231–2255.
41. Dawson TE, Mambelli S, Plamboeck AH, Templer PH, Tu KP (2002) Stable Isotopes in Plant Ecology. Annual Review of Ecology and Systematics 33: 507–559.
42. Vitousek PM, Field C, Matson P (1990) Variation in foliar $\delta^{13}C$ in Hawaiian Metrosideros polymorpha: a case of internal resistance? Oecologia 84: 362–370.
43. Compton JE, Hooker TD, Perakis SS (2007) Ecosystem N distribution and $\delta^{15}N$ during a century of forest regrowth after agricultural abandonment. Ecosystems 10: 1197–1208.
44. Billings S, Richter D (2006) Changes in stable isotopic signatures of soil nitrogen and carbon during 40 years of forest development. Oecologia 148: 325–33.
45. Schedlbauer JL, Kavanagh KL (2008) Soil carbon dynamics in a chronosequence of secondary forests in northeastern Costa Rica. Forest Ecology and Management 255: 1326–1335.
46. Pardo LH, Hemond HF, Montoya JP, Fahey TJ, Siccama TG (2002) Response of the natural abundance of ^{15}N in forest soils and foliage to high nitrate loss following clear-cutting. Canadian Journal of Forest Research 1136: 1126–1136.
47. Bonal D, Born C, Brechet C, Coste S, Marcon E, et al. (2007) The successional status of tropical rainforest tree species is associated with differences in leaf carbon isotope discrimination and functional traits. Annals of Forest Science 64: 169–176.
48. Beekma J, Zonta A, Keijzer B (1996) Base ambiental para el desarrollo del departamento de Pando y la provincia Vaca Diez. La Paz, Bolivia.
49. World Agroforestry Center (2010) Wood Density Database.
50. Nogueira E, Fearnsid P, Nelson B, Franca M (2007) Wood density in forests of Brazil's "arc of deforestation": Implications for biomass and flux of carbon from land-use change in Amazonia. Forest Ecology and Management 248: 119–135.
51. Fearnside P (1997) Wood density for estimating forest biomass in Brazilian Amazonia. Forest Ecology and Management 90: 59–87.
52. Chave J, Andalo C, Brown S, Cairns MA, Chambers JQ, et al. (2005) Tree allometry and improved estimation of carbon stocks and balance in tropical forests. Oecologia 145: 87–99.
53. Chao A, Chazdon RL, Colwell RK, Shen TJ (2005) A new statistical approach for assessing similarity of species composition with incidence and abundance data. Ecology Letters 8: 148–159.
54. Chazdon RL, Chao A, Colwell RK, Lin S-Y, Norden N, et al. (2011) A novel statistical method for classifying habitat generalists and specialists. Ecology 92: 1332–1343.
55. Amundson R, Austin E, Schuur EA, Yoo K, Matzek V, et al. (2003) Global patterns of the isotopic composition of soil and plant nitrogen. Global Biogeochemical Cycles 17: 1031–1042.
56. Toledo M, Poorter L, Peña-Claros M, Alarcón A, Balcázar J, et al. (2011) Climate and soil drive forest structure in Bolivian lowland forests. Journal of Tropical Ecology 27: 333–345.
57. Swenson NG, Enquist BJ (2007) Ecological and evolutionary determinants of a key plant functional trait: wood density and its community-wide variation across latitude and elevation. American Journal of Botany 94: 451–459.
58. Wojciechowski MF, Lavin M, Sanderson MJ (2004) A phylogeny of legumes (Leguminosae) based on analysis of the plastic MATK gene resolves many well-supported subclades within the family. American Journal of Botany 91: 1846–1862.
59. Zerega NJ, Clement WL, Datwyler SL, Weiblen GD (2005) Biogeography and divergence times in the mulberry family (Moraceae). Molecular Phylogenetics and Evolution 37: 402–416.
60. Wikstrom N, Savolainen V, Chase MW (2001) Evolution of the angiosperms: calibrating the family tree. Proceedings of the Royal Society B 268: 2211–2220.
61. Hedges SB, Dudley J, Kumar S (2006) TimeTree: a public knowledge-base of divergence times among organisms. Bioinformatics 22: 2971–2972.
62. Webb CO, Ackerly DD, Kembel SW (2008) Phylocom: software for the analysis of phylogenetic community structure and trait evolution. Bioinformatics 24: 2098–2100.
63. Kembel SW, Cowan PD, Helmus MR, Cornwell WK, Morlon H, et al. (2010) Picante: R tools for integrating phylogenies and ecology. Bioinformatics 26: 1463–1464.
64. Blomberg SP, Garland T, Ives AR (2003) Testing for phylogenetic signal in comparative data: behavioral traits are more labile. Evolution 57: 717–745.
65. Ackerly DD (2000) Taxon sampling, correlated evolution, and independent contrasts. Evolution. International Journal of Organic Evolution 54: 1480–1492.
66. Davies TJ, Kraft NJB, Salamin N, Wolkovich EM (2011) Incompletely resolved phylogenetic trees inflate estimates of phylogenetic conservatism. Ecology 93: 242–247.
67. Schoenholtz SH, Van Miegroet H, Burger JA (2000) A review of chemical and physical properties as indicators of forest soil quality: challenges and opportunities. Forest Ecology and Management 138: 335–56.
68. Prach K, Pyšek P, Šmilauer P (1993) On the rate of succession. Oikos 66: 343–346.
69. Van Breugel M, Bongers F, Martínez-Ramos M (2007) Species dynamics during early secondary forest succession: recruitment, mortality and species turnover. Biotropica 35: 610–619.
70. Gómez-Pompa A, Vázquez-Yanes C (1981) Successional studies of a rainforest in Mexico. In D. C. West, H. H. Shugart, and D. B. Botkin (Eds.). Forest succession, concepts and application, 246–266. Springer-Verlag, New York, New York.
71. Guariguata MR, Chazdon RL, Denslow JS, Dupuy JM, Anderson L (1997) Structure and floristics of secondary and old-growth forest stands in lowland Costa Rica. Plant Ecology 132: 107–120.
72. Norden N, Chazdon RL, Chao A, Jiang YH, Vilchez-Alvarado B (2009) Resilience of tropical rain forests: tree community reassembly in secondary forests. Ecology Letters 12: 385–394.
73. Blundell A, Peart D (2004) Density-dependent population dynamics of a dominant rain forest canopy tree. Ecology 85: 704–715.
74. Canham C (1988) Growth and canopy architecture of shade-tolerant trees: Response to canopy gaps. Ecology 69: 786–795.
75. Silver WL, Ostertag R, Lugo AE (2000) The potential for carbon sequestration through reforestation of abandoned tropical agricultural and pasture lands. Restoration Ecology. 8: 394–407.
76. Saldarriaga J (1985) Forest succession in the Upper Rio Negro of Columbia and Venezuela. Ph.D. dissertation, University of Tennessee, Knoxville, TN, USA.
77. Uhl C, Buschbacher R, Serrao EAS (1988) Abandoned pastures in eastern Amazonia. I. Patterns of plant succession. Journal of Ecology 76: 663–681.
78. Silver WL, Neff J, McGroddy M, Veldkamp E, Keller M, et al. (2000) Effects of soil texture on belowground carbon and nutrient storage in a lowland Amazonian forest ecosystem. Ecosystems 3: 193–209.
79. Guo L, Gifford R (2002) Soil carbon stocks and land use change: a meta-analysis. Global Change Biology 8: 345–360.
80. Feldpausch TR, Rondon MA, Fernandes EC, Riha SJ, Wandelli E (2004) Carbon and nutrient accumulation in secondary forests regenerating on pastures in central Amazonia. Ecological Applications 14: 164–176.
81. Paul K, Polglase P, Nyakuengama J, Khanna P (2002) Change in soil carbon following afforestation. Forest Ecology and Management 168: 241–257.
82. Werner P (1984) Changes in soil properties during tropical wet forest succession in Costa Rica. Biotropica 16: 43–50.
83. Awiti AO, Walsh MG, Kinyamario J (2008) Dynamics of topsoil carbon and nitrogen along a tropical forest–cropland chronosequence: Evidence from stable isotope analysis and spectroscopy. Agriculture, Ecosystems & Environment 127: 265–272.
84. Kauffman JB, Cummings DL, Ward DE, Babbit R (1995) Fire in the Brazilian Amazon: 1. Biomass, nutrient pools, and losses in slashed primary forests. Oecologia 104: 397–408.
85. Keller M, Veldkamp E, Weitz AM, Reiners WA (1993) Effect of pasture age on soil trace-gas emissions from a deforested area of Costa Rica. Nature 365: 244–246.
86. Markewitz D, Davidson E, Moutinho P, Nepstad D (2004) Nutrient loss and redistribution after forest clearing on a highly weather soil in Amazonia. Ecological Applications 14: S177–S199.
87. Cleveland CC, Townsend AR, Schimel DS, Fisher H, Howarth RW, et al. (1999) Global patterns of terrestrial biological nitrogen (N-2) fixation in natural ecosystems. Global Biogeochemical Cycles 13: 623–645.
88. Rastetter EB, Vitousek PM, Field C, Shaver GR, Herbert D, et al. (2001) Resource optimization and symbiotic nitrogen fixation. Ecosystems 4: 369–388.
89. Holland EA, Dentener FJ, Braswell BH, Sulzman JM (1999) Contemporary and pre-industrial global reactive nitrogen budgets. Biogeochemistry 46: 7–43.

90. Handley LL, Raven JA (1992) The use of natural abundance of nitrogen isotopes in plant physiology and ecology. Plant Cell and Environment 15: 965–985.
91. Robinson D (2001) $\delta^{15}N$ as an integrator of the nitrogen cycle. Trends in Ecology & Evolution 16: 153–162.
92. Garten CT Jr (1991) Nitrogen isotope composition of ammonium and nitrate and bulk precipitation and forest throughfall. International Journal of Environmental Analytical Chemistry 87: 33–45.
93. Andreae MO, Browell EV, Garstang M, Gregory G, Harriss R, et al. (1988) Biomass-burning emissions and associated haze layers over Amazonia. Journal of Geophysical Research 93: 1509–1527.
94. Piccolo MC, Neill C, Cerri CC (1994) Natural abundance of ^{15}N in soils along forest-to-pasture chronosequences in the western Brazilian Amazon Basin. Oecologia 99: 1–2.
95. Delwiche C, Zinke P, Johnson C, Virginia RA (1979) Nitrogen isotope distribution as a presumptive indicator of nitrogen fixation. Botanical Gazette 140: 65–69.
96. Yoneyama T, Muraoka T, Murakami T, Boonkerd N (1993) Natural abundance of ^{15}N in tropical plants with emphasis on tree legumes. Plant and Soil 153: 295–304.
97. Hobbie EA, Colpaert JV (2003) Nitrogen availability and colonization by mycorrhizal fungi correlate with nitrogen isotope patterns in plants. New Phytologist 157: 115–126.
98. Hobbie EA, Ouimette AP (2009) Controls of nitrogen isotope patterns in soil profiles. Biogeochemistry 95: 355–371.
99. Piccolo MC, Neill C, Melillo JM (1996) ^{15}N natural abundance in forest and pasture soils of the Brazilian Amazon Basin. Plant Soil 182: 249–258.
100. Houlton B, Sigma DM, Hedin LO (2006) Isotopic evidence for large gaseous nitrogen losses from tropical forests. Proceedings of the National Academy of Sciences 103: 8745–50.
101. Austin AT, Vitousek PM (1998) Nutrient dynamics on a precipitation gradient in Hawai'i. Oecologia 113: 519–529.
102. Nadelhoffer KJ, Fry B (1988) Controls on natural nitrogen-15 and carbon-13 abundance in forest soil organic matter. Soil Science Society of America 52: 1633–1640.
103. Schuur EA, Matson PA (2001) Net primary productivity and nutrient cycling across a mesic to wet precipitation gradient in Hawaiian montane forest. Oecologia 128: 431–442.
104. Wallander H, Morth CM, Giesler R (2009) Increasing abundance of soil fungi is a driver for ^{15}N enrichment in soil profiles along a chronosequence undergoing isostatic rebound in northern Sweden. Oecologia 160: 87–96.
105. Delwiche C, Steyn P (1970) Nitrogen isotope fractionation in soils and microbial reactions. Environmental Science and Technology 11: 929–935.
106. Koerner W, Dambrine E, Dupouey JL, Benoit (1999) $\delta^{15}N$ of forest soil and understory vegetation reflect the former agricultural land use. Oecologia 121: 421–425.
107. Boeckx P, Paulino L, Oyarzun C, van Cleemput O, Godoy R (2005) Soil $\delta^{15}N$ patterns in old-growth forests of southern Chile as integrator for N-cycling. Isotopes in Environmental and Health Studies 41: 249–259.
108. Cochrane MA (2011) The past, present, and future importance of fire in tropical rainforests. In Bush MB, Flenley JR, Gosling WD (Eds.) Tropical rainforest responses to climate change (second edition), 213–240. Springer-Verlag, New York, New York.
109. Myneni RB, Yang W, Nemani RR, Huete AR, Dickinson RE, et al. (2007) Large seasonal swings in leaf area of Amazon rainforests. Proceedings of the National Academy of Sciences 104: 4820–4823.
110. Boutton TW (1991) Stable carbon isotope ratios of natural materials: II. Atmospheric, terrestrial, marine and freshwater environments. In Coleman DC and Fry B (eds.), Carbon Isotope Techniques. San Diego, Academic Press: pgs. 173–185.
111. Schweizer M, Fear S, Cadisch G (1999) Isotopic ((13)C) fractionation during plant residue decomposition and its implications for soil organic matter studies. Rapid Communications in Mass Spectrometry 13: 1284–1290.
112. Agren GI, Bosatta E, Balesdent J (1996) Isotope discrimination during decomposition of organic matter: a theoretical analysis. Soil Science Society of American Journal 60: 1121–1126.
113. Lin G, Enleringer JR, Rygiewicz PT, Johnson MG, Tingey DT (1999) Elevated CO_2 and temperature impacts on different components of soil CO_2 efflux in Douglas-fir terracosms. Global Change Biology 5: 157–168.
114. Ehleringer J, Buchmann N, Flanagan L (2000) Carbon isotope ratios in belowground carbon cycle processes. Ecological Applications 10: 412–422.
115. Klumpp R, Shaufele R, Lotscher M, Lattanzi FA, Feneis W, et al. (2005) C-isotope composition of CO_2 respired by shoots and roots: fractionation during dark respiration? Plant, Cell and Environment 28: 241–250.
116. Harmon ME, Bond-Lamberty B, Tang J, Vargas R (2011) Heterotrophic respiration in disturbed forests: A review with examples from North America. Journal of Geophysical Research 116: 1–17.
117. Schwartz D (1991) Interet de la mesure du $\delta^{13}C$ des sols en milieu naturel equatorial pour la connaissance des aspects pedologiques et ecologiques des relations savane-forest. Cahiers de l'Orstom, série Pédologie. 26: 327–341.
118. Veldkamp E, Weitz AM (1994) Uncertainty analysis of studies. Soil Biology and Biochemistry 26: 153–160.
119. Bernoux M, Cerri CC, Neill C, de Moraes JFL (1998) The use of stable carbon isotopes for estimating soil organic matter turnover rates. Geoderma 82: 43–58.
120. Balesdent J, Girardin C, Mariotti A (1993) Site-related matter in a temperate forest. Ecology 74: 1713–1721.
121. Desjardins T, Andreux F, Volkoff B, Cerri CC (1994) Organic carbon and ^{13}C contents in soils and soil size-fractions, and their changes due to deforestation and pasture installation in the eastern Amazonia. Geoderma 61: 103–118.
122. Feigl BJ, Mellilo J, Cerri CC (1995) Changes in the origen and quality of soil organic matter after pasture introduction in Rondonia (Brazil). Plant Soil 175: 21–29.
123. Peterson BJ, Fry B (1987) Stable isotopes in ecosystem studies. Annual Review of Ecology and Systematics 18: 293–320.
124. López-Ulloa M, Veldkamp E, de Koning GHJ (2005) Soil Carbon Stabilization in Converted Tropical Pastures and Forests Depends on Soil Type. Soil Science Society of America Journal 69: 1110–1117.
125. Finkelstein DB, Pratt LM, Brassell SC (2006) Can biomass burning produce a globally significant carbon-isotope excursion in the sedimentary record? Earth and Planetary Science Letters 250: 501–510.
126. Wright IJ, Reich PB, Westoby M, Ackerly DD, Baruch Z, et al. (2004) The worldwide leaf economics spectrum. Nature 428: 821–827.
127. Wright IJ, Reich PB, Cornelissen JH, Falster DS, Garnier E, et al. (2005) Assessing the generality of global leaf trait relationships. The New Phytologist 166: 485–496.
128. Wright IJ, Westoby M, Reich PB (2001) Strategy shifts in leaf physiology, structure and nutrient content between species of high- and low-rainfall and high- and low-nutrient habitats. Functional Ecology 15: 351–359.
129. He JS, Wang X, Schmid B, Flynn DFB, Li X, et al. (2010) Taxonomic identity, phylogeny, climate and soil fertility as drivers of leaf traits across Chinese grassland biomes. Journal of Plant Research 123: 551–561.
130. Powers JS, Tiffin P (2010) Plant functional type classifications in tropical dry forests in Costa Rica: leaf habit versus taxonomic classifications. Functional Ecology 24: 927–936.
131. Fyllas NM, Patiño S, Baker TR, Bielefeld Nardoto G, Martinelli LA, et al. (2009) Basin-wide variations in foliar properties of Amazonian forest: phylogeny, soils and climate. Biogeosciences 6: 2677–2708.
132. Asner GP, Martin RE, Bin Suhaili A 2012. Sources of canopy chemical and spectral diversity in lowland Bornean forest. Ecosystems 15: 504–517.
133. Chazdon RL, Field CB (1987) Determinants in photosynthetic capacity in six rainforest Piper species. Oecologia 73: 222–230.
134. Kull O, Kruijt B (1999) Acclimation of photosynthesis to light: A mechanistic approach. Functional Ecology 13: 24–36.
135. Niinemets U (2007) Photosynthesis and resource distribution through plant canopies. Plant, Cell and Environment 30: 1052–1071.
136. Binkley D, Giardina C (1998) Why do tree species affect soils? The warp and woof of tree-soil interactions. Biogeochemistry 42: 89–106.
137. Toledo M, Salick J (2006) Secondary succession and indigenous management in semideciduous forest fallows of the Amazon Basin. Biotropica 38: 161–170.
138. Asner GP, Martin RE (2011) Canopy phylogenetic, chemical and spectral assembly in a lowland Amazon forest. New Phytologist 189: 999–1012.
139. Letcher SG (2010) Phylogenetic structure of angiosperm communities during tropical forest succession. Proceedings of the Royal Society - Biological Sciences 277: 97–104.
140. Pringle EG, Adams RI, Broadbent E, Busby PE, Donatti CI, et al. (2010) Distinct leaf-trait syndromes of evergreen and deciduous trees in a seasonally dry tropical forest. Biotropica 43: 299–308.
141. Chazdon RL, Letcher SG, van Breugel M, Martínez-Ramos M, Bongers F, et al. (2007) Rates of change in tree communities of secondary Neotropical forests following major disturbances. Philosophical transactions of the Royal Society of London 362: 273–289.
142. Yan ER, Wang XH, Huang JJ (2006) Shifts in plant nutrient use strategies under secondary forest succession. Plant and Soil 289: 187–197.
143. Huc R, Ferhi A, Guehl JM (1994) Pioneer and late stage tropical rainforest tree species (French Guiana) growing under common conditions differ in leaf gas exchange regulation, carbon isotope discrimination and leaf water potential. Oecologia 99: 297–305.
144. Herbert D, Williams M, Rastetter E (2003) A model analysis of N and P limitation on carbon accumulation in Amazonian secondary forest after alternate land-use abandonment. Biogeochemistry 65: 121–150.
145. Romanya J, Khanna PK, Raison RJ (1994) Effects of slash burning on soil phosphorus fractions and sorption and desorption of phosphorus. Forest Ecology and Management 65: 89–103.
146. Quintero CE, Boschetti NG, Benevidez RA (2003) Effect of soil buffer capacity on soil test phosphorus interpretation and fertilizer requirement. Communications in Soil Science and Plant Analysis. 34: 1435–1450.
147. Vitousek PM, Shearer G, Kohl D (1989) Foliar ^{15}N natural abundance in Hawaiian rainforest: patterns and possible mechanisms. Oecologia 78: 383–388.
148. Wang L, Shaner P-J, Macko S (2007) Foliar ^{15}N patterns along successional gradients at plant community and species levels. Geophysical Research Letters. 34: 1–6.
149. Johnson NC, Zal DR, Tilman D, Pfleger FL (1991) Dynamics of vesicular-buscular mycorrhizae during old field succession. Oecologia 86: 349–358.
150. Hobbie EA, Jumpponen A, Trappe J (2005) Foliar and fungal 15N:14N ratios reflect development of mycorrhizae and nitrogen supply during primary succession: Testing analytical models. Oecologia 122: 273–283.

151. Jackson R, Canadell J, Ehleringer JR, Mooney HA, Sala OE, et al. (1996) A global analysis of root distributions for terrestrial biomes. Oecologia 108: 389–411.
152. Nepstad DC, de Carvalho CR, Davidson EA, Jipp PH, Lefebvre PA, et al. (1994) The role of deep roots in the hydrological and carbon cycles of Amazonian forests and pastures. Nature 372: 666–669.
153. Farquhar GD, Ehleringer JR, Hubick KT (1989) Carbon isotope discrimination and photosynthesis. Annual Review of Plant Physiology and Plant Biology 40: 503–537.
154. Park R, Epstein S (1960) Carbon isotope fractionation during photosynthesis. Geochimica et Cosmochimica Acta 21: 110–126.
155. Farquhar GD, O'Leary MH, Berry JA (1982) On the relationship between carbon isotope discrimination and the intercellular carbon dioxide concentration in leaves. Australian Journal of Plant Physiology 9: 121–137.
156. Yakir D, Sternberg LD (2000) The use of stable isotopes to study ecosystem gas exchange. Oecologia 123: 297–311.
157. Quay P, King S, Wilbur D (1989) $^{13}C/^{12}C$ of atmospheric CO_2 in the Amazon basin: Forest and river sources. Journal of Geophysical Research. 94: 13,327–18,336.
158. Smith BN (1971) Carbon isotope ratios of respired CO_2 from castor bean, corn, peanut, pea, radish, squash, sunflower and wheat seedlings. Plant Cell Physiology. 12: 451–455.
159. Ometto JPH, Flanagan BLB, Martinelli LA, Moreira MZ, Higuchi N, et al. (2002) Carbon isotope discrimination in forest and pasture ecosystems of the Amazon Basin, Brazil. Global Biogeochemical Cycles 16: 1–10.
160. Aranibar J, Berry J, Riley WJ, Pataki DE, Law BE, et al. (2006) Combining meteorology, eddy fluxes, isotope measurements, and modeling to understand environmental controls of carbon isotope discrimination at the canopy scale. Global Change Biology 12: 710–730.
161. Farquhar GD, Sharkey TD (1982) Stomatal conductance and photosynthesis. Annual Review of Plant Physiology 33: 317–345.
162. Schulze ED (1986) Carbon dioxide and water vapor exchange in response to drought in the atmosphere and in the soil. Annual Review of Plant Physiology 37: 247–274.
163. Körner CH, Diemer M (1987) In situ photosynthetic responses to light, temperature and carbon disxide in herbaceous plants from low to high altitude. Functional Ecology 1: 179–194.
164. Domingues F, Berry JA, Martinelli LA, Ometto JP, Ehleringer JR (2005) Parameterization of canopy structure and leaf-level gas exchange for an eastern Amazonian tropical rain forest. Earth Interactions 9: 1–23.
165. Koch GW, Sillett SC, Jennings GM, Davis SD (2004) The limits to tree height. Nature 428: 851–854.
166. Waring RH, Silvester WB (1994) Variation in foliar $\delta^{13}C$ values within the crowns of Pinus radiata trees. Tree physiology 14: 1203–1213.
167. Ehleringer J, Field C, Lin Z, Kuo C (1986) Leaf carbon isotope and mineral composition in subtropical plants along an irradiance cline. Oecologia, 70: 520–526.
168. Ewel J (1976) Litter fall and leaf decomposition in a tropical forest succession in eastern Guatamala. Journal of Ecology 64: 293–308.
169. Huston M, Smith T (1987) Plant succession: life history and competition. American Naturalist 130: 168–198.
170. Kobe RK (1999) Light gradient partitioning among tropical tree species through differential seedling mortality and growth. Ecology 80: 187–201.
171. Bazzaz FA, Pickett STA (1980) Physiological ecology of tropical succession: A comparative review. Annual Review of Ecology and Systematics 11: 287–310.
172. Clark DA, Clark DB (1992) Life History Diversity of Canopy and Emergent Trees in a Neotropical Rain Forest. Ecological Monographs 62: 315–344.
173. Matson P (1990) Plant-soil interactions in primary succession at Hawaii Volcanoes National Park. Oecologia 85: 241–246.
174. Zarin DJ, Davidson EA, Brondizio E, Vieira ICG, Sa T, et al. (2005) Legacy of fire slows carbon accumulation in Amazonian forest regrowth. Frontiers in Ecology 3: 365–369.
175. Vitousek PM, Turner DR, Kitayama K (1995) Foliar nutrients during long-term soil development in Hawaiian montane rain forest. Ecology 76: 712–720.
176. Vitousek PM, Reiner WA (1975) Ecosystem succession and nutrient retention: a hypothesis. BioScience 25: 376–381.
177. Ometto JPH, Ehleringer JR, Tomas DF, Berry JA, Ishida FY, et al. (2006) The stable carbon and nitrogen isotopic composition of vegetation in tropical forests of the Amazon Basin, Brazil. Biogeochemistry 79: 251–274.
178. Martinelli LA, Piccolo MC, Townsend AR, Vitousek PM, Cuevas E, et al. (1999) Nitrogen stable isotopic composition of leaves and soil: Tropical versus temperate forests. Biogeochemistry 46: 45–65.
179. Kuo S (1996) Phosphorus. Methods of soil analysis. Part 3: Chemical methods. (ed D.L. Sparks), 869–919. Soil Science Society of America, Madison, Wisconsin.
180. Elmore AJ, Asner GP (2006) Effects of grazing intensity on soil carbon stocks following deforestation of a Hawaiian dry tropical forest. Global Change Biology 12: 1761–1772.
181. Skjemstad JO, Le Feuvre RP, Prebbie RE (1990) Turnover of soil organic matter under pasture as determined by ^{13}C abundance. Australian Journal of Soil Research 28: 267–276.
182. Kraft NJ, Valencia R, Ackerly DD (2008) Functional traits and niche-based tree community assembly in an Amazonian forest. Science 322: 580–582.
183. Schulze ED, Chapin FS III, Gebauer G (1994) Nitrogen nutrition and isotope differences among life forms at the northern treeline of Alaska. Oecologia 100: 406–412.
184. Xinhua H, Xu M, Guo YQ, Zhou J (2009) Use of ^{15}N stable isotope to quantify nitrogen transfer between mycorrhizal plants. Journal of Plant Ecology 2: 107–118.
185. Barnes VB, Denton SR, Spurr SH, Zak DR (1998) Forest Ecology (4th ed.). John Wiley & Sons Inc. New York.
186. Gilbert GS, Strong DR (2007) Fungal symbionts of tropical trees. Ecology 88: 539–540.
187. Brown S, Lugo AE (1990) Effects of forest clearing and succession on the carbon and nitrogen content of soils in Puerto Rico and US Virgin Islands. Plant and Soil 124: 53–64.
188. Wong SC, Cowan IR, Farquahr GD (1979) Stomatal conductance correlates with photosynthetic capacity. Nature 282: 424–426.
189. John R, Dalling JW, Harms KE, Yavitt JB, Stallard RF, et al. (2007) Soil nutrients influence spatial distributions of tropical tree species. Proceedings of the National Academy of Sciences 104: 864–869.
190. Martinelli LA, Victoria RL, Trivelin PC, Devol AH, Richey JE (1992) ^{15}N natural abundance in plants of the Amazon river floodplain and potential atmospheric N_2 fixation. Oecologia 90: 591–596.

17

Influence of Arbuscular Mycorrhizae on Biomass Production and Nitrogen Fixation of Berseem Clover Plants Subjected to Water Stress

Sergio Saia, Gaetano Amato, Alfonso Salvatore Frenda, Dario Giambalvo, Paolo Ruisi*

Dipartimento di Scienze Agrarie e Forestali, Università degli Studi di Palermo, Palermo, Italy

Abstract

Several studies, performed mainly in pots, have shown that arbuscular mycorrhizal symbiosis can mitigate the negative effects of water stress on plant growth. No information is available about the effects of arbuscular mycorrhizal symbiosis on berseem clover growth and nitrogen (N) fixation under conditions of water shortage. A field experiment was conducted in a hilly area of inner Sicily, Italy, to determine whether symbiosis with AM fungi can mitigate the detrimental effects of drought stress (which in the Mediterranean often occurs during the late period of the growing season) on forage yield and symbiotic N_2 fixation of berseem clover. Soil was either left under water stress (i.e., rain-fed conditions) or the crop was well-watered. Mycorrhization treatments consisted of inoculation of berseem clover seeds with arbuscular mycorrhizal spores or suppression of arbuscular mycorrhizal symbiosis by means of fungicide treatments. Nitrogen biological fixation was assessed using the ^{15}N-isotope dilution technique. Arbuscular mycorrhizal symbiosis was able to mitigate the negative effect of water stress on berseem clover grown in a typical semiarid Mediterranean environment. In fact, under water stress conditions, arbuscular mycorrhizal symbiosis resulted in increases in total biomass, N content, and N fixation, whereas no effect of crop mycorrhization was observed in the well-watered treatment.

Editor: Lam-Son Phan Tran, RIKEN Center for Sustainable Resource Science, Japan

Funding: This study was funded by Regione Siciliana (Italy) – Project SI.CO.BIO.S. The funder had no role in study design, data collection and analysis, decision to publish, or preparation of the manuscript.

Competing Interests: The authors have declared that no competing interests exist.

* E-mail: paolo.ruisi@unipa.it

Introduction

Arbuscular mycorrhizal (AM) fungi are commonly occurring fungi that live in an obligate symbiotic status with the majority of land plants. AM symbiosis has a positive influence on plant growth, which is mainly attributable to the ability of AM fungi to take up nutrients from the soil — especially phosphorus [1,2] and to a lesser extent nitrogen (N) [3,4] — and deliver them to the roots of its host, and also to enhance the health of its host by protecting it from pathogens, pests, and parasitic plants [5]. In addition, in legume crops, AM symbiosis can improve the degree of dependence of the legume on atmospheric N_2 [6] by increasing the uptake of key nutrients to simultaneously feed plants and nodules or through direct effects on the functioning of the legume-*Rhizobium* symbiosis [7]. Moreover, AM symbiosis can mitigate the negative effects of water stress (WS) on plant growth [8], although the underlying mechanisms of this are still unclear [9]. Nevertheless, an extensive review by Augé [10] covering hundreds of studies (which were performed mainly in pots) highlights the fact that mycorrhizal effects on plant-water relationships are often subtle, transient, and probably circumstance- and symbiont-specific.

Berseem clover (*Trifolium alexandrinum* L.) is a drought-tolerant forage legume that is well adapted to semiarid conditions and widely cultivated under rain-fed conditions in the Mediterranean, central Asia, and, most recently, the United States [11]. In Mediterranean areas, it is usually sown in the autumn and grown until early summer as a multicut forage species, thanks to its ability to regrow after cutting [12]. However, during the spring, plant regrowth is often limited by water deficiency, which negatively affects photosynthesis and symbiotic N_2 fixation and leads to a reduction in accumulation and quality of biomass. In berseem clover, experiments performed under controlled conditions have shown that AM symbiosis can play an important role in protecting the host plant from both biotic and abiotic stresses [13,14]. Pellegrino et al. [15], in a study that included both pot and field trials, found that the use of AM fungal inocula may be very effective in improving berseem clover productivity and quality and that the choice of the most appropriate AM fungal inoculum is essential to the success of inoculation practices. Moreover, other studies have found that AM symbiosis plays a role in determining the N_2 fixation activity of berseem clover under low soil phosphorus concentration, at least during the early stages of the *Rhizobium*-legume interaction [16]. However, to our knowledge, no information is available about the effects of AM symbiosis on berseem growth and N_2 fixation under water shortage conditions. Thus, we performed a field experiment to determine whether symbiosis with AM fungi can mitigate the detrimental effects of drought stress (which in Mediterranean environments often occurs during the late period of the growing season) on forage yield, quality, and symbiotic N_2 fixation of berseem clover.

Materials and Methods

Ethics Statement

No specific permits were required for the described field study. The location is not protected in any way.

The experiment did not involve endangered or protected species.

Experimental site

The trial was conducted during the 2007–2008 growing season at the Pietranera farm, which is located about 30 km north of Agrigento, Sicily, Italy (37°30' N, 13°31' E; 178 m asl), on a deep, well-structured, clay soil classified as a Vertic Haploxerept. Soil characteristics are presented in table S1. The climate of the experimental site is semiarid Mediterranean, with a mean annual rainfall of 552 mm, mostly in the autumn and winter (74%) and in the spring (18%). The dry period lasts from May to September. The mean air temperature is 15.9°C in autumn, 9.8°C in winter, and 16.5°C in spring. The average minimum and maximum annual temperatures are 10.0°C and 23.3°C, respectively. The weather data were collected from a weather station located within 500 m of the experimental site.

During the 2007–2008 growing season (from September to May), total rainfall was 516 mm (slightly lower than the long-term average) and it was well distributed during fall and winter (Fig. 1). Total rainfall during the spring (91 mm) was markedly lower than normal (–30%). Average, minimum, and maximum temperatures were similar to long-term averages.

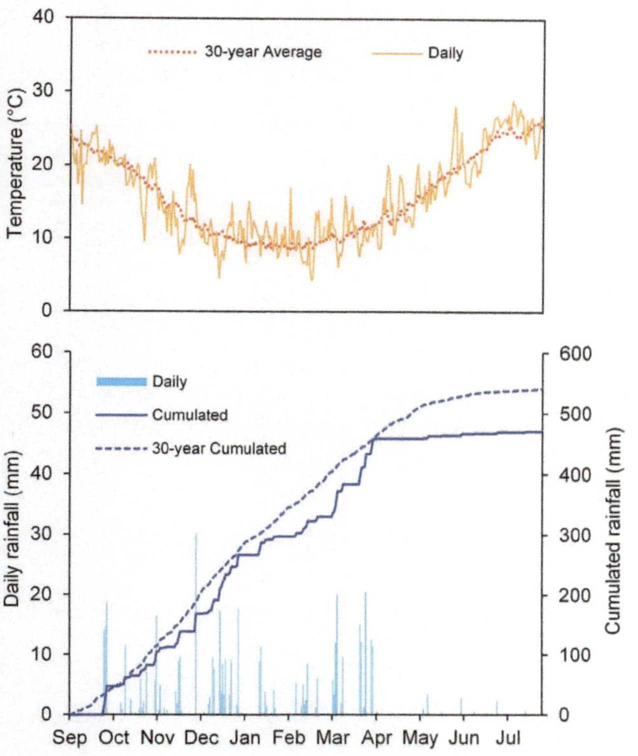

Figure 1. Cumulative daily rainfall and mean air temperatures at the experimental site during the 2007–2008 growing season. 30-year (1980–2010) average cumulative daily rainfall and temperatures are included.

Experimental design and crop management

A split-plot design with four replicates was adopted, with plot sizes of 36 m^2 each (12 rows 12 m long and 0.25 m apart). The main plot was soil water availability: crops were subjected to WS or well-watered (WW) conditions. The subplot was mycorrhization: inoculation of berseem clover seeds with AM spores (+AM) or suppression of AM symbiosis by means of fungicide treatments (–AM). In WS treatment, berseem clover was grown entirely under rain-fed conditions, whereas in WW three sprinkle irrigations, 30 mm each, were done by irrigating when soil water dropped below a value of about 0.7 of the available soil water capacity (116, 125, and 132 days after sowing, DAS). The amount of water applied at each irrigation point was sufficient to bring soil water content to field capacity. Soil moisture content was measured gravimetrically (in the 0–0.40-m layer) once per month from sowing to 110 DAS and every 3 days thereafter.

Natural AM spore population in the field was measured before sowing by the wet-sieving method [17] and consisted of a mycorrhizal consortium formed mainly by *Glomus*-group AM species and *Acaulospora*. Overall AM spore density was 5 spores per 10 g air-dried soil. AM hyphal mass in the soil was not detected. In –AM treatment, suppression of the natural AM symbiosis was achieved by spraying the plots/plants with systemic fungicides: Captan (20 mg a.i. m^{-2}), Carbendazim (20 mg a.i. m^{-2}), and Benomyl (20 mg a.i. m^{-2}) were used at sowing time to suppress AM spore germination; then, Benomyl, Fenpropimorph, and Carbendazim were applied once per month during the growth of the crop, at rates of 10, 5, and 5 mg a.i. m^{-2}, respectively, to reduce AM hyphal growth and symbiotic activity. These fungicides have no toxic effects on plant growth [18,19]. Inoculation with AM fungi involved the application of a commercial AM inoculum at a rate of 12 g kg^{-1} of seed. The inoculum consisted of a mixture of spores of *Rhizophagus irregularis* (formerly *Glomus intraradices*) and *Funneliformis mosseae* (formerly *G. mosseae*), each of which was present at a rate of 1000 spores g^{-1} of inoculum. Berseem clover seed was not inoculated with Rhizobium before planting because prolific inoculation occurs naturally at this site.

Soil, cropped in the previous growing season with durum wheat (*Triticum durum* Desf.), was plowed to a depth of 30 cm in the summer and then shallowly harrowed twice to control weeds and prepare suitable seedbed conditions. No herbicides were applied. Berseem clover (cv Lilibeo) was sown by hand at the end of December at a rate of 1200 viable seeds m^{-2}. Weeds were removed by hand during the growing season.

The entire plot was cut at a 5-cm stubble height at 76 DAS (C1) and again at 116 DAS (C2); at each cut, a subplot (2.25 m^2) was used to determine plant hypogeic (belowground) biomass. Starting at C2, water treatment was established, and crop regrowth was monitored by cutting different subplots (2.25 m^2) after 7 (R7), 14 (R14), 21 (R21), and 28 days (R28).

Symbiotic N_2 fixation was assessed at R14 and R28 using the ^{15}N-isotope dilution technique by applying a ^{15}N-labeled fertilizer ($[NH_4]_2SO_4$ with an isotopic composition of 10 atom% ^{15}N) at 116 DAS, soon after C2. ^{15}N-labeled fertilizer was applied as drench at a rate of 8 kg N ha^{-1} to the subplots assigned to R14 and R28. In all cases, the rest of the plot outside the ^{15}N-labeled areas received a topdressing of ammonium sulfate in an amount equivalent to that in the subplots. Ryegrass (*Lolium multiflorum* var. *Westerwoldicum* cv Elunaria) was chosen as a reference crop because its growth rate and regrowth ability are similar to those of berseem clover [20]; ryegrass received the same treatments as berseem clover.

Table 1. Field response of shoots and roots of berseem clover to AM inoculation at first cut (C1, at 76 days after sowing - DAS) and second cut (C2, at 116 DAS).

		C1 (76 DAS)			C2 (116 DAS)		
		–AM	+AM		–AM	+AM	
Total aboveground biomass	g m^{-2}	159	187	**	548	568	ns
Removed (Rem)	g m^{-2}	113	136	**	393	410	ns
Residual (Res)	g m^{-2}	46	51	*	155	158	ns
Leaves in Rem	% on Rem biomass	74.7	68.2	**	40.6	41.5	ns
Leaves in Res	% on Res biomass	32.2	32.8	ns	28.6	28.5	ns
Total LAI	m^2 m^{-2}	2.96	3.39	*	5.93	6.30	ns
LAI of Res	m^2 m^{-2}	0.44	0.51	*	1.31	1.33	ns
Root biomass	g m^{-2}	35	41	*	36	36	ns

+AM, inoculation with arbuscular mycorrhizal spores; –AM, suppression of arbuscular mycorrhizal symbiosis.
* and ** indicate significant differences at 0.05 and 0.01 probability level, respectively; ns indicates differences not significant.

Plant harvests and analyses

At C1 and C2 the removed biomass (above cutting height) was weighed, and two subsamples of 1 kg of fresh matter each were taken. The first was separated into leaves, stems, and heads, whereas the second was dried at 65°C until a constant weight was reached. Furthermore, on subplots assigned to destructive measures, both residual aboveground biomass (below cutting height) and hypogeic biomass (in the first 20 cm of soil) were measured. The residual aboveground biomass was separated into leaves and stems. All botanical fractions were weighed and dried at 65°C to a constant weight. The leaf area index (LAI) of the residual and removed leaves was immediately measured on a 10-g subsample. On subplots assigned to R7, R14, R21, and R28 cuts, total aboveground biomass (above cutting height + below cutting height) and hypogeic biomass were measured. Samples of total aboveground biomass were separated into botanical fractions, and LAI was measured using the same method described earlier. Biomass samples from R14 and R28 were analyzed for total N and ^{15}N enrichment by means of an elemental analyzer–isotope ratio mass spectrometer (Carlo Erba NA1500).

AM root colonization was measured according to Giovannetti and Mosse [21] on root samples (from five plants randomly chosen from each plot) collected at 55 DAS, at R14, and at R28; roots were stained with 0.05% trypan blue in lactic acid according to Phillips and Hayman [22].

Calculations and statistical analysis

Data on biomass ^{15}N enrichment were used to calculate the percentage of N_2 derived from symbiotic N_2 fixation (%Ndfa) according to Fried and Middelboe [23]:

$$\%\text{Ndfa} = \left(1 - \frac{\text{Atom}\%^{15}\text{N}_{\text{berseem}}}{\text{Atom}\%^{15}\text{N}_{\text{ryegrass}}}\right) \times 100,$$

Table 2. F values and significance levels for water availability (W) and crop mycorrhization (M) on measured traits of regrowth, conducted separately for each sampling date (R7, R14, R21, and R28).

Treatments	Biomass regrowth	Taproot biomass	LAI	Stems %	Heads %	Above ground N		Taproot N		Ndfa		Mycorrhizal infection
	g m^{-2}	g m^{-2}		on biomass regrowth		g kg^{-1}	g m^{-2}	g kg^{-1}	g m^{-2}	%	g m^{-2}	%
Water availabilty (W)												
R7	22.54*	0.14^{ns}	15.30*	0.13^{ns}	†	-	-	-	-	-	-	-
R14	20.03*	8.96^{ns}	262.25***	3.19^{ns}	33.14*	0.27^{ns}	25.67*	4.61^{ns}	50.03**	7.83^{ns}	57.77**	2.52^{ns}
R21	110.99**	2.63^{ns}	48.46**	38.79**	51.06**	-	-	-	-	-	-	-
R28	10.42*	0.03^{ns}	92.74**	233.98***	246.77***	3.81^{ns}	81.93**	0.06^{ns}	0.04^{ns}	16.14*	79.73**	19.87*
Crop mycorrhization (M)												
R7	0.08^{ns}	0.47^{ns}	3.42^{ns}	2.05^{ns}	†	-	-	-	-	-	-	-
R14	1.00^{ns}	2.82^{ns}	0.49^{ns}	0.13^{ns}	0.16^{ns}	0.03^{ns}	1.80^{ns}	0.47^{ns}	2.80^{ns}	0.00^{ns}	0.63^{ns}	1348.18***
R21	22.89**	1.03^{ns}	0.06^{ns}	0.14^{ns}	3.62^{ns}	-	-	-	-	-	-	-
R28	13.08*	0.01^{ns}	4.79^{ns}	1.15^{ns}	0.11^{ns}	0.82^{ns}	0.65^{ns}	0.17^{ns}	0.32^{ns}	14.38**	21.38**	887.47***
W × M												
R7	4.93^{ns}	3.14^{ns}	1.38^{ns}	0.23^{ns}	†	-	-	-	-	-	-	-
R14	0.36^{ns}	0.46^{ns}	0.06^{ns}	5.71^{ns}	0.16^{ns}	0.43^{ns}	2.20^{ns}	0.00^{ns}	0.38^{ns}	1.05^{ns}	4.74^{ns}	0.03^{ns}
R21	6.22*	0.01^{ns}	4.99^{ns}	0.73^{ns}	4.66^{ns}	-	-	-	-	-	-	-
R28	29.78**	1.49^{ns}	0.41^{ns}	0.01^{ns}	4.31^{ns}	0.04^{ns}	7.33*	0.00^{ns}	3.41^{ns}	6.68*	21.55**	14.88**

*, ** and *** indicate significant differences at 0.05, 0.01 and 0.001 probability level, respectively; ^{ns} indicates differences non significant.
†Heads were absent in all the treatments.
-, not detected.

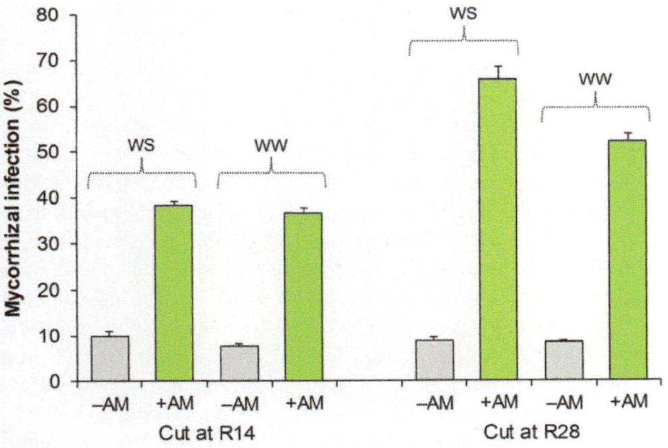

Figure 2. Mycorrhizal infection of berseem clover roots as affected by water availability and mycorrhization treatment. Data are means (n = 4) ± S.E. measured at R14 (14 days after C2) and R28 (28 days after C2). WW, well-watered; WS, water-stressed; +AM, inoculation with arbuscular mycorrhizal spores; –AM, suppression of arbuscular mycorrhizal symbiosis.

where Atom%$^{15}N_{berseem}$ represents the Atom%^{15}N excess of berseem tissues and Atom%$^{15}N_{ryegrass}$ represents the Atom%^{15}N excess of ryegrass tissue. The ^{15}N-natural abundance of the atmosphere (0.3663%^{15}N) was used to calculate the Atom%^{15}N excess of both crops. The amount of N fixed by berseem clover was estimated as:

$$\mathrm{Ndfa} = \mathrm{Total\ berseem\,N} \times \frac{\%\mathrm{Ndfa}}{100}.$$

Analysis of variance [24] was performed separately for each cut according to the experimental design. All measured variables were assumed to be normally distributed. All variables corresponding to proportions were arcsine transformed before analysis to ensure a better fit with the Gaussian law distribution.

Results

Soil moisture did not drop below a value of about 0.7 of the available soil water capacity until C2. As expected, from C2 to R28, soil moisture content was always significantly higher in WW than WS conditions (25.4% vs. 19.4% across all sampling dates; P

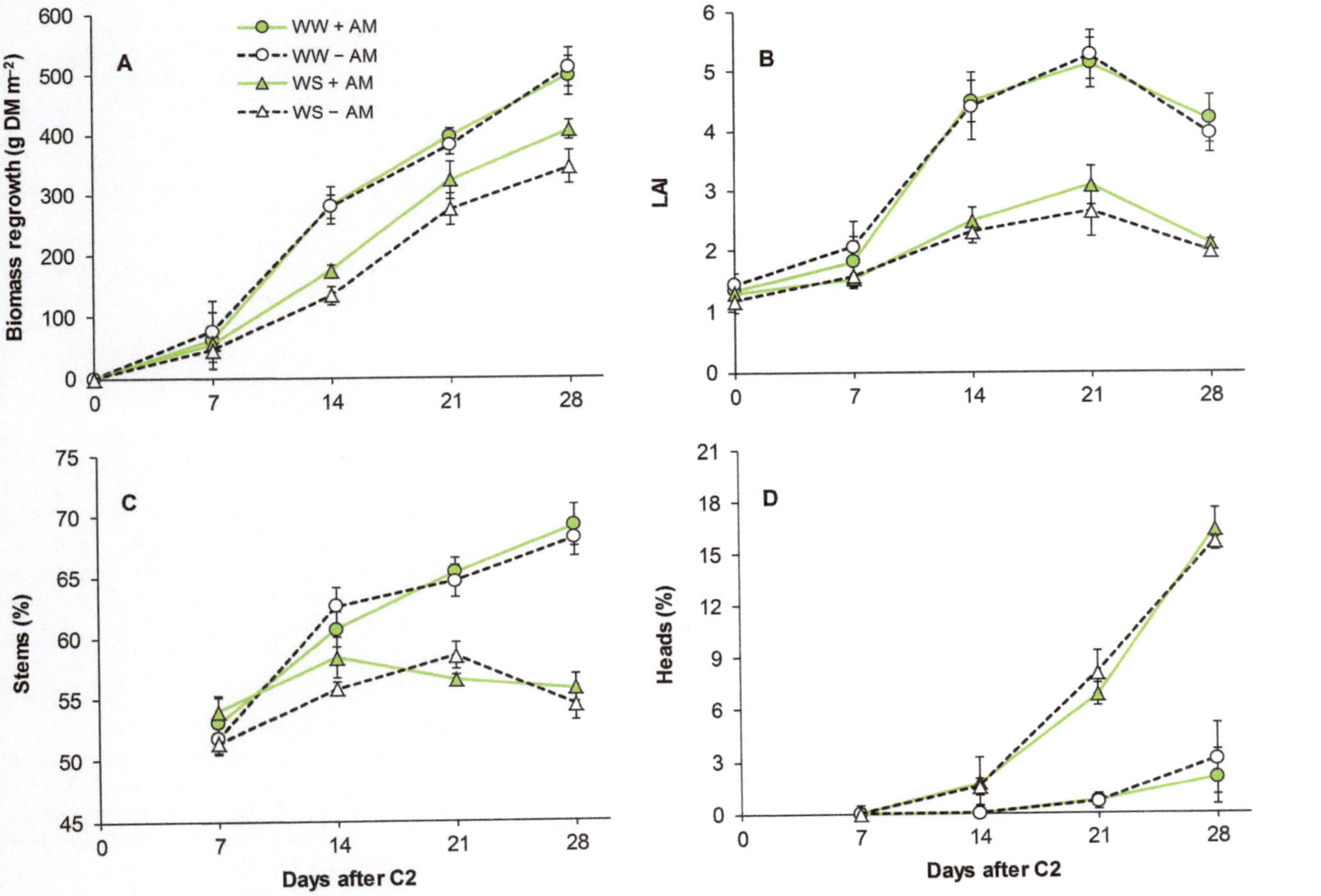

Figure 3. Biomass regrowth (A), LAI (B), and incidence of stems (C) and heads (D) as a percentage of berseem clover biomass as affected by water availability and mycorrhization treatment. Data are means (n = 4) ± S.E. measured after 7, 14, 21 and 28 days from second cut (C2). WW, well-watered; WS, water-stressed; +AM, inoculation with arbuscular mycorrhizal spores; –AM, suppression of arbuscular mycorrhizal symbiosis.

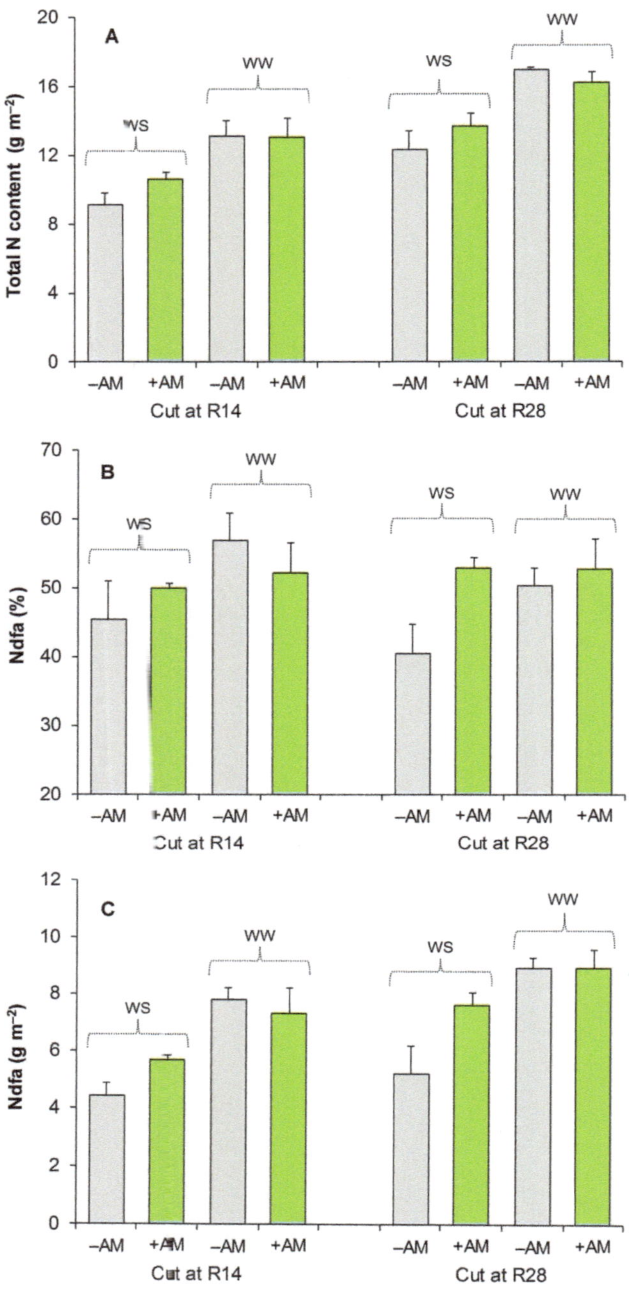

Figure 4. Total N content (A), Nitrogen derived from the atmosphere as a percentage of total N content (B) and total fixed N (C) of berseem clover as affected by water availability and mycorrhization treatment. Data are means (n=4) ± S.E. measured at R14 (14 days after C2) and R28 (28 days after C2). WW, well-watered; WS, water-stressed; +AM, inoculation with arbuscular mycorrhizal spores; −AM, suppression of arbuscular mycorrhizal symbiosis.

< 0.001), and the differences between the two treatments increased with time. Soil moisture content was not affected by crop mycorrhization, and the interaction with soil moisture regimen was never significant.

At 55 DAS, root colonization by AM fungi was significantly lower in the −AM crops than in the +AM crops (7.0% and 32.4%, respectively; $P < 0.001$). At C1, both total and residual biomass of berseem clover, and LAI of total biomass and of residual biomass, were significantly higher in the +AM than the −AM crops (Table 1). Also, root biomass (up to 20 cm depth) was significantly higher in +AM than in −AM. At C2, no significant effect of mycorrhization was observed on biomass accumulation, canopy traits (LAI and percentage of leaves), or root biomass.

Tests of significance for the effect of water treatment and crop mycorrhization on the traits measured during crop regrowth (at R7, R14, R21, and R28) are reported in Table 2. At R14 (14 days after the establishment of water treatment), AM infection was, on average, 8.8% in −AM and 37.7% in +AM (significantly different at $P < 0.001$); at this time, no effect of the soil moisture regime was observed on root colonization by AM fungi (Fig 2). At R28, root colonization by AM fungi was very low in the −AM treatment areas, irrespective of soil moisture regime (average of 8.7%), whereas in +AM treatment areas, root colonization was significantly higher in the WS areas than in the WW areas (66.0% and 52.4%, respectively).

The increase in water availability in WW compared to WS led to significant increases in biomass regrowth starting at R7 (Table 2 and Fig. 3A). The differences between WW and WS increased until R14 and then remained constant until R28. Crop mycorrhization increased biomass regrowth in WS but not in WW and the difference between +AM and −AM increased with time.

Irrigation led to an increase in LAI and stem percentage and a decrease in head percentage with differences between WW and WS increasing with time (Fig. 3B, 3C and 3D). Crop mycorrhization did not affect LAI or the incidence of stems and heads as a percentage of crop biomass in either WS or WW (Table 2 and Fig. 3B, 3C and 3D).

Both water treatment and crop mycorrhization almost never affected taproot biomass (38.4 g m^{-2} averaged across all sampling dates, with slight increases with time).

N concentration did not vary according to the treatment applied (on average 30.5 g kg^{-1} and 24.7 g kg^{-1} in aboveground biomass at R14 and R28, respectively, and 18.6 g kg^{-1} and 17.6 g kg^{-1} in taproot biomass at R14 and R28, respectively). Irrigation increased total N content both at R14 and R28 (Fig. 4A). Crop mycorrhization resulted in an increase in total N content in WS but not in WW.

Irrigation increased total N fixed both at R14 (on average 50% compared to WS) and at R28 (on average 39% compared to WS) (Fig. 4C). At R14, crop mycorrhization did not affect Ndfa (as both a percentage of total N content and the amount of N fixed on an area basis). On the contrary, at R28, Ndfa values were significantly higher in +AM than in −AM in the WS treatment, whereas no effects of crop mycorrhization were observed in the WW treatment.

Discussion

On the whole, when berseem clover did not experience WS (i.e., in WW treatments), aboveground biomass production was about 12 Mg dry matter ha^{-1}; this is similar to the biomass obtained in the same environment when rainfall during the growing season was 30% higher than the long-term average [12,20]. On average, no differences between −AM and +AM conditions were observed in terms of total aboveground biomass production (11.7 vs. 12.0 Mg dry matter ha^{-1}, calculated as a sum of C1+C2+R28 in WW treatments). This result is in contrast with the findings of Pellegrino et al. [15], who observed, in a field trial, an increase of berseem clover productivity (from 6.5 to 10.0 Mg dry matter ha^{-1}) following inoculation with AM fungi in comparison with a non-inoculated control. Because the main recognized plant benefit from AM symbiosis consists of improved access to limited soil resources (i.e., nutrients—especially phosphorus—and water), it is probable that, in our experiment, in contrast to that of Pellegrino et al. [15], the high level of soil fertility and the adequate water availability made mycorrhizal symbiosis superfluous. In such conditions, the only advantage of AM symbiosis that we observed occurred at the first cut (C1, at 76 DAS), at which time mycorrhiza inoculation resulted in increased root and shoot dry matter. According to Miller [25], during the early phase of the crop cycle, when roots are not yet well developed, AM symbiosis can make an important contribution to plant growth by enhancing nutrient acquisition.

In contrast to the WW conditions, under WS conditions, the AM symbiosis enhanced aboveground biomass production of berseem clover. This result is consistent with the findings of other authors who found, in pot-based studies, a beneficial effect of AM symbiosis under water restriction for both legumes [26,27] and non-legume plants [28,29]. The AM fungi contribution to plant drought tolerance is well documented (as reviewed by Augé [10]), but the underlying mechanisms are still unclear. These include hyphal water uptake [28], enhanced nutrient uptake —especially phosphorus— and enhanced plant osmotic adjustment [30,31], which favor plant water uptake from soil.

Numerous pot-based experiments have clearly indicated that AM symbiosis can greatly assist nodulation and N_2 fixation of legumes [32,33]. We observed significant increases in Ndfa (as both percentage of total N content and amount of N fixed on an area basis) only under WS conditions. This result may be attributable to the enhancement of nutrient uptake offered by AM fungi under WS conditions, which allowed the plants to satisfy both the nodule's and their own nutritional needs for growth. In this context, as is well known, an increase in phosphorus acquisition can play an important role in stimulating both nodulation and N_2 fixation [34]. Our research was performed in soil that is very rich in phosphorus; such conditions can strongly reduce the mycorrhizal benefit in terms of plant growth and phosphorus uptake [35]. However, in some cases, under WS conditions, AM plants are able to absorb more phosphorus than non-mycorrhizal plants, even when this element is highly available in the soil [9]; in fact, phosphorus effectively becomes less available with decreasing soil moisture [36,37], so that AM colonization and effective development of external mycelium can be extremely important for nutrient uptake in these situations. Moreover, other authors [27] highlighted that the advantages of the AM symbiosis in WS conditions are related to the defense of plant roots against the oxidative damage generated by drought, which can then ensure a certain level of protection against nodule senescence (with positive effects on the N_2 fixation process).

In WS treatments, the benefits of AM symbiosis were effective only at R28 (i.e., at 28 days after C2), whereas no differences were observed at R14 (i.e., 14 days after C2). This result can be explained by considering that, in legume forage crops subjected to cuts, regrowth ability is highly dependent on the plant's capacity to mobilize carbon and N reserves stored in roots and stubble and on photosynthetic activity of the residual leaf area [12,38,39]. Thus, during the early regrowth phase (from C2 to R14), carbon can represent a limiting factor, and in our experiment, it is likely that the high sink strength of the growing shoots limited the availability of resources for the AM fungus, counterbalancing the potential benefit offered by AM symbiosis; in the late regrowth phase (from R14 to R28), the photosynthetic apparatus reached an extent so great that plants were able to guarantee sufficient availability of the photosynthates to sustain both AM symbiosis and plant growth. The lower percentage of root colonization by AM fungi observed in R14 with respect to R28 seems to confirm the plant's difficulty in sustaining mycorrhizal symbionts during the early regrowth phase.

In conclusion, the results of our study confirm that, in field conditions, AM symbiosis can play an important role in the growth of plants under conditions of abiotic stresses such as WS. In fact, AM symbiosis in berseem clover subjected to WS mitigated the negative effects of drought on plant growth also through the stimulation of N_2 fixation. Moreover, AM symbiosis made an important contribution during the early growth phase, when nutrient uptake is limited by the relatively poorly developed plant root system.

Supporting Information

Table S1 Physical and chemical characteristics of the top layer (0–0.40-m) of soil at the experimental site.

Acknowledgments

The authors wish to thank Vincenzo Cannella and Francesco Cannella for their technical assistance in field management.

Author Contributions

Conceived and designed the experiments: SS GA DG PR. Performed the experiments: SS ASF PR. Analyzed the data: SS ASF DG PR. Contributed reagents/materials/analysis tools: GA ASF DG. Wrote the paper: SS GA DG PR.

References

1. Barea JM, Ferrol N, Azcón-Aguilar C, Azcón R (2008) Mycorrhizal symbioses. In: White PJ, Hammond JP, editors. The Ecophysiology of Plant-Phosphorus Interactions. Dordrecht: Springer Netherlands, Vol. 7. pp. 143–163.
2. Lambers H, Raven JA, Shaver GR, Smith SE (2008) Plant nutrient-acquisition strategies change with soil age. Trends Ecol Evol 23: 95–103.
3. Azcón R, Ruiz-Lozano JM, Rodríguez R (2001) Differential contribution of arbuscular mycorrhizal fungi to plant nitrate uptake (^{15}N) under increasing N supply to the soil. Can J Bot 79: 1175–1180.
4. Saia S, Benítez E, García-Garrido JM, Settanni L, Amato G, et al. (2014) The effect of arbuscular mycorrhizal fungi on total plant nitrogen uptake and nitrogen recovery from soil organic material. J Agric Sci In Press. doi: 10.1017/S002185961300004X. Published on line 07 February 2013
5. Jung SC, Martinez-Medina A, Lopez-Raez JA, Pozo MJ (2012) Mycorrhiza-induced resistance and priming of plant defenses. J Chem Ecol 38: 651–664.
6. Chalk P, Souza R, Urquiaga S, Alves B, Boddey R (2006) The role of arbuscular mycorrhiza in legume symbiotic performance. Soil Biol Biochem 38: 2944–2951.
7. Barea JM, Werner D, Azcón-Aguilar C, Azcón R (2005) Interactions of arbuscular mycorrhiza and nitrogen-fixing symbiosis in sustainable agriculture. In: Werner D, Newton WE, editors. Nitrogen Fixation in Agriculture, Forestry, Ecology, and the Environment. Springer Netherlands. pp. 199–222.
8. Jayne B, Quigley M (2013) Influence of arbuscular mycorrhiza on growth and reproductive response of plants under water deficit: a meta-analysis. Mycorrhiza In press doi: 10.1007/s00572-013-0515-x. Published on line 06 August 2013
9. Smith SE, Facelli E, Pope S, Andrew Smith F (2010) Plant performance in stressful environments: interpreting new and established knowledge of the roles of arbuscular mycorrhizas. Plant Soil 326: 3–20.
10. Augé RM (2001) Water relations, drought and vesicular-arbuscular mycorrhizal symbiosis. Mycorrhiza 11: 3–42.
11. Amato G, Giambalvo D, Ruisi P (2013) Cut and Post-Cut Herbage Management Affects Berseem Clover Seed Yield. Agron J 105: 1222.
12. Giambalvo D, Amato G, Stringi L (2011) Effects of Stubble Height and Cutting Frequency on Regrowth of Berseem Clover in a Mediterranean Semiarid Environment. Crop Sci 51: 1808.
13. Hasan N, Jain RK (1987) Parasitic nematodes and vesicular-arbuscular mycorrhizal (VAM) fungi associated with berseem (*Trifolium alexandrinum* L.) in Bundelkhand region. Indian J Nematol 17: 184–188.

14. Shokri S, Maadi B (2009) Effects of arbuscular mycorrhizal fungus on the mineral nutrition and yield of *Trifolium alexandrinum* plants under salinity stress. J Agron: 79–83.
15. Pellegrino E, Bedini S, Avio L, Bonari E, Giovannetti M (2011) Field inoculation effectiveness of native and exotic arbuscular mycorrhizal fungi in a Mediterranean agricultural soil. Soil Biol Biochem 43: 367–376.
16. Patterson NA, Chet I, Kapulnik Y (1990) Effect of mycorrhizal inoculation on nodule initiation, activity and contribution to legume productivity. Symbiosis 8: 9–20.
17. Gerdemann JW, Nicolson TH (1963) Spores of mycorrhizal Endogone species extracted from soil by wet sieving and decanting. Trans Br Mycol Soc 46: 235–244.
18. Udaiyan K, Manian S, Muthukumar T, Greep S (1995) Biostatic effect of fumigation and pesticide drenches on an endomycorrhizal-Rhizobium-legume tripartite association under field conditions. Biol Fertil soils 20: 275–283.
19. Schweiger P, Spliid N, Jakobsen I (2001) Fungicide application and phosphorus uptake by hyphae of arbuscular mycorrhizal fungi into field-grown peas. Soil Biol Biochem 33: 1231–1237.
20. Giambalvo D, Ruisi P, Di Miceli G, Frenda AS, Amato G (2011) Forage production, N uptake, N_2 fixation, and N recovery of berseem clover grown in pure stand and in mixture with annual ryegrass under different managements. Plant Soil 342: 379–391.
21. Giovannetti M, Mosse B (1980) An evaluation of techniques for measuring vesicular arbuscular mycorrhizal infection in roots. New Phytol 84: 489–500.
22. Phillips JM, Hayman DS (1970) Improved procedures for clearing roots and staining parasitic and vesicular-arbuscular mycorrhizal fungi for rapid assessment of infection. Trans Br Mycol Soc 55: 158–IN18.
23. Fried M, Middelboe V (1977) Measurement of amount of nitrogen fixed by a legume crop. Nitrogen Isot Tech 47: 713–715.
24. SAS Institute (2008) SAS/STAT® 9.2. User's Guide.
25. Miller MH (2000) Arbuscular mycorrhizae and the phosphorus nutrition of maize: A review of Guelph studies. Can J Plant Sci 80: 47–52.
26. Azcon R, El-Atrach F, Barea J (1988) Influence of mycorrhiza vs. soluble phosphate on growth, nodulation, and N_2 fixation (^{15}N) in alfalfa under different levels of water potential. Biol Fertil soils 7: 28–31.
27. Porcel R, Ruiz-Lozano JM (2004) Arbuscular mycorrhizal influence on leaf water potential, solute accumulation, and oxidative stress in soybean plants subjected to drought stress. J Exp Bot 55: 1743–1750.
28. Ruiz-Lozano JM, Azcon R (1995) Hyphal contribution to water uptake in mycorrhizal plants as affected by the fungal species and water status. Physiol Plant 95: 472–478.
29. Marulanda A, Azcon R, Ruiz-Lozano JM (2003) Contribution of six arbuscular mycorrhizal fungal isolates to water uptake by *Lactuca sativa* plants under drought stress. Physiol Plant 119: 526–533.
30. Goicoechea N, Szalai G, Antolín MC, Sánchez-Díaz M, Paldi E (1998) Influence of arbuscular mycorrhizae and *Rhizobium* on free polyamines and proline levels in water-stressed alfalfa. J Plant Physiol 153: 706–711.
31. Kubikova E, Moore JL, Ownley BH, Mullen MD, Auge RM (2001) Mycorrhizal impact on osmotic adjustment in Ocimum basilicum during a lethal drying episode. J Plant Physiol 158: 1227–1230.
32. Antunes PM, de Varennes A, Rajcan I, Goss MJ (2006) Accumulation of specific flavonoids in soybean (Glycine max (L.) Merr.) as a function of the early tripartite symbiosis with arbuscular mycorrhizal fungi and *Bradyrhizobium japonicum* (Kirchner) Jordan. Soil Biol Biochem 38: 1234–1242.
33. Olesniewicz KS, Thomas RB (1999) Effects of mycorrhizal colonization on biomass production and nitrogen fixation of black locust (*Robinia pseudoacacia*) seedlings grown under elevated atmospheric carbon dioxide. New Phytol 142: 133–140.
34. Ganry F, Diem HG, Dommergues YR (1982) Effect of inoculation with *Glomus mosseae* on nitrogen fixation by field grown soybeans. Plant Soil 68: 321–329.
35. Lekberg Y, Koide RT (2005) Is plant performance limited by abundance of arbuscular mycorrhizal fungi? A meta-analysis of studies published between 1988 and 2003. New Phytol 168: 189–204.
36. Viets FG (1972) Water deficits and nutrient availability. In: Kozlowski TT, editor. Water deficits and plant growth. Academic Press New York, NY, Vol. 3. pp. 217–239.
37. Gahoonia TS, Raza S, Nielsen NE (1994) Phosphorus depletion in the rhizosphere as influenced by soil moisture. Plant Soil 159: 213–218.
38. Avice J, Lemaire G, Ourry A, Boucaud J (1997) Effects of the previous shoot removal frequency on subsequent shoot regrowth in two *Medicago sativa* L. cultivars. Plant Soil 188: 189–198.
39. Meuriot F, Avice J-C, Simon J-C, Laine P, Decau M-L, et al. (2004) Influence of initial organic N reserves and residual leaf area on growth, N uptake, N partitioning and N storage in alfalfa (*Medicago sativa*) during post-cutting regrowth. Ann Bot 94: 311–321.

18

Combining Forces - The Use of Landsat TM Satellite Imagery, Soil Parameter Information, and Multiplex PCR to Detect *Coccidioides immitis* Growth Sites in Kern County, California

Antje Lauer[1]*, Jorge Talamantes[2], Laura Rosío Castañón Olivares[3], Luis Jaime Medina[2], Joe Daryl Hugo Baal[1], Kayla Casimiro[1], Natasha Shroff[1], Kirt W. Emery[4]

1 Department of Biology, California State University, Bakersfield, California, United States of America, **2** Department of Physics & Engineering, California State University, Bakersfield, California, United States of America, **3** Laboratorio de Micología Médica, Facultad de Medicina, Universidad Nacional Autónoma de México, Mexico City, Mexico, **4** County of Kern Public Health Services Department, Bakersfield, California, United States of America

Abstract

Coccidioidomycosis is a fungal disease acquired through the inhalation of spores of *Coccidioides* spp., which afflicts primarily humans and other mammals. It is endemic to areas in the southwestern United States, including the San Joaquin Valley portion of Kern County, California, our region of interest (ROI). Recently, incidence of coccidioidomycosis, also known as valley fever, has increased significantly, and several factors including climate change have been suggested as possible drivers for this observation. Up to date details about the ecological niche of *C. immitis* have escaped full characterization. In our project, we chose a three-step approach to investigate this niche: 1) We examined Landsat-5-Thematic-Mapper multispectral images of our ROI by using training pixels at a 750 m×750 m section of Sharktooth Hill, a site confirmed to be a *C. immitis* growth site, to implement a Maximum Likelihood Classification scheme to map out the locations that could be suitable to support the growth of the pathogen; 2) We used the websoilsurvey database of the US Department of Agriculture to obtain soil parameter data; and 3) We investigated soil samples from 23 sites around Bakersfield, California using a multiplex Polymerase Chain Reaction (PCR) based method to detect the pathogen. Our results indicated that a combination of satellite imagery, soil type information, and multiplex PCR are powerful tools to predict and identify growth sites of *C. immitis*. This approach can be used as a basis for systematic sampling and investigation of soils to detect *Coccidioides* spp.

Editor: Vishnu Chaturvedi, California Department of Public Health, United States of America

Funding: The work was supported by National Science Foundation grant HRD-0331537, and support from Chevron (Research Experience Vitalizing Science Undergraduate Program (REVS-UP) at CSUB. The funders had no role in study design, data collection and analysis, decision to publish, or preparation of the manuscript.

Competing Interests: The authors have the following interest: This study was partly funded by Chevron (Research Experience Vitalizing Science Undergraduate Program [REVS-UP]) at CSUB. There are no patents, products in development or marketed products to declare. This does not alter the authors' adherence to all the PLOS ONE policies on sharing data and materials, as detailed online in the guide for authors.

* Email: alauer@csub.edu

Introduction

Valley fever research has predominantly focused on the medical and epidemiological aspects of *Coccidioides immitis* and *Coccidioides posadasii*, the fungi that cause coccidioidomycosis ([1,2], and references therein). *Coccidioides* spp. can have a complete life cycle as soil dwelling organisms but if the soil is disturbed, their arthroconidia can become air-borne and are able to infect a host via the respiratory tract. About 60% of infected patients report no symptoms [3]; about 25% exhibit severe flu-like symptoms, such as cough, sputum, fever, and muscle aches; the remaining 15% become very ill with pneumonia-like symptoms (e.g. pleurisy and heavier sputum) requiring medication and bed rest. In a small number of cases (about 0.5–1%), the disease disseminates beyond the lungs to e.g. the skin, bones, and/or meninges of the brain, and the disease can be fatal. Certain sectors of the population seem to be more susceptible to infection, such as the very young, persons newly arrived to the endemic areas (since immunity develops with infection), field-, and construction workers, and those with impaired immune systems [4].

Coccidioides spp. are endemic in the southern part of the San Joaquin Valley in California, southern California, the southern part of Arizona, New Mexico and Texas, most of northern Mexico, and some areas in Guatemala, Honduras, Venezuela, northeastern Brazil, Argentina, and Paraguay [5,6]. Given its geographic distribution, it is evident that *C. posadasii* is able to flourish in desert regions of the Americas (besides California), in contrast to its close relative *C. immitis* which seems to be restricted to areas in California. Of the two fungal species, it is *C. immitis* which afflicts the San Joaquin Valley portion of Kern County, California [7,8] which is the Region of Interest (ROI) of this study. However, population genomic sequencing of *Coccidioides* spp.

revealed recent hybridization between both species [9], and nothing is known about the distribution and ecology of these hybrids.

It is reasonable to expect that climatic fluctuations might affect the rate at which humans become infected [4]. For example, an extended drought might decimate less heat tolerant, non-sporeforming soil microorganisms that had acted as natural antagonists to the pathogen in its natural environment. A wetter than-normal rainy season could help *Coccidioides* spp. bloom, and windy spells might facilitate the dispersal of its arthroconidia. The "grow and blow" hypothesis has first been introduced by Comrie and Glueck [10]. It has long been surmised that *Coccidioides* spp. are generally poor competitors [11], but that they are more heat-resistant than competing microorganisms – thus, it can be expected that hot summers might favor its presence or dominance. Indeed, anecdotal evidence to these effects is well documented in the literature [12–20]. There have been a number of attempts at demonstrating this connection quantitatively with various degrees of success [10,11,21–25]. Yet, despite extensive study, there is currently no ecologically consistent link identified between the environment and coccidioidomycosis rates [26]. The predicted warming of the climate in California will add another piece of the puzzle in the already complicated interrelationships of environmental factors that might support or suppress the growth of pathogens with environmental reservoirs [27,28]. However, occasionally, the pathogen was detected in other regions, such as the recent detection of *Coccidioides immitis* in soils of Eastern WA [29].

There also have been several attempts to characterize the ecological niche of *Coccidioides* spp. in more detail [4,31–33], but we still do not have a complete description of this niche. To date, Fisher et al. [31] present the most comprehensive review of this subject. We need to direct attention to a few fundamental points about what is known in regards to this niche. First, it is important to realize that *C. immitis* and *C. posadasii* do not grow in disturbed soils [4,30,31] such as cultivated fields, gardens, etc. Second, whereas it was initially thought that *Coccidioides* spp. ecological niche corresponds to the Lower Sonoran Life Zone (as defined and described by Merriam [34]), or similar environments [14,35,36]. Later research [30] showed that this is not quite correct, and indeed more recent works [31,32,37,38] suggested that the fungus's niche corresponds more closely with thermic and hyperthermic soils in which temperatures can reach or exceed 22°C in 50 cm depth. Fisher et al. [31] described sites where *Coccidioides* spp. were suspected to have been present because humans or animals were reported to have been infected at these sites. Fisher et al. [31] also made the general observations that the vegetation at those sites ranged from sparse to relatively thick cover in lower Sonoran Deserts, Chaparral-upper Sonoran brush and grasslands, as well as Mediterranean savannas and forested foothills. Furthermore, they stated that the temperature regimes, climate conditions in general, and soil textures are the only indicative variables of the presence of *Coccidioides* spp. Microbial diversity in soils is highly influenced by the habitat's chemical and physical parameters. But biotic soil factors such as plant and microeukaryote diversity influence fungal and bacterial soil communities as well through root exudation (additional available nutrients), microbial antagonism (antibiotic production) and synergism, as well as through selective grazing by microeukaryotes [32,39–40]. It is currently being discussed that the pathogen is in fact not very competitive as a soil saprophyte because it has lost the ability to produce a variety of enzymes that are involved in important biodegradation processes of soil organic matter, which might explain the difficulty to detect it in bulk soil [41].

There are few published data available about the distribution of *C. immitis* growth sites in Kern County, California [42] most probably because it has been very difficult in the past to isolate and identify *Coccidioides* spp. from soil and dust samples [4,14,30]. Recently, first attempts using molecular biological techniques to identify *C. immitis* in bulk soil samples from Kern County, predominantly around Bakersfield, have been performed [32]. Based on that study, it appeared that *C. immitis* is likely to be found in the Bakersfield area at locations that are non-agricultural and have about equal parts of sand, clay, and silt (clay loam), a pH between 7.8 and 8.5, an available water capacity of about 0.15–0.2 cm/cm, a water content of about 30% (1/3 bar), an available water supply (0–25 cm) of 4–5 cm, and a Cation Exchange Capacity (CEC7) of over 20 milliequivalents per 100 grams.

The idea of using remote sensing (RS) techniques to piece together environmental characteristics, environmental change, and their relationship to disease transmission has been used extensively in connection with other diseases such as malaria [43], cholera [44], and African trypanosomiases [45]. Even though the ecological niche of *C. immitis* is not well characterized, we present here a RS technique that allows the mapping of sites around Bakersfield, California, where the pathogen is suspected to grow based on data obtained in a previous study by Lauer et al. [32]. Our method utilized a location well-known for being a *C. immitis* growth site (Sharktooth hill [STH], Bakersfield, California) as a basis, and then examined satellite images of the ROI to find all locations with similar spectral signatures. This is similar to characterizing the growth sites by the vegetation that tends to grow in the same environment as *C. immitis*, using the vegetation type as a marker. This is reasonable because the vegetation type closely reflects the co-variation of the relevant physical and chemical parameters such as clay and sand content, temperature, pH, nutrients, water content, etc., and also affects the development of the microbial diversity in the soils [46].

To validate our approach, we investigated if a combination of remote sensing and soil parameter information can predict locations which might be suitable to support the growth of *C. immitis*, followed by a molecular biological approach to detect the fungus in these soils with a culture independent polymerase chain reaction (PCR) based method [32,47].

Material and Methods

No specific permissions were required for the soil sampling. Our field study did also not involve endangered or protected species.

Multispectral image analysis

Landsat-5 Thematic Mapper (TM) L1G corrected multispectral images were downloaded from the United States Geological Survey archive (http://EarthExplorer.usgs.gov). The satellite relayed a continuous data stream which was then framed into individual scenes each 23.92 sec (see, e.g., http://landsat.gsfc.nasa.gov/about/wrs.html). The images for path 42, rows 35 and 36: Worldwide Reference System to cover our ROI were downloaded, and then the two images were mosaicked. Most of the analysis that is presented here was performed on a spatial subset of this mosaicked image. This subset corresponds to an area approximately one million hectares that covers the San Joaquin Valley portion of Kern County. Our work mainly focused on a multispectral image taken on April 20, 2008 at 10:23 PM local time. This image was chosen because it was obtained at a date (during spring) were microbial activity and biomass in the soil is generally considered high, because of supportive environmental parameters, such as moderate temperatures and increased water

Table 1. Location and description of sampling sites used as test data for the remote sensing approach.

sampling sites and year sampled	coordinates	soil type (map unit symbol)	GS or AS of the pathogen	rodent activity*
Bakersfield city				
1. CSUB Children Center ('08, '09)	119° 06′ 29.0′′ W, 35° 20′ 57.0′′ N	Wasco sandy loam (243)	AS	yes
2. Belle Terrace/P Str. ('11)	119° 00′ 37.2′′ W, 35° 20′ 49.8′′′ N	Kimberlina Urban land, Cajon-complex (180)	GS	no
3. Belle Terrace/Gay Str. ('11)	118° 59′ 22.7′′ W, 35° 20′ 40.0′′′ N	Kimberlina Urban land, Cajon-complex (180)	GS	yes
4. Marella Way ('11)	118° 63′ 15.0′′ W, 35° 21′ 40.0′′′ N	Kimberlina Urban land, Cajon-complex (180)	NS	no
5. Flood Plain CSUB ('08, '09)	119° 06′ 05.0′′ W, 35° 21′ 16.0′′ N	River Wash (229)	AS	no
SW Bakersfield				
6. Bike Path West ('08, '09)	119° 15′ 06.0′′ W, 35° 18′ 20.0′′ N	Cajon sandy loam (125)	NS	yes
7. Lake Webb ('08, '09)	119° 16′ 27.0′′ W, 35° 13′ 53.0′′ N	Zalvidea sandy loam (240)	AS	no
8. Cole's Levee Rd. I ('08, '09, '11)	119° 13′ 60.0′′ W, 35° 14′ 08.0′′ N	Garces loam (180)	GS	yes
9. Cole's Levee Rd. II ('11)	119° 13′ 65.3′′ W, 35° 14′ 09.7′′ N	Garces loam (180)	GS	yes
10. Olen Avenue ('11)	119° 14′ 50.0′′ W, 35° 14′ 72.0′′ N	Garces loam (180)	GS	yes
11. Valley Street Field ('08, '09)	118° 52′ 18.0′′ W, 35° 24′ 29.0′′ N	Delano sandy loam (139)	AS	no
NE Bakersfield				
12. Across CALM ('11)	118° 53′ 14.1′′ W, 35° 25′ 50.3′′ N	Chanac Clay Loam (130)	GS	yes
13. Ant Hill Oil Field ('08, '09, '11)	118° 51′ 25.0′′ W, 35° 23′ 50.0′′ N	Chanac Clay Loam (131)	GS	yes
14. Round Mt. Rd. I ('08, '09)	118° 52′ 20.0′′ W, 35° 27′ 10.0′′ N	Xeric Torriorthents-Calcic Haploxerept association (174)	AS	yes
15. Round Mt. Rd. II ('08, '09)	118° 53′ 30.0′′ W, 35° 28′ 42.0′′ N	Xeric Torriorthents-Calcic Haploxerept association (174)	NS	yes
16. Sharktooth hill I	118° 55′ 03.4′′ W, 35° 27′ 44.5′′ N	Chanac Pleito Premier Association (305)	nd	yes
17. Sharktooth hill 2	118° 54′ 37.0′′ W, 35° 28′ 20.0′′ N	Pleito Trigo Chanac Complex (205)	GS	yes
18. Sharktooth hill 3	118° 54′ 33.0′′ W, 35° 28′ 21.3′ N	Pleito Trigo Chanac Complex (205)	GS	yes
NW Bakersfield				
19. Acari Rd. ('11)	119° 15′ 26.8′′ W, 35° 23′ 16.1′′ N	Garces silt loam (156)	NS	no
20. Elementary Lne. ('11)	119° 15′ 16.1′′ W, 35° 25′ 20.5′′ N	Panoche clay loam (211)	GS	no
21. Beech Str. ('11)	119° 15′ 43.5′′ W, 35° 26′ 39.6′′ N	Garces silt loam (156)	GS	yes
Wasco				
22. Gun Club Rd.('11)	119° 29′ 54.0′′ W, 35° 39′ 34.9′′ N	Garces silt loam (156)	NS	yes
23. McCoy Rd. ('11)	119° 31′ 34.3′′ W, 35° 37′ 24.8′′ N	Garces silt loam (156)	NS	yes
Arvin				
24. Di Giorgio Rd. ('11)	118° 57′ 28.7′′ W, 35° 15′ 06.6′′ N	Garces loam (180)	GS	yes
25. Bear Mt. Rd. ('11)	118° 57′ 05.9′′ W, 35° 12′ 30.0′′ N	Garces loam (180)	GS	yes

Growth sites (GS), accumulation sites (AS) and negative sites (NS) were determined by multiplex PCR results, nd: not determined.
* Proof of rodent activity was observed in the immediate neighborhood of the sampling site. Soil disturbing activity was also observed by burrowing owls, coyotes, kit foxes, spiders or large ants at some locations. The dominant rodents observed were ground squirrels, kangaroo rats and hares.

content. Furthermore, this image had 0% cloud cover. Our analysis started by defining a 25 pixel ×25 pixel area centered at latitude 35° 28′ 20.29′′ N, and longitude 118° 54′ 37.04′′ W. This location is at STH, an area where *C. immitis* has been repeatedly detected ([30,31,48], this study). These 625 pixels were used to train the algorithm, and thus, they define a spectral class which is referred to in what follows as the "STH-vegetation class". To implement the Maximum Likelihood Classification (MLC) method distributed (Richards & Jia, 2006), TM bands 1 (0.45–0.52 µm, blue-green), 2 (0.52–0.60 µm, green), 3 (0.63–0.69 µm, red), 4 (0.76–0.90 µm, near infrared), 5 (1.55–1.75 µm, mid infrared), and 7 (2.08–2.35 µm, mid infrared) were used. Band 6 (10.40–12.50 µm, thermal infrared) was not used in our MLC scheme because the resolution was 60 m instead of the 30 m (as it is for the other bands). However, this band was used to compute surface temperatures as described in more detail below. Our MLC scheme then entailed computing, for each of the pixels in the ROI, the probability that it belonged to the STH-vegetation class. This probability was assumed to be normally distributed [49], and thus is given by

$$p(x) = \frac{1}{(2\pi)^{N/2}|\Sigma|^{1/2}} e^{-\left[\frac{1}{2}(x-m)^T \Sigma^{-1}((x-m))\right]}$$

where $\boldsymbol{x}$ is a vector location in pixel space, $N = 6$ is the dimensionality of pixel space, $\boldsymbol{\Sigma}$ the covariance matrix of the distribution, and $\boldsymbol{m}$ is the mean position of the spectral class. ($\boldsymbol{m}$ and $\boldsymbol{\Sigma}$ are computed from the training pixels). A threshold value was set at p_0, meaning that if a pixel had a probability $p \geq p_0$ of being in the STH-vegetation class, then the pixel was put into this

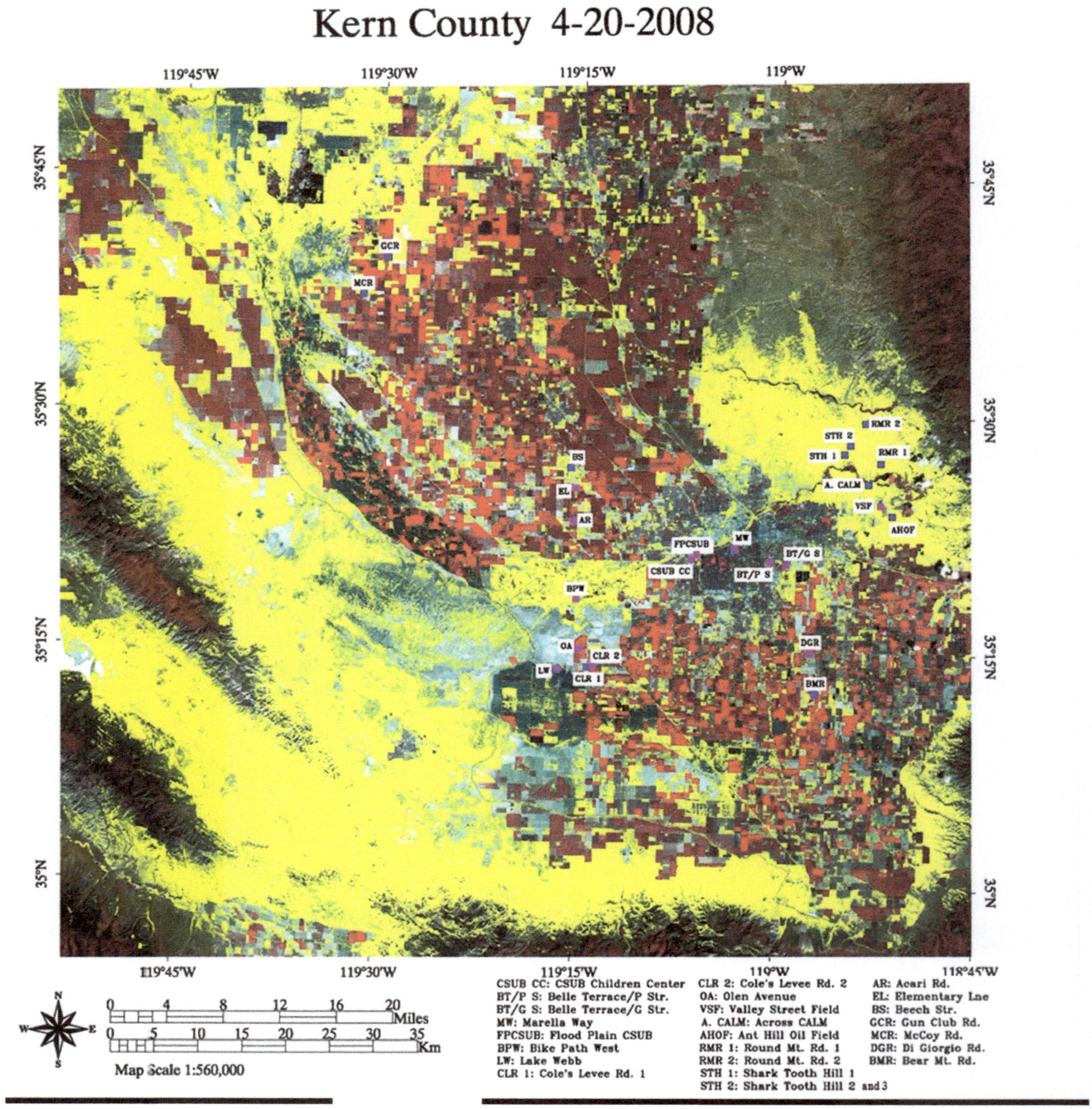

Figure 1. False color image of the ROI on April 20, 2008. Yellow pixels indicate locations in the STH-vegetation class, $p_0 = 0.95$ and $f = 0.32$.

spectral class. Otherwise ($p < p_0$) the pixel was simply left unclassified. Clearly, as the parameter p_0 decreased, the fraction f of pixels in the ROI which belong to the STH-vegetation class increased. This is because pixels which are less and less like the training pixels get included into this class.

It was also investigated how much p_0 needed to be reduced from 1 until the sites which tested positive for *C. immitis* came into the STH-vegetation class. This served to calibrate the method and as a validation step. Clearly, if the *C. immitis*-positive sites get included in the STH-vegetation class for $p_0 \lesssim 1$, then our method is robust. However, if our *C. immitis*-positive sites remain unclassified until $p_0 \gtrsim 0$, then our spectral class is poorly defined.

Lastly, we determined the area (km^2) that was characterized by vegetation that belonged into the STH-vegetation class over the sampling period and until early 2014 using landsat images and the software ENVI 5.1+IDL 8.3.

Surface temperatures

Surface temperature variations across the ROI were of interest as well. In addition to utilizing a vegetation class to assess potential

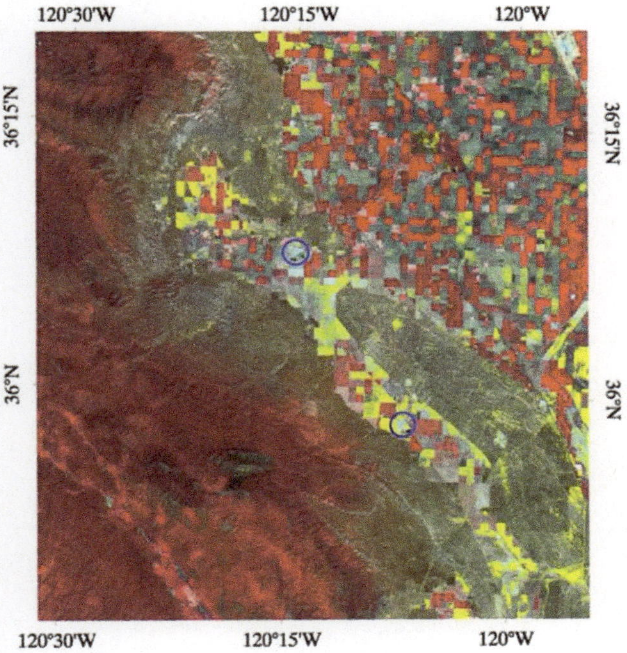

Figure 2. False color image of two San Joaquin Valley prisons. STH-vegetation class pixels are shown in yellow. The circles indicate the location of the prisons. Upper left: Pleasant Valley State Prison in Fresno County, California. Lower right: Avenal State Prison in Kings County, California. Images were taken on April 20, 2008. Maximum Likelihood Classification scheme was used with $p_0 = 0.95$.

sites for *C. immitis* growth, surface temperature variations across the ROI may also help to characterize the niche of this fungus. Landsat-5 TM-6 is an infrared band. This band (from the same April 20, 2008 image) was used as follows to examine the thermal landscape of our ROI. The same area in STH was taken as training pixels, and their average $\bar{x}$ and standard deviation σ were computed. Then, a simple parallelepiped method [49] was used to find other locations in the ROI with similar values. Thus, all pixels whose value was in the ranges$(x-n\sigma, x+n\sigma)$, $n = 1,2,3$ were put into this spectral class, and we referred to this as the "STH-thermal" class. All other pixels were left unclassified. The image's digital numbers were converted to temperatures by applying the procedure described in the National Aeronautics and Space Administration's Landsat 7 Science Data Users Handbook (http://landsathandbook.gsfc.nasa.gov/). See also Chander & Markham [50] for details. As a result, a map was obtained where the surface temperature was close to STH at the time the image was taken.

Weather data

Precipitation data for the Southern San Joaquin Valley was obtained from the California Data Exchange Center (http://cdec.water.ca.gov/snow_rain.html). The cumulative monthly precipitation (inches) over time was assembled from 5 stations (Calaveras Big Trees [CVT], Hetch Hetchy [HTH], Yosemite HQ [YSV], North Fork RS (NFR), and Huntington Lake (HNT]). A more detailed analysis of the weather data for Bakersfield in particular was not the focus of this study.

Physical and chemical soil parameters

To determine physical and chemical soil parameters of all soil samples, the websoilsurvey database of the United States Department of Agriculture (http://websoilsurvey.nrcs.usda.gov/) was used. Furthermore, all sampling sites were characterized by using the soil series extent mapping tool from the website of the Center of Environmental Informatics (CEI) (http://www.cei.psu.edu/cei_wp/). Thus, through agricultural and environmental support tools available from the USDA and CEI websites, our sampling sites were further characterized in regard to land use and vegetation. Additional geological information was obtained as well, such as the distribution of certain soil types and series in California. By using the soil series extent mapping tool, our soil samples were linked to known soil series and soil groups that are characteristic for the Southern San Joaquin Valley and beyond.

Soil sampling sites

Soil physical and chemical parameters that could likely support the growth of the pathogen based on results of the study by Lauer et al. [32] were used to choose 13 new sites that were investigated in winter and spring 2011 (Jan–Apr). Additionally, two sites that were found to be strong growth sites of the pathogen in 2008/2009 were also investigated again in 2011. Six sites were the pathogen was not detected were included in this study as well. Sampling sites included in this study were all non-agricultural silt, clay or sandy loams that differed in regard to physical and chemical parameters. All sites were located within the Central Valley Portion of Kern County. Overall, 23 sites were investigated in this study by satellite imagery and multiplex PCR. Two additional sites from STH were investigated by satellite imagery only (reference sites). Based on information from the USDA websoilsurvey database, the soils belonged to 13 different soil map units. Samples were taken each month in 2008, 2009 and 2011 (some sites were not sampled in 2011) from three different depths (0–2 cm, 5–7 cm, and 18–20 cm), placed on ice during transport to the lab, and frozen at −80°C when not processed immediately. See table 1 for detailed information about all sites, including exact location, soil type, observed rodent activity and indication of the presence or absence of *C. immitis*. Also see the first column of table two for the year they were investigated. Our sampling sites were not chosen based on Landsat imagery. They were chosen mainly based on the percentage of clay in the soil as indicated by the USDA websoilsurvey database. About 30% of clay had been indicative of a potential *C. immitis* positive site based on previous research [32]. After results from the multiplex PCR approach became available, we evaluated if sites where *C. immitis* was detected correlate with sites indicated by Landsat imagery to fall into STH vegetation sites.

DNA extraction and multiplex Polymerase Chain Reaction (PCR)

DNA was extracted from well-mixed soil samples (two replicates) using the MoBio PowerSoil DNA Isolation Kit (MoBio Laboratories, Solana Beach, CA) following the manufacturer's protocol. The multiplex PCR approach developed by Greene et al. [47] and optimized for the detection of *C. immitis* from soil DNA by Lauer et al. [32] was used to determine the presence of fungi in general and specifically *C. immitis* in all soil samples with two primer pairs. Primer pair ITSC1A/ITS C2 (18S ribosomal intertranscribed spacer [ITS] region, 223 bp), which is specific for *C. immitis*, was used in combination with primer pair RDS478/RDS482 (18S ribosomal gene, 650 bp) which amplifies 18S rDNA from all fungi. The ITS region was chosen due to its high

Table 2. Probability that the sites fall in the STH-vegetation class, as predicted by Landsat data.

sampling sites and year sampled	p_0=**0.90**				***C. immitis* growth site (multiplex PCR)**	p_0=**0.95**			
Bakersfield city	4/20/08	4/23/09	4/26/10	4/29/11		4/20/08	4/23/09	4/26/10	4/29/11
1. CSUB Children Center ('08, '09)	N	N	N	N	accumulation site*	N	N	N	N
2. Belle Terrace/P Str. ('11)	**Y**	N	N	N	growth site*	N	N	N	N
3. Belle Terrace/Gay Str. ('11)	N	N	**Y**	N	growth site	N	N	**Y**	N
4. Marella Way ('11)	N	N	N	N	negative site*	N	N	N	N
5. Flood Plain CSUB ('08, '09)	N	N	N	N	accumulation site	N	N	N	N
SW Bakersfield									
6. Bike Path West ('08, '09)	N	N	N	N	negative site	N	N	N	N
7. Lake Webb ('08, '09)	N	N	N	N	accumulation site	N	N	N	N
8. Cole's Levee Rd. I ('08, '09, '11)	**Y**	**Y**	**Y**	**Y**	growth site	**Y**	N	N	N
9. Cole's Levee Rd. II ('11)	**Y**	**Y**	**Y**	**Y**	growth site	**Y**	**Y**	**Y**	N
10. Olen Avenue ('11)	N	N	N	N	growth site	N	N	N	N
11. Valley Street Field ('08, '09)	N	N	N	N	accumulation site	N	N	N	N
NE Bakersfield									
12. Across CALM ('11)	**Y**	**Y**	N	N	growth site	**Y**	**Y**	N	N
13. Ant Hill Oil Field ('08, '09, '11)	**Y**	**Y**	**Y**	N	growth site	**Y**	**Y**	**Y**	N
14. Round Mt. Rd. I ('08, '09)	**Y**	**Y**	**Y**	N	accumulation site	**Y**	**Y**	**Y**	N
15. Round Mt. Rd. II ('08, '09)	**Y**	**Y**	N	N	negative site	**Y**	**Y**	N	N
16. Sharktooth hill I	**Y**	**Y**	**Y**	**Y**	nd*	**Y**	**Y**	**Y**	**Y**
17. Sharktooth hill 2	**Y**	**Y**	**Y**	**Y**	nd**	**Y**	**Y**	**Y**	**Y**
18. Sharktooth hill 3 ('11)	**Y**	**Y**	**Y**	**Y**	growth site	**Y**	**Y**	**Y**	**Y**
NW Bakersfield									
19. Acari Rd. ('11)	N	N	N	N	negative site	N	N	N	N
20. Elementary Lne. ('11)	N	N	N	N	growth site	N	N	N	N
21. Beech Str. ('11)	N	**Y**	N	**Y**	growth site	N	**Y**	N	**Y**
Wasco									
22. Gun Club Rd.('11)	**Y**	**Y**	**Y**	**Y**	negative site	**Y**	**Y**	N	**Y**
23. McCoy Rd. ('11)	**Y**	**Y**	**Y**	**Y**	negative site	**Y**	N	**Y**	**Y**
Arvin									
24. Di Giorgio Rd. ('11)	N	**Y**	N	N	growth site	N	**Y**	N	N
25. Bear Mt. Rd. ('11)	**Y**	N	N	N	growth site	**Y**	N	N	N
Fraction of area covered by class (%)	35	42	24	28		32	39	22	26

(**Y** = in class [indicated in bold], N = not in class).
nd* = not determined in this study.
nd** = not determined in this study, but confirmed as growth site by Swatek (1970).
growth site* = sites were *C. immitis* was detected at least twice in a deeper soil layer during the late winter/spring (February-May).
accumulation site* = the pathogen could only be detected on the surface of the sampling site and never in a deeper soil layer over a several year period.
negative site* = the pathogen could not be detected in any of the soil samples using the multiplex PCR method as described in this study.

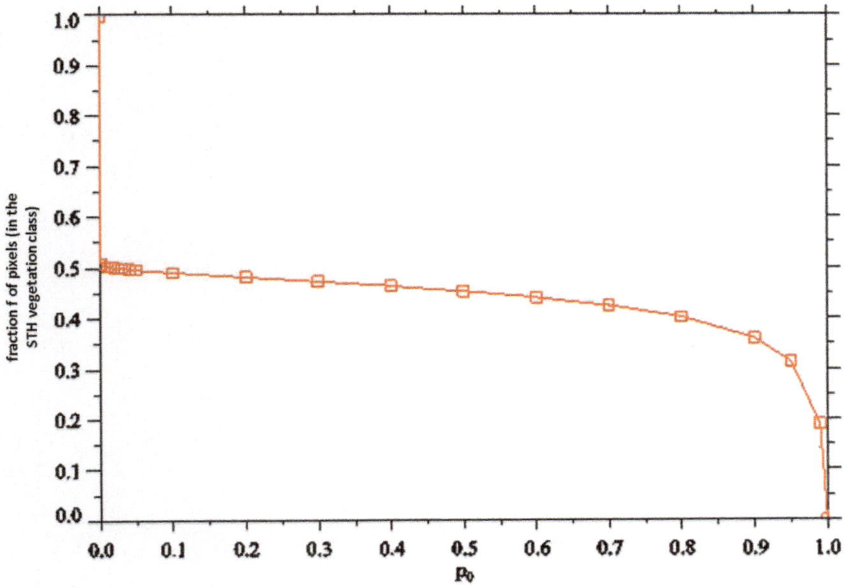

Figure 3. Plot of the fraction *f* of pixels in the STH-vegetation class vs. the threshold value p_0. Pixels whose probability of being in the STH-vegetation class is $p < p_0$ are left unclassified. Pixels with $p \geq p_0$ are put in the class. In this plot, $f = 0.51$ for $p_0 = 10^{-4}$, and $f = 1.0$ for $p_0 = 10^{-5}$.

nucleotide variability. Amplified ITS fragments were extracted from the 2% Agarose Gel, extracted with the Zymo Clean Gel DNA Recovery kit (ZymoResearch, Irvine, CA), and subsequently sequenced to confirm the presence of *C. immitis*. Extracted DNA from a *C. immitis* isolate (M39), obtained from the Laboratory of Medical Mycology at the Universidad Nacional Autónoma de México was used as positive control. Negative controls and positive controls were included in all PCR's to detect contamination and to verify the amplification of a PCR product of the desired size.

Sites were *C. immitis* was detected at least twice in a deeper soil layer during the late winter/spring (February–May) when the soil is moist and the soil temperature increased were referred to as 'growth sites' of the pathogen in this study, assuming that the soil parameters likely supported the growth of the fungus and thus, the pathogen could be detected consecutively over a several year period in the same location, over several growth seasons. Based on the definition provided by Fisher et al. [31], growth sites are sites where physical, chemical, and biological conditions are suitable for completion of the entire growth cycle required by the organism. Thus, it could be assumed that if the pathogen finds supportive environmental conditions, it would likely expand into deeper soil layers, and not just remain on the surface which can be more hostile due to desiccation and increased uv-radiation. In fact, the majority of the soil samples that contained the pathogen in deeper soil layers also contained the pathogen in surface layers. To the contrary, sites were termed 'accumulation sites' in this study when the pathogen could only be detected occasionally on the surface of the sampling site and never in a deeper soil layer over a several year period. This made it likely that arthroconidia had been transported to this location by the wind, but the pathogen was never able to complete its life cycle because of non-supportive environmental conditions. 'Accumulation sites' were also never positive in consecutive years in contrast to 'growth sites'. Fisher et al. [31] defined 'accumulation sites' as sites where arthroconidia of *Coccidioides* may have been deposited on or near the soil surface after being transported from growth sites by wind, water, organisms, or anthropogenic means. We are aware that we did not investigate the activity of the pathogen in the soil or verify its growth, and that finding the pathogen in the surface layer of the soil does not mean that it cannot grow there at all. Therefore, we have to consider that some of our results might have been false negatives.

Results

Remote Sensing Approach

A false color map of our ROI for April 28, 2008 was generated and is presented in figure 1, with indication of all sampling sites. Sites which were similar in vegetation to site STH, a confirmed growth site of *C. immitis*, were indicated in yellow, whereas sites that are characterized by different vegetation types appeared in various shades of green and red (agricultural fields, housing developments with gardens, higher elevated mountain slopes etc.). Results by Landsat imagery indicated large areas west of Bakersfield as potential growth sites of the pathogen, in addition to the STH area east of Bakersfield. The city of Taft southwest of Bakersfield was completely surrounded by vegetation that is similar to the vegetation type that characterizes STH. Landsat imagery furthermore indicated small pockets of potential growth sites of *C. immitis* scattered throughout the Southern San Joaquin Valley and around and within the city of Bakersfield. Overall, the yellow colors indicate that about 15% of the landscape visible in the satellite image was covered with vegetation that has the same reflection pattern as the STH vegetation.

We also applied Landsat imagery to an area northwest of Bakersfield where two prisons are located near the cities of Avenal and Coalinga (Kings County and Fresno County) (Fig. 2). In this area, the incidence of coccidioidomycosis has been observed to be large among prison inmates, so one might hypothesize that *C. immitis* could be present in the neighboring environments. And indeed, yellow areas in the immediate neighborhood of the prisons, as presented by Landsat imagery, indicated the presence of potential growth sites of the pathogen.

Table 3. Probability that the sites fall in the STH-thermal class, as predicted by Landsat data.

sampling sites and year sampled	n = 3				*C. immitis* **growth site (multiplex PCR)**	n = 2			
Bakersfield city	4/20/08	4/23/09	4/26/10	4/29/11		4/20/08	4/23/09	4/26/10	4/29/11
1. CSUB Children Center ('08, '09)	N	N	N	N	accumulation site	N	N	N	N
2. Belle Terrace/P Str. ('11)	N	N	**Y**	**Y**	growth site	N	N	N	**Y**
3. Belle Terrace/Gay Str. ('11)	N	**Y**	**Y**	N	growth site	N	**Y**	**Y**	N
4. Marella Way ('11)	N	N	N	N	negative site	N	N	N	N
5. Flood Plain CSUB ('08, '09)	N	N	N	N	accumulation site	N	N	N	N
SW Bakersfield									
6. Bike Path West ('08, '09)	N	**Y**	**Y**	**Y**	negative site	N	N	N	N
7. Lake Webb ('08, '09)	N	N	N	N	accumulation site	N	N	N	N
8. Cole's Levee Rd. I ('08, '09, '11)	N	**Y**	**Y**	**Y**	growth site	N	**Y**	N	**Y**
9. Cole's Levee Rd. II ('11)	N	**Y**	**Y**	N	growth site	N	**Y**	**Y**	N
10. Olen Avenue ('11)	N	N	N	N	growth site	N	N	N	N
11. Valley Street Field ('08, '09)	N	N	**Y**	N	accumulation site	N	N	N	N
NE Bakersfield									
12. Across CALM ('11)	**Y**	N	**Y**	N	growth site	**Y**	N	N	N
13. Ant Hill Oil Field ('08, '09, '11)	**Y**	**Y**	**Y**	N	growth site	**Y**	**Y**	N	N
14. Round Mt. Rd. I ('08, '09)	**Y**	**Y**	**Y**	N	accumulation site	**Y**	**Y**	**Y**	N
15. Round Mt. Rd. II ('08, '09)	**Y**	**Y**	**Y**	N	negative site	**Y**	**Y**	N	N
16. Sharktooth hill I	**Y**	**Y**	**Y**	**Y**	nd*	**Y**	**Y**	**Y**	**Y**
17. Sharktooth hill 2	**Y**	**Y**	**Y**	N	nd**	**Y**	**Y**	N	N
18. Sharktooth hill 3 ('11)	**Y**	**Y**	**Y**	N	growth site	**Y**	**Y**	N	N
NW Bakersfield									
19. Acari Rd. ('11)	N	N	N	N	negative site	N	N	N	N
20. Elementary Lne. ('11)	N	N	N	N	growth site	N	N	N	N
21. Beech Str. ('11)	N	**Y**	**Y**	N	growth site	N	**Y**	**Y**	N
Wasco									
22. Gun Club Rd.('11)	N	N	**Y**	**Y**	negative site	N	N	**Y**	**Y**
23. McCoy Rd. ('11)	**Y**	**Y**	**Y**	**Y**	negative site	**Y**	N	**Y**	**Y**
Arvin									
24. Di Giorgio Rd. ('11)	**Y**	**Y**	**Y**	**Y**	growth site	N	**Y**	**Y**	**Y**
25. Bear Mt. Rd. ('11)	N	N	N	N	growth site	N	N	N	N
Fraction of area covered by class (%)	35	42	24	28		32	39	22	26

(**Y** = in class [indicated in bold], N = not in class).
nd* = not determined in this study.
nd** = not determined in this study, but confirmed as growth site by Swatek (1970).

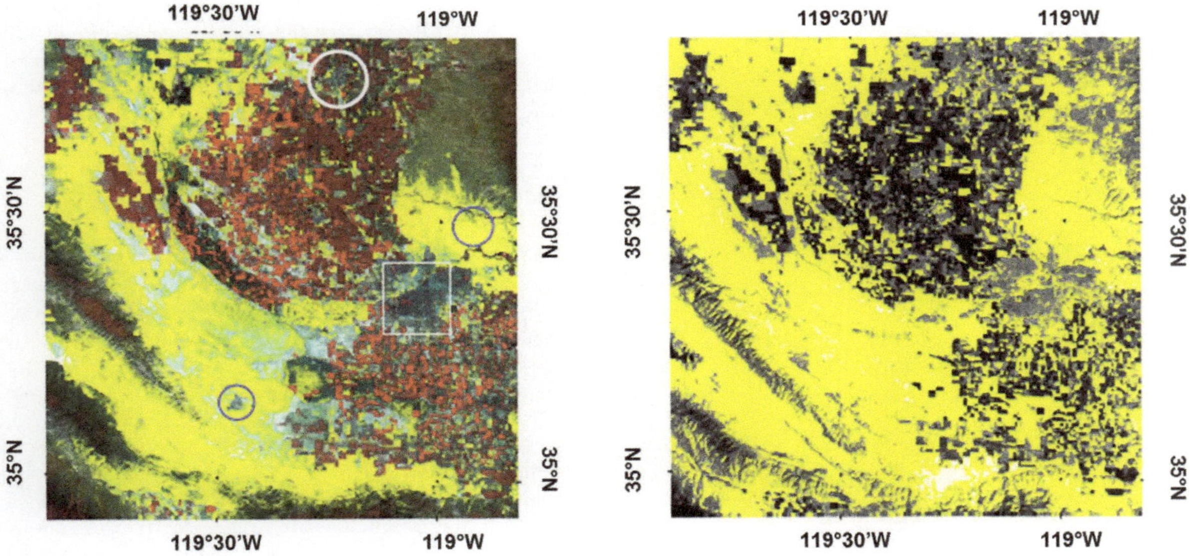

Figure 4. Left: False color image of the ROI on April 20, 2008. Yellow pixels indicate locations in the STH-vegetation class. $p_0 = 0.95$ and $f = 0.32$. The square denotes the location of the city of Bakersfield, the circle on the top indicates the city of Delano, the circle on the right indicates the location of STH, and the circle on the left indicates the location of the city of Taft. **Right**: Spectral class comparison. STH-thermal class is shown in yellow for the same April 20, 2008 image. Parallelepiped scheme was used with thresholds $\bar{x} \mp 20$, with $\bar{x}(=36.4^\circ C)$ the average surface temperature on the STH training pixels, and $\sigma(=1.9^\circ C)$ the corresponding standard deviation.

Overall, 4 false color maps were generated for our ROI showing results for April 2008 to April 2011 (one for each April, see Table 2). For each satellite picture we obtained one STH vegetation profile. Therefore, it did not matter if variation in the vegetation occurred. The April profiles were considered the most definitive for our work, because in early spring, the climate (soil and air temperature and humidity) still support the growth of the vegetation, and grasses and herbs which are characteristic for the STH- vegetation profile have not dried up yet, compared to the summer months. A validation of our approach is shown in figure 3. In this figure, the corresponding plot of f vs. p_O is presented. This figure also presents the dependence of the fraction f on p_O. It can be pointed out that when p_O drops from 1, f changes from zero to 0.36 at $p_0 = 0.90$, but then increases slowly with decreasing p_O until $p \approx 0.04$, where $f \approx 0.50$, and jumps to $f = 1$ for $p_0 = 10^{-5}$. Most likely, this is due to the STH-vegetation class being quite distinct from all other possible spectral classes in the ROI, with a large distance (in pixel space) from those classes. Otherwise, discrete increases in f with decreasing p_O would be expected as other surface types get merged into the STH-vegetation class.

As a further consistency check, we also wanted to examine the extent to which the STH-vegetation and STH-thermal classes overlapped and if growth sites of the pathogen could be predicted by soil thermal data (tables 2 and 3, figure 4). Thus, we obtained the STH-thermal class by implementing a parallelepiped scheme as described before (see methods). For $n = 1$, only sites 13, 14, 16 and 21 were included in the STH-thermal class (April 2008, data not shown). Sites 8, 9, 12–18, 21, and 22–24 were added when $n = 2$. Site 11 was included when $n = 3$. Site 25 (one of the strongest growth sites of the pathogen) was never included. Figure 4 (right) shows our results for $n = 2$. We present this figure here because site 8, which we identified as a strong growth site of the pathogen, came into the STH-thermal class for this value of n (but site 8 was not in the class for $n = 1$). By evaluating the agreement between satellite imagery (STH-vegetation class and STH-thermal class, data from 4 consecutive years), we found that both data sets disagreed in 12% (sites 6, 11 and 25). Almost all sites that fell into the STH-vegetation class also fell into the STH-thermal class (see tables 2 and 3).

Soil series and soil parameters

By using the soil series extent mapping tool, we found that the soil series and soil groups in which the pathogen was detected around Bakersfield, California, belonged to the Garces (Natragid, sites CLR, Bear Mt. Rd.), Chanac (Haploxerept, site AHOF), and Pleito (Haploxeroll, sites STH1 and 2 and 3) series. These soil series are not restricted to the Southern San Joaquin Valley. See figure 5 for a distribution of these soil series in California. Soils that belonged to the Chanac soil series can also be found in western Arizona and southern Nevada. All soils were of mixed mineralogy, had a superactive cation exchange capacity, and were thermic soils with predominantly fine loamy particles. Soils that belonged to these soil series are among the dominant soils in the Southern San Joaquin Valley, especially Kern County and Kings County, but can also be found in northern and western California. The use of software such as the USDA websoilsurvey database, as well as tools available at the Center of Environmental Informatics (CEI) have been found to be very valuable in obtaining information about physical and chemical parameters of soils that could support the growth of *C. immitis*. Detailed information about soil type, landform, dominant parent material of the soil, as well as soil physical and chemical parameters are listed for all sampling sites in table 4. Using these tools, information about land use, vegetation, mean annual soil temperatures, and geographic setting was accessed as well and is summarized in table S1 in file S1.

Detection of *C. immitis* by multiplex PCR

In addition to the two growth sites of the pathogen that were detected in 2008 and 2009 (Cole's Levee Rd. [CLR], Ant Hill Oil

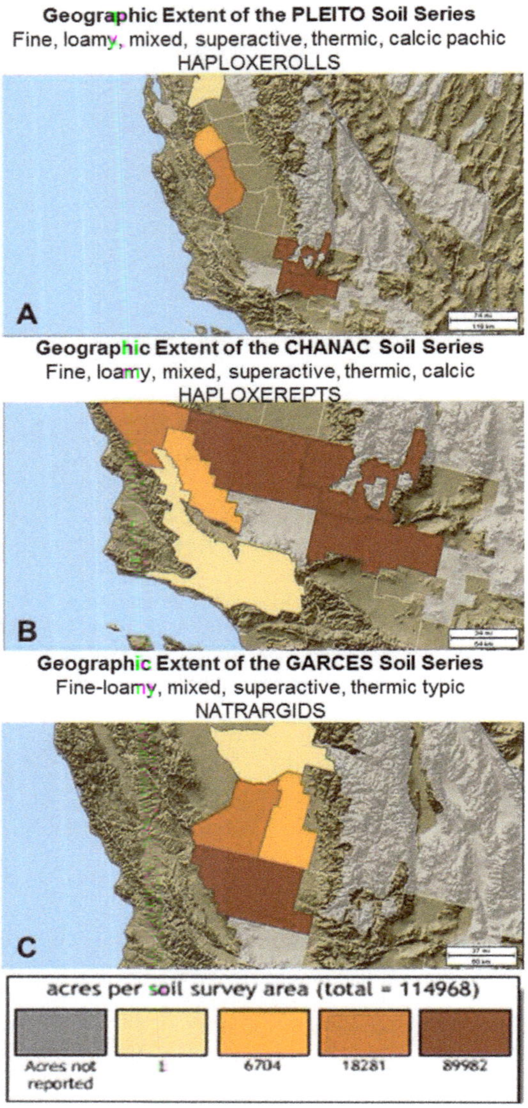

Figure 5. Extend of soil series in the San Joaquin Valley, CA, which can support the growth of *C. immitis*. A: Pleito (brown: SE and NE Kern County, dark orange: W Fresno County, light orange: W Merced County, tan: San Joaquin County) **B**: Chanac (brown: SE, NE and NW Kern County, dark orange: San Louis Obispo County [Paso Robles area], light orange: San Luis Obispo County, [Carrizo Plains]), and **C**: Garces soil series (brown: NW Kern County, dark orange: Kings County, light orange: W Tulare County, tan: E Fresno Area), Center for Environmental Informatics at Pennsylvania State University (CEI), http://www.cei.psu.edu/soiltool/semtool.html.

Field [AHOF], see [32]), we were able to detect additional "hot spots" of the pathogen in 2011. These sites were located within Bakersfield city, in the southwest, northeast and northwest of Bakersfield, and near Arvin, California (see table 1). Sites where the pathogen was detected more than once in a deeper soil layer were considered growth sites, whereas sites where the pathogen was detected occasionally in the surface layer only were considered accumulation sites. Sites were *C. immitis* was never detected were considered negative sites and included areas within Bakersfield city, an area northwest of Bakersfield, and 2 sites near Wasco, California. Fungal DNA could be detected in all soil samples. Of all sites investigated in 2011 (two growth sites investigated in 2008/09 [CLR and AHOF], and 14 new sites, out of 16 sites altogether for 2011), 4 sites (25%) were found negative, and 12 sites (75%) were confirmed as growth sites of the pathogen. No new accumulation sites were discovered in 2011. The site at Bear Mt. Road was the strongest growth site of *C. immitis* for the 2011 sampling set (positive for *C. immitis* from Jan–Apr). For an example of multiplex PCR results, see figure 6.

Correlation between Landsat imagery and multiplex PCR results

By comparing results obtained by satellite imagery (STH-vegetation class) and multiplex PCR, we found that in ~74% (17 out of 23 tested sites) the satellite imagery results and the results obtained by multiplex PCR agreed at least in one year out of four years ($p_0 = 0.90$). When $p_0 = 0.95$, the agreement was ~70% (16 sites). When satellite imagery based on STH-thermal class was compared with multiplex PCR results, we found that both methods agreed only in 61% ($n=3$) or 65% ($n=2$) (table 5). Tables 2 and 3 show the probability that the sites fall in the '*C. immitis* growth area' based on Landsat data in comparison to results obtained by multiplex PCR. We set Red, Green and Blue (RGB) to TM bands 4, 3, and 2 respectively. With this choice, the different depths of red indicated different plant associations. These maps were the result of implementing the MLC method with $p_0 = 0.90$and$p_0 = 0.95$. Two of the three sites at STH listed in table 1 were confirmed growth sites of *C. immitis*. STH site 2 was confirmed as a growth site by Frank Swatek (Fisher FS, personal communication based on [30]), and STH site 3 was confirmed as a growth site by multiplex PCR in this study. Sampling site 8 (Cole's Levee Rd. I), which was determined as a strong growth site of the pathogen by multiplex PCR in every year, became included in the '*C. immitis* growth area' based on Landsat data when p_0 was reduced to 0.95. Sampling site 7 (Lake Webb, accumulation site, located less than 1 mile west of CLR) was added when p_0 was reduced to 0.10. Sampling sites 1, 5, 6, and 11, (near Children Center, Flood Plain, Bike Path West, and Valley Street Field) never got added for $p_0 > 10^{-5}$. This was consistent with results obtained by multiplex which confirmed the absence of the pathogen (site 6, negative site), or which detected the pathogen occasionally in surface samples only (sites 1, 5 and 11). Sites, 1, 5 and 11 were termed accumulation sites, where the arthroconidia had been likely transported to by the wind, and where the presence of the pathogen could not be detected in deeper layers by multiplex PCR. We interpret this to mean that STH was quite representative of the *C. immitis* ecological niche within our ROI. However, in some occasions the prediction made by satellite imagery to indicate soils that could potentially harbor the pathogen could not be confirmed by multiplex PCR. Of all 25 sites included in this study, only sites 8 and 9 (Cole's Levee Rd. I and II) fell in the STH-vegetation class in each year when$p_0 = 0.90$. These sites were confirmed as positive for *C. immitis* by multiplex PCR. Other sites that were confirmed as growth sites of the fungus by our culture independent approach fell in this class at least on one occasion out of four when $p_0 = 0.90$(sites 2, 3 [Belle Terrace/P Str. and Belle Terrace/Gay Str.], 12 [Across CALM], 13 [AHOF], 21 [Beech Str.], 24 [Di Giorgio Rd.], and 25 [Bear Mt. Rd.]). These sites were still included in the STH-vegetation site when p_0 was increased to$p_0 = 0.95$, with the exception of site 25. Of all sites that were found to be *C. immitis* growth sites by multiplex PCR, two sites were never indicated as a potential growth site by the MLC method (site 10 [Olen Ave.] and site 20 [Elementary Lne.]), but the pathogen was present in soil samples from both sites as

Table 4. Detailed physical and chemical information obtained from the USDA websoilsurvey database for all sites included in this study.

soil	sampling sites					
parameters soil parameters	**Elementary Lne.**	**across CALM Ant Hill Oil Field**	**Bear Mt. Rd. Di Georgio Rd. Olen Ave. Cole's Levee Rd.**	**Sharktooth hill**	**Belle Terrace/Gay Str. Belle Terrace/P Str. Marella Way**	**McCoy Rd. Gun Club Rd. Acari Rd. Beech Str.**
soil type	Panoche clay loam	Chanac clay loam	Garces loam	Pleito-Trigo-Chanac complex	Kimberlina-Urban land Cajon complex	Garces silt loam
landform	alluvial fans	fan remnants	Alluvium derived from granitoid	Fan remnants, stream terraces	alluvial fans	rims on basin floors
parent material	alluvium derived from igneous and sedimentary rock	alluvium derived from mixed	alluvium derived from granitoid	Alluvium derived from mixed	alluvium derived from igneous and sedimentary rock	alluvium derived from granite
(map unit symbols)	211	130/131	180	205	180	156
Physical parameters						
Surface texture	clay loam	clay loam	clay loam	clay loam	loamy sand	silt loam
% clay	31	31	25.5	30	12	26.8
% sand	35.4	35.4	38	33.5	71.3	34.2
% silt	33.6	33.6	36.5	36.5	16.7	39.1
Available water capacity (cm/cm)	0.17	0.17	0.21	0.16	0.12	0.11
Available water supply (0–25 cm)	4.25	4.25	5.04	3.69	2.64	2.7
Organic matter	0.25	0.75	0.98	1.5	0.75	0.06
Water content (15 bar)	18.9	18.2	16.7	17.2	8.7	16.2
Water content (1/3 bar)	32	30.1	30.9	27.8	17.7	30.2
Sat. hydraulic conductivity (Ksat) (micrometers/s)	9	9	8.37	2.82	28	0.8362
Chemical parameters						
pH	7.9	7.9	8.5	7.8	7.5	8.9
CaCO3	3	3	3	0	3	3
Cation Exchange Capacity (CEC7)	15	24.4	20.6	24.3	7.5	13.1
Gypsum	0	0	0	0	0	0
Sodium adsorption ratio (SAR)	0	0	2	0	0	14
Electrical conductivity (EC)	1	0	5	0.5	1	10.2

Indicated in cursive are the parameters which seemed to be most important to distinguish *C. immitis* growth sites from negative sites.

confirmed by multiplex PCR (see Discussion). Furthermore, two sites that were indicated as potential growth sites of the pathogen by Landsat imagery at all times when $p_0 = 0.90$, could not be confirmed by multiplex PCR to harbor the pathogen. These sites were located near Wasco, California (NW of Bakersfield), (sites 22 [Gun Club Rd.] and 23 [McCoy Rd.]).

Changes in extend of areas (km^2) that fell into the STH-vegetation class were observed for the sampling period until early 2014 and are displayed in table S2 in file S1. The year with the highest precipitation (2011) had the lowest area of vegetation that belonged into the STH-vegetation class in comparison to the years 2008 and 2009 which were characterized by a significantly reduced amount of precipitation and showed an increased area of vegetation that belonged into the STH-vegetation class (see figure S1).

Discussion

The purpose of our study was to identify soil types in Kern County that could support the growth of *C. immitis* by combining Landsat imagery (based on vegetation and soil temperature), and soil parameter information (from 25 sites) with a culture independent PCR-based method to detect the pathogen. We showed that satellite imagery, combined with soil parameter information, can provide a map of locations within our ROI, where *C. immitis* might reasonably be expected to be found. We were able to verify the presence of the pathogen by a multiplex PCR method in about 74% ($p = 0.90$), when soil samples were investigated over a 4 year period. However, for about a quarter of our sites (26%), results obtained by Landsat imagery and multiplex PCR did not correlate. The reasons for this observation could be multifold. Some main factors to be considered are: 1) The amount of *Coccidioides* DNA extracted from the soil might have been under the detection limit of our PCR based methods (sites 22 and

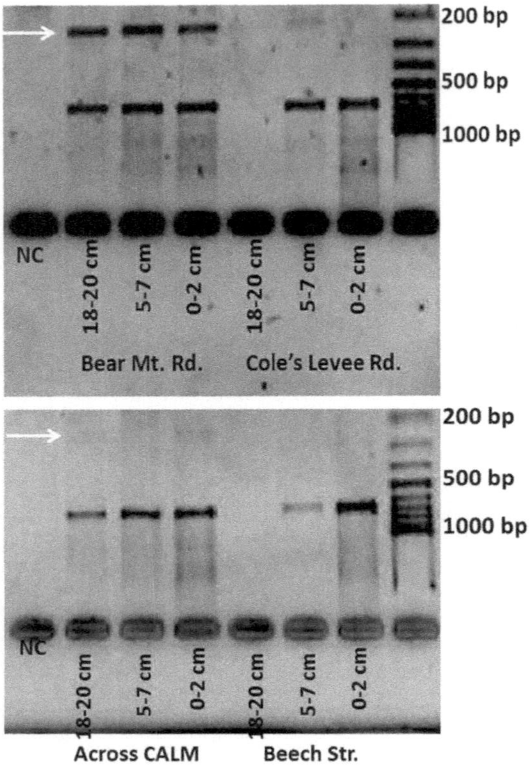

Figure 6. Example of multiplex PCR results. White arrows point on a 223 bp fragment that represents *C. immitis*. Site Bear Mt. Rd. shows the strongest ITS amplicons in all soil layers, whereas sites Cole's Levee Rd. and site Across CALM gave a weaker signal in some soil layers, and site Beech Str. was negative. NC = negative control. Bands that indicate the presence of the pathogen in the 2% Agarose gel were confirmed to origin from *C. immitis* by sequencing.

23), or 2) the resolution of the satellite imagery might not have been detailed enough (site 10, a small site of only 10 m^2), and 3) the distribution of *Coccidioides* in the soil might have been spotty, and the positive site was missed (sites that were not positive for the pathogen in all sampling years). A closer look at sites where satellite imagery and soil parameter data indicated potential growth sites for *C. immitis* also revealed that these sites were not uniform in regard to plant coverage, distribution, and diversity, thus, generating microhabitats for soil microorganisms that most likely would be quiet distinct, especially in and around the rhizosphere [32]. Other factors, such as fluctuation in climate and pollution of the soil might have had an impact on our analyses as well. Furthermore, it has to be considered that *C. immitis* might be able to persist in soils that have been converted to agricultural fields for an unknown amount of time, but its arthroconidia might never germinate and grow into vegetative hyphae. These sites could be termed dormant sites (e.g. site 20, an orchard with young almond trees). To assess these impacts on our results was not the focus of our work, but we are aware of these limitations. In previous research we have investigated the limitation of the multiplex PCR approach to detect *C. immitis*, (see [33] for results of primer efficiency). Briefly, we found that the sensitivity of the diagnostic PCR (ITS primer pair) was reduced compared to the primer pair that amplifies 18S rDNA fragments of all fungi (RDS primer pair).

In previous work [23–25], it was suggested that some environmental fluctuations are a fundamental link missing from coccidioidomycosis incidence statistical modeling schemes. One important aspect to investigate is whether fluctuations in the STH-vegetation class can provide this connection, and be statistically linked to the observed variations in incidence of valley fever. In this regard, one effect to consider is the extent to which this RS approach continues to be valid through the seasons. In the spring, when plants are blooming, the different vegetation types have different spectral signatures. As the weather dries and plants wither, the spectral signatures of the relevant vegetation types may become less distinct. Thus, the vegetation on STH may not be as good a marker for *C. immitis* in the fall, as it is in the spring. The implicit assumption in this study is that the STH environment is the only type of environment which harbors *C. immitis* within the San Joaquin Valley area of Kern County. We presented in this paper arguments to support this assumption; nevertheless, it would be useful to find more similarly suitable test sites to further corroborate our findings, or to find slightly different ecotypes that support *C. immitis*, beside of those detected in this study.

Table 5. Agreement between multiplex PCR and MLC for the STH vegetation class and the STH-thermal class to predict growth sites of *C. immitis* (to agree a prediction by either multiplex PCR or MLC must be confirmed at least once for the four years by the other method).

	STH-vegetation class	
	p_O = 0.90	p_O = 0.95
multiplex PCR and MLC agree	17 (74%)	16 (70%)
multiplex PCR predicts growth site and MLC disagrees	2 (9%)	3 (13%)
MLC predicts growth site and multiplex PCR disagrees	4 (17%)	4 (17%)
	STH-thermal class	
	n = 3	**n = 2**
multiplex PCR and MLC agree	14 (61%)	15 (65%)
multiplex PCR predicts growth site and MLC disagrees	3 (13%)	4 (17.5%)
MLC predicts growth site and multiplex PCR disagrees	6 (26%)	4 (17.5%)

From altogether 25 sites, only 23 were considered, because no multiplex PCR results were obtained for STH sites I and II.

We also observed changes in the extent of the STH-vegetation class over time. A comparative analysis of precipitation between 2008 and 2011 (up to early 2014) suggest that years with a reduced precipitation (drought) favor plants of the STH-vegetation class, but other factors likely play a role as well, such as development and changes in land use (see figure S1 and table S1 in File S1), which was not assessed in this study.

Compared to the STH-vegetation class data, the STH-thermal data showed considerably more variation for the four different years, as expected. The vegetation on a certain day in each year may be very similar, but soil temperatures might be more variable in different years (data for $n=2$ and $n=3$ can be seen in table 3, no data is shown for$n=1$). Other strong growth sites of the pathogen (sites 8, 9, and 13) were also not consistently included in all years, not even with$n=3$. It should also be noted that sites 6 (negative site) and 11 (accumulation site) were never included into the STH-vegetation class by satellite imagery, but were included in the STH-thermal class when$n=3$. We concluded therefore, that the STH-thermal classes alone might not be sensitive enough to predict growth sites of *C. immits*. Site 25 was never included (thermal class) maybe because of the limited resolution of the satellite imagery, as discussed earlier. It is important to keep in mind that the TM-6 image tells us surface temperatures. It may very well be that what matters is the temperature below the surface [31]. For our ROI, all landcovers were similar on the macroscale; therefore, apparent temperatures were appropriate for comparison purposes.

To improve the value of satellite imagery data, the actual spectral reflectance profiles of various soil components could be included to complement the satellite data (for details see http://www.africasoils.net/data/ldsf-description) in future studies. A time series analysis could also be considered, if feasible. Phenology development throughout the year can make the analysis more specific to a particular vegetation type. Niche modeling rather than automated classification could be considered as well to obtain a richer output that indicates variables of importance. However, a large dataset would be necessary that would include presence and also absence data of the pathogen in a certain type of soil at a certain time with presence or absence of a certain type of vegetation. Furthermore, it should be considered that broad band signatures over larger geographic areas and ecotones might not be precise enough to be useful in predicting growth sites of a pathogen, especially when the pathogen could be adapted to grow in a variety of different ecosystems.

Our results indicated that strong growth sites of the pathogen were likely associated with 3 different USDA soil map units (180, 131, and 205), which were all loamy sands. Several sites around Bakersfield, California, that fell into one of these map units were indeed growth sites of *C. immitis*, as confirmed by multiplex PCR, and were similar in vegetation compared to the STH area. These types of soils are not restricted to the Southern San Joaquin Valley, but can be found in other areas of California as well. One could hypothesize that with a drier and warmer climate, as it is predicted for California in the near future [51,52], *C. immitis* might be able to expand its current range. In our study, we focused on soil samples from only one County, the above mentioned Kern County in the Southern San Joaquin Valley of California, a highly endemic area for *C. immitis*. The soil types investigated here did not comprise all types that can be found in our ROI. Even though Kern County is a hot spot of *C. immitis* with the highest incidence of coccidioidomycosis documented for as long as incidence data is recorded in California, we cannot conclude that soils that predominate in this area are the ones that also predominantly support the growth of the pathogen. A more rigorous sampling framework should be attempted in the future that would include locations beyond Kern County covering as wide a range of habitats as possible to correctly determine growth sites of *C. immitis*, as well as determining sites that are not supporting the growth of the pathogen. Developing such a sampling plan should include stratification, replicate sampling, and determination of important chemical and physical soil parameters, including investigations in other countries where coccidioidomycosis occurs would be of value as well. The ultimate goal would then be to generate a U.S. or America-wide database of occurrence and absence of *Coccidioides* spp. Such a database could be useful for characterizing the ecological niche for both *Coccidioides* species, and could indicate a variety of supporting ecosystems, as well as being an advisory public health tool, to reduce incidence of coccidioidomycosis in Kern County and elsewhere.

In conclusion, the combination of the methods used in our research can be used to generate maps that indicate potential growth sites of *C. immitis*, and thus serve as a tool to further investigate the ecological niche occupied by the pathogen in the Southern San Joaquin Valley and beyond in more detail. Recent advances in computer processing and geographic information system and global positioning system technologies facilitate integration of remote sensing data, such as environmental parameters with disease incidence data, so that models for disease surveillance and control can be developed [53,54].

Supporting Information

Figure S1 Cumulative monthly precipitation (inches) over time for the Southern San Joaquin Valley, assembled from 5 stations (Calaveras Big Trees [CVT], Hetch Hetchy [HTH], Yosemite HQ [YSV], North Fork RS (NFR), and Huntington Lake (HNT]) obtained from the California Data Exchange Center at http://cdec.water.ca.gov/snow_rain.html).

File S1 Supporting tables. Table S1. Detailed soil series descriptions of sites which were found to be growth sites of *C. immitis*. **Table S2.** Extend of STH-vegetation class in our ROI between 2008 and 2011 based on satellite imagery.

Acknowledgments

The authors gratefully acknowledge Tom Mace, retired Senior Science Advisor at NASA Armstrong Flight Research Center (Science Mission Directorate, Airborne Sciences), and to Ned Horning of the Biodiversity Informatics Facility at the American Museum of Natural History's Center for Biodiversity and Conservation (New York, NY) for valuable technical advice on satellite image analysis. The authors also thank Kathleen Szick (CSUB) for helpful discussions and comments to improve this manuscript.

Author Contributions

Conceived and designed the experiments: AL JT. Performed the experiments: AL JT LM JB KC NS. Analyzed the data: AL JT. Contributed reagents/materials/analysis tools: LC KE. Wrote the paper: AL JT LC KE.

References

1. Hector RF, Rutherford GW III, Tsang CA, Erhart LM, McCotter O et al. (2011) The public health impact of coccidioidomycosis in Arizona and California. Int J Environ Res Public Health 8: 1150–1173.
2. Nguyen C, Barker M, Hoover S, Nix DE, Ampel NM et al. (2013) Recent advances in our understanding of the environmental, epidemiological, immunological, and clinical dimensions of coccidioidomycosis, Clin Microbiol Rev 26: 505–525.
3. Lee RV (1944) Coccidioidomycosis: In the western flying training command. Cal West Med 61: 133–134.
4. Pappagianis D (1988) Epidemiology of coccidioidomycosis. Current Topics in Mycology 2: 199–238.
5. Pappagianis D (1994) Marked increase in cases of coccidioidomycosis in California: 1991, 1992, and 1993. Clin Infect Dis 19 (Suppl 1), S14–S18.
6. Negroni R (2008) Evolución de los conocimientos sobre aspectos clinic epidemiológicos de la coccidioidomycosis en las Américas. Rev Argent Microbiol 40: 246–256.
7. Fisher MC, Koenig GL, White TJ, San-Blas G, Negroni R et al. (2001) Biogeographic range expansion into South America by *Coccidioides immitis* mirrors new world patterns of human migration. Proc Nat Acad Sci 98: 4558–4562.
8. Fisher MC, Koenig GL, White TJ, Taylor JW (2002) Molecular and phenotypic description of *Coccidioides posadasii* sp. nov., previously recognized as the non-California population of *Coccidioides immitis*. Mycologia 94: 73–84.
9. Neafsey DE, Barker BM, Sharpton TJ, Stajich JE, Park DJ et al. (2010) Population genomic sequencing of *Coccidioides* fungi reveals recent hybridization and transposon control. Genome Res 20: 938–946.
10. Comrie AC, Glueck MF (2007) Assessment of climate coccidioidomycosis model: Model sensitivity for assessing climatologic effects on the risk of acquiring coccidioidomycosis. Ann NY Acad Sci 1111: 83–95.
11. Kolivras KM, Johnson PS, Comrie AC, Yool SR (2001) Environmental variability and coccidioidomycosis (valley fever). Aerobiologia 17: 31–42.
12. Smith CE, Beard RR, Rosenberger HG, Whiting EG (1946) Effect of season and dust control on coccidioidomycosis. J Am Med Assoc 132: 833–838.
13. Hugenholtz PG (1957) Climate and coccidioidomycosis. In: Proceedings of symposium on coccidioidomycosis, Phoenix, AZ. Public Health Service, Publication no. 575. Atlanta: US Public Health Service, pp. 136–43.
14. Maddy KT (1957) Ecological factors possibly relating to the geographic distribution of *Coccidioides immitis*. In: Proceedings of the Symposium on Coccidioidomycosis (pp. 144–157). Comunicable Disease Center, Atlanta, GA. PHS Pub. No. 575.
15. Pappagianis D, Einstein H (1978) Tempest from Tehachapi takes toll or *Coccidioides* conveyed aloft and afar. West J Med 129: 527–530.
16. Jinadu BA, Welch G, Talbot R, Caldwell J, Johnson R et al. (1994) Update: coccidioidomycosis in California 1991–1993. MMWR 43: 421–423.
17. Mosley D, Komatsu K, Vaz V, Vertz D, England B et al. (1996) Coccidioidomycosis in Arizona 1990–1995. MMWR 45: 1069–1073.
18. Kirkland TN, Fierer J (1996) Coccidioidomycosis: A reemerging infectious disease. Emerg Infect Dis 2: 192–199.
19. Schneider E, Hajeh RA, Spiegel RA, Jibson RW, Harp EL et al. (1997) A coccidioidomycosis outbreak following the Northridge, Calif. earthquake. J Am Med Assoc 277: 904–908.
20. Komatsu K, Vaz V, McRill C, Colman T, Comrie A et al. (2003) Increase in coccidioidomycosis in Arizona 1998–2001. MMWR 52: 109–112.
21. Kolivras KM, Comrie AC (2003) Modeling valley fever (coccidioidomycosis) incidence on the basis of climate conditions. Int J Biometeor 47: 87–101.
22. Comrie AC (2005) Climate factors influencing coccidioidomycosis seasonality and outbreaks. Environ Health Perspect 113: 688–692.
23. Zender CS, Talamantes J (2006) Climate controls on valley fever incidence in Kern County, California. Int J Biometeorol 50: 174–182.
24. Talamantes J, Behseta S, Zender CS (2007a) Fluctuations in climate and incidence of coccidioidomycosis in Kern County, California: a review. Ann NY Acad Sci 1111: 73–82.
25. Talamantes J, Behseta S, Zender CS (2007b) Statistical modeling of valley fever data in Kern County, California. Int J Biometeorol 51: 301–315.
26. Tamerius JD and Comrie AC (2011) Coccidioidomycosis incidence in Arizona predicted by seasonal precipitation. PLOS One 6: 1–7.
27. Cayan D, Maurer E, Dettinger M, Tyree M, Hayhoe K (2008) Climate change scenarios for the California region. Climate Change 87: 21–42.
28. Greer A, Ng V, Fisman D (2008). Climate change and infectious diseases in North America: the road ahead. CMAJ 11: 178–186.
29. Marsden-Haug N, Goldoft M, Ralston C, Limaye AP, Chua J et al. (2013) BRIEFREPORT - Coccidioidomycosis acquired in Washington State. CID 2013: 56.
30. Swatek FE (1970) Ecology of *Coccidioides immitis*. Mycopath Mycol Applic 40: 3–12.
31. Fisher FS, Bultman MW, Johnson SM, Pappagianis D, Zaborsky E (2007) *Coccidioides* niches and habitat parameters in the southwestern United States: a matter of scale. Ann NYAcad Sci 1111: 47–72.
32. De Boer, Folman LB, Summerbell RC, Boddy L (2005) Living in a fungal world: impact of fungi on soil bacterial niche development. FEMS Microbiol Rev 29: 795–811.
33. Lauer A, Baal JCH, Baal JDH, Verma M, Chen JM (2011). Detection of *Coccidioides immitis* in Kern County, California, by multiplex PCR. Mycologia. doi:10.3852/11-127.
34. Merriam CH (1898) Life zones and crop zones of the United States. Bulletin 10 Division of Biological Survey, United States Department of Agriculture. Washington, DC.
35. Maddy KT (1958) The geographic distribution of *Coccidioides immitis* and possible ecologic implications. Ariz Med 15: 178–188.
36. Maddy KT (1959) Coccidioidomycosis in animals. Vet Med 54: 233–242.
37. Baptista-Rosas RC, Catalan-Dibene J, Romero-Olivares AL, Hinojosa A, Cavazos T et al. (2012) Molecular detection of *Coccidioides* spp. form environmental samples in Baja California: linking Valley Fever to soil and climate conditions. Fung Ecol 5: 177–190.
38. Barker BM, Tabor JA, Shubitz L, Perrill R, Orbach MJ (2012) Detection and phylogenetic analysis of *Coccidioides posadasii* in Arizona soil samples, Fungal Ecol. 5: 163–176.
39. Zak DR, Holmes WE, White DC, Peacock AD, Tilman D (2003) Plant diversity, soil microbial communities, and ecosystem function: Are there any links? Ecology 84: 2042–2050.
40. Nannipieri P, Ascher J, Ceccherini MT, Landi L, Pietramellara G, et al. (2003) Microbial diversity and soil function. Europ J Soil Sci 54: 655–670.
41. Sharpton TJ, Stajich JE, Rounsley SD, Gardner MJ, Wortman JR et al. (2009) Comparative genomic analyses of the human fungal pathogens *Coccidioides* and their relatives. Genome Res. doi:10.1101/gr.087551.108.
42. Egeberg RO, Ely F (1956) *Coccidioides immitis* in the soil of the southern San Joaquin Valley. Am J Med Sci 231: 151–154.
43. Rogers DJ, Randolph SE, Snow RW, Hay SI (2002) Satellite imagery in the study and forecast of malaria. Nature 415: 710–715.
44. Koelle K, Rodó X, Pascual M, Yunus M, Mostafa G (2005) Refractory periods and climate forcing in cholera dynamics. Nature 436: 696–70.
45. Rogers DJ (2000) Satellites, space, time, and the African trypanosomiases. Adv Parasitol 47: 128–171.
46. Goetz S, Prince S, Small J (2000) Advances in satellite remote sensing of environmental variables for epidemiological applications. Adv Parasitol 47: 289–307.
47. Greene DR, Koenig G, Fisher MC, Taylor GW (2000) Soil isolation and molecular identification of *Coccidioides immitis*. Mycologia 92: 406–410.
48. Kemp R (1974) The distribution of *Coccidioides immitis* at Sharktooth hill, Oildale, California [MS Thesis].. Long Beach, California: California State College. 62 p.
49. Richards JA, Jia X (2006) Remote sensing digital image analysis, An Introduction. Springer-Verlag, Germany.
50. Chander G, Markham B (2003) Revised Landsat-5 TM radiometric calibration procedures and postcalibration dynamic ranges. IEEE Trans Geosci Remote Sens 41: 2674–2677.
51. Cayan D, Maurer E, Dettinger M, Tyree M, Hayhoe K (2008) Climate change scenarios for the California region. Climate Change 87: 21–42. Clim Change 87: 21–42.
52. Greer A, Ng V, Fisman D (2008). Climate change and infectious diseases in North America: the road ahead. CMAJ 11: 178–186.
53. Nuckols JR, Ward MH, Jarup L (2004) Using geographic information systems for exposure assessment in environmental epidemiology studies. Environ Health Perspect 11: 1007–1015.
54. Craglia M, Goodchild MF, Annoni A, Camara G, Gould M et al. (2008) Next-Generation Digital Earth. A position paper from the Vespucci Initiative for the Advancement of Geographic Information Science. Int J Spat Data Infrastruct Res 3: 146–167.

19

Linking Bovine Tuberculosis on Cattle Farms to White-Tailed Deer and Environmental Variables Using Bayesian Hierarchical Analysis

W. David Walter[1]*, Rick Smith[2], Mike Vanderklok[2], Kurt C. VerCauteren[3]

1 U.S. Geological Survey, Pennsylvania Cooperative Fish and Wildlife Research Unit, Pennsylvania State University, University Park, Pennsylvania, United States of America, **2** Animal Industry Division, Michigan Department of Agriculture and Rural Development, Lansing, Michigan, United States of America, **3** United States Department of Agriculture, Animal and Plant Health Inspection Services, Wildlife Services, National Wildlife Research Center, Fort Collins, Colorado, United States of America

Abstract

Bovine tuberculosis is a bacterial disease caused by *Mycobacterium bovis* in livestock and wildlife with hosts that include Eurasian badgers (*Meles meles*), brushtail possum (*Trichosurus vulpecula*), and white-tailed deer (*Odocoileus virginianus*). Risk-assessment efforts in Michigan have been initiated on farms to minimize interactions of cattle with wildlife hosts but research on *M. bovis* on cattle farms has not investigated the spatial context of disease epidemiology. To incorporate spatially explicit data, initial likelihood of infection probabilities for cattle farms tested for *M. bovis*, prevalence of *M. bovis* in white-tailed deer, deer density, and environmental variables for each farm were modeled in a Bayesian hierarchical framework. We used geo-referenced locations of 762 cattle farms that have been tested for *M. bovis*, white-tailed deer prevalence, and several environmental variables that may lead to long-term survival and viability of *M. bovis* on farms and surrounding habitats (i.e., soil type, habitat type). Bayesian hierarchical analyses identified deer prevalence and proportion of sandy soil within our sampling grid as the most supported model. Analysis of cattle farms tested for *M. bovis* identified that for every 1% increase in sandy soil resulted in an increase in odds of infection by 4%. Our analysis revealed that the influence of prevalence of *M. bovis* in white-tailed deer was still a concern even after considerable efforts to prevent cattle interactions with white-tailed deer through on-farm mitigation and reduction in the deer population. Cattle farms test positive for *M. bovis* annually in our study area suggesting that the potential for an environmental source either on farms or in the surrounding landscape may contributing to new or re-infections with *M. bovis*. Our research provides an initial assessment of potential environmental factors that could be incorporated into additional modeling efforts as more knowledge of deer herd factors and cattle farm prevalence is documented.

Editor: Joao Inacio, National Institute for Agriculture and Veterinary Research, IP (INIAV, I.P.), Portugal

Funding: Funding for this research was provided by the National Wildlife Research Center of the United States Department of Agriculture, Animal and Plant Health Inspection Service, Wildlife Services. The funders had no role in study design, data collection and analysis, decision to publish, or preparation of the manuscript.

Competing Interests: The authors have declared that no competing interests exist.

* E-mail: wdwalter@psu.edu

Introduction

Bovine tuberculosis (bTB) is a bacterial disease (*Mycobacterium bovis*) in livestock and wildlife that results in United States Department of Agriculture-mandated depopulation of cattle herds costing farmers millions in lost revenue throughout the world [1,2]. Preliminary efforts by the Michigan Department of Agriculture-Animal Industry Division (MDA) have created protocols that farmers could follow to reduce potential for *M. bovis* infection of cattle in Michigan's Modified Accredited Zone (MAZ) [3]. Basic risk-assessment efforts were needed, however, to address the spatial context of disease epidemiology (i.e., infection probability if a farm is adjacent to a bTB-infected farm) and dynamics of primary reservoirs in the MAZ (i.e., white-tailed deer [*Odocoileus virginianus*]).

The influence of wildlife activity on transmission of *M. bovis* depends on possible hosts and their ability to transmit disease [4–6]. Direct observation of farms in Michigan, USA documented that indirect interactions between cattle and white-tailed deer were dominated by use of pastures and silage storage areas but deer fed from hay racks or silage troughs on only one occasion [7]. Visitation of farm yards and cattle-use areas by sixteen GPS-collared white-tailed deer was documented in Michigan's MAZ and deer were documented using confined feeding areas, water tubs, and pastures [8]. Prevalence of *M. bovis* in deer was as high as 10–12% in some townships but currently can range from 2 to ≥5% in some townships due to changes in management regulation for deer and feeding on some cattle farms [3,9,10].

Reoccurrence of *M. bovis* in farms depopulated of cattle in Michigan would suggest an environmental or mammalian host source of re-infection as several farms have become re-infected with *M. bovis* on ≥2 separate occasions often spanning 3–7 years between re-infection [3,11]. Under natural shaded conditions on pastures, survival of *M. bovis* in cattle feces was documented to span up to 5 months post-application during winter but only up to 2 months during spring and summer [12]. Effluent plots tested positive for *M. bovis* for up to 29, 13, and 35 days post application for soil, radishes, and lettuce, respectively, in a study in raised garden plots (lined plywood boxes) [13]. Although environmental and anthropogenic variables that influence odds of contracting a disease have been addressed in North America [3,14], only recently has the spatial matrices incorporating proximity to adjacent infected individuals been successfully modeled in disease epidemiology research with advances in software (i.e., WinBUGS; [15–17]). Understanding the spatial dynamics of *M. bovis* will increase our ability to predict future spread or occurrences and variables influencing these occurrences across the MAZ in the northern, lower peninsula of Michigan.

To incorporate spatially explicit data, likelihood of infection probabilities within a geographically designed grid can be determined for cattle herds that tested positive for *M. bovis* and incorporated into a Bayesian hierarchical framework. Although on-farm management practices are believed to influence *M. bovis* transmission, consensus on the most important farm-level factor responsible for transmission is absent and varied across studies in Europe and North America ([11,18,19] but see [3] for a detailed summary). Spatially explicit data on environments that cattle farms occupy is often lacking for researchers attempting to understand the underlying distribution of disease in the landscape and has not been modeled in this system since discovery of *M. bovis* in a free-ranging white-tailed deer in 1975. Our objectives were to model odds of infection with *M. bovis* in cattle farms at the herd level using Bayesian hierarchical analysis by incorporating prevalence of *M. bovis* in the deer population, environmental variables, spatial structure, and unstructured spatial heterogeneity across the MAZ in Michigan. An understanding of conditions that sustain survival of *M. bovis* in the environment would be valuable to our ability to focus surveillance for the disease and predict future spread or occurrences outside of the MAZ in Michigan.

Materials and Methods

Study area

We conducted our study in the northern, lower peninsula of Michigan in the MAZ. The 8,062 km^2 study area included the entirety of Alcona, Alpena, Montmorency, Oscoda, and Presque Isle counties (Fig. 1). The area encompassed the majority of the cattle farms where *M. bovis* has been found in Michigan. Our study area surrounds Deer Management Unit 452 that has been defined as the bovine tuberculosis core area by the Michigan Department of Natural Resources (MDNR) due to the high prevalence of *M. bovis* in free-ranging deer and the presence of *M. bovis*-positive cattle on farms (Fig. 1; [20,21]). Vegetation categories present in our study area included: *developed* that included roads, development, and barren land; *grass* that included pasture/hay fields and native grasses; *agriculture* that included crops other than forage; *forest* that included upland hardwood stands (*Quercus alba*, *Acer rubrum*, and *A. saccharum*), aspen stands (*Populus tremuloides* and *P. grandidentata*), hardwood/aspen mixed stands, upland conifer stands (*Pinus glauca*, *P. banksiana*, and *P. resinosa*), and hardwood/conifer mixed stands, and swamp that included lowland conifer forests/swamps (*P. glauca*, *P. mariana*, *Thuja occidentalis*, *Abies balsamea*. and *Latrix laricinea*). Elevations in the area ranged from 150–390 m above sea level and the mean annual temperature was 6.6°C, the mean rainfall was 72.5 cm, and there was a mean snowfall of 175 cm [22].

To link the disease status (positive or negative) of each farm in the sample to deer herd and environmental-level predictors, we first overlaid a 5×5 km square grid having a resolution of 25 square kilometers (hereafter referred to as *grid cell*), which is equal to a quarter township in size. We selected quarter townships as the proper resolution given that township would likely be too coarse a scale and section would be too fine a resolution for model convergence based on previous research with Bayesian hierarchical models [17]. There were a total of 368 grid cells covering the MAZ and we assigned each farm in our study to its appropriate grid cell.

Observation component

Cattle farm data. Our data included 762 cattle farms of known infection status (*observation component*) provided by the Michigan Department of Agriculture and Rural Development (MDA) from mandatory testing of cattle on an annual basis. Based on current knowledge that indirect transmission (i.e., environmental source) of *M. bovis* to cattle may be important, we used replicates for each farm that tested positive on ≥1 occasion over the 11 year span of our study (58 positive, 704 negative) with positive farms coded as 1 and negative farms coded as 0. We included a farm each time it tested positive for *M. bovis* and this was deemed warranted because conditions of that farm or host characteristics in the area were responsible for continued infections of *M. bovis* and replicates would weigh environmental characteristics of farms that tested positive on >1 occasion. Because farms that tested positive for *M. bovis* were depopulated of cattle then repopulated prior to subsequently testing positive, each farm was considered an independent observation for the purposes of our study design. Each farm was tested annually for *M. bovis* but we did not include additional negatives as replicates because that would have likely masked the effects of the positives that we were attempting to model for odds of infection.

We included all cattle farms in this region because we wanted to determine the environmental drivers of disease that were not associated with farm practices and remained unaltered when cattle farms were depopulated or permanently closed (e.g., surrounding habitats, soil composition). Furthermore, we did not include any farm-level covariates (e.g., herd size, feeding practices) in our models because farm mitigation strategies were initiated by the MDA during our study [3], would be difficult to quantify and standardize across the study region, and would only be considered a contamination source (e.g., cattle fed in deer habitat) but would not influence environmental persistence or survival of *M. bovis* in the landscape.

Process component

Host-level variables. The MDNR provided section-level data on deer prevalence for *M. bovis* from 1995 to 2009. We limited our analysis to white-tailed deer prevalence for 2005 to 2009 because deer herd management, ban on baiting deer, and mitigation of on-farm practices indicated that deer prevalence has stabilized within the past 5 years thus, more reflective of current deer prevalence [10,23]. Apparent prevalence of *M. bovis* in deer was determined for each grid cell by dividing the total number of deer testing positive by the total number of deer tested resulting in percent prevalence that was entered into models. Annual deer densities were provided by the MDNR from 2005 to 2009 at the county level for the MAZ based on Sex-Age-Kill reconstruction technique or additional methods if available [24,25]. We averaged deer density over the time period to match deer prevalence (2005–2009) that resulted in a single estimate of deer density per grid cell. Due to the logistical constraints of estimating deer densities at a fine scale, such as to the section-level, we used the only available data to represent deer densities in our study site as deer per square kilometer. We did not select mean deer prevalence or mean deer density for the entire span of sampling of cattle farms because we were interested in modeling the effects of more recent deer

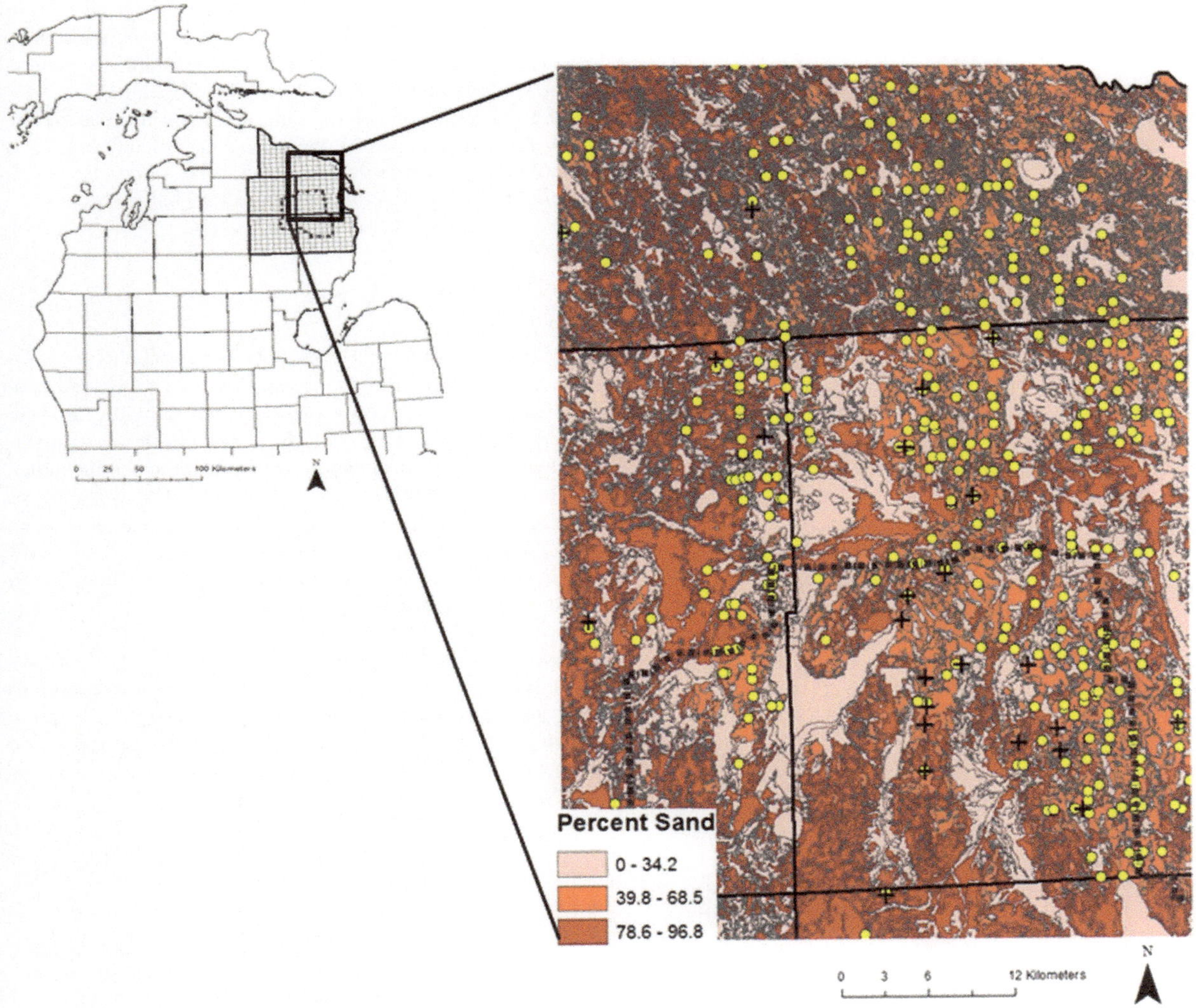

Figure 1. Sampling grid (25 km^2 cells) that contained all cattle farms tested for bovine tuberculosis within the Modified Accredited Zone (5 counties in bold) in the upper, lower peninsula of Michigan. Deer Management Unit 452 (dashed polygon) is considered the bovine tuberculosis core area for surveillance in white-tailed deer (Outset). Cattle farms that tested negative (yellow circles) and positive (black crosses) for *Mycobacterium bovis* used in Bayesian Hierarchical models overlayed on percent sand within a portion of the study area in the upper, lower peninsula of Michigan (Inset).

prevalence that likely would be influenced by initiation of deer and on-farm management practices in 1996.

Environmental-level variables. We hypothesized *a priori* that *M. bovis* patterns on farms were structured in part by spatial heterogeneities in features of the landscape, therefore, we identified four environmental-level predictors of infection based on optimal survival characteristics of *M. bovis* identified in the literature [3]. Proportion of sand content in the soil was selected because dry sandy loam soils at the proper pH and moisture promoted bacteria growth [26,27]. Proportion of the landscape ponding frequently and proportion of swamp/wetland were selected because the duration of standing water occurring in non-wetlands (ponding) and soil characteristics of inundated areas (wetlands) were conducive to long-term survival of *M. bovis* [27,28]. Mean soil pH was selected because soil pH from 5.8 to 6.9 was conducive to culture of *M. bovis* at the optimum temperature (37°C) for survival [27,29]. Sand (where "sand" was defined as soil particles with size >2 μm), landscape ponding frequently, and soil pH was determined using the Advanced Mode of the Soil Data Viewer available through the National Resources Conservation Service of the US Department of Agriculture in ArcMap 9.x (ArcMap; Environmental Systems Research Institute, Redlands, CA, USA). Soil Data Viewer provides interactive mapping software to query the Soil Survey Geographic (SSURGO) database and descriptive characteristics for each soil type. Sand was expressed as the percent sand within a soil type polygon (~2 ha resolution) for each soil map unit [30]. Each grid cell was thus potentially composed of multiple soil type polygons with varying sand contents. Therefore, we calculated the mean percent sand for each 25 km^2 grid cell using a weighted average based on the area of the various soil type polygons and their associated sand content. Similarly, in the Advanced Mode of the Soil Data Viewer, we identified the proportion of each grid cell that ponded frequently with frequently defined as "ponding occurs, on the average, more than once in 2 years and the chance of ponding is >50% in any year [30]." Similar to sand and ponding frequently, we calculated the mean soil pH for each 25 km^2 grid cell using a weighted average based on the area of the various pH polygons and their associated pH value.

We used the National Land Cover Database of 2006 (NLCD) that was created from Landsat 7 imagery to determine the proportion of swamp/wetland across the study site (MRLC 2007). To standardize analyses across the MAZ, we reclassified land cover from the NLCD into 8 categories used in Kaneene et al. [11]: hardwood forest, coniferous forest, mixed forest, open areas and shrubs, wetland/swamp, agricultural use, open water, and other (industrial, residential). We extracted the proportion of wetland/swamp from NLCD within each grid cell with all environmental-level variables, except soil pH, presented as a percentage in a grid cell in modeling efforts. Skewness of data and correlation among covariates was assessed but data transformations and exclusions were not considered necessary prior to entering into models.

Statistical analysis

We used a Bayesian hierarchical model structure [31,32] with logistic regression models (described below) to examine how recent TB prevalence at the deer herd-level and landscape factors influenced the probability of a farm being infected, while adjusting for the other covariates and spatial structure in the data [33]. To adjust for latent spatial effects we included two types of random effects that captured both the influence of the local neighborhood (i.e., cells sharing a border or vertex with each 25-km^2 grid cell; CAR) in determining spatial clustering of *M. bovis*, as well as any spatially independent influences occurring at the 25-km^2 spatial resolution of our grid (HET). Our models were constructed hierarchically to accommodate the fact that information from multiple levels (i.e., fixed-effects and spatial random effects) was being used to estimate individual-level infection probabilities. Taking a Bayesian approach, we used Markov Chain Monte Carlo (MCMC) simulation methods available within the program WinBUGS [34] to produce the unnormalized joint posterior density for the parameters of interest across all models examined based on the product of the data likelihood and the prior densities for each parameter. We used this approach to estimate the posterior marginal probability distributions for the parameters governing the influence of the host- and environmental-level, and spatial random effects predictors on the probability of infection. For each model we ran three independent Markov chains with varying initial values for 350,000 iterations and discarded the first 100,000. We thinned the Markov chains by keeping every twentieth iteration for inference. To determine if the three Markov chains used for each model had converged on the same posterior distribution, we used the statistical program R with the package boa [35] and employed several graphical and quantitative diagnostics, including autocorrelation plots, trace plots, and univariate corrected scale reduction factors for each parameter. To assess simultaneous convergence of all parameters for the top models, we calculated the multivariate potential scale reduction factor [32,36,37]. All inferences were based on the mean of each parameter's marginal posterior distribution.

Likelihood functions

The observation component of the data likelihood specifies each farm's observed *M. bovis* infection status as a Bernoulli random variable with parameter φ_{ij}:

$$Y_{ij}|\varphi_{ij} \sim \text{Bernoulli}(\varphi_{ij}),$$

where Y_{ij} is the infection status of the ith farm for $i = 1, \ldots, n$ from the jth grid cell for $j = 1, \ldots, m$, and φ_{ij} represents the probability of infection. Thus, given the probability of infection we assume each farm's infection status is conditionally independent.

The process component of the data likelihood models, via the logit link function, defined as the probability of infection as a function of individual, environmental and spatial covariates as well as random effects that account for spatial variability:

$$\text{logit}(\varphi_{ij}) = \mu + x_{ij}'\beta + \theta_j + \varphi_j, \quad (1)$$

where μ defines the baseline *M. bovis* infection probability, x_{ij}' is the transpose of a $k \times 1$ matrix of covariates for the ith farm from the jth grid cell, β is a $k \times 1$ vector of parameter estimates for these covariates, θ_j is a random effect for the jth grid cell capturing extra-binomial variability over the entire study region at the individual quarter township scale (HET; i.e., 25 km^2), and ϕ_j is a random effect term for the jth grid cell that models the extra-binomial variability associated with local disease clustering (CAR; i.e., grid cells closer together will have similar infection probabilities due to proximity of cattle farms and pastures).

Prior distributions

We assumed non-informative $\mathcal{N}(0, 100{,}000)$ prior distributions for each of the β parameters, and an improper (flat) prior over the entire real line for μ. For the random effect describing region-wide heterogeneity (HET), we assumed the following:

$$\theta_j \overset{iid}{\sim} N\left(0, \sigma_h^2\right). \quad (2)$$

To describe the spatial structure, we assumed an intrinsic Gaussian conditional autoregressive prior with a sum to zero constraint [31] for the local clustering random effect (CAR):

$$\varphi_j|\varphi_{j+} \overset{iid}{\sim} N\left(\frac{1}{n_{j+}}\sum_{i\text{in}j+}\varphi_i, \frac{\sigma_c^2}{n_{j+}}\right), \quad (3)$$

where n_{j+} is the number of grid cells that share a border or vertex with the jth grid cell. Thus, the random effect of the jth grid cell is conditional on the values of its n_{j+} (usually = 8) neighboring cells. Adjacency matrices were created with the Adjacency for WinBUGS Tool in ArcMap that provides a matrix relating one areal unit to a collection of neighboring areal units in text files for use in WinBUGS.

Because of the marginal specification for σ_h^2 and conditional specification for σ_c^2 of the random effects, we generated prior distributions for the precisions (i.e., $\frac{1}{\sigma_h^2}$ and $\frac{1}{\sigma_c^2}$) using simulations in WinBUGS where we varied the values of the parameters and determined the parameter values that created an expectation of ~0.5 for the *psi* metric [38], where *psi* is as follows:

$$psi = \frac{\sigma_c}{\sigma_c + \sigma_h} \quad (4)$$

These simulations attempted to ensure an equal emphasis on the priors of the standard deviations of the random effects [31]. Based on our simulation results we defined

$$\frac{1}{\sigma_h^2} \sim \text{Gamma}(10.368,3.22) \text{ and } \frac{1}{\sigma_c^2} \sim \text{Gamma}(1.0,1.0).$$

Model Selection

To test our original hypothesis that environmental variables would influence the odds of *M. bovis* infection, our set of candidate models for logistic regression consisted of 12 different structures with strictly additive effects on the logit scale (Table 1). The 12 logistic regression models represented all possible combinations of deer and environmental variables, as well as inherent regional and local spatial structure of the data. Viewing these 12 models as competing hypotheses, we used deviance information criterion (DIC) [15,33,39] to compare the models' respective fits to the data from sampled farms, and then estimated parameters and examined goodness-of-fit and other metrics for the top models. For model comparison we used DIC weights [33], which allow for an intuitive comparison of the evidence in the data for each candidate model. The weights are considered a measure of the strength of evidence in the data for i^{th} model being the "best" model of those within the candidate set, and therefore provide a measure of model selection uncertainty [33,39].

We used parameter estimates from the top model to calculate odds ratios for the effect of variables on *M. bovis* infection odds among farms in that area. Model averaging was not appropriate because DIC, unlike BIC and AIC, is not based on any assumption of a "true" model and is primarily concerned with short-term predictive ability [33]. We treated host-level (deer prevalence, deer density) and environmental-level predictors (percent sand, soil pH, proportion of wetland/swamp, and area that ponded frequently) as a group of variables, such that they were all entered or removed from the models together (Table 1). To examine the goodness-of-fit of the top model from our candidate set we conducted a numerical posterior predictive check [32]. We examined correlation and trace plots, as well as the estimates of the corrected scale reduction factor for each parameter and multivariate potential scale reduction factors and determined that that the three chains for each model had converged (data not shown). For each dataset, the top models selected via our model selection procedures and their corresponding estimates were similar regardless of prior specification (data not shown; [33]).

Table 1. Model selection results for the candidate set of models investigating the effect of covariates on the probability of bovine tuberculosis infection from 2005–2010 in Modified Accredited Zone in Michigan, USA using non-informative *N* (0, 0.00001) prior distributions for the fixed effects parameters and diffuse gamma priors for the random effects with farm-level factors removed.

Model Terms	Dbar	Dhat	pD	DIC	ΔDIC	Weights
Deer + ------ + HET + Envir	272.9	226.4	46.5	319.4	0.0	0.6386
Deer + ------ + HET + ------	276.2	231.5	44.7	320.9	1.5	0.3073
------ + ------ + HET + Envir	275.9	227.5	48.4	324.4	4.9	0.0540
Deer + CAR + ------ + ------	289.0	235.5	53.5	342.4	23.0	0.0000
------ + CAR + ------ + Envir	290.5	231.7	58.8	349.3	29.8	0.0000
Deer + CAR + ------ + Envir	298.1	232.0	66.1	364.2	44.7	0.0000
Deer + CAR + HET + ------	302.3	224.8	77.5	379.8	60.4	0.0000
------ + CAR + HET + Envir	309.3	226.2	83.1	392.3	72.9	0.0000
Deer + ------ + ------ + Envir	387.9	380.9	7.0	394.9	75.5	0.0000
Deer + ------ + ------ + ------	394.2	391.2	2.9	397.1	77.6	0.0000
------ + ------ + ------ + Envir	405.5	400.7	4.8	410.3	90.8	0.0000
Deer + CAR + HET + Envir	317.1	222.1	95.0	412.0	92.6	0.0000

"Deer" represents deer herd factors: apparent prevalence of deer and deer density. "Envir" represents the environmental variables: percent sand, percent ponding frequently, percent swamp/wetland, and mean soil pH in each sampled farms quarter township grid cell. "HET" represents the random effect capturing region-wide heterogeneity and "CAR" is the random effect capturing local clustering.

Results

Out of the 762 cattle farms tested on an annual basis, 704 were negative while 37, 9, and 1 tested positive for *M. bovis* on 1, 2, and 3 occasions, respectively. Of our 12 models determined *a priori*, the top two models combined to account for over 95% of the summed weights of all models considered. Models weights of 95% provided strong assurance that some combination of these 2 models and their parameters reflected the underlying infection-generating process far better than other models in the candidate set (Table 1). Parameters in the top model included deer herd factors and local landscape features suggesting that these factors increased the odds of *M. bovis* infection to cattle in the northern, lower peninsula of Michigan (Table 2). As documented since initial diagnosis of a positive white-tailed deer in 1975, deer apparent prevalence ranged from 0.0% to 5.2% and was the most supported variable in the top model (odd ratio = 1.004, 95% CI = 1.001 to 1.007; Table 2). Sand within the vicinity of sampled farms was by far the most supported environmental variable and ranged from 37% to 79% on farms that test positive whereas sand ranged from 17% to 88% for cattle farms that tested negative for *M. bovis* (Fig. 1). The odds of infection for *M. bovis* increased by about 4% for every 1% increase in sand in the area (odd ratio = 1.036, 95% CI = 1.01 to 1.07; Table 2).

Our analyses also identified that an unstructured random effect (HET) dominated over spatial structure (CAR) in influencing the odds of *M.bovis* infection (odd ratio = 3.36, 95% CI = 1.69 to 10.91; Table 2). Inclusion of the unstructured random effect in both our top models would suggest that additional covariates are driving odds of *M. bovis* infection and not spatial occurrence of *M. bovis*-positive farms in our study area.

Discussion

Our findings support the premise that deer herd-related factors play an important role in sustaining *M. bovis* presence in the northern, lower peninsula of Michigan similar to that found in previous research [11,40]. Because mitigation measures have been implemented to reduce deer access to feeding and cattle use areas [3], we focused our analysis simply on deer density and deer prevalence in the area without reference to farm practices. Farm practices have been the primary focus of most efforts to control *M. bovis* transmission between reservoirs and hosts [11,18,41,42] but are difficult to standardize and document for inclusion in modeling efforts. Bayesian hierarchical models provide the ability to assess spatially the influence of region-wide cattle farm and host-level variables while adjusting for additional covariates [31,43]. The bias that may have been introduced by entering 36% of cattle farms more than once into our models was deemed warranted to achieve our objectives of assessing environmental factors that may lead to continued presence of *M. bovis* on cattle farms. Furthermore, we don't deny that farm-management practices

Table 2. Mean parameter estimates, standard deviation (SD), Monte Carlo error (MC error), odds ratios (OR), and 95% credible intervals for best-fitting model investigating the effect of covariates on the probability of bovine tuberculosis infection from 2005–2010 in Modified Accredited Zone in Michigan, USA.

Parameter	Mean	SD	MC error	2.50%	Median	97.5%	OR	95% CI
Intercept	−3.401	2.886	0.02	−9.219	−3.317	2.031	0.0333	0.00 to 7.622
Deer density	−0.2219	0.1454	0.00	−0.515	0.219	0.056	0.8001	0.60 to 1.06
Deer prevalence	0.4147	0.1412	0.00	0.137	0.414	0.697	1.004	1.001 to 1.007
Percent wetland	−0.0209	0.0262	0.00	−0.074	−0.020	0.029	0.9793	0.93 to 1.03
Percent sand	0.0357	0.0152	0.00	0.007	0.035	0.067	1.0363	1.01 to 1.07
Soil pH	0.04212	0.3368	0.00	−0.591	0.035	0.732	1.0430	0.55 to 2.08
Percent ponding	0.0240	0.035	0.00	−0.043	0.023	0.095	1.0243	0.96 to 1.10
HET	1.213	0.4819	0.00	0.524	1.128	2.39	3.3636	1.69 to 10.91

"HET" represents the random effect capturing region-wide.

are important in *M. bovis* infection on cattle farms, however, it is very difficult to accurately represent or measure one of these farm practices in a standardized format for hundreds of farms to include in our models. For this reason, we selected objectives that would look at determining new facet in understanding *M. bovis* infection (i.e., environmental variables) as opposed to conducting another study that suggested a different farm practice was responsible for bTB infection in cattle farms as documented in previous research [11,18,41,42].

Although deer densities in the area have been reduced to about 10–15 deer/km^2 since 1994 and deer prevalence has remained at just below 2% overall at the DMU 452-level [10], the role of a primary host for *M. bovis* in this region is still influencing transmission of *M. bovis* based on our study. Including mean deer prevalence for the duration of the study (i.e., 1998–2009) in our models would likely have yielded similar results to our current modeling effort but the role of the host would still be supported nonetheless. Due to public pressure about low deer densities and continued occurrence of cattle positive for *M. bovis* on an annual basis, reducing deer densities further is likely not possible. Relatedly, even with our conservative estimate of prevalence of *M. bovis* in deer (i.e., 2005–2009), deer prevalence was in the most supported model even though the odds ratio would likely have been greater if we included 1998–2009.

Our study identified environmental variables that were not possible to assess in previous modeling efforts that can further assist agencies in their attempts to eradicate *M. bovis* in Michigan. Environmental variables have been documented to contribute to presence or viability of infectious agents of disease in several areas in North American and Europe [17,44,45] although environmental sampling has yet to identify *M. bovis* on cattle farms in this region [46,47]. Variables to consider should be based on *a priori* knowledge of the disease agent studied and mechanisms that may hinder or promote survival. Survival of *M. bovis* has been linked to moist, humid environments that maintain the proper soil type and pH [48–50]. Our top model indicated that a combination of landscape variables played an important role in determining infection probability for *M. bovis* on farms which was the impetus for us to select a combination of covariates that were conducive to moist, humid environments with low sunlight exposure on the landscape (e.g., wetlands, ponding frequency, soil types). Percent sand was a significant predictor and increased the odds of *M. bovis* infection by 3.6% for every 1% increase in proportion of sandy soil in the local area (Table 2). Survival of *M. bovis* has been linked to soil type, temperature and pH [3,29,48] but research in natural settings has limited the advancement of knowledge in this area.

The exact composition of sandy soils or functional role of these soils that make them conducive to the survival of *M. bovis* likely requires further research. Sandy soils are defined as having loose particle sizes (> 2 μm) with structural integrity during desiccation that may promote survival of *M. bovis* when associated with wetlands and areas that routinely have standing water. Moraines that have steep slopes and sandy, well-drained soils dominated by northern hardwood forests were linked to infection of white-tailed with *M. bovis* [51] but are confounded by the fact that they characterize preferred deer habitat in years with heavy oak mast production. Based on prevalence studies on white-tailed deer and our current study of farms positive for *M. bovis*, environmental or landscape-specific characteristics would appear to be a logical focus of future studies and potential assessment of viable *M. bovis* detection in soils or water. Areas that contain these sandy, well-drained soils in northern hardwood forests likely provide moist, humid microclimates conducive to survival of *M. bovis* for extended periods of time and should be considered for focused surveillance for *M. bovis* in deer and cattle as well as the focus of future on-farm mitigation measures.

Since initial detection of *M. bovis* in farms over a decade ago, farms have tested positive on an annual basis. Even after considerable efforts have been implemented to reduce deer densities, limiting or preventing aggregation of deer (e.g., ban on baiting of deer), and limiting deer-cattle interactions through on-farm mitigation measures, Michigan still does not have *M. bovis*-free status. Although modeling efforts are unable to include movements of cattle between farms and its influence on movements and spread of *M. bovis*, repeated positive tests have occurred on numerous farms since the first farm tested positive in 1998 [28]. Repeated positive tests would suggest that potentially an environmental source conducive to survival of the bacteria may be responsible for maintaining *M. bovis* in the region. Mycobacteria have waxy, lipid-rich cell walls that are relatively resistant to biocides used in decontamination procedures thus complicating management of the disease.

Unlike previous work in this region, we were able to assess spatial processes that may be influencing the transmission or presence of *M. bovis* using a Bayesian hierarchical modeling framework [31,43]. We identified that unstructured spatial heterogeneity (HET) was included in the top two models explaining infection of *M. bovis* on farms. If probability of infection was driven by spatial structure or clustering of disease (i.e., contiguous grid cells more alike than 2 arbitrary grid cells) at our site, we would have expected spatial structure (CAR) to be included in our top models but the opposite occurred [43]. Unstructured spatial heterogeneity would suggest that additional covariates that we may not have accounted for in our models were also influencing the disease across our study region. As stated previously, movement of cattle between farms and additional on-farm practices is controlled or mitigated to some extent for some farms but is very difficult to enforce and has been documented to be the cause of contamination of several cattle farms within and outside of the MAZ [21]. Our unstructured spatial heterogeneity could simply be on-farm practices not included in our modeling effort or additional environmental covariates conducive to survival or destruction of *M. bovis* such as slope/aspects conducive to direct exposure to ultraviolet light.

Although host prevalence in DMU 452 and clustering of cattle farms were considered important in *M. bovis* infection in previous research [11,52], our results suggest that less spatially structured components in the landscape are influencing continued occurrence of *M. bovis* in cattle on farms in Michigan. Considering the logistics of managing the host and reservoir through the northern, lower peninsula of Michigan, management efforts could focus on environmental and landscape characteristics that are potentially supporting the continued presence of *M. bovis* in Michigan. Landscape characteristics that are conducive to survival of *M. bovis* such as habitat inundated with standing water, soil composition, and prime deer habitat should be the focus of future management and research. On-farm mitigations should focus efforts in areas at high risk for continued survival of *M. bovis* prior to mandating complete risk mitigation of farms for an entire area such as DMU 452 (1,479 km^2) or the 5 county study area (8,062 km^2). Focused surveillance and management of on-farm practices in reservoirs for disease in domestic livestock would provide a logistically feasible approach to combating a disease in an endemic area as well as new areas that the disease has recently been introduced or spread from the endemic area.

Acknowledgments

We thank the Michigan Department of Natural Resources for deer-related data used in this manuscript. We thank Daniel J. O'Brien, Michigan Department of Natural Resources, for a helpful review of an earlier draft of this manuscript. Any use of trade, firm, or product names is for descriptive purposes only and does not imply endorsement by the U.S. Government.

Author Contributions

Conceived and designed the experiments: WDW RS MV KCV. Performed the experiments: WDW RS. Analyzed the data: WDW. Wrote the paper: WDW RS MV KCV.

References

1. de Lisle GW, Bengis RG, Schmitt SM, O'Brien DJ (2002) Tuberculosis in free-ranging wildlife: detection, diagnosis and management. Review Science Technology, OIE 21: 317–334.
2. Kaneene J, Thoen CO (2004) Tuberculosis. Journal of the American Veterinary Medical Association 224: 685–691.
3. Walter WD, Anderson CW, Smith R, Vanderklok M, Averill JJ (2012) On-farm mitigation of transmission of tuberculosis from white-tailed deer to cattle: literature review and recommendations. Veterinary Medicine International 2012: 1–15.
4. Clifton-Hadley RS, Wilesmith JW, Stuart FA (1993) Mycobacterium bovis in the European badger (Meles meles): epidemiological findings in tuberculosis badgers from a naturally infected population. Epidemiology and Infection 111: 9–19.
5. Griffin JM, Hahesy T, Lynch K, Salman MD, McCarthy J, et al. (1993) The association of cattle husbandry practices, environmental factors and farmer characteristics with the occurrence of chronic bovine tuberculosis in dairy herds in the Republic of Ireland. Preventive Veterinary Medicine 17: 145–160.
6. Corner LAL (2006) The role of wild animal populations in the epidemiology of tuberculosis in domestic animals: How to assess the risk. Veterinary Microbiology 112: 303–312.
7. Hill JA (2005) Wildlife-cattle interactions in northern Michigan: implications for the transmission of bovine tuberculosis Thesis. Logan: Utah State University. 1–58 p.
8. Berentsen AR, Miller RS, Misiewicz R, Malmberg JL, Dunbar MR (2013) Characteristics of white-tailed deer visits to cattle farms: implications for disease transmission at the wildlife-livestock interface. European Journal of Wildlife Research: In press.
9. O'Brien DJ, Schmitt SM, Fierke JS, Hogle SA, Winterstein SR, et al. (2002) Epidemiology of Mycobacterium bovis in free-ranging white-tailed deer, Michigan, USA, 1995-2000. Preventive Veterinary Medicine 54: 47–63.
10. O'Brien DJ, Schmitt SM, Fitzgerald SD, Berry DE (2011) Management of bovine tuberculosis in Michigan wildlife: Current status and near term prospects. Veterinary Microbiology 151: 179–187.
11. Kaneene JB, Bruning-Fann CS, Granger LM, Miller R, Porter-Spalding BA (2002) Environmental and farm management factors associated with tuberculosis on cattle farms in northeastern Michigan. Journal of the American Veterinary Medical Association 221: 837–842.
12. Williams RS, Hoy WA (1930) The viability of B. tuberculosis (bovinus) on pasture land, in stored faeces and in liquid manure. Journal of Hygeine 30: 413–419.
13. Van Donsel DJ, Larkin EP (1977) Persistence of Mycobacterium bovis BCG in soil and on vegetables spray-irrigated with sewage effluent and sludge. Journal of Food Protection 40: 160–163.
14. Carstensen M, DonCarlos MW (2011) Preventing the establishment of a wildlife disease reservoir: a case study of bovine tuberculosis in wild deer in Minnesota, USA. Veterinary Medicine International 2011: 1–10.
15. Farnsworth ML, Hoeting JA, Hobbs NT, Miller MW (2006) Linking chronic wasting disease to mule deer movement scales: a hierarchical bayesian approach. Ecological Applications 16: 1026–1036.
16. Osnas EE, Heisey DM, Rolley RE, Samuel MD (2009) Spatial and temporal patterns of chronic wasting disease: fine-scale mapping of a wildlife epidemic in Wisconsin. Ecological Applications 19: 1311–1322.
17. Walter WD, Walsh DP, Farnsworth ML, Winkelman DL, Miller MW (2011) Soil clay content underlies prion infection odds. Nature Communications 2: 1–6.
18. Hutchings MR, Harris S (1997) Effects of farm management practices on cattle grazing behaviour and the potential for transmission of bovine tuberculosis from badgers to cattle. The Veterinary Journal 153: 149–162.
19. Mathews F, Lovett L, Rushton S, Macdonald DW (2006) Bovine tuberculosis in cattle: reduced risk on wildlife-friendly farms. Biology Letters 2: 271–274.
20. O'Brien DJ, Schmitt SM, Berry DE, Fitzgerald SD, Vanneste JR, et al. (2004) Estimating the true prevalence of Mycobacterium bovis in hunter-harvested white-tailed deer in Michigan. Journal of Wildlife Diseases 40: 42–52.
21. Okafor CC, Grooms DL, Bruning-Fann CS, Averill JJ, Kaneene JB (2011) Descriptive epidemiology of bovine tuberculosis in Michigan (1975–2010): lessons learned. Veterinary Medicine International 2011.
22. Hughey BD (2003) Are there "hot spots" of bovine tuberculosis in the free-ranging white-tailed deer (Odocoileus virginianus) herd of northeastern Michigan? Thesis. East Lansing: Michigan State University. 1–86 p.
23. Okafor CC, Grooms DL, Bruning-Fann CS, Averill JJ, Kaneene JB (2011) Descriptive epidemiology of bovine tuberculosis in Michigan(1975-2010): lessons learned. Veterinary Medicine International 2011: 874924.
24. Eberhardt LL (1960) Estimation of vital characteristics of Michigan deer herds. East Lansing: Department of Conservation.
25. Creed WA, Haberland F, Kohn BE, McCaffery KR (1984) Harvest management: the Wisconsin experience. In: Halls LK, editor. White-tailed deer ecology and management. Harrisburg.
26. Phillips CJC, Foster CRW, Morris PA, Teverson R (2002) Genetic and management factors that influence the susceptibility of cattle to Mycobacterium bovis infection. Animal Health Research Reviews 3: 3–13.
27. Mitscherlich E, Marth EH (1984) Microbial survival in the environment: bacteria and rickettsiae important in human and animal health. Berlin: Springer-Verlag. 1–802 p.
28. Miller R, Kaneene JB (2006) Evaluation of historical factors influencing the occurrence and distribution of *Mycobacterium bovis* infection among wildlife in Michigan. American Journal of Veterinary Research 67: 604–615.
29. Phillips CJC, Foster CRW, Morris PA, Teverson R (2003) The transmission of Mycobacterium bovis infection to cattle. Research in Veterinary Science 74: 1–15.
30. USDA NRCS (2007) Soil Data Viewer 5.2 User Guide. United States Department of Agriculture, Natural Resources Conservation Science.
31. Banerjee S, Carlin BP, Gelfand AE (2004) Hierarchical modeling and analysis for spatial data. New York: Chapman and Hall/CRC. 1–448 p.
32. Gelman A, Carlin JB, Stern HS, Rubin DB (2004) Bayesian data analysis. New York: Chapman and Hall/CRC. 1–696 p.

33. Spiegelhalter DJ, Best NG, Carlin BP, van der Linde A (2002) Bayesian measures of model complexity and fit. Journal of the Royal Statistical SocietySeries B (Statistical Methodology) 64: 583–639.
34. Spiegelhalter D, Thomas A, Best N, Lunn D (2003) WinBUGS Version 1.4 user manual. Cambridge: MRC Biostatistics Unit. 1–60 p.
35. Smith BJ (2007) boa: an R package for MCMC output convergence assessment and posterior inference. Journal of Statistical Software 21: 1–37.
36. Gelman A, Rubin DB (1992) Inference from iterative simulation using multiple sequences. Statistical Science 7: 457–472.
37. Brooks SP, Gelman A (1998) General methods for monitoring convergence of iterative simulations. Journal of Computational and Graphical Statistics 7: 434–455.
38. Eberly LE, Carlin BP (2000) Identifiability and convergence issues for Markov chain Monte Carlo fitting of spatial models. Statistics in Medicine 19: 2279–2294.
39. Burnham KP, Anderson DR (2002) Model selection and multimodel inference: a practical information-theoretic approach. New York: Springer-Verlag. 1–488 p.
40. Miller R, Kaneene JB, Fitzgerald SD, Schmitt SM (2003) Evaluation of the influence of supplemental feeding of white-tailed deer (Odocoileus virginianus) on the prevalence of bovine tuberculosis in the Michigan wild deer population. Journal of Wildlife Diseases 39: 84–95.
41. Knust BM, Wolf PC, Wells SJ (2011) Characterization of the risk of deer-cattle interactions in Minnesota by use of an on-farm environmental assessment tool. American Journal of Veterinary Research 72: 924–931.
42. Brook RK (2010) Incorporating farmer observations in efforts to manage bovine tuberculosis using barrier fencing at the wildlife-livestock interface. Preventive Veterinary Medicine 94: 301–305.
43. Besag J, York J, Mollie A (1991) Bayesian image restoration, with two applications in spatial statistics. Annals of the Institute of Statistical Mathematics 43: 1–59.
44. Imrie CE, Korre A, Munoz-Melendez G (2009) Spatial correlation between the prevalence of transmissible spongiform diseases and British soil geochemistry. Environmental Geochemistry and Health 31: 133–145.
45. Fine AE, Bolin CA, Gardiner JC, Kaneene JB (2011) A study of the persistence of Mycobacterium bovis in the environment under natural weather conditions in Michigan, USA. Veterinary Medicine International 2011: 1–12.
46. Witmer G, Fine AE, Gionfriddo J, Pipas M, Shively K, et al. (2010) Epizootiological survey of bovine tuberculosis in northern Michigan. Journal of Wildlife Diseases 46: 368–378.
47. Fine AE, O'Brien DJ, Winterstein SR, Kaneene JB (2011) An effort to isolate Mycobacterium bovis from environmental substrates during investigations of bovine tuberculosis transmission sites (cattle farms and wildlife areas) in Michigan, USA. ISRN Veterinary Science.
48. Duffield BJ, Young DA (1985) Survival of Mycobacterium bovis in defined environmental conditions. Veterinary Microbiology 10: 193–197.
49. Young JS, Gormley E, Wellington EMH (2005) Molecular detection of Mycobacterium bovis and Mycobacterium bovis BCG (Pasteur) in soil. Applied and Environmental Microbiology 71: 1946–1952.
50. Jackson R, de Lisle GW, Morris RS (1995) A study of the environmental survival of Mycobacterium bovis on a farm in New Zealand. New Zealand Veterinary Journal 43: 346–352.
51. Miller R, Kaneene JB, Schmitt SM, Lusch DP, Fitzgerald SD (2007) Spatial analysis of Mycobacterium bovis infection in white-tailed deer (Odocoileus virginianus) in Michigan, USA. Preventive Veterinary Medicine 82: 111–122.
52. O'Brien DJ, Schmitt SM, Fitzgerald SD, Berry DE, Hickling GJ (2006) Managing the wildlife reservoir of Mycobacterium bovis: The Michigan, USA, experience. Veterinary Microbiology 112: 313323.

Conservation of Sandy Calcareous Grassland: What Can Be Learned from the Land Use History?

Anja Madelen Ödman*, Pål Axel Olsson

Biodiversity, Department of Biology, Lund University, Lund, Sweden

Abstract

Understanding the land use history has proven crucial for the conservation of biodiversity in the agricultural landscape. In southern Sweden, very small and fragmented areas of the disturbance-dependent habitat xeric sand calcareous grassland support a large number of threatened and rare plants and animals. In order to find out if historical land use could explain variation in present-day habitat quality, the land use on eight such sites was traced back to the 18th century and compared with key factors such as the amount of bare sand, lime content and P availability. There was no support for the common explanation of the decline in xeric sand calcareous grassland being caused by abandonment of agricultural fields during the last century. Instead, fertilization history was the main explanation for the difference in depletion depth of $CaCO_3$ seen between the sites. The decline in xeric sand calcareous grassland since the 18th century is most probably the result of the drastic changes in land use during the 19th century, which put an end to the extensive sand drift. Since cultivation was shown to have played an important role in the historical land use of xeric sand calcareous grassland, grazing alone may not be the optimal management option for these grasslands. Instead more drastic measures are needed to restore the high calcium content and maintain proper disturbance levels.

Editor: Peter Shaw, Roehampton university, United Kingdom

Funding: Financial support from The County Administrative Board, The Swedish Research Council, and the Gyllenstiernska Krapperup Foundation made this study possible. The funders had no role in study design, data collection and analysis, decision to publish, or preparation of the manuscript.

Competing Interests: The authors have declared that no competing interests exist.

* E-mail: Anja.Odman@biol.lu.se

Introduction

There is a growing awareness among conservation biologists regarding the importance of land use history and local management practices when implementing conservation and restoration measures for biodiversity and threatened species [1,2,3]. It is also recognised that land use changes long ago may have caused the vegetation patterns observed today [4,5]. For example, Eberhardt et al. [2] found that large areas of the sand forests at Cape Cod had been agricultural fields in the middle of the 19th century. They concluded that it was important to "mimic past agricultural practices in order to maintain and restore important sand-plain habitatsÓ. In particular for sandy areas, erosion and vegetation burial have been an important part of habitats history, and plants are adapted to this [6,7].

Throughout Europe, sandy grassland habitats support a large number of threatened species. Perhaps the best example of such sandy habitats is the threatened xeric sand calcareous grassland (Natura 2000 code 6120, 2002/83/EC Habitat Directive), which is home to several endangered species of vascular plants [8,9,10], bryophytes [11], fungi [12] and invertebrates [13]. These habitats are dry, open grasslands on calcareous, more or less humus-free, nutrient-poor and well-drained sandy soils with a discontinuous vegetation cover. The xeric sand calcareous grassland specialists are adapted to high pH [14] and low nutrient content [8,15], and many of the species depend on the open sand for regeneration [8,9]. In eastern Skåne (southern Sweden), these grasslands developed on lime-rich glaciofluvial sand [8,16], which was deposited by glacial meltwater when the ice retreated around 14,000 years ago, and later re-deposited by wind and waves [16]. Sites with natural erosion such as river banks, steep slopes, sand dunes and dune slacks have been proposed as sites where the type of species found in dry grasslands could have occurred before deforestation [17].

Before 3000 BC, eastern Skåne (Fig. 1) was predominantly covered by forest [18]. After 3000 BC, coppice agriculture led to an opening of the landscape, and there was a concentration of settlements to the flat coastal areas, and around 800 BC large areas of the coastal plains were deforested [18]. From 800 BC, permanent settlements replaced the mobile settlements and fields previously practiced [18]. At this time, intensified land use also opened up the landscape on other places in the Baltic region [19]. An infield/outland system with fertilisation was introduced in the area and there was an expansion of cereal cultivation, which also increased erosion. The traditional method of farming was extensive and with long (up to 30 years) periods of fallow when the arable fields were used for grazing [20,21,22]. During the 17th and 18th centuries the human population in southern Sweden grew, which caused further deforestation and intensified agricultural practices [20,22], and in the 18th century, Linné [23] noted an extensive sand drift and the occurrence of sand fields all over eastern Skåne. The problem with sand drift was for the most part solved at the end of the 18th century and during the 19th century by the introduction of ley plants and more efficient use of fertilisers, which increased the production per surface area and thereby stopped over-exploitation. At the same time, the afforestation of Skåne started [22] and pine was planted to

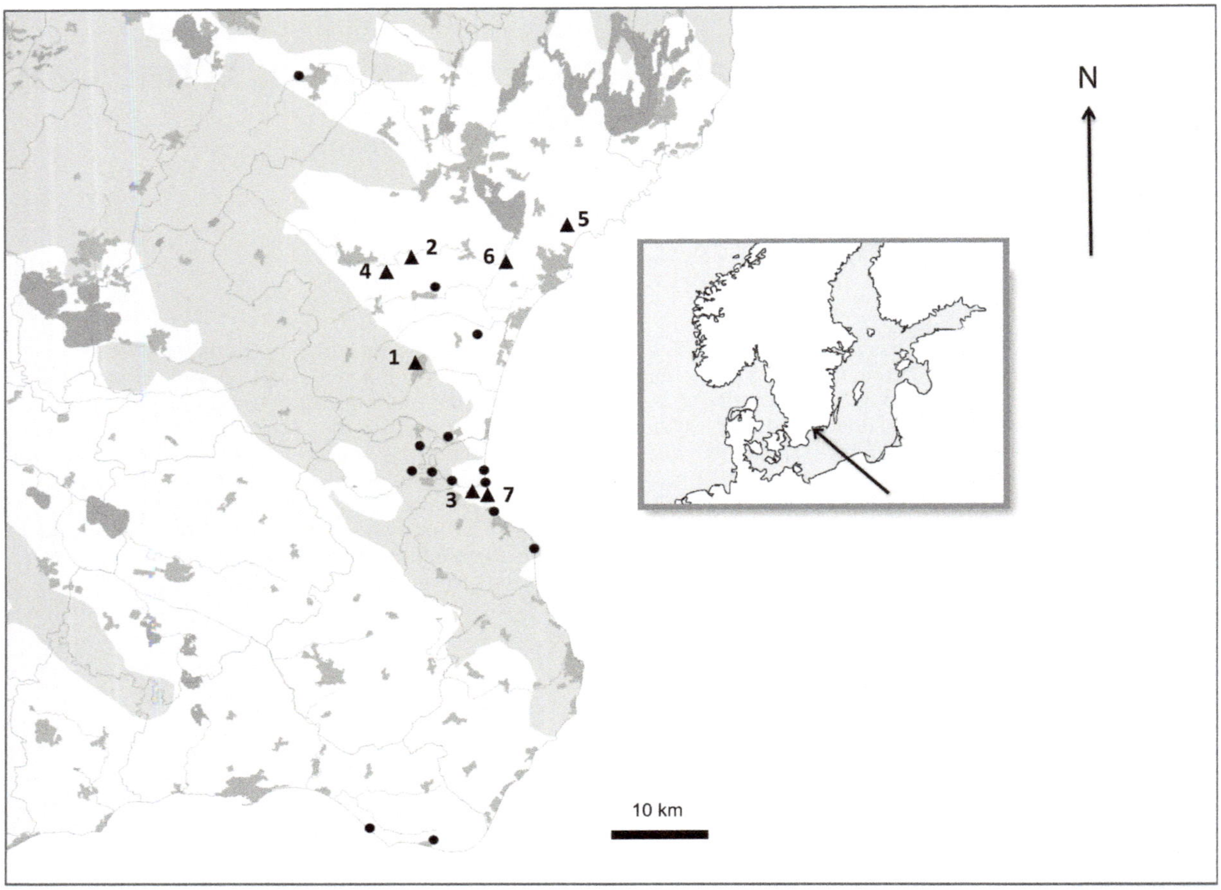

Figure 1. Map showing the existing xeric sand calcareous grassland sites in Skåne. Sites included in the study are represented by triangles and circles represent sites not included. Grey shading indicates forested land and unshaded represents open land. Scale bar represent 10 km. Enclosed is an overview map for southern Scandinavia, where an arrow points out the study area. The sites chosen for the study were: (1) Degeberga, (2) Everöd, (3) Klammersbäck, (4) Lyngsjö, (5) Rinkaby, (6) Ripa and (7) Vitemölla.

prevent wind erosion. Similar trends with deforestation until the middle of the 19th century, and then afforestation again has also been found in other places in Sweden [24] as well as in New England, USA [25].

Today, the old agricultural practices in sandy grasslands have been abandoned, and the arable fields have been turned into pine plantations or pastures [26,27], leaving the vegetation cover to close and thereby ending sand drift [28]. A natural accumulation of nutrients and organic matter follows [10], which can be further accelerated by atmospheric nitrogen deposition [29]. In addition, there is a threat of acidification of the topsoil [8]. The remaining areas of xeric sand calcareous grassland are small and fragmented, which poses a further threat to the rarest species (Fig 1).

Linné [23] described a number of xeric sand calcareous grassland species growing in eastern Skåne. In particular *Dianthus arenarius* ssp *arenarius*, which is a key species in these grasslands and protected by the Natura 2000 network (EU Habitat Directive, Annex II), was mentioned several times and was said to grow on all the sand fields, almost like a weed. In fact, all xeric sand calcareous grassland specialists but two, *Silene conica* and *Alyssum alyssoides*, which came to Skåne in the 19th century, were present in Skåne in 1749 [23,30,31]. Observations by Linné [23] and Campbell [20] suggest that xeric sand calcareous grassland was widespread in eastern Skåne during the 18th century, but there are no records of the actual area before the 1970s. Since the 1970s, reported areas range between 30 and 50 ha, with no consistent trend between the different studies [26, 27 32]. However, there is a strong indication of decline in xeric sand calcareous grassland species since the 18th century, which is reflected in the high density of nationally red-listed (and thereby declining) plant species [8].

Changes in land use during the last century, i.e. that the agricultural practices on these grasslands were abandoned, has been the common explanation for the decline of xeric sand calcareous grassland in Skåne [26 27, 33]. However, this hypothesis had never been tested prior to this study. The aim of this study was to improve the conservation of dry calcareous grassland by increasing the understanding of how xeric sand calcareous grasslands came to be and why they have disappeared. Land use change and decalcification were compared at eight sites in eastern Skåne in areas with a range of different quantities and qualities of xeric sand calcareous grassland. The amount of bare sand and the $CaCO_3$ content of the soil were used as proxies for favourable conditions and, in the case of open sand, as a proxy for erosion. Historical maps were used to determine the former land use and aerial photographs were used to map the extent of bare sand during the last 70 years. By analysing soil profiles, possible links between land use history and decalcification in the studied grasslands were investigated at specific sites within each studied landscape. Extractable P concentrations were also measured as a proxy for fertilisation history, since P remains in the soil for a longer period than N and causes problems during restoration [34,35].

Materials and Methods

Site descriptions

The sites studied were situated in the coastal area of eastern Skåne in southernmost Sweden. The area is 80 km long and 35 km wide, situated between 6172130 and 6204470 N, and 441330 and 458270 E (SWEREF99 TM). Sites were selected on the basis that they contained at least some remnants of xeric sand calcareous grassland vegetation. They should also be, or be part of, a larger open area. Since the focus of the study was the effects of agriculture, sites known to have been created by sand pits or other digging activities were excluded. Still, the majority of landscapes with xeric sand calcareous grassland in the area were included. The sites chosen for the study were: 1. Degeberga (6188081, 442627), 2. Everöd (6198995, 443126), 3. Klammersbäck (61749171, 448414), 4. Lyngsjö (6198615, 442108), 5. Rinkaby (6202616, 455831), 6. Ripa (6198003, 452133) and 7. Vitemölla (6173465, 449873) (Fig 1). The sites are listed in Table 1, where information about their topography, present day management and the amount and quality of the remaining xeric sand calcareous grassland vegetation can be found. These sites are all under nature conservation management, and the conducted study was made in collaboration with the county administrative board in Skåne and followed site-specific regulations. The Everöd, Klammersbäck, Lyngsjö, Rinkaby and Ripa sites are all more or less flat areas. The vegetation at these sites predominantly consists of Fennoscandian lowland species-rich, dry to mesic grassland (N6270, EU Habitat Directive) as well as small areas of heath-like vegetation resembling continental dunes with open Corynephorus and Agrostis grasslands (N2330, EU Habitat Directive) [8,28]. All of the flat sites contain patches of xeric sand calcareous grassland of varying sizes, depending on the level of degradation. Degeberga and Vitemölla are hilly sites with large presence of xeric sand calcareous grassland vegetation on the south facing slopes, and with Fennoscandian lowland species-rich, dry to mesic grassland on flatter areas.

Soil sampling and analyses

Soil was sampled at the following locations: Everöd (6199099, 443104), Klammersbäck (6174279, 448413), Lyngsjö (6198665, 442138), Rinkaby (6202600, 455808), Ripa pasture (6198191, 452028), and Ripa airfield (6197740, 452277) (SWEREF99 TM). At each of these sites a smaller area was selected that was assumed to consist of degraded xeric sand calcareous grassland. The vegetation at most sampling sites was either Fennoscandian dry to mesic grassland or dry Corynephorus grassland. The soil was sampled on 8-18th June 2009 and soil cores of 20 cm length (Ø 15 cm) were taken down to 1.6 m using a soil auger. Three profiles were sampled at each site, located approximately 10 m from each other in a triangle. At the Ripa airfield site, the soil could only be sampled down to 1.2 m for two of the three profiles because bedrock or boulders were encountered at this depth.

For pH measurements, 10 g of soil was mixed with 50 ml distilled water, and the pH was measured electrometrically in supernatants obtained by 2-h extraction in a rotator. The total amounts of Ca in the samples were measured by digesting the soil in concentrated HNO_3, followed by ICP-AES analysis. One g of soil at field moisture and 50 ml HNO_3 were placed in a 300 ml flask, covered with a watch glass and heated on a hot plate for

Table 1. A summary of the characteristics of the eight sites analyzed in this study, including the total area analysed and the area of xeric sand calcareous grassland (Natura 2000 habitat 6120) according to Olsson [32].

Site	Topography	Present day management/Disturbance	Area (ha)		Vegetation (vascular plants)				
			Total	Habitat type 6120	Total cover (%)	Specialists cover (%)	Red-listed cover (%)	Red-listed (Nr/m^2)	N
1. Degeberga	Hilly	Ungulate grazing	13.5	2.6	47±20	23±13	17±14	2.7±0.6	3
2. Everöd	Flat	Ungulate grazing	15	4.9	57±23	14±16	9.2±15	1.0±0.7	10
3. Klammersbäck	Flat	Ungulate grazing	38	0.09	34±16	4±1.4	2.5±3.5	0.5±0.7	2
4. Lyngsjö	Flat	Horse grazing	3	0.1	58±23	12±16	10±16	0.6±0.7	17
5. Rinkaby	Flat	Ungulate grazing/military training	155	4.5	64±20	21±15	14±12	1.4±0.8	25
6a. Ripa airfield	Flat	Fallow/wild grazers (rabbits)	14	0.2	71±23	20±26	17±29	0.3±0.6	3
6b. Ripa pasture	Flat	Ungulate grazing	22	3.0	52±6.4	7.0±5.7	5.5±6.4	1.0±0.0	2
7. Vitemölla	Hilly	Fallow/recreation, wild grazers (rabbits)	38	4.8	59±22	23±21	16±18	1.7±1.3	12

Vegetation data comes from a survey in 2004-2005 [9] and includes the total vegetation cover, as well as the cover of specialist species and nationally red-listed species as defined in Olsson and Ödman [55]. In addition, the number of red-listed species per square meter (Nr/m^2) is given. Means are displayed (± SD), and N denotes the number of analysed squares (1 m^2) at each site. The dominant red-listed and specialist species were *Koeleria glauca, Dianthus arenarius* ssp *arenarius* and *Alyssum alyssoides.*

72 h. The solution was evaporated to 2 ml and diluted to 25 ml with distilled water. Plant-available (extractable) ortho-phosphate (P_{ext}) in the top 20 cm was determined using flow injection analysis, with stannous chloride and ammonium molybdate as reactant (measured at 720 nm, method ISO/FDIS 15681-1). Ten g fresh soil was extracted for 1 min with Bray-1 solution [36].

Soil in Degeberga and Vitemölla was not sampled during this study and samples collected for earlier studies by Olsson et al. [8] and Bahr et al. [37] respectively were used instead. At Degeberga, two samples were used: one from Degeberga south slope (xeric sand calcareous grassland) and one from Degeberga north slope (Fennoscandian dry to mesic grassland), which had been sampled down to 1.5 m. At Vitemölla, four samples were used from a mosaic of xeric sand calcareous grassland and dry Corynephorus grassland. The samples were taken down to 0.6 m at three sites and 0.5 m for one of the sites. Data for soil pH and Ca content comes from data collected for earlier studies by Olsson et al. [8] and Bahr et al. [37]. At Vitemölla, Ca concentration values were only available for the top 0.1 m. Additional measurements of P_{ext} were performed on the soil samples for this study, using the methods described above. The samples from the top 0.1 m for all samples at Vitemölla had unfortunately been used up in the previous study, and the 0.1–0.2 m layer was used for the P_{ext} analysis.

Analysis of land use history

Information on land use in the past was obtained through visual analysis of historical maps and aerial photographs. The area at each site was delimited so that it enclosed the soil sampling plots, had clearly defined borders (arable field margins, roads, water etc.) and only included open land without trees in the 2007 aerial photograph. The sizes of the delimited areas are presented in Table 1. Historical maps were obtained from Lantmäteriet (the Swedish land surveying agency), thru a thorough search in their archives [38]. The historical maps used dated from the middle of the 18th century (oldest maps available from Lantmäteriet) until 1974 and consisted of economic and property redistribution maps (Table 2). The analyses used arable field margins and other objects to locate the sites, and explanatory text or legends were used to determine land use. The land use was either printed on the map or found in a supplementary document accompanying the map. The land uses recorded were: arable field, pasture, meadow, forest and other land uses of interest such as military training or recreation. Areas named "utmark" (outland), "fälad" (old word for common pasture in southern Sweden) and "allmänning" (common land) were classified as pastures, and areas named "wång" (old word for common arable fields in southern Sweden) were classified as arable fields. In the economic map from 1974, only arable fields were mapped and all other land uses were treated as one class (other). The aerial photographs used dated from the 1940s until 2007 (Table 2) and were obtained from Lantmäteriet [39]. The resolution varied between 0.2 m^2 and 1 m^2. The analysis, which was done visually, was based on the following criteria: forest (>50% tree cover), arable field (open, with furrows) or pasture (open, without furrows). Furrows were interpreted as signs of ploughing. The map and aerial photograph analyses were performed at each site, both for the whole area (landscape level) and at the exact sites (plot level) where the soil was sampled. The results from the analysis were used to calculate the number of years since the area was last used for cultivation. The number of years was counted from the last year the area was recorded as having been cultivated. In the case of Degeberga and Vitemölla, which had never been cultivated, the number of years from the first available record of land use was used.

Analysis of the amount of bare sand and forest

The aerial photographs described above were used to analyse how the amount of bare sand changed between the 1940s and 2007. The aerial photographs were rectified and classified using ArcGIS Desktop 9.3 (ESRI 2008). The delimited areas were the same as those described above for the historical land use analysis. The amount of bare sand was estimated by dividing the surface features shown on the aerial photographs into two classes: bare sand and other (anything which was not bare sand). The classes were identified separately for each photograph using supervised classification where areas of bare sand were identified based on colour (and previous knowledge of the areas for the 2007 photographs), and then used as training areas for the software. The delimited areas were then classified using maximum likelihood classification. The classification for 2007 was evaluated in the field by comparing classified and real patches, and the older ones by comparing the classified patches of bare sand with the visual ones in the aerial photographs. The analysis identified patches of bare sand, but did not consider the overall density of the vegetation. Consequently, only patches with close to 100% bare sand could be identified, and areas with partly bare sand of varying amounts were classified as "other" since they could not be distinguished.

The amount of forest (>50% tree cover) in the surrounding landscape was also analysed from the aerial photographs. The size of the aerial photographs varied considerably between sites and years, and four different areas around the sites, 1×1, 1.5×1.5, 2×2 and 3×3 km, were used. The Everöd and Lyngsjö sites are situated within the same area and were analysed together. Analysis showed that, in those areas where squares larger than 1×1 km could be fitted, the amount of forest for all years increased slightly with increasing size of areas (although not significantly so). However, differences between the years were constant and so only results from the analyses using the 1×1 km squares are presented. Polygons were drawn manually around all forest and the percentage of forest in the total area was calculated using ArcGIS.

Statistical analyses

Correlations between land use history, amount of forest, amount of bare sand, depletion depth for $CaCO_3$, maximum $CaCO_3$, pH and P_{ext} content were performed using Pearson correlation or Spearman's correlation when parameters were not normally distributed. For the amount of forest and amount of bare sand, correlations were also performed with year, using the same method as described above. Differences between sites, land use history and depths in terms of the amount of $CaCO_3$ and P were tested using a general linear model followed by Tukey's post hoc test. All statistical analyses were performed using IBM SPSS Statistics (IBM Corporation, New York, US). An Arcsine square root transformation of the parameter amount of bare sand was performed to obtain normal distribution. For the correlation between pH and $CaCO_3$, the $CaCO_3$ values were log transformed. Results presented are mean values and standard deviations.

Results

Historical land use changes

The results from the analysis of land use change at the soil sampling plots showed that the last record of cultivation at the six flat sites was between 1940 and 1957, except for the Ripa pasture, where indications of cultivation could be seen also in aerial photographs from 1985 (Table 2). At the sampling plots at the two

Table 2. Results from visual analysis of land use history through historical maps (economic and property redistribution maps, with original names in Swedish) and aerial photographs for all eight soil-sampling sites.

	1. Degeberga	2. Everöd	3. Klammersbäck	4. Lyngsjö	5. Rinkaby	6a. Ripa airfield	6b. Ripa pasture	7. Vitemölla
Aerial photographs								
2007	Pasture	Pasture	Pasture	Pasture	Military training/Pasture	Pasture	Pasture	Pasture/recreation
2001					Military training/Pasture			
1999	Pasture	Pasture	Pasture	Pasture		Pasture	Pasture	Pasture/recreation
1985	Pasture	Pasture	Pasture	Pasture	Military training/Pasture	Airfield/Pasture	Arable field?	Pasture/recreation
1970					Military training/Pasture	Airfield/Pasture	Arable field?	
1969	Pasture	Pasture	Pasture	Arable field				Pasture/recreation
1957	Pasture	Arable field		Arable field		Pasture?	Arable field	
1956			Pasture		Arable field			Pasture/recreation
1940	Pasture	Arable field	Arable field	Arable field	Arable field	Arable field	Arable field	Pasture/recreation
Historical maps								
1974, Ekonomiska kartan	Other	Other	Other	Other	Other	Other	Other	Other
1926-34, Härads ekonomiska	Pasture	Arable field	Arable field	Arable field	Arable field	Arable field	Arable field	Pasture
1845, Laga skifte					Plantation			Meadow, drift sand
1826, Enskifte		Arable field		Arable field				
1822, Enskifte			Arable field					
1818, Storskifte						Arable field	Arable field	
1811, Enskifte	Pasture							
1803, Avmätning		Arable field		Arable field				
1751, Ägodelning					Arable field			Pasture
Years since cultivation	196	50	67	38	51	67	22	256

"Other" means an area not classified as arable field or real estate in the economic maps (Ekonomiska kartan). Maps and aerial photographs were obtained from Lantmäteriet [38,39].

hilly sites, none of the maps or aerial photographs showed signs of agricultural cultivation (Table 2).

At the landscape level, the six flat areas, Everöd, Lyngsjö, Klammersbäck, Ripa pasture, Ripa airfield and Rinkaby, have all been cultivated but, according to the maps, cultivation ceased between 1940 and the 1990s. They have gone through a historical land use change from a mixture of cultivation and grazing and mowing, to a mixture of forest and grazing. During the 19th century arable fields were common, but at Everöd, Rinkaby and Klammersbäck, pastures occurred as well, and these pastures were managed as commons. At the two coastal sites Rinkaby and Klammersbäck, the common pastures occurred closest to the sea, while fields occurred inland. During the 20th century, the agricultural fields were gradually abandoned and turned into pastures. During this time, forests started to emerge at the Everöd, Rinkaby and Ripa sites.

The two hilly sites, Vitemölla and Degeberga, have a rather similar history. In 1756, the northern part of the site at Vitemölla was classified as very poor arable fields spoiled by sand drift and therefore unusable. The southern part was classified as useless and left as a common in 1778. In 1828 and 1845, the southern part of the area was described as sandy soil and drift sand, and the lower parts close to the sea as useless. According to the maps from 1926-34 and onward, the area was used as pasture and for recreation. The Degeberga site has been used as pasture continually from the first map in 1811 until today. Consequently, according to the maps used in this study, the two hilly sites were not cultivated during the analysed period. However, the note from 1756 about parts of Vitemölla being arable field spoiled by sand drift indicates that some parts had previously been cultivated.

Forest cover and amount of bare sand from 1940 to 2007 at landscape level

There was no significant temporal trend in the mean forest cover between 1940 and 2007, when analysed in 1×1 km squares within each landscape. In three areas (Ripa, Everöd/Lyngsjö and Rinkaby) there was an increase in forest cover, while the forest cover decreased in Klammersbäck, and almost no changes could be seen in the other two areas (Fig 2a). The area with the highest mean forest cover over the years ($n = 6$) was Klammersbäck (50±6%), followed by Rinkaby (44±4%), Degeberga (28±1%), Ripa (20±21%), Everöd/Lyngsjö (18±6%) and Vitemölla (11±2%).

The proportion of bare sand was analysed for eight sites, as Ripa airfield and Ripa pasture were treated separately, ranging in size between 3 and 155 ha. At Vitemölla, Rinkaby and Degeberga, the analysis showed a decrease in the amount of bare sand between 1940 and 2007 (Fig 2b). However, the general trend for all sites was that changes since the 1940s were not drastic, and that the mean amount of bare sand at the studied sites was never more than about 3% of the total study area. There was no significant change in the mean amount of bare sand between 1940 and 2007. The largest change was found at Vitemölla, where the amount of bare sand decreased from 11.2% to 4.4%. There was no

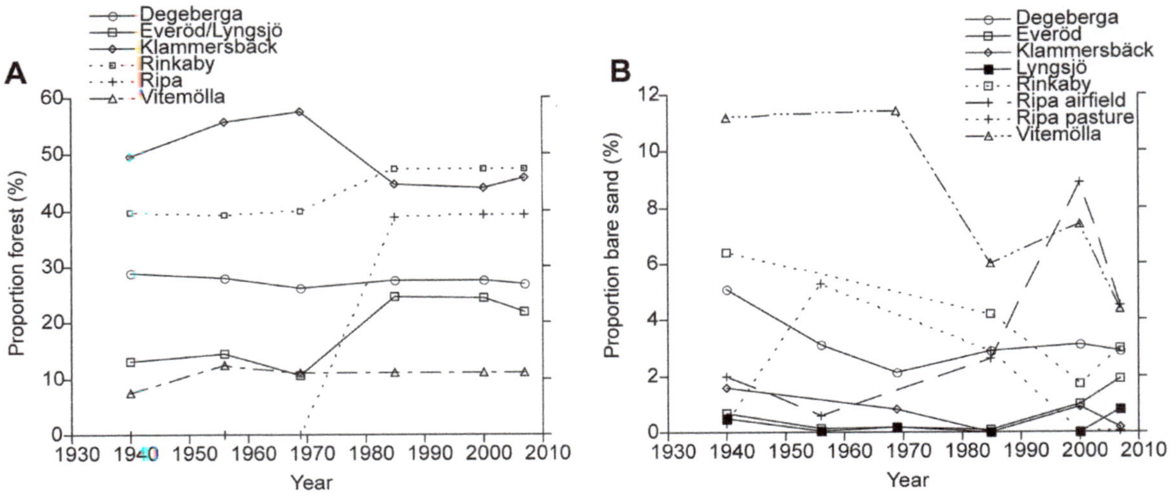

Figure 2. The proportion of land area occupied by forest (a) and bare sand (b) between 1940 and 2009 as analysed from aerial photographs (for simplification we did not differ between 1956/1957 and 1969/1970 in the graph). The proportion of forest was analysed for a 1×1 km square around the sites (Everöd/Lyngsjö and Ripa airfield/Ripa pasture were each contained within the same square), and the proportion of bare sand for the whole area of each of the eight sites.

correlation between the amount of bare sand and the forest cover or the number of years since cultivation.

Historical land use and soil properties

Extractable P (P_{ext}) showed large variation between sites. Lyngsjö had the highest mean ($n = 3$) concentration of P_{ext} (79 ± 18 μg g^{-1} soil), followed by Ripa pasture (51 ± 4.4 μg g^{-1} soil), Ripa airfield (48 ± 4.4 g^{-1} soil) and Everöd (47 ± 6.8 μg g^{-1} soil). Degeberga north (26 μg g^{-1} soil) and the four sampling points at Vitemölla (13 ± 8.2 μg g^{-1} soil) had intermediate concentrations, while the sites at Rinkaby (6.5 ± 4.4 μg g^{-1} soil), Degeberga south (6.3 μg g^{-1} soil) and Klammersbäck (5.1 ± 1.6 μg g^{-1} soil) showed considerably lower P_{ext} values than the other sites. Lyngsjö had significantly higher P_{ext} values than all the other sites ($p < 0.05$) and the Everöd, Ripa pasture and Ripa airfield sites had significantly higher P_{ext} values than Degeberga, Klammersbäck, Rinkaby and Vitemölla ($p < 0.05$). No significant differences were found between the other sites. There was a positive correlation between the P_{ext} and the depletion depth in the sites that had been cultivated ($N = 18$, $R^2 = 0.53$, $p < 0.001$, Fig. 3). There was no significant correlation between the P_{ext} and the number of years since cultivation or the amount of bare sand.

None of the sites had an enrichment of $CaCO_3$ at any particular depth. Most of the topsoils were depleted but, when the lime horizon was reached, the $CaCO_3$ content remained rather constant further down (Fig. 4). The lowest $CaCO_3$ values were found at Lyngsjö ($0.2 \pm 0.03\%$), which was also the only site where the $CaCO_3$ content never exceeded 10%, even in the deepest layer. The only sites with a $CaCO_3$ content in the topsoil above 2% were Rinkaby ($13.1 \pm 3.4\%$) and Vitemölla (5.4% in sampling point 2), and at 0.2–0.4 m only Everöd, Klammersbäck and Rinkaby had $CaCO_3$ concentrations above 2%. The pH values were strongly correlated with the $CaCO_3$ values ($N = 156$, $R^2 = 0.93$, $p < 0.001$).

The maximum $CaCO_3$ concentration for each profile was identified as well as the depletion depth (the first depth where the $CaCO_3$ concentrations exceeded 2%). The maximum $CaCO_3$ value was significantly higher at Rinkaby and Degeberga than at Everöd, Klammersbäck, Lyngsjö and Ripa pasture ($p < 0.05$). The maximum $CaCO_3$ value at Ripa airfield was significantly higher than that of Lyngsjö ($p < 0.001$), but did not differ from that of Rinkaby and Degeberga. Lyngsjö had a significantly lower maximum $CaCO_3$ value than all other sites except Klammersbäck ($p < 0.001$).

There was a positive correlation between the maximum $CaCO_3$ and the mean amount of open sand ($N = 7$, $R^2 = 0.73$, $p < 0.01$). The maximum $CaCO_3$ value and the depletion depth were not significantly correlated but there was a weak negative correlation between maximum $CaCO_3$ and the depth at which the maximum $CaCO_3$ was found ($N = 20$, $R^2 = 0.30$, $p < 0.01$). The maximum $CaCO_3$ and the pH in the top 20 cm were also positively correlated ($N = 20$, $R^2 = 0.28$, $p < 0.01$). There was no correlation between the number of years since cultivation and the depletion depth, or between the mean amount of bare sand and the depletion depth.

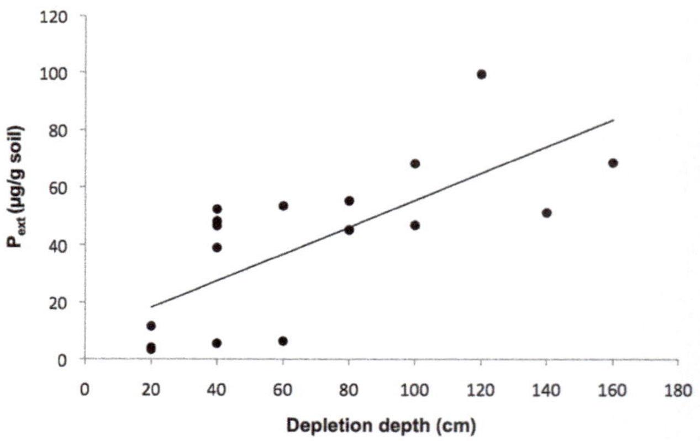

Figure 3. The correlation between P_{ext} (μg g-1 soil) and depletion depth (cm) at the previously cultivated sites ($N = 18$, $R^2 = 0.53$, $p < 0.001$). The number of years since cultivation was counted from the last year the area was recorded as having been cultivated. In the case of Degeberga and Vitemölla the number of years from the first available record of land use was used.

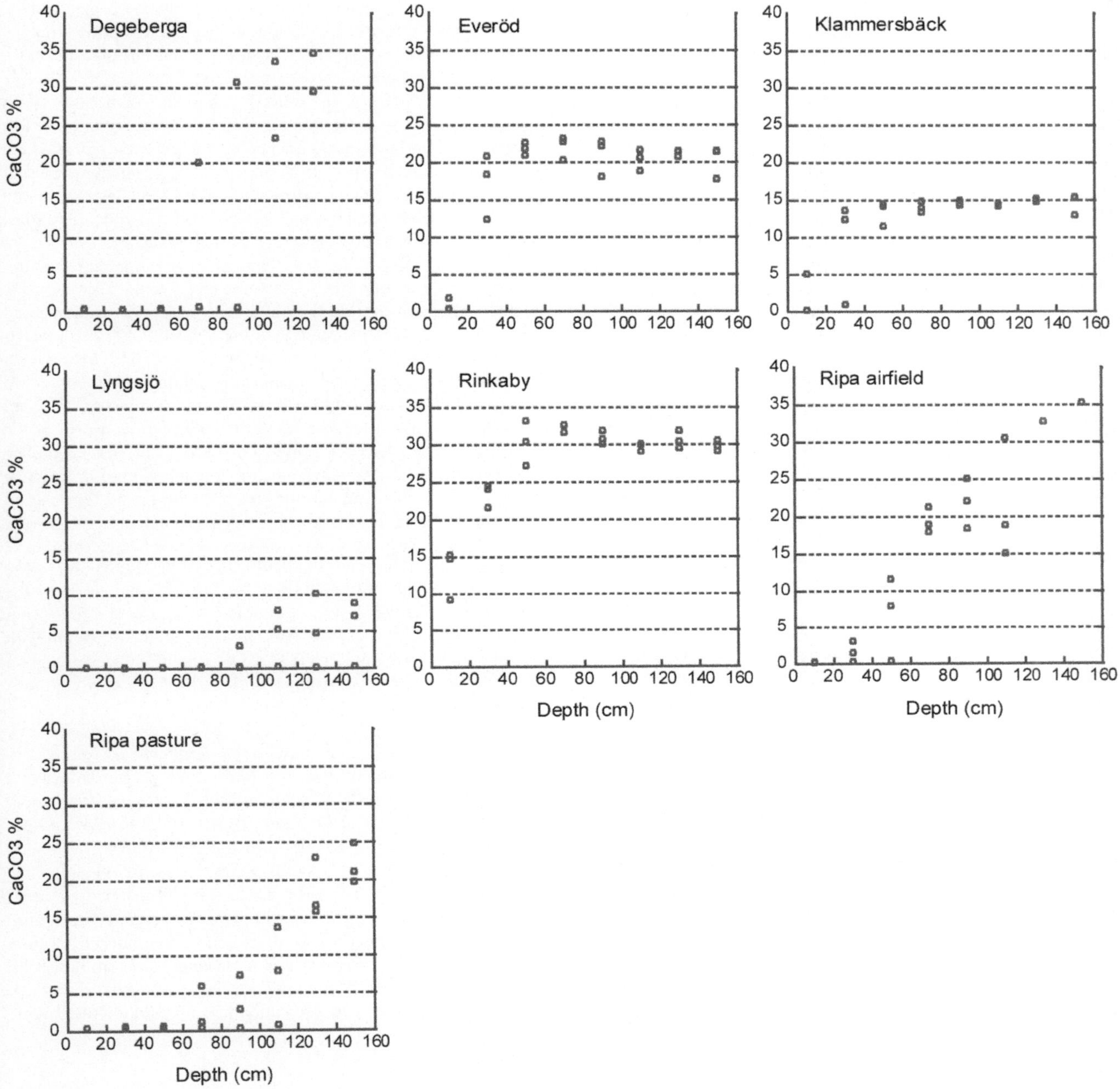

Figure 4. The $CaCO_3$ content as percentage of the soil weight for 20 cm at a time down to 160 cm. Vitemölla was not included since samples were only available for the top 20 cm.

Discussion

Most of the remaining xeric sand calcareous grasslands in Skåne are found on eroded sites [27], where the underlying lime-rich soil is exposed. This gives an indication about the great importance of erosion and other soil disturbances in the restoration and preservation of these grasslands. The importance of sand pits and military training areas for the conservation of biodiversity and threatened species has also been demonstrated in other parts of Europe [40,41,42,43]. An important finding of this study was that all the flat sites with xeric sand calcareous grassland included in the study had been cultivated in the past, which suggests that former agricultural practices were important in creating the habitat where natural erosion does not occur. Similarly, the chalk grasslands of Salisbury Plain in the UK, which were previously only thought to have a history as sheepwalks, were found to have a history of cultivation [44]. In other types of semi-natural grasslands, grazing continuity is the most important factor for diversity of grazing-dependent species [45,46]. In the case of sandy grasslands it may instead be that continuous traditional low-intensity cultivation is a very important factor in creating the proper conditions [47], particularly on flat areas where natural erosion is low when covered with vegetation and trampling is not enough to create bare soil. Therefore, the result that disturbance-dependent species were found in areas with long-term continuous soil disturbance created by cultivation should not be surprising, nor should the decline of dry sandy grassland species as soil disturbance decreased.

The cultivation of the soil is thought to have prevented the vegetation cover in the xeric sand calcareous grassland from closing and the $CaCO_3$ from being depleted by mixing the soil [26]. The highest average amount of bare sand during the period 1940 to 2007 was just over 3%, and there was no drastic change in the amount of bare sand during this period. The descriptions of Linné [23] and Campbell [20] indicate that there was much more bare sand in the area in the 18th century, and the most important decline probably occurred during the 19th century. Although there was a difference between the sites in amount of bare sand, there was no negative correlation between the amount of bare sand and the number of years since cultivation, which would be expected if lack of cultivation is an important factor in explaining recent declines in xeric sand calcareous grassland species. In fact, one of the sites not cultivated during the last 260 years had the highest amount of bare sand even in the 1940s. It must be acknowledged that there is a problem when comparing flat and hilly sites, as sloping ground will naturally be more prone to soil erosion. However, even when the two hilly sites were removed from the analysis, no correlation between bare sand and number of years since cultivation could be seen. There was a possible constraint to the method used in this study to detect the amount of bare sand. Assuming that cultivation was the important factor and it only occurred every 8–20 years, the analysis made using aerial photographs from just one date every fifteen years in this study could have failed to spot the bare sand produced. Completely bare sand might have been visible only in the first years, although the vegetation cover could have been sparse for many more. This might still have been enough of disturbance to sustain xeric sand calcareous grassland, and the failure to detect semi-bare sand was a weakness in the method used.

Another explanation for the decline in xeric sand calcareous grassland species could be nutrient enrichment. In the present study, some of the more recently cultivated sites had much higher P_{ext} value than sites that were cultivated a long time ago or not at all. However, Rinkaby and Klammersbäck stood out with their low P_{ext} values, suggesting that they have not been fertilized despite their recent cultivation history. The higher P_{ext} values found at some of the more recently cultivated plots could be a reason for the lack of correlation between the amount of bare sand and the number of years since cultivation, since high P values could speed up succession [48], thereby counteracting the effects of soil disturbance. In addition, the fertilisers would also have contained N, which is known to accelerate the decalcification process [49,50,51]. This was supported by the positive correlation between P_{ext} and depletion depth of $CaCO_3$, which suggests that fertilization could be the main explanation for the difference in depletion depth seen between the sites.

The depletion depth of $CaCO_3$ was not correlated with the time since cultivation, which would be expected if the important factor for retention of $CaCO_3$ in the topsoil was indeed cultivation. No correlation could be found between the mean amount of bare sand and the depletion depth, indicating a lack of connection between depletion and the amount of soil disturbance during the last century. The positive correlation between P and depletion depth of $CaCO_3$ would in fact suggest the opposite, i.e. that cultivation in the last 70 years, if including fertilization, might have had a negative effect. The maximum $CaCO_3$ value and the depth at which it was found were negatively correlated, although the correlation was weak. Still, this suggests that the depletion depth of $CaCO_3$ partly depended on the soil type. Different glaciofluvial deposits have different $CaCO_3$ content depending on the lithology of the local bedrock and the previously deposited soil types [16], which would explain the observed variation in $CaCO_3$ content between sites. None of the above-mentioned explanations seems entirely satisfactory in explaining differences in depletion depth between sites and they could not explain the presence of $CaCO_3$ in the topsoil of some sites. It is important to keep in mind that depletion has been going on throughout the 14,000 years since the soil was deposited by glacial melt-water. Consequently, there must have been a process counteracting the depletion.

The results discussed above indicate that ploughing was not directly preventing the $CaCO_3$ from being depleted. However, it is probable that the wind erosion resulting from the cultivation could have counteracted depletion by exposing the $CaCO_3$ rich sand. It seems to be a likely scenario, since wind erosion has already been found to remove fertile topsoil and cause organic matter content to decline [52,53]. Some sand drift still occurs in southern Sweden [22], but it is limited compared to the extensive sand drift of the 18th century described by Linné [23] and Campbell [20]. The most important factor determining the amount of sand drift is how much bare sand is available for erosion. During the time period from the 1940s until today, the amount of bare sand has been very low, but in the 18th century extensive areas of bare sand were observed due to the intensified cultivation [20,22,23]. One other factor that would affect wind erosion is the forest cover in the surroundings. Although the amount of forest at some sites increased during the last 70 years, no correlation was found with the amount of bare sand or the $CaCO_3$ content. However, a great change in forest cover took place in the 19th century, when forest was planted to bind the sand [22].

The results strongly suggest that the causes of present-day decline in threatened species may have origins far back in time and that conservation should take this into account. Human activity, such as expansion of arable fields and intensive livestock grazing, also caused sand drift in Germany [9,10], the UK [54] and the Netherlands [52]. In Germany and the Netherlands, dry sandy grasslands, including xeric sand calcareous grassland and Corynephorus grassland, are believed to have developed in areas where arable farming or livestock grazing created bare sand, which was then exposed to wind erosion and resulted in drifting sand [9,52]. Today, these sandy grasslands, in common with similar areas in Sweden, are suffering from changes in land use including afforestation, intensification and abandonment [9,52].

Conclusions

There was no support for the common explanation of the decline in xeric sand calcareous grassland being caused by abandonment of agricultural fields during the last century, i.e. that the old agricultural practices in these grasslands were abandoned. Instead, fertilization history (as indicated by soil P levels) was the main explanation for the difference in depletion depth of $CaCO_3$ seen between the sites. The decline in xeric sand calcareous grassland since the 18th century is most probably the result of the drastic changes in land use during the 19th century, which put an end to the extensive sand drift. The importance of sand movement is shown by the high density of red-listed vascular plant species at the hilly sites (Table 1). Since cultivation played a very important role in the creation of xeric sand calcareous grassland, grazing alone is not a good conservation option for these grasslands [28,55]. Instead more drastic measures are needed to restore the high calcium content, remove the accumulated nutrients and create bare sand for regeneration

[56]. An example of such a restoration measure is topsoil removal, where the decalcified and nutrient rich topsoil is removed much like it would be during sand drift [57]. Not surprisingly, topsoil removal has been shown to be a successful method for restoring xeric sand calcareous grasslands [55].

Acknowledgments

The authors would like to thank Merit Kindström and Micael Runnström for input on the methods used in the map and aerial photograph analysis, and for helpful comments on the manuscript.

Author Contributions

Conceived and designed the experiments: AMÖ PAO. Performed the experiments: AMÖ. Analyzed the data: AMÖ PAO. Contributed reagents/materials/analysis tools: AMÖ PAO. Wrote the paper: AMÖ PAO.

References

1. Poschlod P, WallisDeVries MF (2002) The historical and socioeconomic perspective of calcareous grasslands – lessons from the distant and recent past. Biol Conserv 104: 361–376.
2. Eberhardt RW, Foster DR, Motzkin G, Hall B (2003) Conservation of changing landscapes: vegetation and land-use history of Cape Cod National Seashore. Ecol Appl 13: 68–84.
3. Antrop M (2005) Why landscapes of the past are important for the future. Landscape Urban Plan 70: 21–34.
4. Motzkin G, Foster D, Allen A, Harrod J, Boone R (1996) Controlling site to evaluate history: vegetation patterns of New England sand plain. Ecol Monogr 66: 345–365.
5. Lunt ID, Spooner PG (2005) Using historical ecology to understand patterns of biodiversity in fragmented agricultural landscapes. J Biogeogr 32: 1859–1873.
6. Kent M, Owen NW, Dale P, Newnham RM, Giles TM (2001) Studies of vegetation burial: a focus for biogeography and biogeomorphology? Progress in Physical Geography 25: 455–482.
7. Owen NW, Kent M, Dale MP (2004) Plant species and community responses to sand burial on the machair of the Outer Hebrides, Scotland. J Veg Sci 15: 669–678.
8. Olsson PA, Mårtensson L-M, Bruun HH (2009) Acidification of sandy grasslands – consequences for plant diversity. Appl Veg Sci 12: 350–361.
9. Poschlod P, Baumann A, Biedermann H, Bugla B, Neugebauer K (2009) Dry sandy grasslands in Southern Germany – a case study of how to re-develop remnants to achieve the former status of a high nature value landscape. In Veen P, Jefferson R, de Smidt J, van der Straaten J, editors. Grasslands in Europe of High Nature Value. KNNV Publishing, the Netherlands. pp. 113–121.
10. Eichberg C, Storm C, Stroh M, Schwabe A (2010) Is the combination of topsoil replacement and inoculation with plant material an effective tool for the restoration of threatened sandy grassland? Appl Veg Sci 13: 425–438.
11. Tyler T (2005) The Bryophyte flora of Scanian sand-steppe vegetation and its relation to soil pH and phosphate availability. Lindbergia 30: 11–20.
12. Olsson PA, Schnoor TK, Hanson S-Å (2010) pH preferences of red-listed gasteromycetes in calcareous sandy grasslands: Implications for conservation and restoration. Fungal Ecol 3: 357–365.
13. Ljungberg H (1999) Skalbaggar och andra insekter på sandstäppslokaler i östra Skåne (Beetles and other insects on xeric sand calcareous grassland sites in eastern Scania). Länsstyrelsen i Skåne län, Malmö, 1–56.
14. Ellenberg H (1991) Indicator values of vascular plants in Europe 1. Indicator values of plants not including *Rubus*. Scripta Geobot 18: 9–166
15. Mårtensson L-M, Olsson PA (2010) Soil chemistry of local vegetation gradients in sandy calcareous grasslands. Plant Ecol 206: 127–138.
16. Persson KM (2000) Description to the quaternary map 2D Tomelilla NO. The Swedish Geological Survey, Uppsala, Sweden.
17. Pigott CD, Walters SM (1954) On the interpretation of the discontinuous distributions shown by certain British species of open habitats. J Ecol 42: 95–116.
18. Berglund BE, Larsson L, Lewan N, Olsson EGA, Skansjö S (1991) Ecological and social factors behind the landscape changes, in: Berglund BE (Ed.), The cultural landscape during 6000 years in southern Sweden. Ecol Bull 41: 425–444.
19. Poska A, Saarse L (1999) Holocene vegetation and land-use history in the environs of Lake Kahala, northern Estonia. Veget Hist Archaeobot 8: 185–197.
20. Campbell Å (1929) Skånska bygder under förra hälften av 1700-talet. Etnografisk studie över den skånska allmogens äldre odlingar, hägnader och byggnader (The Scanian countryside during the first half of the 18th century). Thesis, Uppsala University, Sweden.
21. Dahl S (1989) Studier i äldre skånska odlingssystem (Studies of the old Scanian agricultural systems). Meddelande serie B 69, Kulturgeografiska institutionen, Stockholms Universitet. Stockholm, Sweden.
22. Emanuelsson U (2002) Det skånska kulturlandskapet (The Scanian Cultural Landscape). Naturskyddsföreningen i Skåne, Lund.
23. von Linné C (1751) Carl Linnaei skånska resa på höga öfwerhetens befallning förrättad år 1749 (Carl Linnaeus's Journey to Scania as ordered by the high authorities in 1749). Stockholm, Sweden.
24. Björkman L, Sjögren P (2003) Long-term history of land-use and vegetation at Ire, an agriculturally marginal area in Blekinge, south Sweden. Veget Hist Archaeobot 12: 61–74.
25. Foster DR (1992) Land-use history (1730-1990) and vegetation dynamics in central New England, USA. J Ecol 753–772.
26. Mattiasson G (1974) Sandstäpp - Vegetation, dynamik och skötsel (Xeric sand calcareous grassland – vegetation, dynamics and management). Meddelanden från Avdelningen för Ekologisk Botanik, Lunds universitet 2: 1–40.
27. Tyler T (2003) The state of the sand-steppe vegetation of eastern Scania (Skåne) in early spring 2003. Bot Notiser 136: 1–22.
28. Schnoor TK, Olsson PA (2010) Effects of soil disturbance on plant diversity of calcareous grasslands. Agr Ecosyst Environ 139: 714–719.
29. Roem WJ, Berendse F (2000) Soil acidity and nutrient supply ratio as possible factors determining changes in plant species diversity in grassland and heathland communities. Biol Conserv 92: 151–161.
30. Snogerup S, Jörgensen M (2000) Från skånska resa till Skånes flora (From the journey to Scania to the Scanian flora). Lunds Botaniska Förening, BTJ Tryck AB, Lund, Sweden.
31. Tyler T, Olsson K-A, Johansson H, Sonesson M (2007) *Floran i Skåne: arterna och deras utbredning (The Scanian flora: the species and their distribution)*. Lunds Botaniska Förening, Grahns Tryckeri AB, Lund, Sweden.
32. Olsson K-A (1994) Sandstäpp i Skåne – ett upprop (Xeric sand calcareous grassland in Scania – a proclamation). Lunds Botaniska Förenings Medlemsblad 1994: 4–7.
33. Ljungberg KO, Löfroth M, Nitare J (1994) Åtgärdsprogram för Sandstäpp 1994-1996 (Action Plan for Xeric Sand Calcareous Grassland 1994-1996). Report no. 390-2540-94. Swedish Environmental Protection Agency, Stockholm, Sweden.
34. Wassen MJ, Venterink HO, Lapshina ED, Tanneberger F (2005) Endangered plants persist under phosphorus limitation. Nature 437: 547–550
35. Smits NAC, Willems JH, Bobbink R (2008) Long-term after-effects of fertilization on the restoration of calcareous grasslands. Appl Veg Sci 11: 279–286.
36. Bray RH, Kurtz LT (1945) Determination of total, organic, and available forms of phosphorus in soils. Soil Sci 59: 39–45.
37. Bahr A, Ellström M, Schnoor TK, Påhlsson L, Olsson PA (2012) Long-term changes in vegetation and soil chemistry in a calcareous and sandy semi-natural grassland. Flora 207: 379–387.
38. Lantmäteriet (2014a) (Swedish land surveying agency), http://historiskakartor.lantmateriet.se/arken/s/search.html?locale = en_US. Accessed 2014 Feb 15.
39. Lantmäteriet (2014b) (Swedish land surveying agency), http://www.lantmateriet.se/Kartor-och-geografisk-information/Flyg--och-satellitbilder/Flygbilder/ Accessed 2014 Feb 15.
40. Gazenbeek A (2005) LIFE, Natura 2000 and the Military. LIFE Focus. European Commission, Luxembourg.
41. Warren SD, Holbrook SW, Dale DA, Whelan NL, Elyn M, et al. (2007) Biodiversity and the heterogeneous disturbance regime on military training lands. Restor Ecol 15: 606–612.
42. Řehounková K, Prach K (2008) Spontaneous vegetation succession in gravel-sand pits: a potential for restoration. Restor Ecol 16: 305–312.
43. Jentsch A, Friedrich S, Steinlein T, Beyschlag W, Nezadal W (2009) Assessing conservation action for substitution of missing dynamics on former military training areas in Central Europe. Restor Ecol 17: 107–116.
44. Wells TCE, Sheail J, Ball DF, Ward LK (1976) Ecological studies on the Porton Ranges: relationships between vegetation, soils and land-use history. J Ecol 64: 589–626.
45. Dahlström A, Cousins SAO, Eriksson O (2006) The history (1620–2003) of land use, people and livestock, and the relationship to present plant species diversity in a rural landscape in Sweden. Environment and History 12: 191–212.
46. Johansson LJ, Hall K, Prentice H, Ihse M, Reitalu T, et al. (2008) Semi-natural grassland continuity, long-term land-use changes and plant species richness in an agricultural landscape on Öland, Sweden. Landscape Urban Plan 84: 200–211.
47. Kent M, Dargie T, Reid C (2003) The management and conservation of machair vegetation. Bot J Scot 55: 161–176.
48. Carroll JA, Caporn SJM, Johnson D, Morecroft MD, Lee JA (2003) The interactions between plant growth, vegetation structure and soil processes in semi-natural acidic and calcareous grasslands receiving long-term inputs of simulated pollutant nitrogen deposition. Environ Pollut 121: 363–376.
49. Oenema O (1990) Calculated rates of soil acidification of intensively used grassland in the Netherlands. Fert Res 26: 217–228.

50. Bolan NS, Adriano DC, Curtin D (2003) Soil acidification and liming interactions with nutrient and heavy metal transformation and bioavailability. Adv Agron 78: 216–272.
51. Stroia C, Morel C, Jouany C (2010) Nitrogen fertilization effects on grassland soil acidification: consequences on diffusive phosphorus ions. Soil Sci Soc Am J 75: 112–120.
52. Riksen M, Ketner-Oostra R, van Turnhout C, Nijssen M, Goossens D, et al. (2006) Will we lose the last inland drift sands of Western Europe? The origin and development of the inland drift-sand ecotype in the Netherlands. Landscape Ecol 21: 431–447.
53. Poortinga A, Visser SM, Riksen MJPM, Stroosnijder L (2011) Beneficial effects of wind erosion: Concepts, measurements and modelling. Aeolian Research 3: 81–86.
54. Bateman MD, Godby SP (2004) Late-Holocene inland dune activity in the UK: a case study from Breckland, East Anglia. The Holocene 14: 579–588.
55. Olsson PA, Ödman AM (2014) Natural establishment of specialist plant species after topsoil removal and soil perturbation in degraded calcareous sandy grassland. Restor Ecol 22: 49-56.
56. Ödman AM, Schnoor TK, Ripa J, Olsson PA (2012) Soil disturbance as a restoration measure in dry sandy grasslands. Biodiv Cons 21: 1921–1935.
57. Riksen M, Spaan W, Stroosnijder L (2008) How to use wind erosion to restore and maintain the inland drift-sand ecotype in the Netherlands? J Nat Conserv 16: 26–43.

21

Chlorophyll *a* Fluorescence in Evaluation of the Effect of Heavy Metal Soil Contamination on Perennial Grasses

Grzegorz Żurek[1¶], Krystyna Rybka[2*¶], Marta Pogrzeba[3], Jacek Krzyżak[3], Kamil Prokopiuk[1]

1 Department of Grasses, Legumes and Energy Plants, Plant Breeding and Acclimatization Institute IHAR-PIB, Radzików, Błonie, Poland, **2** Department of Plant Physiology and Biochemistry, Plant Breeding and Acclimatization Institute IHAR-PIB, Radzików, Błonie, Poland, **3** Department of Environmental Biotechnology, Phytoremediation Team, Institute for Ecology of Industrial Areas, Katowice, Poland

Abstract

Chlorophyll *a* fluorescence gives information about the plant physiological status due to its coupling to the photosynthetic electron transfer chain and to the further biochemical processes. Environmental stresses, which acts synergistically, disturbs the photosynthesis. The OJIP test, elaborated by Strasser and co-workers, enables comparison of the physiological status of plants grown on polluted vs. control areas. The paper shows that the Chl *a* measurements are very useful tool in evaluating of heavy metal ions influence on perennial grasses, tested as potential phytoremediators. Among 5 cultivars tested, the highest concentration of Cd and Zn ions, not associated with the yield reduction, was detected in the biomass of tall fescue cv. Rahela. Chl *a* fluorescence interpreted as double normalized curves pointed out Rahela as the outstanding cultivar under the HM ions stress.

Editor: Rajagopal Subramanyam, University of Hyderabad, India

Funding: The work was financed by the Ministry of Agriculture and Rural Development, Poland, as grant PW3-3-00-0-03. The funders had no role in study design, data collection and analysis, decision to publish, or preparation of the manuscript.

Competing Interests: The authors have declared that no competing interests exist.

* E-mail: k.rybka@ihar.edu.pl

¶ These authors are co-first authors on this work.

Introduction

Phytoremediation requires plants that maintain good fitness on contaminated soils in parallel with the highest possible concentration of pollutants in aerial parts [1]. For evaluation of species and varieties as potential phytoremediators in breeding programs, low-cost, easy to carry and high throughput techniques are required. Chlorophyll *a* (Chl *a*) fluorescence measurements fulfill those conditions and in parallel with detection of heavy metal (HM) ions concentration represent a good screening tool for evaluation of plant suitability to grow on polluted lands.

Chl *a* fluorescence is a natural phenomenon, characteristic for all photosynthetic organisms, since the sun light energy absorbed by chlorophyll molecules, if not used to drive photosynthesis, is dissipated as heat radiation or is re-emitted as light photons. Even though the amount of fluorescence usually makes no more than 1–8% of total light absorbed, the fact that light induced processes are as communicating vessels and the increase in the efficiency of one of them results in the decrease in the yield of the other two, measurement of Chl *a* fluorescence gives information about changes in the efficiency of photochemistry and heat scattering [2,3]. Since such a measurement is convenient, non-invasive, highly sensitive, rapid, reliable and gives quantitative probe of oxygenic photosynthesis, it is in common use as suitable tool for determination of plant physiological status [4]. It can be read in the way as the electrocardiograms are (Goltsev, personal communication, 2012). Plants subjected to HM ions stress show disturbances in photosynthesis due to single or cumulative phenomena of: *i*/direct interaction of HM ions with protein thioyl-, histidyl- and carboxyl-groups; *ii*/formation of reactive oxygen species (ROS); and/or *iii*/displacement of essential cations in protein active centers. Some ions such as Hg, Cu, Cd, Ni or Zn may substitute the central Mn ion in chlorophyll molecule, forming chlorophyll-metal complexes and thus lower PSII quantum efficiency [5,6,7]. These circumstances affect most of detected parameters [8]. Beside the comparison of particular parameters, among which the F_V/F_0 and F_V/F_m are the most widely known and used, the interpretation of double normalized curves, based on JIP test is an increasingly popular approach in applied, environmental studies [9]. Plots are drown from the data recorded with high sampling rate within a second after the exposure of dark adapted leaf to the light, on logarithmic timeline as the independent variable [10]. On such graphs points of inflections (J-I-P), noted on the increasing part of the registered fluorescence, are the base of inference about the structure and function of the photosynthetic apparatus, with the obligatory assumption, that the efficiency of chlorophyll antenna in light energy capture is the only cause of detected fluctuations and that the only limitation of the photochemical conversion in PSII, is plastoquinone Q_A, the electron acceptor [11,12]. Even though the accurate interpretations of Chl *a* fluorescence measurements, which are the result of in-depth knowledge accumulated over the years, still needs the further studies [13] both: the knowledge and the modern technology have turned measurements of Chl *a* fluorescence into convenient and reliable tool for outdoor, environmental studies [9].

In the field studies, the experimental system is complex. All environmental stimuli: contaminants in parallel with weather fluctuations act synergistically, so Chl *a* fluorescence reflects theirs cumulative effect. For phytoremediation - revealing plants of better physiological activity on contaminated soils, in parallel with higher concentration of HM ions in aerial parts, is the goal. However it should be considered that good physiological state of the plant detected in contaminated rural areas can be misleading, since health risk for organisms in the food chain. The phytotoxicity of some metals has been long established [14].

The aim of our research was to evaluate the usefulness of Chl *a* fluorescence as a method for phytoremediation studies in field experiments on five perennial grass species grown on naturally HM contaminated soils. The acquired knowledge is to be used in planning of biomass processing systems and, in parallel, in phytoremediation of polluted soils, using cultivation of energy plants as a main concept.

Materials and Methods

Field Experiment

The experiment was located on contaminated soil on Silesia, southern Poland (50.4°N/018.9°E). Plots were established in the vicinity of a closed-down cadmium/lead/zinc/ore mine and processing plant, which operated for more than 100 years and had significant impact on local soils. Reference site (uncontaminated soil) was located in a distance of 25 km North from above, (50.6°N/018.88°E) on field with similar forecrop history. Permission was not required for research activities on both locations. No endangered or protected species were involved in our field studies.

Plant Materials

Five perennial grass cultivars: tall oat grass, *Arrhentherum elatius* L. (Ae.) 'Wiwena'; rescuegrass, *Bromus carinathus* Hooker et Arn (Bc.) 'Broma'; tall wheat grass, *Elytrigia elongata* Nevski (Ee.) 'Bamar'; smooth brome grass, *Bromus inermis* Leyss (Bi.) 'Brudzynska' and tall fescue, *Festuca arundinacea* Schreb. (Fa.) 'Rahela' were sown at the end of June 2010 on 1×1 m plots, in 3 replications per variety on both locations. No further treatments such as: fertilization, herbicides or watering were applied, despite of collecting aboveground biomass at the end of growing seasons in 2010 and 2011. All observations and measurements: plant height [PH, cm], generative shoots abundance [GS, %] per plot area and yield [BY, $t{\cdot}ha^{-1}$] as well as Chl *a* fluorescence were conducted at the first year after the sowing (2011) at both locations: all parameters were measured in June, beside the biomass yield which was measured in October.

Soil Analysis

Five representative soil samples were collected from 0–30 cm horizon at each site for chemical analysis. Large chunks of soil were spread out to dry in a forced air oven at 35°C and then were grounded to pass a 60 mesh screen.

Soil pH was measured in 1 M KCl and H_2O (1:2.5) suspension (combination glass/calomel electrode and pH meter) and electrical conductivity (EC) e in water extract (1:2.5) after 24 h of equilibration; the fraction of bioavailable metals was obtained by extraction of 3 g of air-dried soil (ground to <0.25 mm) with 30 ml 0.01 M $CaCl_2$ for 5 h; total metal concentrations by extraction of air-dried soil ground to <0.25 mm with concentrated $HClO_4$ and HF.

Concentrations of metals in soil extracts were detected using inductive coupled plasma spectrometry (ICP-AES) (Spectro Analytical Instruments GmbH, Kleve, Germany).

Determination of HM Ions Concentration in Plant Biomass

Plant material was washed with tap water and then with deionised water in an ultrasonic washer to remove all soil particles followed by drying at 70°C for 3 days. One g of dried, ground material was wet-ashed using concentrated nitric acid (Merck) in a microwave system (MDS 2000, CEM, USA). Concentration of lead, cadmium and zinc ions in plant material was also detected using inductive coupled plasma spectrometry (ICP-AES).

Measurements of Chl *a* Fluorescence Transient

Chl *a* fluorescence was measured using PocketPEA portable fluorometers (Hansatech Instruments, King's Lynn, Norfolk, UK). Recordings were performed simultaneously on both locations at 10–12 a.m. UMT on the first fully developed leaf, (3 leaves per 5 plants per replication). Dark adaptation (0.5 hour) of the middle part of abaxial leaf blade preceded the measurement. Fluorescence was induced by saturating, red actinic light with energy of 3.500 $\mu mol\ m^{-2}\ s^{-1}$, and first 3 seconds of transient fluorescence, covering more than its exponential growing part, was registered with time intervals increasing from 10 µs within first 300 µs of the measurement up to 100 ms intervals for times longer than 0.3 sec. Additionally, collected data points were double normalized between points O-I, O-J, O-K and a kinetic differences ΔW_{OI}, ΔW_{OJ} and ΔW_{OK} were calculated and analyzed as graphs on logarithmic time scale [10].

List of measured parameters:

$\mathbf{F_O \approx F_{50\ \mu s}}$ [minimal fluorescence];

$\mathbf{F_1, F_2, F_3, F_4, F_5}$ [fluorescence at times: 0.05, 01, 03, 2.0 and 30 ms after the start of actinic illumination F_O. Values at 0.3, 2.0 and 30 ms responds to fluorescence at K, J, I points of inflections, on fluorescence transient curve];

$\mathbf{F_M = F_P}$ [maximal recorded fluorescence];

$\mathbf{t_{FM}}$ [time (in ms) to reach the maximal fluorescence, F_M];

Area [total complementary area between the fluorescence induction curve and F_M of OJIP curve].

Calculated parameters:

a) listed by PocketPEA software:

$\mathbf{F_V}$ [maximal variable fluorescence calculated as F_M–F_O];

$\mathbf{F_V/F_M}$ [force of the light reactions];

RC/ABS [the amount of active reaction centers per absorption];

$\mathbf{(1-V_J)/V_J}$ [measure of forward electron transport];

$\mathbf{PI_{ABS}}$ [performance index];

b) not listed by PocketPEA software:

$\mathbf{W_{OI}}$, $\mathbf{W_{OJ}}$ and $\mathbf{W_{OK}}$ double normalized fluorescence readings at points O-I, O-J, O-K [ratio of variable fluorescence (F_t-F_O) to the amplitude: (F_I-F_O), (F_J-F_O) or (F_K-F_O)];

$\mathbf{\Delta W_{OI}}$, $\mathbf{\Delta W_{OJ}}$ and $\mathbf{\Delta W_{OK}}$ [difference of: [W_{OI}(reference)−W_{OI}(control)], [W_{OJ}(reference)−W_{OJ}(control)] and [W_{OK}(reference)−W_{OK}(control)], respectively]. Calculated separately for each cultivar.

Climatic Data

Monthly precipitation and air temperatures for both localities were estimated on the basis of free climatic service (http://klimat.icm.edu.pl/serv_climate.php), where air temperature and precipitation data for current and past periods can be projected for given geographic coordinates in Poland (Table 1).

Statistical Calculations

Statistical calculations were performed using SAS statistical package. The least significant differences (LSD) between means

Table 1. Monthly precipitation and mean temperatures during the course of experiment in polluted and reference site.

Year	Month	Precipitation (mm)		Temperature (°C)	
		Site type:			
		polluted	reference	polluted	reference
2010	1	43.4	40.3	−6.2	−6.6
	2	28.0	25.2	−1.3	−1.5
	3	31.0	31.0	3.8	3.5
	4	45.0	39.0	8.9	8.8
	5	232.5	217.0	12.8	12.6
	6	66.0	60.0	17.0	16.8
	7	108.5	108.5	20.4	20.3
	8	102.3	102.3	18.6	18.6
	9	99.0	87.0	12.5	12.4
	10	6.2	6.8	6.3	6.2
	11	69.0	72.0	6.5	6.2
	12	49.6	46.5	−5.2	−5.5
2011	1	28.2	27.9	−0.6	−0.6
	2	12.0	11.2	−2.2	−2.5
	3	37.5	31.0	4.1	4.0
	4	24.3	21.0	10.7	10.7
	5	72.5	62.0	13.6	13.6
	6	51.3	54.0	18.3	18.1

were calculated according to the Duncan honesty test and significance of difference between means was tested with the probability of 95%. Divergence among tested cultivars was estimated using Principal Component Analysis (PCA) analysis, based on correlation matrix algorithm for Chl *a* fluorescence traits.

Results

In order to examine, to what extent the Chl *a* fluorescence was useful in studies of the effect of HM polluted soil in natural agro-ecosystem, localizations with the most similar soil composition (beside HM ions concentration) were chosen to run the experiment. Soils did not differ much in organic matter content: 4% on contaminated area vs. 3.5% on reference place. The pH of contaminated soil was nearly neutral (6.8), whereas the reference one was slightly acidic (5.2). Although the content of silt and clay were on similar level, the reference soil was 1.6 times more sandy than contaminated soil. The amount of HM ions in polluted soil exceeded by almost 2 orders of magnitude the concentration of Cd and Zn ions and about 50 times of Pb ions when compared to reference place (Table 2).

Climatic data for both localities were also very similar (Table 1). Reference site received similar amount of rain and heat during vegetation in 2010 as well as in 2011.

It is very likely that the elevated concentration of Pb, Cd and Zn ions in the soil affected plant growth (Table 3). Biomass yields of plants grown on the polluted site ranged from 1.7 (cv. Broma) to 3.7 (cv. Wiwena & Rahela) $t{\cdot}ha^{-1}$ of dry mass and it was ca. 0.8 $t{\cdot}ha^{-1}$ lower than the yield of grasses grown on the reference site. The highest reduction of the yield were noted for brome grasses: cv. Brudzynska (2.2 $t{\cdot}ha^{-1}$) and cv. Broma (1.1 $t{\cdot}ha^{-1}$), grown on polluted site. High biomass yields of cv. Wiewena and cv. Rahela were associated with the significantly lower GS value, as compared to the reference site. No significant differences in GS values were detected for other cultivars grown on both sites. Cultivar Bamar grown on the polluted soil did not differ significantly from that grown on the reference one, beside minor disparity in the Chl *a* fluorescence parameters.

The differences in HM ions sequestration were detected as well (Table 4). The highest concentrations of Cd and Zn ions were found in the biomass of cv. Rahela whereas the lowest concentration of Cd was detected in cv. Bamar and Broma. The least Zn ions were collected by the cv. Wiwena. Rahela accumulated 3.9 fold more of Cd ions and 2.2 more of Zn ions. No statistically important differences in Pb ions accumulation were noted.

Inhibition of the plant growth as a reaction on HM ions stress was the result of disturbances in the plant cell metabolism affecting also the chloroplasts and reflected in changes of the fluorescence parameters (Table 5). Fluorescence intensities at different time points distinguished cultivars. At points F_M, F_V and F_5 the recorded data were the smallest for cv. Rahela, whereas for shorter times (from F_1 to F_4) cv. Bamar. was characterized by the weakest signal. Also the F_V/F_M ratio as well as the other calculated parameters were the lowest for cv. Rahela.

Principal Component Analysis (PCA) enabled reduction in dimension of multivariate fluorescence data set. The first two components, accounted about 86% of total variation: 1st –54.6% of total variation explained, and was highly correlated with measured values: F_O, F_M, F_V, F_1–F_5; 2nd explaining 31.7% of total variation, was highly correlated with parameters: RC/ABS, $(1-V_j)/V_j$ and P_I (Table 6). The multi-trait analysis visually presents the overall magnitude of changes related to the plant growth in polluted and unpolluted site across all Chl *a* parameters (Fig. 1). Cultivars tested in unpolluted *vs.* polluted site could be easily

Table 2. The physico-chemical characteristics of the contaminated and the reference soil, where, under the natural conditions, the tested plants were cultivated.

Soil property	Soil type:	
	contaminated	reference
pH (KCl)	6.79±0.01	5.17±0.02
pH (H_2O)	7.01±0.03	5.96±0.02
EC (μS cm^{-1})	98.23±7.01	63.90±9.59
organic matter content (%)	4.0±0.03	3.5±0.03
Soil composition		
clay (<0.002 mm). %	16.1±2.0	12.7±1.5
silt (0.002–0.05 mm). %	55.9±3.1	40.0±2.2
sand (0.05–1 mm). %	28.0±2.0	47.3±1.1
Total heavy metal concentration (extraction with aqua regia)		
Pb (mg kg^{-1})	547.0±27.92	16.76±0.50
Cd (mg kg^{-1})	20.84±1.17	0.74±0.04
Zn (mg kg^{-1})	2174±103.0	35.06±0.69
$CaCl_2$ extractable metal fraction		
lead [Pb^{+2}] (mg kg^{-1})	16.76±0.50	0.019±0.01
cadmium [Cd^{+2}] (mg kg^{-1})	0.74±0.04	0.016±0.01
zinc [Zn^{+2}] (mg kg^{-1})	35.06±0.70	0.533±0.02

distinguished on the basis of location on *x*-axis (measured parameters of Chl *a* fluorescence).

Splits of the cultivars into groups on PCA diagrams induced the idea for calculation of mean values of fluorescence data points for each of plant group: those from polluted soil and those from reference one, in order to present the average values on logarithmic time scale. The fluorescence transient of plants grown on HM ion contaminated soil were lower than those of plants grown on the reference site. Points of inflection: J at 2.0 ms and I at 30 ms, important for the results interpretation, are visible there (Fig. 2a). Other points of local inflections: L at 150 μs, K at 300 μs, identified and described by Strasser [15] are so tiny that can be visible as local extreme points only after additional mathematical operations; here, the differences of double normalized transients measured for plants grown on polluted and reference sides are presented, as described in Materials and Methods. Double normalization of the data at points O and I, reflects all local disturbances originating from the physical and biochemical events occurring along the electron transport chain (ETC) in chloroplasts: from the electron trapping through the plastoquinone reduction, and further towards the sink at PSI acceptor side [15]. Curves drown for all cultivars, but not cv. Wiwena, have positive values (Fig. 2b). Point of inflection J, at about 2 ms, is detectable for all cultivars. Point K, at about 0.3 ms, is sharp for cv. Wiwena. Point L, at 0.150 ms, is not visible on the Fig. 2b. For all cultivars extremes not discussed by Tsimilli-Michael and Strasser [15] at

Table 3. The agronomic characteristics of grass cultivars grown on the contaminated agricultural soil in comparison to the plants cultivated on the reference soil: yield of the biomass (BY) [t ha^{-1}] and abundance of generative stems (GS) [%].

Grass cultivar	Trait	Site type:		Significance of difference
		polluted	reference	
(Ae.) Wiwena	BY [t ha^{-1}]	3.7±0.1	4.3±0.7	ns.
	GS [%]	8.3±0.2	18.3±0.9	***
(Ee.) Bamar	BY [t ha^{-1}]	2.0±0.1	1.7±0.2	ns.
	GS [%]	1.0±0.1	1.3±0.2	ns.
(Bc.) Broma	BY [t ha^{-1}]	1.7±0.2	2.8±0.9	***
	GS [%]	86.7±8.1	90.0±7.1	ns.
(Bu.) Brudzynska	BY [t ha^{-1}]	1.8±0.2	4.0±0.9	***
	GS [%]	15.0±1.1	16.7±2.1	ns.
(Fa.) Rahela	BY [t ha^{-1}]	3.7±0.9	4.3±1.0	ns.
	GS [%]	9.0±0.8	26.7±3.7	**

Plant height (PH) [cm] values are not shown since it did not differentiate studied cultivars. ns – not significant; **significant for P>95%; ***significant for P>99%.

Table 4. The concentration of HM in plant biomass.

Grass cultivar	Soil type	HM concentration [mg kg^{-1}]		
		Cd	Pb	Zn
(Ae.) Wiwena	reference	0.11±0.02	0.31±0.08	20.45±1.0
	polluted	3.0±0.1	64.0±4.5	185.5±6.4
(Ee.) Bamar	reference	0.14±0.01	0.25±0.01	18.45±2.1
	polluted	2.2±0.2	31.0±4.2	253.0±2.8
(Bc.) Broma	reference	0.08±0.01	0.35±0.02	16.7±1.8
	polluted	2.2±0.6	51.0±9.9	197.0±43.8
(Bu.) Brudzynska	reference	0.18±0.03	0.37±0.02	25.25±5.4
	polluted	3.5±0.4	59.5±12.0	256.0±4.2
(Fa.) Rahela	reference	0.19±0.02	0.73±0.1	19.5±2.0
	polluted	8.5±0.9	50.0±11.3	417.0±50.2
LSD (P>95%)	reference	n.s.	n.s.	n.s.
	polluted	1.33	n.s.	113.5

n.s. – not significant.

about 5 ms (the minimum) and at about 10 ms (the maximum) are visible. Shape of the curve drown for cv. Rahela is noticeably different, more flat, then curves of the other ones.

Normalization procedure at points O and J, and differentiation of results from polluted and reference soils, visualized the K-band, the most expressed with minus sign, for cv. Wiwena and the weakest for cultivars: Brudzynska (with the negative sign) and Rahela (with positive one) (Fig. 2c). Normalization at points O and K visualized L-band (Fig. 2d). All curves, but not that drawn for cv. Rahela, turned out to be almost identical, with the minus sign. The plot for the cv. Rahela was close to zero.

Discussion

The detected differences in soil composition of contaminated and reference site resulted from outdoor natural environments of rural areas, where the experiments were run. The fact that the vegetation seasons prior to Chl *a* fluorescence measurements were rainy and water availability did not limited the plant growth, opened the opportunity to interpret the differences in grasses performance as the response to HM. More sand in soil at reference site might negatively affect the plant growth and performance but mostly during the low precipitation periods. In March, April and May, three months prior to Chl *a* measurements, polluted and reference site received 107% and 91% of rainfall long term average (1971–2000), respectively, which periods are defined as to average precipitation periods in Poland [16]. Lower EC and pH at reference site could be the indirect effect of lower level of HM concentration, so it was assumed, that detected differences in plant performance resulted mostly from high HM level in soil. The HM ion concentration in polluted site exceeded four fold the limits of Cd and Pb ions concentration for arable soils and six fold the limits of Zn, which are: 12.7 mg·kg^{-1} Cd, 304.5 mg·kg^{-1} Pb, and of 1677.5 mg·kg^{-1} Zn (Table 2). Plants can tolerate relatively large amount of zinc in the soil, because it is an essential micronutrient, which deficiency as well as excess is harmful [17].

Contamination of agricultural land by trace metals is the side effect of industrialization. In regions of long history of industrial emission elevated levels of cadmium, lead, zinc and other ions emitted in the course of coal and ore mining and processing are detected. Since the maintenance of agricultural areas in those regions is important from ecological and sociological standpoint, the alternative farming activities are needed. Perennial grass biomass production for energy purposes is one of such solution [18,19]. However, caution should be made on practical use of biomass with high amounts of heavy metals. Considering its chemical composition it is not a substrate for direct energy production but a hazardous waste. Contaminated crop disposal process must be always planned in any phytoremediation activities, especially when large quantities of biomass are to be generated [20].

Many plant species has been already screened for their ability to accumulate HM. For this purpose two strategies have been tested: *i*/the cultivation of hyperaccumulators on polluted sides (usually plants of relatively low aboveground biomass, accumulating high amounts of one or more toxic ions); *ii*/growing plants which produce high biomass, despite lower ability of HM sequestration [21]. The latter seems much promising, because there are numbers of species available from many families, from which willows and perennial grasses are the most interesting [22]. Although the willow produces the high biomass with the effective nutrient uptake as well as the clone specific capacity for HM uptake, high demand for water and the specific harvest equipment required limit its usefulness to the large and flat fields rich in water, what is not frequent in the case of polluted areas [23]. Considering perennial grasses, there are no such limitations. A number of perennial grasses are used increasingly in Europe and North America as renewable bioenergy sources, as they can be grown with low maintenance on marginal soils and harvested to produce large volumes of biomass [24].

Dynamics of HM ions absorption, transport and neutralization influence theirs toxicity. It was shown, that dose dependent relative cadmium toxicity is 4-fold higher than zinc and 3-fold higher than lead, for the onion epidermal cells used as a model. The way of HM inactivation affects theirs toxicity: HM ions can be stored in vacuoles, Cd ions can form the complexes with the cell wall polymers, Pb can be locked in endoplasmic reticulum vesicles [25]. Non neutralized HM ions induce formation of an excess of free radicals and slowdown turnover of chloroplasts proteins, mostly D1 protein [17,26]. Chloroplasts development in young leaves as

Table 5. The parameters of Chl *a* fluorescence of the grass cultivars grown on the polluted and the reference soil.

Chl *a* fluorescence parameters	Soil type	Grass cultivars				
		(Ae.) Wiwena	(Ee.) Bamar	(Bu.) Brudz.	(Bc.) Broma	(Fa.) Rahela
measured parameters:						
$F_1 = F_{50\mu s} = F_O (AU \times 10^3)$	reference	6.9	6.9	8.0	7.2	7.2
	polluted	5.0 ab	4.6 b	4.9 ab	5.3 ab	5.5 a
	sign. of diff.	**	*	***	**	***
$F_2 (AU \times 10^3)$	reference	7.7	7.3	8.7	7.9	8.0
	polluted	5.4 ab	5.0 b	5.4 ab	5.8 ab	6.2 a
	sign. of diff.	***	*	***	**	**
$F_3 (AU \times 10^3)$	reference	10.6 ab	9.1 b	12.1 a	11.0 ab	11.2 ab
	polluted	7.2 bc	6.4 c	8.0 ab	8.4 ab	8.7 a
	sign. of diff.	***	*	**	**	**
$F_4 (AU \times 10^3)$	reference	16.7 ab	14.3 b	19.9 a	17.2 ab	17.7 ab
	polluted	13.3 b	11.5 c	14.0 ab	15.0 a	13.6 ab
	sign. of diff.	**	n.s.	**	***	***
$F_5 (AU \times 10^3)$	reference	25.7	26.1	28.6	26.5	25.7
	polluted	20.1 ab	18.7 ab	19.2 ab	20.4 a	18.3 b
	sign. of diff.	**	**	***	**	***
$F_M [AU \times 10^3]$	reference	31.7	33.9	34.5	31.6	30.5
	polluted	24.8 a	22.5 ab	22.4 ab	24.0 a	20.2 b
	sign. of diff.	**	**	***	**	***
$F_V [AU \times 10^3]$	reference	25.5	27.5	27.4	25.2	24.1
	polluted	20.3 a	18.3 ab	18.2 ab	19.3 ab	15.3 c
	sign. of diff.	**	***	***	***	***
Tfm	reference	750 a	733 a	550 ab	385 b	550 ab
	polluted	500	650	450	500	850
	sign. of diff.	****	n.s.	n.s.	n.s.	n.s.
Area $[AU x 10^3]$	reference	977 b	1223 a	930 bc	652 d	721 cd
	polluted	651 a	551 ab	455 b	433 bc	280 c
	sign. of diff.	**	***	***	*	***
calculated paramaters:						
F_V/F_M	reference	0.806	0.812	0.794	0.798	0.791
	polluted	0.818 a	0.811 ab	0.811 ab	0.804 b	0.760 c
	sign. of diff.	**	n.s.	**	n.s.	***
RC/ABS	reference	1.15 b	2.31 a	1.12 b	1.1025 b	0.985 b
	polluted	1.53 a	1.72 a	0.98 b	1.058 b	0.761 c
	sign. of diff.	**	n.s.	n.s.	n.s.	*
PI	reference	2.80 b	7.38 a	2.32 b	2.48 b	1.97 b
	polluted	3.90 b	4.46 a	1.94 c	2.03 c	1.04 d
	sign. of diff.	**	n.s.	n.s.	*	***
(1–Vj)/Vj	reference	0.59 b	0.72 a	0.53 b	0.57 b	0.53 b
	polluted	0.57 b	0.61 a	0.46 cd	0.47 c	0.43 d
	sign. of diff.	n.s.	**	*	**	***

The cultivars were compared within soil types (reference and polluted) by the LSD test (1) as well as the influence of the soil type on the cultivar was tested by LSD test (2). Values with the same small letters are not different significantly within soil type (rows). *- p>95%; **– P>99%; ***- P>99.9%; n.s. - not significant.

well as chlorophyll and β-carotene content and Chl *a*/*b* ratio is also affected by Cd ions. Zinc and cobalt divalent cations have the ability to replace Mn^{+2} or Fe^{+2} in chloroplast proteins active centers, decreasing theirs functional activity [27,28].

Thus the HM impact on the plant interferes with photosynthesis and the measurement of Chl *a* fluorescence reveals the disturbances. At the beginning, the fluorescence of dark-adapted leaf rises rapidly from the minimal level (O), typically recorded at

Table 6. Eigenvector values of three components calculated after Principal Component Analysis (PCA) performed on Chl *a* fluorescence parameters.

Chl *a* fluorescence parameters	Component number:		
	#1	#2	#3
F_O	0.98	0.13	−0.13
F_M	0.92	0.39	0.09
F_V	0.88	0.45	0.14
F_V/F_M	−0.16	0.66	0.71
T_{FM}	−0.11	0.18	−0.94
AREA	0.66	0.70	−0.06
F_1	0.99	0.03	−0.13
F_2	0.99	−0.04	−0.14
F_3	0.96	−0.27	−0.07
F_4	0.93	−0.30	0.15
F_5	0.97	0.21	0.07
RC/ABS	−0.04	0.97	0.04
V_J	0.30	0.93	−0.02
P_I	0.02	0.98	0.01
Eigen values	8.19	4.75	1.71
Variaton explained (%)	54.6	31.7	11.4

50 microsecond, to the maximum (marked as 'M' or 'P'), usually appearing at about one second. Minimal fluorescence reflects the oxidized state of quinone Q_A, whereas the maximal one – the completely reduced state of the whole plastoquinone pool (PQ) [15]. This increasing part of the fluorescence transient shows the dynamics of light induced electron flow through the electron transfer chain (ETC) in chloroplasts: pheophytin → plastoquinone pool (PQ) → cytochrome b6f complex (cyt-bf) → plastocyanin (PC) up to the end electron acceptors on PSI acceptor side. ETC damage and sink dependent slowdown of electron flow are visible directly on fluorescence curves drawn on logarithmic time abscissa or on double normalized curves as well as can be recognized by parameters calculated using JIP-test and its applications [3].

Fluorescence values F_2, F_3, F_4 and F_5 at times: 0.15, 0.30, 2.0 and 30 ms responds to L, K, J, I points of inflections, on fluorescence transient curve. Usually under the stress conditions

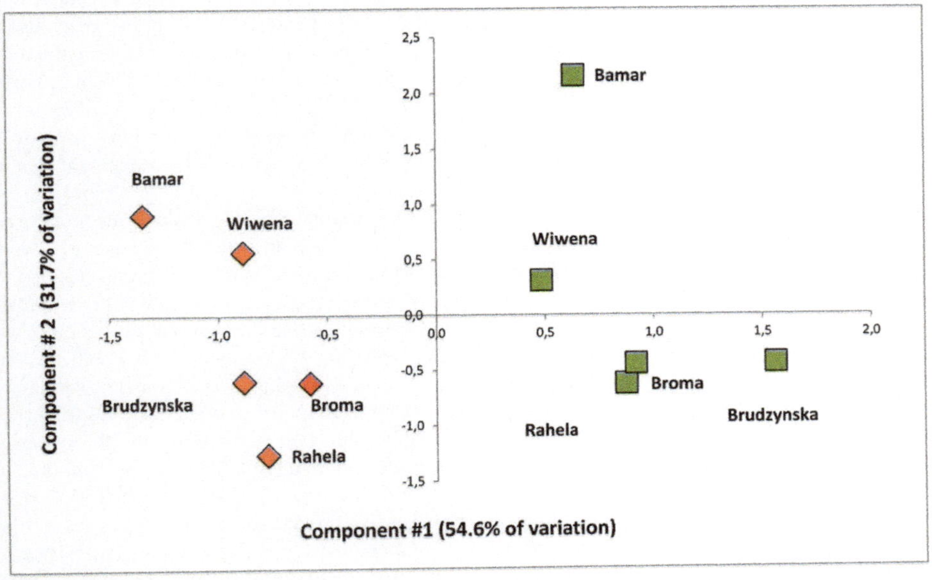

Figure 1. The distribution of tested cultivars in PCA plot created on the basis of Chl *a* fluorescence data. The first two components, accounted about 86% of total variation: 1st was related to measured values (FM, FV, F1-F5), 2nd was highly correlated with calculated parameters (RC/ABS, (1-Vj)/Vj and PI. Cultivars tested in unpolluted site- green boxes; polluted site- red boxes.

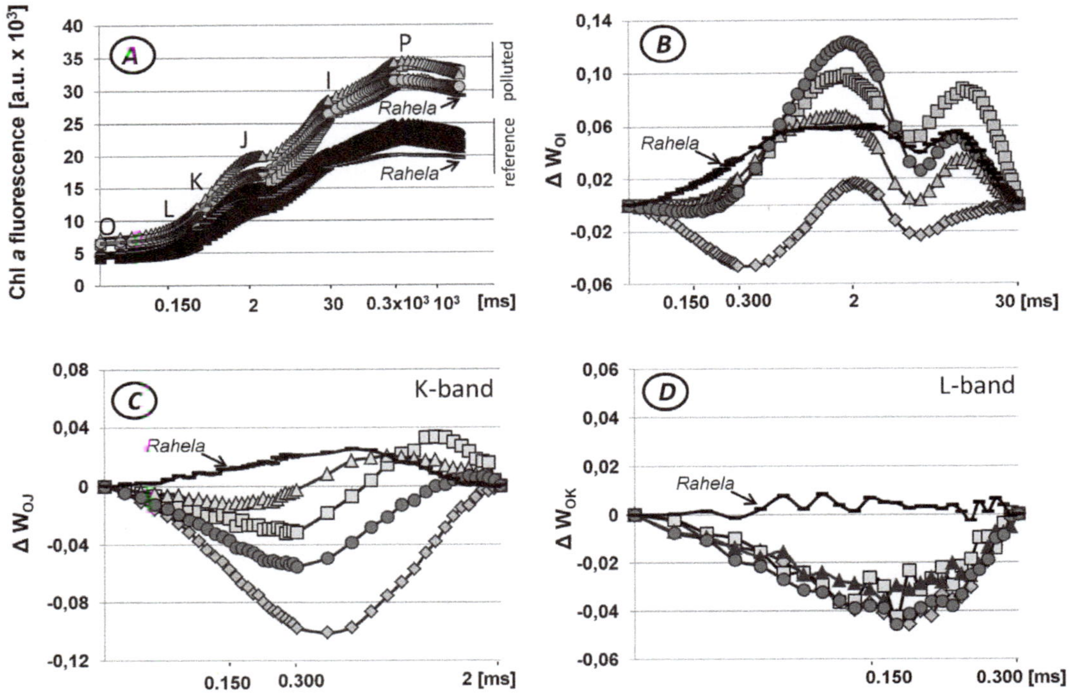

Figure 2. The chlorophyll a fluorescence transients (OJIP) from dark adapted leaves of five perennial grass cultivars given on a logarithmic time scale from 50 µs to 1 s. A/The fluorescence registered for cultivars grown on the reference and the HM ions polluted soil plotted on a logarithmic time scale. The time points for the calculation of structural and functional parameters of the JIP test are marked: O – the fluorescence intensity at 50 µs, L – at 150 µs, K – at 300 µs, J – at 2 ms and I – at 30 ms, P – the maximal fluorescence intensity at the time about 1 s, denoted as t_{FM}; **B/**ΔW_{OI} – the difference of double normalized data at points F_O and F_I for each cultivar grown on polluted and reference soil where $W_{OI} = [(F_t - F_O)/(F_I - F_O)]$; bands in the time range 0.050–30 ms; **C/**ΔW_{OJ} – the difference of double normalized data at points F_O and F_J for each cultivar grown on polluted and reference soil where $W_{OJ} = [(F_t - F_O)/(F_J - F_O)]$; K band; D/$\Delta W_{OK}$ – the difference of double normalized data at points F_O and F_K for each cultivar grown on polluted and reference soil where $W_{OK} = [(F_t - F_O)/(F_K - F_O)]$; L-band. Cultivars are marked as follow: diamonds – (Ae.) 'Wiwena'; squares – (Ee.) 'Bamar'; triangles – (Bi.) 'Brudzynska'; circles – (Bc.) 'Broma'; black dashes – (Fa.) 'Rahela'.

the fluorescence transient curve becomes flatter, i.e.: the F_M decreases [29]; almost all Chl *a* fluorescence induction parameters (F_o, F_v, F_m) were found to be affected by HM concentration in different plant species [8]. The curve deformation depends on both: the resistance of the plant as well as on the time of exposure to the stress; in outdoor studies- to stresses/environmental conditions acting synergistically. The change of F_M entails changes of the other F's, which is well documented and as a consequence affects the values of calculated parameters [30,31].

Our data distinguished cv. Rahela as the cultivar which reached the highest HM ions concentration in aerial parts and its photosynthesis was the most influenced by them. The maximum quantum yield of PSII photochemical reaction described by parameter *Fv*/*Fm* measured on the reference area was slightly lower than respected for the healthy plants (0.83 vs. 0.79) and decreased upon reference side conditions (0.76). It indicated a decline of the maximum efficiency at which light absorbed by plants grown on the polluted soil is used for the reduction of Q_A [12,27]. Other cultivars did not show differences of F_v/F_m parameter, which can happen [32] and can be explained by much lower concentration of sequestrated HM ions. Kalaji *et al.* [12] discussed the dynamics of J-I-P parameters reflecting the stress resistance: the time to reach F_M value (T_{FM}), reduced plastoquinone pool size (Area) and the vitality index of PSII (PI) were more sensitive for action of the stress than commonly used parameters of maximal quantum yield of PSII: F_V/F_M and F_V/F_O.

In our studies the first component of PCA analysis depended on measured values: F_M, F_V, F_1–F_5, while the second on calculated parameters: RC/ABS, (1–Vj)/Vj and PI. Grouping of cultivars calculated on bases of 2 principal components which explained 85% of differentiation, gave logical results. Cultivars were grouped in sub-classes as "polluted" and "reference" (Fig. 1). The other statistical method, presented by Goltsev *et al.* [33] is neutral network. We have found the PCA analysis as a convenient tool in plant grouping.

Since the first PCA variable was dependent on measured parameters (Fig. 1) and also the points of inflection J and I were detectable on the logarithmic time scale (Fig. 2a) it was decided to use double normalized induction curves to compare the dynamics of electron transfer processes in various objects and thereby to estimate, which variety tolerates stresses better [10,34,35].

Analysis of normalized curves distinguish cv. Rahela. as the only outstanding cultivar with the shape of particular curves, different on all drown figures. The inflection point K on the fluorescence curve is associated with water photolysis and the amount of electron transported along Electron Transfer Chain (ETC); high relative fluorescence means slowing of electron flow from the chlorophyll Reaction Centers (RC) of Photosystem II (PSII) as a result of lower activity of PS II Oxygen Evolving Complex (OEC) and slower water splitting [36,37]. The negative peak on ΔW_{OJ} curve, calculated as the difference between fluorescence of plants grown on the polluted area minus values of plants grown on the reference site, means that the relative slowing of the electron flow in chloroplasts of plants from polluted area was smaller than in chloroplasts from plants grown on the reference area. It can be hypothesized that in such a situation the ascorbate can be the electron donor in chloroplasts of plants from the polluted area [36,37,38,39]. Again the cv. Rahela stood out, as on the Fig. 2b. Normalization at points O and K visualized L-band (Fig. 2d). A negative L-band indicates higher grouping of PSII antennas as compared to reference sample [35,38,40], which is the fact for all cultivars, beside cv. Rahela. The similarities of the shape of all normalized curves for all cultivars beside cv. Rahela may suggest the different physiological response to stress of that one. The y-axis shift is detected on all graphs. Cv. Wiwena. is at the bottom of Fig. 2b, Fig. 2c and Fig. 2d while cv. Rahela is at the top. Obtained results can be interpreted as a mark of cultivars reaction on HM ion stress. It looks that all cultivars except cv. Rahela. took from the soil only such amount of HM, which is not too stressful for them. Those cultivars, under the influence of HM ions stress accelerated metabolism, which is a typical physiological response of plants slightly resistant or non-resistant to the stress. Resistant plants typically launch additional defense mechanisms and rebuild its metabolism [41,42]. Contrary to expectation, cv. Rahela which sequestrated the highest concentration of HM ions in the aerial part, did not show PSII antennas grouping under the stress, as other cultivars did. It should be considered that such cultivars which survive the stress by accelerating photosynthesis as well as the whole metabolism could be less resistant in front of additional environmental stresses such as i.e. drought or cold and could react more rapidly on them. *Festuca arundinacea* cv. Rahela in fact deals the best with the HM stress. Its antennas are not damaged, might be due to redox guard in the form of ascorbate which in other cultivars were not induced strongly enough to be visible on double normalized graphs.

Conclusions

1. Chl *a* fluorescence is a good tool in phytoremediation studies.
2. The combined effects of elevated concentration of lead, cadmium and zinc in soil was manifested in alteration of some parameters of Chl *a* fluorescence as well as plant growth.
3. Heavy metal soil contamination resulted in biomass yield reduction of cultivars of *Bromus inermis* 'Brudzynska' and *B. carinathus* 'Broma'.
4. *Festuca arundinacea* 'Rahela', may absorb relatively high amount of HM ions from the soil, without significant reduction of biomass yield, what is expected in the case of bioremediation practices. Mentioned cultivar could be regarded as a good species for the phytoextraction of Cd-contaminated soil.

Author Contributions

Conceived and designed the experiments: GZ KR MP JK KP. Performed the experiments: GZ KR MP JK KP. Analyzed the data: GZ KR. Contributed reagents/materials/analysis tools: GZ KR MP JK KP. Wrote the paper: GZ KR MP.

References

1. Nedunuri K, Lowell C, Meade W, Vonderheide A, Shann J (2010) Management practices and phytoremediation by native grases. Int J Phytorem 12: 200–214.
2. Baker N, Oxborough K (2004) Chlorophyll fluorescence as a probe of photosynthetic productivity. In: Papageorgiou G, Govindjee, editors. Chlorophyll *a* Fluorescence: A Signature of Photosynthesis. pp. 65–82.
3. Stirbet A, Govindjee, Strasser BJ, Strasser RJ (1998) Chlorophyll *a* Fluorescence Induction in Higher Plants: Modelling and Numerical Simulation. J Theor Biol 193: 131–151.
4. Murchie EH, Lawson T (2013) Chlorophyll fluorescence analysis: a guide to good practice and understanding some new applications. J Exp Bot 64: 3983–3998.
5. Dietz K-J, Pfannschmidt T (2011) Novel Regulators in Photosynthetic Redox Control of Plant Metabolism and Gene Expression. Plant Physiol 155: 1477–1485.
6. Hall JL (2002) Cellular mechanisms for heavy metal detoxification and tolerance. J Exp Bot 53: 1–11.
7. Sharma SS, Dietz K-J (2009) The relationship between metal toxicity and cellular redox imbalance. Trends Plant Sci 14: 43–50.
8. Joshi MK, Mohanty P (2004) Chlorophyll *a* fluorescence as a probe of heavy metal ion toxicity in plants. In: Papageorgiou G, Govindjee, editorss. Chlorophyll *a* Fluorescence: A Signature of Photosynthesis. pp. 637–661.
9. Stirbet A, Govindjee. (2011) On the relation between the Kautsky effect (chlorophyll *a* fluorescence induction) and Photosystem II: Basics and applications of the OJIP fluorescence transient. J Photochem Photobiol, B 104: 236–257.
10. Oukarroum A, Madidi SE, Schansker G, Strasser RJ (2007) Probing the responses of barley cultivars (*Hordeum vulgare*, L.) by chlorophyll *a* fluorescence OLKJIP under drought stress and re-watering. Environ Exp Bot 60: 438–446.
11. Stirbet A, Govindjee (2012) Chlorophyll *a* fluorescence induction: A personal perspective of the thermal phase, the J-I-P rise. Photosynth Res 113: 15–6.
12. Kalaji HM, Goltsev V, Bosa K, Allakhverdiev SI, Strasser RJ, Govindjee (2012) Experimental *in vivo* measurements of light emission in plants: A perspective dedicated to David Walker. Photosynth Res 114: 69–96.
13. Schansker G, Toth SZ, Holzwarth AR, Garab G (2013) Chlorophyll *a* fluorescence: beyond the limits of the Q model. Photosynth Res 1: 1doi: 10.1007/s11120-01309806-01309805.
14. Valko M, Morris H, Cronin MTD (2005) Metals, Toxicity and Oxidative Stress. Curr Med Chem 12: 1161–1208.
15. Tsimilli-Michael M, Strasser R (2008) *In vivo* assessment of plants' vitality: applications in detecting and evaluating the impact of mycorrhization on host plants. In: Varma A, editor. Mycorrhiza: State of the Art, Genetics and Molecular Biology, Eco-Function, Biotechnology, Eco-Physiology, Structure and Systematics. pp. 679–703.
16. Bak B, Labedzki L (2002) Assesing drought severity with the relative percipitation index (RPI) and the standardised percipitation index (SPI). J Water Land Develop 6: 29–49.
17. Lin YF, Aarts MGM (2012) The molecular mechanism of zinc and cadmium stress response in plants. Cell Mol Life Sci 69: 3187–3206.
18. Kucharski R, Sas-Nowosielska A, Małkowski E, Japenga J, Kuperberg JM, et al. (2005) The use of indigenous plant species and calcium phosphate for the stabilization of highly metal-polluted sites in southern Poland. Plant Soil 273: 291–305.
19. Bryson GM, Barker AV (2007) Phytoextraction of zinc by Indian Mustard and Tall Fescue. Commun Soil Sci Plant Anal 38: 315–335.
20. Sas-Nowosielska A, Kucharski R, Malkowski E, Pogrzeba M, Kuperberg JM, et al. (2004) Phytoextraction crop disposal– an unsolved problem. Environmen Pollut 128: 373–379.
21. Tlustos P, Szakova J, Hruby J, Hartman I, Najmanova J, et al. (2006) Removal of As, Cd, Pb, and Zn from contaminated soil by high biomass producing plants. Plant Soil Environ 52: 413–423.
22. Borjesson P (1999) Environmental effects of energy crop cultivation in Sweden. I: Identification and quantification. Biomass Bioenergy 16: 137–154.
23. Pulford ID, Watson C (2003) Phytoremediation of heavy metal-contaminated land by trees– a review. Environ Int 29: 529–540.
24. Gibson DJ (2009) Grasses and grassland ecology. New York: Oxford University Press Inc. 315 p.
25. Wierzbicka MH, Przedpelska E, Ruzik R, Ouerdane L, Polec-Pawlak K, et al. (2007) Comparison of the toxicity and distribution of cadmium and lead in plant cells. Protoplasma 231: 99–111.
26. Siedlecka A, Krupa Z (1999) Cd/Fe interaction in higher plants– its consequences for the photosynthetic apparatus. Photosynthetica 36: 321–331.
27. Molins H, Michelet L, Lanquar V, Agorio A, Giraudat J, et al. (2013) Mutants impaired in vacuolar metal mobilization identify chloroplasts as a target for cadmium hypersensitivity in *Arabidopsis thaliana*. Plant Cell Environ 36: 804–817.
28. Mysliwa-Kurdziel B, Prasad MNV, Strzałka K (2004) Photosynthesis in heavy metal stressed plants. In: Prasad M, editor. Heavy Metals Stress in Plants. From Biomolecules to Ecosystems pp. 146–182.
29. Kalaji HM, Loboda T (2007) Photosystem II of barley seedlings under cadmium and lead stress. Plant Soil Environ 53: 511–516.
30. Buonasera K, Lambreva M, Rea G, Touloupakis E, Giardi MT (2011) Technological applications of chlorophyll *a* fluorescence for the assessment of environmental pollutants. Anal Bioanal Chem 401: 1139–1151.
31. Krupa Z, Moniak M (1998) The stage of leaf maturity implicates the response of the photosynthetic apparatus to cadmium toxicity. Plant Sci 138: 149–156.
32. Caldelas C, Araus J, Febrero A, Bort J (2012) Accumulation and toxic effects of chromium and zinc in *Iris pseudacorus*, L. Acta Physiol Plant 34: 1217–1228.

33. Goltsev V, Zaharieva I, Chernev P, Kouzmanova M, Kalaji HM, et al. (2012) Drought-induced modifications of photosynthetic electron transport in intact leaves: Analysis and use of neural networks as a tool for a rapid non-invasive estimation. Biochim Biophys Acta 1817: 1490–1498.
34. Guha A, Sengupta D, Reddy AR (2013) Polyphasic chlorophyll *a* fluorescence kinetics and leaf protein analyses to track dynamics of photosynthetic performance in mulberry during progressive drought. J Photochem Photobiol, B 119: 71–83.
35. Gururani MA, Upadhyaya CP, Strasser RJ, Woong YJ, Park SW (2012) Physiological and biochemical responses of transgenic potato plants with altered expression of PSII manganese stabilizing protein. Plant Physiol Biochem 58: 182–194.
36. Srivastava A, Guisse B, Greppin H, Strasser RJ (1997) Regulation of antenna structure and electron transport in Photosystem II of *Pisum sativum* under elevated temperature probed by the fast polyphasic chlorophyll *a* fluorescence transient: OKJIP. Biochim Biophys Acta, Bioenerg 1320: 95–106.
37. Tóth SZ, Schansker G, Garab G, Strasser RJ (2007) Photosynthetic electron transport activity in heat-treated barley leaves: The role of internal alternative electron donors to photosystem II. Biochim Biophys Acta, Bioenerg 1767: 295–305.
38. Strasser RJ, Tsimilli-Michael M, Srivastava A (2004) Analysis of the Chlorophyll *a* Fluorescence Transient. In: Papageorgiou G, Govindjee, editors. Chlorophyll *a* Fluorescence: A Signature of Photosynthesis. pp. 321–362.
39. Tóth SZ, Schansker G, Garab G (2013) The physiological roles and metabolism of ascorbate in chloroplasts. Physiol Plant 148: 161–175.
40. Janik E, Maksymiec W, Mazur R, Garstka M, Gruszecki WI (2010) Structural and functional modifications of the major light-harvesting complex II in cadmium or copper-treated *Secale cereale*. Plant Cell Physiol 51: 1330–1340.
41. Foyer C, Neukermans J, Queval G, Noctor G, Harbinson J (2012) Photosynthetic control of electron transport and the regulation of gene expression. J Exp Bot 63: 1637–1661.
42. Zagdanska B (1995) Energy metabolism in plants under water deficits. Acta Biochim Polon 42: 281–289.

22

Evaluation of Dissolved Organic Carbon as a Soil Quality Indicator in National Monitoring Schemes

David L. Jones[1]*, Paul Simfukwe[1], Paul W. Hill[1], Robert T. E. Mills[2], Bridget A. Emmett[3]

1 School of the Environment, Natural Resources & Geography, Bangor University, Bangor, Gwynedd, United Kingdom, **2** Ecole Polytechnique Fédérale de Lausanne (EPFL), School of Architecture, Civil and Environmental Engineering (ENAC), Laboratory of Ecological Systems (ECOS), Lausanne, Switzerland, **3** Centre for Ecology and Hydrology, Environment Centre Wales, Bangor, Gwynedd, United Kingdom

Abstract

Background: Monitoring the properties of dissolved organic carbon (DOC) in soil water is frequently used to evaluate changes in soil quality and to explain shifts in freshwater ecosystem functioning.

Methods: Using >700 individual soils (0–15 cm) collected from a 209,331 km^2 area we evaluated the relationship between soil classification (7 major soil types) or vegetation cover (8 dominant classes, e.g. cropland, grassland, forest) and the absorbance properties (254 and 400 nm), DOC quantity and quality (SUVA, total soluble phenolics) of soil water.

Results: Overall, a good correlation ($r^2 = 0.58$) was apparent between soil water absorbance and DOC concentration across the diverse range of soil types tested. In contrast, both DOC and the absorbance properties of soil water provided a poor predictor of SUVA or soluble phenolics which we used as a measure of humic substance concentration. Significant overlap in the measured ranges for UV absorbance, DOC, phenolic content and especially SUVA of soil water were apparent between the 8 vegetation and 7 soil classes. A number of significant differences, however, were apparent within these populations with total soluble phenolics giving the greatest statistical separation between both soil and vegetation groups.

Conclusions: We conclude that the quality of DOC rather than its quantity provides a more useful measure of soil quality in large scale surveys.

Editor: Dafeng Hui, Tennessee State University, United States of America

Funding: This work was funded by the United Kingdom Department for Environment, Food and Rural Affairs. The funders had no role in study design, data collection and analysis, decision to publish, or preparation of the manuscript.

Competing Interests: The authors have declared that no competing interests exist.

* E-mail: d.jones@bangor.ac.uk

Introduction

Measurement of dissolved organic matter (DOM) concentrations have provided a sensitive indicator for assessing large scale geographical and temporal (e.g. decadal) changes in water quality [1]. In contrast, while DOM is frequently measured in soil studies, it has rarely been used for evaluating changes in soil quality in large scale (e.g. national) monitoring programmes [2]. To some extent, the DOM concentration of freshwaters should reflect that of the surrounding catchment area suggesting that an assessment of soil DOM might provide a sensitive soil quality indicator. This would also match with studies investigating the progressive increase in DOM within rivers, a widespread trend observed in both Europe and North America [3–4]. An investigation of the spatial or temporal changes in soil DOM within different soil types could therefore prove useful for explaining these trends in catchments containing a complex mixture of soils. One potential drawback, however, is that determination of DOM in soil can be very time consuming and expensive when large numbers of samples need to be processed. One potential solution is to use the spectral properties of soil water which provides a rapid proxy for estimating dissolved organic C (DOC) concentration [5–10]. For a single geographical location, strong relationships are often apparent between DOC concentration and UV absorbance [11–13]. However, these relationships are often non-linear and the mathematical function describing the relationship can vary with season or changes in hydrological flow patterns [14–16]. Further, whilst differences in the quality and quantity of DOC are frequently reported for contrasting soil and vegetation types, whether these dissimilarities remain valid over large geographical scales remains uncertain. In this context, we undertook a national large scale survey of the quality and quantity of DOM across Great Britain to evaluate its potential for evaluating changes to soil quality in national soil quality monitoring programmes. Our first aim was to critically evaluate the relationship between direct and indirect methods for DOC quantification. Secondly, we aimed to evaluate whether different soil classes and vegetation cover types possessed unique DOM signatures in terms of both quantity and quality and whether these might be useful for explaining changes in DOM occurring at a national scale.

Methods

Soil Sampling

To encompass all the major soil and land use types, a total of 702 soil samples were collected throughout Great Britain in June-

July, 2007 (area 209,331 km^2; ca. 300 km^2 sample^{-1}) as part of the Centre for Ecology and Hydrology Countryside Survey (CS) [17]. Samples were selected, based on a stratified random sample of 1 km squares at gridpoints on a 15 km grid using the Institute of Terrestrial Ecology (ITE) Land Classification as the basis of the stratification (Fig. S1). At each grid intersection, a 1 km^2 sample area was selected. Within the 1 km^2 sample area, a 5×5 m^2 plot was randomly located and replicate 15 cm long ×4 cm diameter soil cores were collected. Topsoils (0–15 cm) were only sampled to reflect past and current (1978-present) standard practice in UK national monitoring [18–20]. This depth was originally selected by the UK Government's Department for Environment, Food and Rural Affairs as it was hypothesized that it would show the greatest change in response to environmental perturbation (including changes in land use, climate, agronomic management or atmospheric deposition). The soil horizons sampled included H, O and A horizons with *a*, *e*, *i*, *h*, *g*, *k* and *p* sub-designations [21]. Across all land use categories, the dominant eight soil groupings (% of total) were: Brown soils (31%), Podzolic soils (15%), Surface water (SW) gley soils (18%), Peat soils (13%), Groundwater (GW) gley soils (12%), Lithomorphic soils (8%), and Pelosol soils (3%) [22]. The FAO World Reference Base Classification equivalent categories for these soil groups are presented in Table S1 and their major chemical, physical and biological properties presented in Table 1. Vegetation cover at each sampling point was classified into eight aggregated vegetation classes (AVC) with the following groupings (% of total): Cropland (15%), Tall grass and herbs (4%), Fertile grassland (19%), Infertile grassland (21%), Lowland woodland (3%), Upland woodland (8%), Moorland grassland mosaic (11%), and Heathland and bog (19%) [17]. Aggregate vegetation classes were derived by cluster analysis of the mean DECORANA scores for 100 smaller classes obtained by TWINSPAN analysis of plant species data in each sample plots [37]. Further descriptions of the vegetation types can be found in Table S2.

Soil Water Collection and Analysis

Soil water was obtained by adding artificial rainwater to the intact soil columns and collecting the leachate as detailed in [23–24]. The absorbance of the soil leachate water was measured at 254 and 400 nm on a Synergy 96 well plate spectrophotometer (BioTek Instruments Inc., Winooski, VT) using Falcon flat-bottom UV well plates (BD Biosciences, Franklin Lakes, NJ). DOC concentrations were measured with a TOC-V analyser (Shimadzu Corp., Kyoto, Japan). Total dissolved phenolics and tannins were assayed colorimetrically using the Folin-Ciocalteu reagent (F9252; Sigma-Aldrich Inc.) according to [25] using gallic acid as a standard. Specific ultraviolet absorbance (SUVA) was calculated by dividing the absorbance at 254 nm (cm^{-1}) by the DOC concentration (mg l^{-1}).

Background Soil Analysis

Background soil characteristics were measured on replicate cores matching those used in the analysis above. After collection from the field, the soil was extruded from the core, roots and stones removed and the soil homogenised and dried (105 °C). Total soil C and N were analysed on an Elementar Vario-EL elemental analyser (Elementar Analysensysteme GmbH, Hanau, Germany) using the UKAS accredited method SOP3102 [23]. Bulk density was calculated as mass/volume after the removal of stones (>2 mm) and accounting for their volume [23]. Soil organic matter was determined by loss-on-ignition (LOI) by first drying soil (10 g) at 105°C and then measuring the mass loss after further heating at 375 °C for 16 h. Available soil P was measured using

Table 1. Properties of the major soil groups.

Soil group	Soil pH	Soil organic matter (%)	C-to-N ratio	Olsen P (mg kg^{-1})	Bulk density (g cm^{-3})	Exchangeable Ca (mmol kg^{-1})	Exchangeable Al (mmol kg^{-1})	Soil respiration (mg C cm^{-3} h^{-1})	Soil respiration (g C g SOC^{-1} h^{-1})
Pelosol	6.8±1.4^{a}	7.3±3.3^{cd}	10.7±0.8^{de}	28±22^{abc}	1.10±0.21^{a}	62±40^{a}	1.3±4.8a^{b}	10.2±26.4^{bc}	4.7±3.4^{a}
Brown	6.5±1.2^{a}	10.0±12.5^{d}	12.2±4.1^{e}	32±27^{abc}	1.04±0.33^{a}	39±27^{a}	2.1±4.8^{b}	8.6±11.0^{c}	7.3±5.6^{a}
Groundwater gley	6.6±1.2^{a}	10.9±13.5^{d}	13.4±4.4^{cde}	32±34^{abc}	0.98±0.36^{a}	42±31^{a}	2.3±5.4^{ab}	13.2±24.1^{bc}	8.1±6.4^{a}
Surface water gley	5.9±1.3^{a}	24.9±30.0^{bc}	14.2±5.5^{cd}	25±28^{bc}	0.78±0.43^{b}	54±49^{a}	6.3±14.6^{ab}	14.9±19.2^{bc}	6.6±5.0^{ab}
Podzolic	4.9±0.9^{b}	28.8±27.8^{b}	17.9±7.0^{b}	21±20^{c}	0.57±0.33^{c}	20±29^{b}	7.1±8.9a^{b}	17.9±32.9^{bc}	5.1±3.9^{b}
Lithomorphic	6.0±1.4^{a}	35.1±32.8^{b}	16.2±6.4^{bc}	41±69^{ab}	0.56±0.38^{c}	42±39^{a}	3.4±5.6^{ab}	20.8±33.0^{b}	6.4±5.6^{ab}
Peat	4.6±0.6^{b}	75.9±27.9^{a}	24.1±8.5^{a}	44±51^{a}	0.20±0.21^{d}	46±50^{a}	9.6±38.0^{a}	43.1±37.1^{a}	4.5±3.8^{b}
ANOVA	***	***	***	**	***	***	**	***	***

Values are expressed on a dry weight basis and represent mean ± SD.
** and *** indicate significant differences between soil groups at the $P<0.01$ and $P<0.001$ levels respectively. Superscript letters indicate significant between soil groups at the $P<0.05$ level.

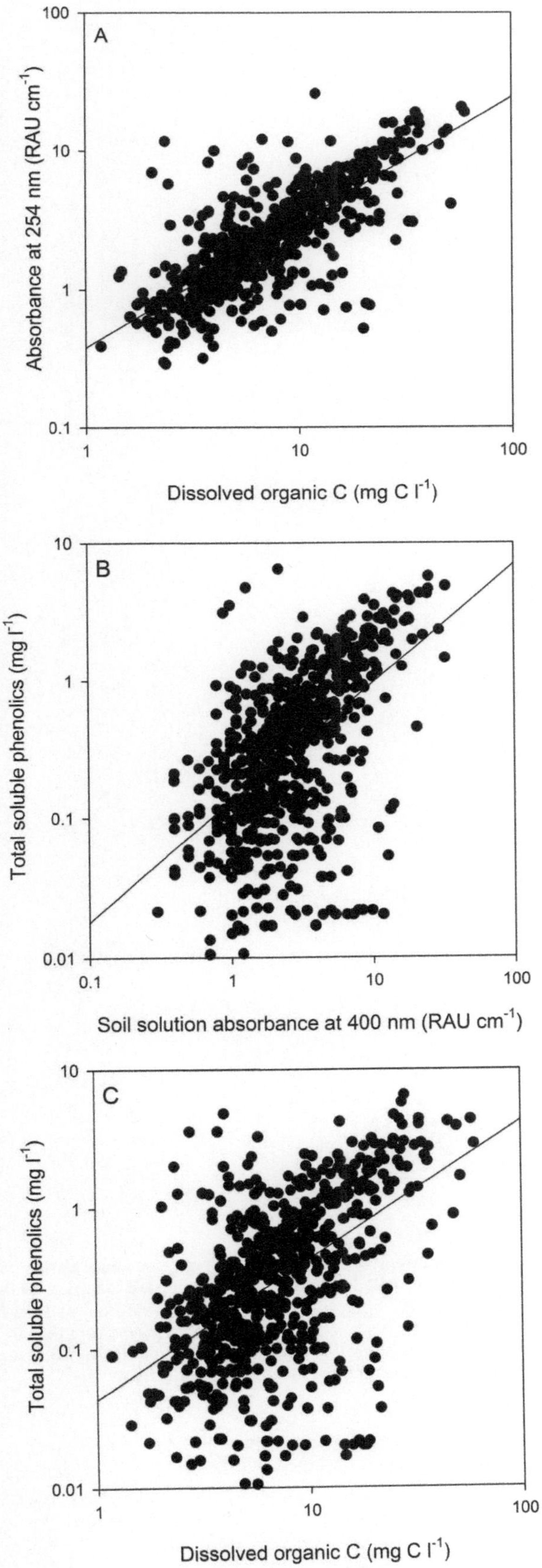

Figure 1. Relationship between the absorbance of soil water from 702 individual soils measured in the UV (254 nm) range and DOC concentration (Panel A), the relationship between total soluble phenolics and either solution absorbance at 400 nm (Panel B) or DOC concentration (Panel C). The lines are linear regression for all the data in the plots (Panel A, $r^2=0.579$; y = 0.387+0.301x; Panel B, $r^2=0.432$; y = 0.12+0.139x; Panel C, $r^2=0.375$; y = 0.025+0.068x).

the Olsen method whereby 5 g of soil was extracted with 100 ml of 0.5 M sodium bicarbonate (pH 8.5). The P in the extract was determined colorimetrically using the molybdate blue method (880 nm) using a continuous flow analyser [23]. Soil pH was measured by equilibrating 10 g of field-moist soil with 25 ml of deionised water. Exchangeable Ca and Al were determined by shaking 5 g of soil with 25 ml of 1.0 M NH_4Cl (250 rev min^{-1}, 60 min). Subsequently, the extracts were centrifuged (5000 *g*, 10 min) and the supernatant recovered for analysis. Ca in the extracts was determined by atomic absorption spectrometry on a Perkin Elmer Analyst 400 Atomic Absorption Spectrometer (PerkinElmer, Waltham, MA, USA). The extracts were diluted with $LaCl_3$ (0.5% w/v) prior to Ca determination. Al concentration in the extracts was determined using the modified catechol violet method [35]. The absorbance of the solution was measured at 580 nm using a PowerWave XS scanning microplate spectrophotometer (BioTek Instrument, Winooski, VT). Basal soil respiration was determined on one replicate core. The cores were wet to field capacity as described previously, placed in a sealed chamber (1250 cm^3 head space). The soils were then incubated at 10°C (average UK air temperature) for 1 h (at which linearity was known to be established following testing on selected cores which covered the range of soil types sampled). Subsequently, the head space gas was analysed for CO_2 concentration using a Clarus 500 Gas Chromatograph (PerkinElmer). SR was determined as the change in CO_2 concentration before and after incubation corrected for soil dry weight and soil organic matter content.

Statistical Analysis

Linear and stepwise regression analysis was undertaken using Minitab v16 (Minitab Inc., State College, PA). When the solution DOC, absorbance, soluble phenolic and SUVA values were grouped according to soil and vegetation type the data failed normality testing (Shapiro-Wilk). Consequently, the data were log transformed, normality verified and an ANOVA performed with Tukey-pairwise comparisons ($P<0.05$ cut-off) using Minitab v16.

Results

Soil Water Properties

The relationship between the absorbance of soil water in the UV (254 nm) and visible (400 nm) range across 702 individual sites showed a strong linear correlation ($r^2=0.931$; $P<0.001$; Fig. S2). A strong positive correlation was also observed between the DOC concentration in soil water and both absorbance at 254 nm ($r^2=0.58$; $P<0.001$; Fig. 1) and 400 nm ($r^2=0.47$; $P<0.001$). The prediction of DOC concentration using both absorbance values in a stepwise regression model did not result in a significantly better fit ($r^2=0.61$). In contrast, the absorbance characteristics of soil water proved to be a less reliable predictor of total soluble phenolics (254 nm $r^2=0.41$; $P<0.001$; 400 nm $r^2=0.43$; $P<0.001$; Fig. 1). Similarly, DOC concentration either with or without inclusion of spectral properties into the regression model, proved to be the least reliable predictor of total soluble phenolics ($r^2=0.38$; $P<0.001$). SUVA proved to be the worst predictor of

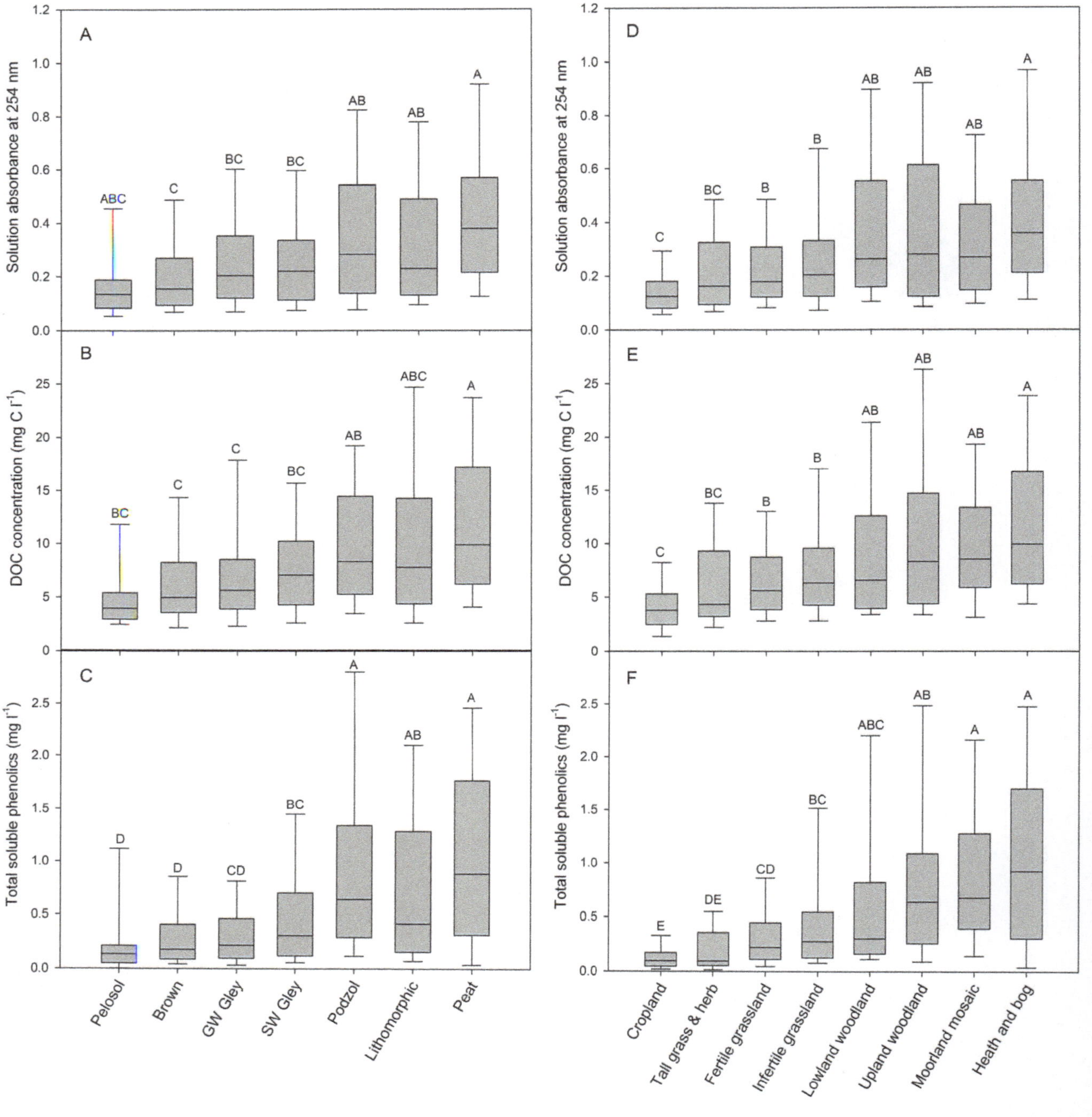

Figure 2. Box plots showing the influence of soil class (panels A–C) and vegetation cover type (panels D–F) on the absorbance of soil solutions at 254 nm, their DOC and total soluble phenolic concentration from 702 individual soils sampled as part of a nationwide soil quality assessment. The boundary of the box closest to zero indicates the 25th percentile, the line within the box marks the median, and the boundary of the box farthest from zero indicates the 75th percentile. Whiskers (error bars) above and below the box indicate the 90th and 10th percentiles respectively. Different letters indicate significant difference between individual groups ($P<0.05$). Soils are ranked in organic matter (OM) content from left (low OM) to right (high OM) while vegetation is broadly ranked according to agricultural productivity.

total soluble phenolics with either linear ($r^2 = 0.13$) or non-linear models (data not presented).

Is Soil or Vegetation a Better Predictor of DOC Quantity and Quality?

The quantity and quality of DOC as a function of different soil class and vegetation types is shown in Figure 2. Overall, there was a large similarity in the measured ranges of DOC, absorbance at 254 nm and soluble phenolics between the different groupings for both soil and vegetation. Despite this, ANOVA revealed significant differences between groups (all $P<0.001$) with response gradients apparent when the groups were ranked according to either organic matter content (soils) or an intensification/altitude scale (vegetation)(Fig. 2). Of the parameters measured, soluble phenolic content gave the greatest separation between groups. In contrast, SUVA revealed no significant differences between groupings for both soil ($P = 0.483$) and vegetation ($P = 0.819$; Fig. S3).

Discussion

Method of Soil Sampling

The sampling undertaken here formed part of a nationwide monitoring programme which is carried out each decade to assess changes in soil and water quality, landscape features and aquatic and terrestrial biodiversity [19,26]. Whilst the sampling depth (0–15 cm) is highly suited to agricultural soils which show a high degree of vertical homogeneity (due to cultivation), we acknowledge that the sampling regime may be less suited to highly stratified soils [27]. In these horizonated soils, differential amounts of O, E, A and possibly B horizons could be sampled by coring from 0–15 cm, depending on their thickness. This has led to many studies, particularly in forests, where soil quality is measured on organic and mineral horizons separately [27–28]. In our case, the historical legacy is such that the sampling protocols will not be changed for the foreseeable future (on scientific, socio-political and economic grounds) and therefore our results should be viewed in the context of this.

DOC and Soil Quality Assessment

Overall, our results show that in large geographical scale soil quality assessments across a broad range of vegetation and soil types, soil water DOC concentrations can be rapidly and cheaply estimated from their spectral properties. This is particularly relevant as DOC in itself can be used to predict the movement of both organic and inorganic pollutants in soil and is often seen as a pollutant itself when entering freshwaters [9,29–30]. Disappointingly, however, we found that all soil and vegetation types had a wide variability in their DOC concentrations. This indicates that the quantity of DOC may not represent a sensitive indicator for monitoring the stability of ecosystems when faced with anthropogenic perturbation (e.g. land use change). This is in contrast to [24] who found different rates of N mineralization ($NH_4^+ + NO_3^-$) in the same soils alongside other soil quality indicators as shown in Table 1. While SUVA is frequently used for assessing the quality of DOC in freshwaters [30], our results indicated that it was incapable of separating between land uses and soil classes over a wide geographical range and is therefore probably unsuited to large scale soil quality assessments. In contrast, total soluble phenols gave the best separation between ecosystem types. This was a surprising result as we had assumed that SUVA and soluble phenols would be highly correlated. This disagreement could be due to interference in both the determination of total phenolics (e.g. SO_2, DOC; [31–32]) or SUVA (e.g. pH, Fe^{3+}; [12]). However, more likely it is due to the natural variability in the total phenol content of humic substances originating from different soil classes and vegetation cover types [33–34]. As expected, the quality and quantity of DOC were dependent on both vegetation and soil type. This is to some extent expected considering that certain soil types favour certain vegetation covers, however, it does indicate that 2-way stratification by soil and vegetation type may prove more useful for revealing unique DOC signatures.

With respect to the measurement of DOC quality, we acknowledge that recent methodological advancements are increasing our potential to characterise the many thousands of compounds that comprise DOC [36]. It is highly likely that these new analytical approaches will reveal compounds that are specific to different functional soil or plant types (e.g. keystone compounds) improving our capacity to use DOC as a soil quality indicator. In addition, this analysis may also directly support the interpretation of temporal changes in other soil quality indicators (e.g. soil organic matter quality and quantity, soil biodiversity).

Supporting Information

Figure S1 Map of the UK showing the individual soil sampling locations used in the study. The total land area is 209,331 km2.

Figure S2 Relationship between the absorbance of soil water from 702 individual soils measured in either the UV (254 nm) or visible (400 nm) range. The line is a linear regression for all the data in the plot ($r^2 = 0.931$; $y = -0.01 + 0.123x$).

Figure S3 Box plots showing the influence of soil type (panel A) and vegetation cover (panel B) on the specific UV absorbance (SUVA) values from 702 individual soils sampled as part of a nationwide soil quality assessment. The boundary of the box closest to zero indicates the 25th percentile, the line within the box marks the median, and the boundary of the box farthest from zero indicates the 75th percentile. Whiskers (error bars) above and below the box indicate the 90th and 10th percentiles respectively. No significant differences were apparent between treatments.

Table S1 Comparable classification of the UK soil groups with those in the FAO World Reference Base Classification (WRB, 2006).

Table S2 Summary descriptions of the eight aggregate vegetation classes represented. Table adapted from Smart et al. (2003).

Acknowledgments

We would like to thank Ed Rowe at the Centre for Ecology and Hydrology for help with the experiments and data handling. Countryside Survey (CS) is funded by a consortium of UK government funded bodies led by the Natural Environment Research Council (NERC) and the Department for Environment, Food and Rural Affairs (Defra), which includes the Centre for Ecology and Hydrology, Countryside Council for Wales, Forestry Commission, Natural England, the Northern Ireland Environment Agency, the Scottish Government, Scottish Natural Heritage, and the Welsh Assembly Government. The completion of CS was only possible with the support and advice of many dedicated individuals from these and other organizations. The authors would like to thank the land owners, farmers, and other land managers who gave permission for the field surveyors to collect data and samples from their land; without such cooperation scientific field studies like CS would not be possible. Additionally, the authors would especially like to acknowledge the major contribution made by the team of laboratory staff who processed all the soils and undertook the chemical analyses.

Author Contributions

Conceived and designed the experiments: DLJ BAE. Performed the experiments: PS RTEM. Analyzed the data: PS PWH DLJ RTEM. Wrote the paper: DLJ PWH PS BAE.

References

1. Camarero L, Rogora M, Mosello T, Anderson NJ, Barbieri A, et al. (2009) Regionalisation of chemical variability in European mountain lakes. Freshwater Biol 54: 2452–2469.
2. Filep T, Rekasi M, Kincses I (2008) Effect of soil quality on dissolved organic matter (DOM) content of soils. Cereal Res Commun 36: 667–670.
3. Evans CD, Monteith DT, Cooper DM (2005) Long-term increases in surface water dissolved organic carbon: Observations, possible causes and environmental impacts. Environ Pollut 137: 55–7.
4. Clutterbuck B, Yallop AR (2010) Land management as a factor controlling dissolved organic carbon release from upland peat soils. 2 Changes in DOC productivity over four decades. Sci Total Environ 408: 6179–6191.
5. Mattson JS, Smith CA, Jones TT, Gerchako SM, Epstein BD (1974) Continuous monitoring of dissolved organic-matter by UV-visible photometry. Limnol Oceanogr 19: 530–535.
6. Dittman JA, Shanley JB, Driscoll CT, Aiken GR, Chalmers AT, et al. (2009) Ultraviolet absorbance as a proxy for total dissolved mercury in streams. Environ Pollut 157: 1953–1956.
7. Brandstetter A, Sletten RS, Mentler A, Wenzel WW (1996) Estimating dissolved organic carbon in natural waters by UV absorbance (254 nm). J Soil Sci Plant Nutr 159: 605–607.
8. Jaffrain J, Gerard F, Meyer M, Ranger J (2007) Assessing the quality of dissolved organic matter in forest soils using ultraviolet absorption spectrophotometry. Soil Sci Soc Am J 71: 1851–1858.
9. Amery F, Degryse F, Cheyns K, De Troyer I, Mertens J, et al. (2008) The UV-absorbance of dissolved organic matter predicts the fivefold variation in its affinity for mobilizing Cu in an agricultural soil horizon. Eur J Soil Sci 59: 1087–1095.
10. Hansson K, Kleja DB, Kalbitz K, Larsson H (2010) Amounts of carbon mineralised and leached as DOC during decomposition of Norway spruce needles and fine roots. Soil Biol Biochem 42: 178–185.
11. Dilling J, Kaiser K (2002) Estimation of the hydrophobic fraction of dissolved organic matter in water samples using UV photometry. Water Res 36: 5037–5044.
12. Weishaar JL, Aiken GR, Bergamaschi BA, Fram MS, Fujii R, et al. (2003) Evaluation of specific ultraviolet absorbance as an indicator of the chemical composition and reactivity of dissolved organic carbon. Environ Sci Technol 37: 4702–4708.
13. Tipping E, Corbishley HT, Koprivnjak JF, Lapworth DJ, Miller MP, et al. (2009) Quantification of natural DOM from UV absorption at two wavelengths. Environ Chem 6: 472–476.
14. Wheeler JR (1977) Dissolved organic-carbon - spectral relationships in coastal waters. Limnol Oceanogr 22: 573–575.
15. Edwards AC, Cresser MS (1987) Relationships between ultraviolet absorbency and total organic-carbon in 2 upland catchments. Water Res 21: 49–56.
16. Izbicki JA, Pimentel IM, Johnson R, Aiken GR, Leenheer J (2007) Concentration, UV-spectroscopic characteristics and fractionation of DOC in stormflow from an urban stream, Southern California, USA. Environ Chem 4: 35–48.
17. Emmett BA, Reynolds B, Chamberlain PM, Rowe E, Spurgeon D, et al. (2010) Countryside Survey: Soils Report from 2007. CS Technical Report No. 9/07, Centre for Ecology and Hydrology, Wallingford, UK.
18. Bellamy PH, Loveland PJ, Bradley RI, Lark RM, Kirk GJD (2005) Carbon losses from all soils across England and Wales 1978–2003. Nature 437: 245–248.
19. Norton LR, Maskell LC, Smart SS, Dunbar MJ, Emmett BA, et al. (2012) Measuring stock and change in the GB countryside for policy - Key findings and developments from the Countryside Survey 2007 field survey. J Environ Manage 113: 117–127.
20. Reynolds B, Chamberlain PM, Poskitt J, Woods C, Scott WA, et al. (2013) Countryside Survey: National "soil change" 1978-2007 for topsoils in Great Britain-acidity, carbon, and total nitrogen status. Vadose Zone J 12: vzj2012.0114
21. FAO (2006) Guidelines for Soil Description. Fourth Edition.Food and Agriculture Organization of the United Nations, Rome, Italy.
22. Avery BW (1990) Soils of the British Isles. Cambridge University Press, Cambridge.
23. Emmett BA, Frogbrook ZL, Chamberlain PM, Griffiths R, Pickup R, et al. (2008) Countryside Survey. Soils Manual. NERC/Centre for Ecology & Hydrology, (CS Technical Report No. 3/07, CEH Project Number: C03259), 180 pp.
24. Rowe EC, Emmett BA, Frogbrook ZL, Robinson DA, Hughes S (2012) Nitrogen deposition and climate effects on soil nitrogen availability: Influences of habitat type and vegetation. Sci Total Environ 434: 62–70.
25. Velioglu YS, Mazza G, Gao L, Oomah BD (1998) Antioxidant activity and total phenolics in selected fruits, vegetables and grain products. J Agric Food Chem 46: 4113–4117.
26. Smart SM, Clarke RT, van de Poll HM, Robertson EJ, Shield ER, et al. (2003) National-scale vegetation change across Britain; an analysis of sample-based surveillance data from the Countryside Surveys of 1990 and 1998. J Environ Manage 67: 239–254.
27. Lawrence GB, Fernandez IJ, Richter DD, Ross DS, Hazlett PW, et al. (2013) Measuring environmental change in forest ecosystems by repeated soil sampling: a North American perspective. J Environ Qual 42: 623–639.
28. O'Neill KP, Amacher MC, Palmer CJ (2005) Developing a national indicator of soil quality on U.S forestlands: Methods and initial results. Environ Monit Assess 107: 59–80.
29. Steinberg CEW, Hoss S, Kloas W, Lutz I, Meinelt T, et al. (2003) Hormonelike effects of humic substances on fish, amphibians, and invertebrates. Environ Toxicol 19: 409–411.
30. Zhang Q, Kuang WF, Liu LY, Li KX, Wong KH, et al. (2009) Trihalomethane, haloacetonitrile, and chloral hydrate formation potentials of organic carbon fractions from sub-tropical forest soils. J Hazard Mater 172: 880–887.
31. Matilainen A, Gjessing ET, Lahtinen T, Hed L, Bhatnagar A, et al. (2011) An overview of the methods used in the characterisation of natural organic matter (NOM) in relation to drinking water treatment. Chemosphere 83: 1431–1442.
32. Ohno T, First PR (1998) Assessment of the Folin and Ciocalteu's method for determining soil phenolic carbon. J Environ Qual 27: 776–782.
33. Sharma OK, Krishnan PS (1966). Colorimetric estimation of humic acid with phenol reagent of Folin and Ciocalteu. Anal Biochem 14: 11–17.
34. Thorn KA, Folan DW, MacCarthy P (1989) Characterization of the International Humic Substances Society standard and reference fulvic and humic acids by solution state carbon-13 (^{13}C) and hydrogen-1 (^{1}H) nuclear magnetic resonance spectrometry. U.S. Geological Survey, Water-Resources Investigations Report 89-4196, Denver, CO, USA.
35. Dougan WK, Wilson AL (1974) The absorptiometric determination of aluminium in water. A comparison of some chromogenic reagents and the development of an improved method. Analyst 99: 413–30.
36. Osborne DM, Podgorski DC, Bronk DA, Roberts Q, Sipler RE, et al. (2013) Molecular-level characterization of reactive and refractory dissolved natural organic nitrogen compounds by atmospheric pressure photoionization coupled to Fourier transform ion cyclotron resonance mass spectrometry. Rapid Comm. Mass Spectrom. 27: 851–858.
37. Bunce RGH, Barr CJ, Gillespie MK, Howard DC, Scott RA, et al. (1999) Vegetation of the British Countryside—the Countryside Vegetation System. Department of the Environment, London.

23

Monitoring Urban Greenness Dynamics Using Multiple Endmember Spectral Mixture Analysis

Muye Gan[1], Jinsong Deng[1,2]*, Xinyu Zheng[1], Yang Hong[2,3], Ke Wang[1]*

1 College of Environmental and Resource Sciences, Zhejiang University, Hangzhou, Zhejiang Province, China, **2** School of Civil Engineering and Environmental Sciences and School of Meteorology, University of Oklahoma, Norman, Oklahoma, United States of America, **3** State Key Laboratory of Hydroscience and Engineering, Department of Hydraulic Engineering, Tsinghua University, Beijing, China

Abstract

Urban greenness is increasingly recognized as an essential constituent of the urban environment and can provide a range of services and enhance residents' quality of life. Understanding the pattern of urban greenness and exploring its spatiotemporal dynamics would contribute valuable information for urban planning. In this paper, we investigated the pattern of urban greenness in Hangzhou, China, over the past two decades using time series Landsat-5 TM data obtained in 1990, 2002, and 2010. Multiple endmember spectral mixture analysis was used to derive vegetation cover fractions at the subpixel level. An RGB-vegetation fraction model, change intensity analysis and the concentric technique were integrated to reveal the detailed, spatial characteristics and the overall pattern of change in the vegetation cover fraction. Our results demonstrated the ability of multiple endmember spectral mixture analysis to accurately model the vegetation cover fraction in pixels despite the complex spectral confusion of different land cover types. The integration of multiple techniques revealed various changing patterns in urban greenness in this region. The overall vegetation cover has exhibited a drastic decrease over the past two decades, while no significant change occurred in the scenic spots that were studied. Meanwhile, a remarkable recovery of greenness was observed in the existing urban area. The increasing coverage of small green patches has played a vital role in the recovery of urban greenness. These changing patterns were more obvious during the period from 2002 to 2010 than from 1990 to 2002, and they revealed the combined effects of rapid urbanization and greening policies. This work demonstrates the usefulness of time series of vegetation cover fractions for conducting accurate and in-depth studies of the long-term trajectories of urban greenness to obtain meaningful information for sustainable urban development.

Editor: Clinton N. Jenkins, Instituto de Pesquisas Ecológicas, Brazil

Funding: This work was supported by the Survey and Evaluation of National Ecological Environment (2000-2010) using Remote Sensing Program of Chinese Ministry of Environmental Protection (No. STSN-05-11). The funders had no role in study design, data collection and analysis, decision to publish, or preparation of the manuscript.

Competing Interests: The authors have declared that no competing interests exist.

* Email: kwang@zju.edu.cn (KW); jsong_deng@zju.edu.cn (JSD)

Introduction

The rapid urbanization in China in recent decades, accompanied by the continuous increase in urban population and the unprecedented growth of cities, has put significant pressure on urban environments. This pressure has led to severe environmental issues, such as urban heat islands, air pollution, and urban flooding [1]. Moreover, improved economic conditions have lead residents to demand better living environments and a better quality of life.

Urban greenness, represented in this study by all vegetation cover in and around cities, e.g., street plantation, lawns, parks, gardens, crops, wetlands, and forests, contributes valuable ecosystem services and plays an irreplaceable role in the improvement of the urban environment [2]. Urban vegetation improves air quality [3], intercepts storm water runoff [4], reduces energy demands [5], and provides numerous psychological benefits, aesthetic views and restorative opportunities for city dwellers who may otherwise have limited exposure to natural environments [6,7].

Rapid urbanization has resulted in significant alterations in the quality and quantity of urban greenness [8]. In a large-sample survey involving 386 European cities, green space was found to increase with the city area and slightly decline with the population density [9]. Dallimer et al. indicated that nine out of thirteen cities in England exhibited declining urban green spaces between 2000 and 2008 [10]. Infill development during urbanization is considered to be a major cause of green space loss. Similarly, a statistically significant decline in tree cover was found in many major U.S. cities, along with increased population densities and likely increased development pressure between 2003 and 2009 [11]. Unlike the declining trend reported by the above studies, green space coverage may increase with urbanization intensity. Green space coverage in the built-up areas of Chinese cities, for example, has increased steadily during the last two decades [12]. Yang et al. confirmed the greening trend in larger cities, and they note that rapid urbanization caused a dramatic turnover in vegetation cover [13]. Accurate and timely monitoring of the status of urban greenness is essential for the protection and management of urban environments. Furthermore, characterizing and understanding the trends in urban vegetation cover change provide insights into the relationship between urban greenness and urbanization and can help guide sustainable urban development [14].

As an important source of urban information, remote sensing data provide a spatially consistent coverage of large areas with both high spatial detail and high temporal frequency [15]. With the increased availability of current and historical remotely sensed data and new analytical techniques, it is now possible to quantify urban greenness in a timely and cost-effective manner [16,17]. Despite the increasing application of high-resolution images, medium-resolution imagery (e.g., Landsat and SPOT) is still the preferred data source. Specifically, medium-resolution imagery is globally available, and it uniquely composes the only long-term, consistent digital dataset. However, commonly used detection techniques, such as the calculation of classification and vegetation indices from medium spatial resolution imagery, may be ineffective in quantifying physically fine-resolution information related to urban greenness. The urban environment is highly heterogeneous and complex, resulting in the mixed pixel problem [18,19]. Because the status and variation of urban greenness typically occur at finer spatial scales than most moderate resolution imagery, the vegetation in a pixel can only be recorded as either present or absent with traditional hard classifiers, resulting in a loss of information on isolated vegetation and within-class variation [20,21]. However, the status of and subtle change in urban greenness are highly related to the urban environment and significantly affect local residents' quality of life. The specific characteristics of the urban landscape require further consideration regarding the capabilities and limitations of remote sensing data and the use of appropriate analysis techniques [22,23].

A variety of approaches has been developed to overcome the mixed pixel problem in urban landscapes. Spectral mixture analysis techniques (SMA and MESMA) [24] have been widely utilized to estimate the proportion of representative urban land cover within each pixel because of their ability to support the repeatable and accurate extraction of quantitative subpixel information with physical meaning [25,26]. To date, a considerable number of studies have utilized this method to characterize urban compositions [18,27,28], map urban vegetation [16,19,29], and monitor urban changes [30,31,32,33]. However, only a few multi-temporal studies have employed these techniques to monitor long-term changes in urban greenness [34]. Given the sharp growth in urban areas in China in terms of area and intensity, it is imperative to analyze the pattern of vegetation cover change within Chinese cities.

Taking Hangzhou, one of the most rapidly urbanizing cities in eastern China, this paper proposes a feasible and cost-effective greenness change detection method in highly-fragmented urban environment by integrating multi-date remote sensing, MESMA and GIS spatial analysis. Specifically, this paper explores the spatiotemporal dynamics and evolution of greenness in response to the rapid urbanization process in the past two decades (1990–2010) aiming to improve the understanding of the effects of urbanization on greenness patterns and provide basic information for appropriate decision-making towards urban sustainable development.

Methods

Study site

Hangzhou, the capital of Zhejiang Province, is located near the eastern edge of the Qiantang River in the southern Yangtze River Delta (Figure 1). This study was conducted in the area within the administrative boundaries of Hangzhou, the city proper, which covers an area of 728 km^2 and has approximately 3.56 million registered permanent residents. At the end of 2010, the local GDP per capita was approximately 109,708 RMB Yuan (equivalent to 17,981 US dollars) [35]. As a notable international garden city, as well as a vigorous and economically competitive key city, Hangzhou has experienced significant economic development and tremendous population growth over the past decades [15]. The urbanization process has significantly modified the vegetation cover in the wetland, forest, and agriculture ecosystems in and around the city.

Data and preprocessing

The Landsat TM images (Path: 119, Row: 39) used to conduct the study were obtained on October 8, 1990, September 23, 2002, and May 24, 2010 (i.e., a total of 20 years). The level 1T data were carefully selected from the USGS GLOVIS portal [36], with special consideration of phenology and cloudlessness. Thermal bands were excluded from the analysis. FLAASH (Fast Line-of-sight Atmospheric Analysis of Spectral Hypercubes) was used for atmospheric correction of these TM data. The study area was then extracted using the boundaries of Hangzhou.

In addition to Landsat imagery, the following ancillary data were collected for analytical purposes: (1) two historical true color aerial photographs with a 0.5-m resolution, acquired on September 26, 2002 and May 20, 2011; and (2) land-use maps derived from the National Detailed Land-use Inventory for 2010. All the auxiliary data were registered to the same projection as the TM images: WGS 84, UTM map projection Zone 50, with a root mean squared error (RMSE) of less than 15 m.

Greenness mapping requires special consideration of vegetation phenology. Generally, images from the same season with full vegetation growth are the most appropriate for detecting changes. However, because a high-quality image was not available for the late summer of 2010, a TM image acquired on May 24 was used. No significant bias in the vegetation cover was observed in late May or late September (except for cropland), according to our pre-experiment using a MODIS NDVI time series and a comparison between the fraction results from a Landsat 5 TM image acquired on May 24, 2010 and a Landsat 7 ETM+ SLC-off image acquired on September 21, 2010. To minimize the information error caused by the crop phenology, a common cropland mask was developed based on the 2010 land-use map and visual interpretation of three TM images. The common cropland pixels for the three dates were extracted, and the vegetation fraction values were set to a consistent value of 1 (Figure 2c, yellow circles). Thus, we assumed that the phenological effect of the vegetation in our study had no effect on our results during processing.

Multiple endmember spectral mixture analysis

The SMA approach assumes that a landscape is formed from continuously varying proportions of idealized land cover types or spectrally 'pure' materials called endmembers. In linear SMA, the reflectance P' measured at pixel i represents a linear combination of N endmembers, weighted by their areal fractions f_{ki}, within the pixel [24,25,33] as follows:

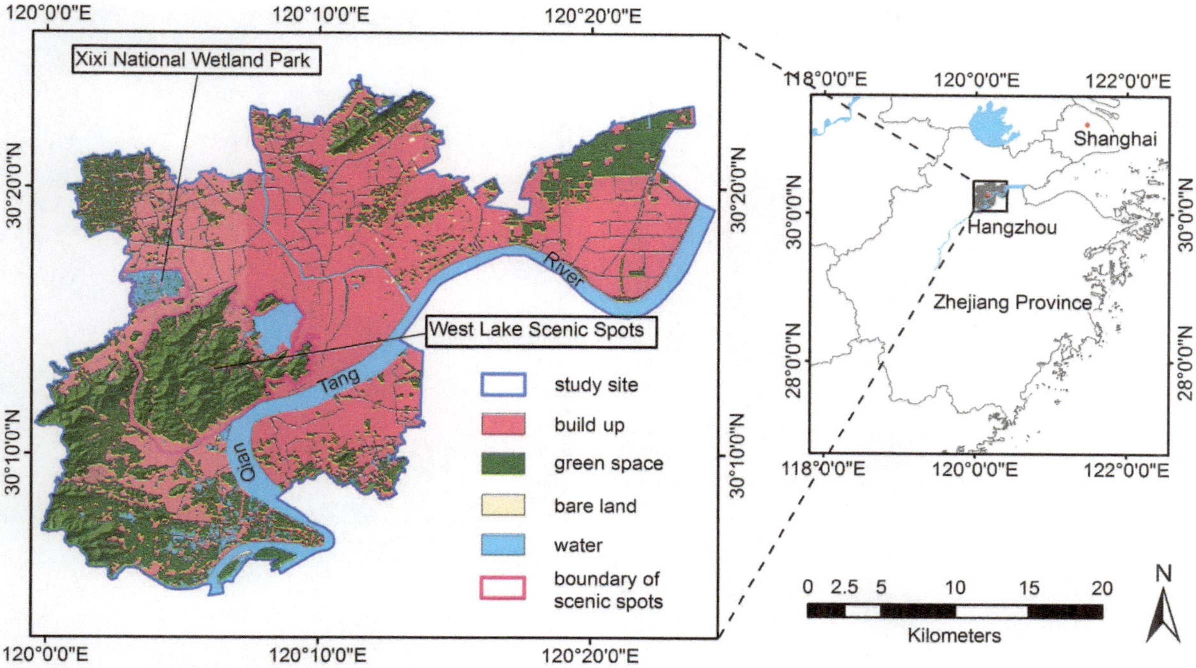

Figure 1. A map of the study area, i.e., Hangzhou city proper.

$$P'_{i\lambda} = \sum_{k=1}^{N} f_{ki} \times P_{k\lambda} + e_{i\lambda} \text{ and } \sum_{k=1}^{N} f_{ki} = 1 \quad (1)$$

where $P_{k\lambda}$ is the reflectance of endmember k for a specific band (λ) and $e_{i\lambda}$ is a residual term indicating the unmodeled portions of the spectrum. The model fit is assessed via an RMSE metric [24,37], calculated as follows:

$$\text{RMSE}_i = \sqrt{\sum_{\lambda=1}^{M} (e_{i\lambda})^2 / M} \quad (2)$$

where M is the number of bands [38]. The standard SMA has two major limitations. First, only one spectrum is allowed for each endmember, i.e., every image pixel is interpreted using the same endmember spectra. Therefore, the method is unable to account for within-class spectral variability. This is particularly problematic in Chinese cities where a large variation of material spectra exists due to rapid development and ubiquitous urban reconstruction. Second, the standard SMA is most often implemented with a fixed number of endmembers for the entire image, regardless of whether the ground components represented by the endmembers are present in the pixel. This may result in the decreased accuracy of the estimated fractions [18,25,39]. MESMA, which is an extension of the standard SMA approach, allows the number and type of endmembers to vary on a per pixel basis to better represent the spectral variability of land cover; thus, it is more suitable for urban landscapes [25,27].

In this study, MESMA was conducted according to the following steps: (1) endmember selection, (2) spectral unmixing modeling, and (3) accuracy assessment. These steps are each detailed in turn. MESMA was implemented using VIPER Tools Version 1.5, which is a plug-in to the ENVI software package [40].

Endmember selection. The careful selection of representative endmembers is essential to all applications of SMA [20,41,42]. In this study, the endmembers were extracted from the 1990, 2002, and 2010 Landsat TM images. When building a spectral library for MESMA, the number of spectra needed for adequately representing the spectral variation in the materials and for attaining computational efficiency should be considered [25,43,44]. Therefore, the endmembers were organized into five groups: vegetation, impervious surfaces, soil, water, and shade. Each group included several subsets, e.g., vegetation (urban forest, primary forest, grass, and crops), impervious surfaces (urban center and suburbs), soil (mine soil, farmland soil, and construction sites), and water (lakes and rivers). An endmember for shade (zero value in all bands) was included in all of the models [41,45].

First, the spectrally "pure" pixels were collected from the image based on the pixel purity index (PPI) [46] and high-resolution aerial photographs to build a candidate spectral library for each endmember. Pixels with high PPI scores, representing pure land cover in the high-resolution aerial photograph and having reasonable spectral signatures, were chosen as candidate spectra and grouped into the corresponding subclasses. The representative spectra for each subclass were then selected from the candidate spectra using the endmember average root mean squared error (EAR) [41], count-based endmember selection (CoB) [47], and minimum average spectral angle (MASA) [38]. The optimal set of spectra for each endmember was iteratively selected by adding spectra with a high CoB index, low EAR, or low MASA to the library and assessing the model performance using RMSE images and a visual comparison with the high-resolution aerial photographs [16]. Ultimately, libraries with 47, 53, and 69 spectra were used for the 1990, 2002, and 2010 images, respectively, to run the unmixing models.

Spectral unmixing modeling. Previous research has shown that natural systems are usually best modeled by two endmembers, while disturbed regions require three and urban landscapes

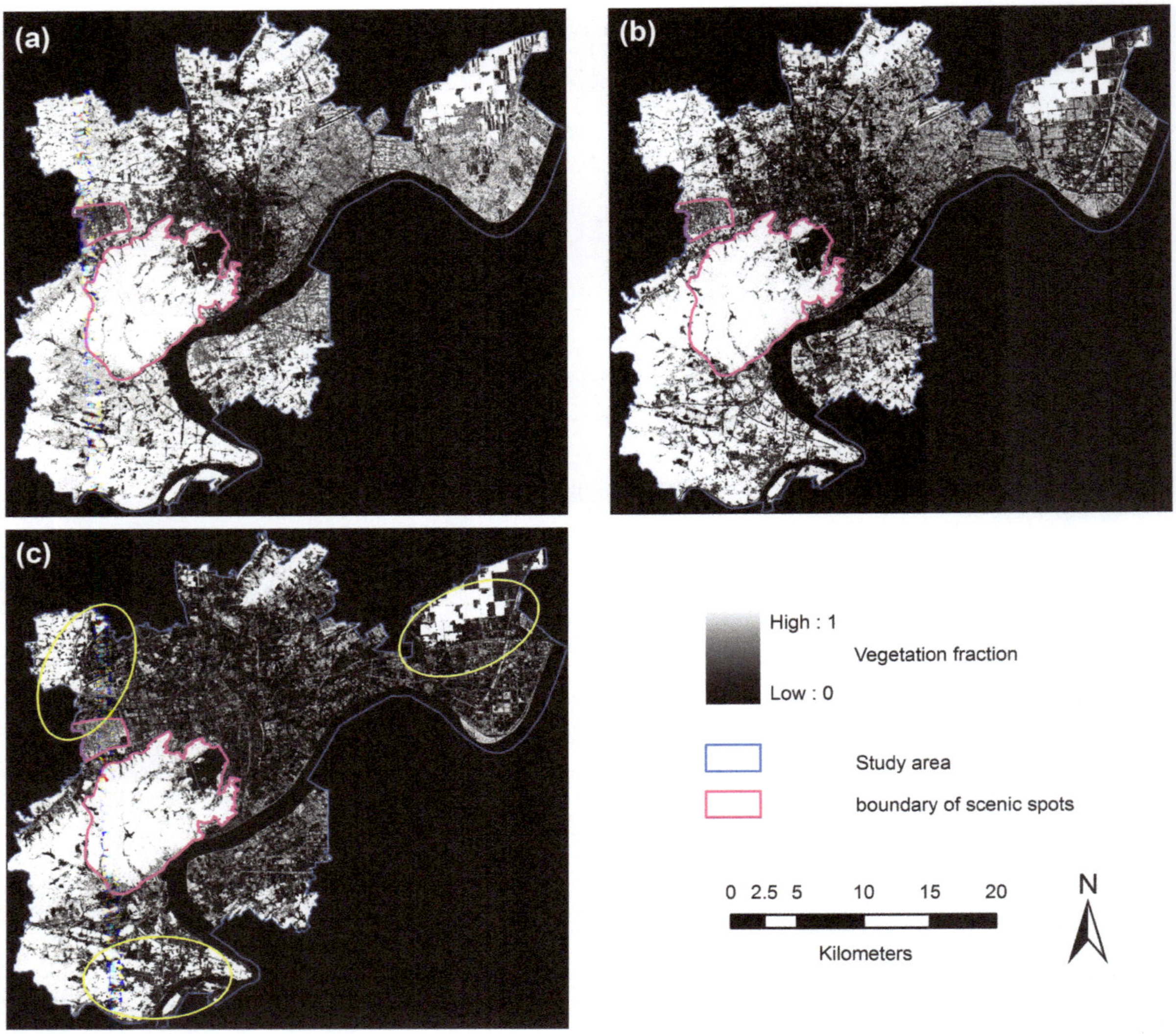

Figure 2. Vegetation fraction maps of Hangzhou for (a) 1990, (b) 2002, and (c) 2010. The vegetation fractions are indicated by the grayscale, where 1 represents pure vegetation (light gray) and 0 represents no vegetation (dark gray) in each pixel.

require four [48]. Therefore, two-endmember models with all the spectra were first applied to map each pixel in the image (Table 1). Then, only spectra that effectively represented the pure land cover within the urban area in the two-endmember models were retained to run three- and four-endmember models. Intra-class spectral mixtures were only considered for impervious surfaces [48]. Every model was evaluated for every pixel in the image. Constraints were determined based on iterative experimentation and the findings of previous studies [37,41,45]. The non-shade fractions were confined to the range of −0.05 to 1.05, the maximum allowable shade fraction was set to 0.5, and a maximum RMSE of 2.5% in reflectance was applied.

After the models with the different levels of spectral complexity (i.e., two-, three-, and four-endmember models) were applied, the model with the lowest RMSE at each level was compared to determine the best model for each pixel. Generally, a lower RMSE in a more complex model does not necessarily indicate better modeling of the true land cover component [28,49]. In addition, we noticed that the introduction of a water endmember significantly improved the model performance in the water-containing and water-edge pixels (e.g., wetlands), but it also caused a mis-unmixing between water and shade classes in pixels affected by topographic shadowing within mountainous regions [50]. Based on previous findings, simpler endmember models are preferred over more complex endmember models, except in cases in which adding an endmember significantly improves the RMSE beyond a specified threshold [30,47,50]. The RMSE threshold value between the two- and three-endmember models was 0.2% for the reflectance, the RMSE threshold between the three- and four-endmember models was 0.4%, and the RMSE threshold between the two- and four-endmember models was 0.6%. When a water endmember was used, the corresponding threshold values were set to 0.7%, 0.4% and 1.1%, respectively, for the reflectance.

The fractions produced by the optimal models were then shade-normalized [24], rescaled [26], and combined to generate fractional maps of each component. Vegetation cover fraction (VCF) maps were retained to investigate the vegetation cover change.

Accuracy assessment. Evaluating vegetation fraction estimates can be challenging due to the difficulty of obtaining

Table 1. Combinations of endmember models by land cover class.

2-endmember	3-endmember	4-endmember
Imp+SHD	Imp_a+Imp_b+SHD	Imp+V+S+SHD
V+SHD	Imp+V+SHD	Imp+V+W+SHD
S+SHD	Imp+S+SHD	S+V+W+SHD
W+SHD	Imp+W+SHD	$Imp_a+Imp_b+V+SHD$
	V+S+SHD	$Imp_a+Imp_b+S+SHD$
	V+W+SHD	$Imp_a+Imp_b+W+SHD$
	S+W+SHD	

V refers to green vegetation, Imp to impervious surface, S to soil, W to water, and SHD to shade. Endmembers with subscripts (a, b) were used multiple times. The same endmember spectra were not used in one combination.

reference data, especially for historical datasets. Because of the lack of high-quality reference images from the 1990s, accuracy assessments for only two dates were performed using the aerial photographs acquired in 2002 and 2011 for the relevant TM images. We assumed that the result of the 1990 fraction image would have acceptable accuracy if the vegetation fraction images in 2002 and 2010 were reasonably accurate. The lack of an accuracy assessment for the 1990 fraction image is a limitation of this study.

Two hundred samples were used to assess the modeled VCF after removing the samples with apparent land cover changes. Each sample had a window size of 3×3 pixels to reduce the impact of geometric errors between the aerial photographs and the TM images [27,51]. The 'true' proportion of vegetation was then calculated by dividing the area of vegetation cover measured through visually interpreting the aerial photograph by the total sampling area (i.e., 8,100 m^2). The accuracy of the vegetation cover fraction was assessed using three metrics: the coefficient of determination (R^2), the root mean squared error (RMSE), and the systematic error (SE) [52] as follows:

$$\mathrm{RMSE}=\sqrt{\sum_{i=1}^{S}(\hat{V}_i-V_i)^2/S} \quad (3)$$

$$\mathrm{SE}=\sum_{i=1}^{S}(\hat{V}_i-V_i)/S \quad (4)$$

where $\hat{V}_i$ is the modeled fractional value of vegetation measured at sample i, V_i is the reference fractional value, and S is the number of samples. The RMSE is a measure of the overall estimation accuracy for all samples, and the SE is a measure of the effects of systematic errors (e.g., overestimation or underestimation) [53].

RGB-vegetation fraction model

An RGB-vegetation fraction model based on additive color theory was applied to the multi-temporal vegetation fraction images to interpret change events in our study area. Using this technique, multiple dates of vegetation fraction images could be viewed at one time, and the changes were highlighted. Vegetation fraction images in 1990, 2002, and 2010 were placed chronologically by red, green, and blue colors, respectively. Any combination of primary colors of similar brightness produces a complementary color that can identify the direction and magnitude of the vegetation cover change.

Change intensity analysis

To study the spatial distribution of the vegetation change intensity, fraction change images from 1990 to 2002 (T1) and from 2002 to 2010 (T2) were produced by subtracting two sequential fraction images. The intensity of the fraction change was divided into 4 intensity categories: extreme decrease (−1 to −0.5), moderate decrease (−0.5 to −0.2), moderate increase (0.2 to 0.5) and extreme increase (0.5 to 1). The relationship between the change intensity and the distance to the urban center was analyzed by counting the percentage of pixels of a certain intensity category at a certain distance.

Concentric analysis

A concentric analysis technique was used to obtain insight into the relationship between urban vegetation status and urban expansion. Twenty concentric belts were identified, radiating from the urban center to the fringe (Figure 3). Each belt spanned 0.5 km, approximately 16 pixels in the TM image, which is approximately two times the minimum resolution of assessment suggested by Powell et al. [18]. To account for the multi-nuclei expansion pattern in Hangzhou [54], we only applied the concentric analysis to the main urban area rather than to the entire study area. Furthermore, two famous tourist attractions, the West Lake Scenic Spots (WLSS) and the Xixi National Wetland Park (XNWP), were excluded from the concentric analysis because they were subject to a different land use policy than the rest of the study area.

MESMA provided subpixel information with physical meaning, where a fraction value of 1 represented full coverage of vegetation within a pixel. The vegetation fraction images were reclassified into 4 density categories: low coverage (0 to 0.2), middle coverage (0.2 to 0.5), middle-high coverage (0.5 to 0.8), and high coverage (0.8 to 1). The overall average vegetation fraction and the area percentages of the pixels within the 4 density categories were calculated for each belt to estimate the general pattern of urban greenness.

Results

Accuracy assessment

The R^2, RMSE, and SE were calculated for the 2002 and 2010 vegetation fraction images to evaluate the modeled results quantitatively. Scatter plots of the accuracy assessment results are shown in Figure 4. The RMSE was 7.78% for the 2010 image

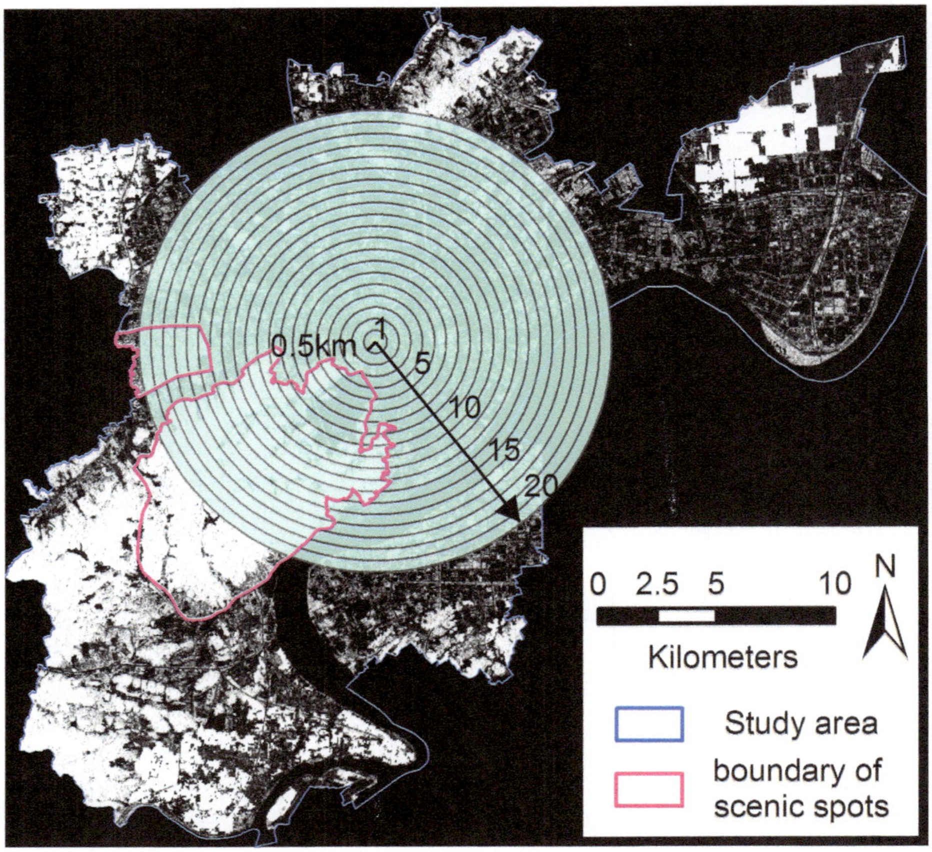

Figure 3. The layout of the concentric belts. Twenty concentric belts radiating from the city center, where each belt spans 0.5 km.

and 10.38% for the 2002 image. The R^2 values were relatively high, ranging from 0.929 to 0.881, indicating a good fit between the 'true' fraction and the fraction derived from MESMA.

The analysis of the residuals, in terms of the SE (−0.800% for 2010 and 0.320% for 2002), indicated that there was no significant underestimation or overestimation relative to the reference fraction.

Spatiotemporal analysis of urban greenness change

To construct vegetation cover images for corresponding years, the amount of vegetation in each pixel was represented by the assigned gray value. Brighter areas signify higher proportions of vegetation, and darker areas signify lower proportions of vegetation.

Overall, the urban greenness of Hangzhou has decreased steadily over the past two decades (Figures 2a, b, and c). In 1990, most of the high-density green space was distributed in natural land-use areas (i.e., agricultural areas, primary forests, and wetlands) around the built-up area, and the vegetation cover within the city was relatively sparse. Along with the urban expansion, an obvious loss of greenness occurred in the urban fringe and the suburban area. The urban greenness became more fragmented and isolated, except in mountainous areas and in the remaining cropland. The total percentage of the vegetated area of Hangzhou decreased greatly from 64.61% to 42.76% (Table 2), especially in T2, nearly twice the rate observed in T1.

Using the RGB-vegetation fraction model, distinctly different patterns were demonstrated among the various segments of the urban area (Figure 5). The core built-up area exhibited an increasing percentage of vegetation cover in T1 and T2, dominated by cyan and blue colors (Figure 5c). The other colors, such as red, yellow and green, revealed a complex reconfiguration of urban greenness in the core built-up area over the past two decades due to urban reconstruction and the urban greening project. In contrast, the urban peripheral areas showed dramatic declines in vegetation cover in the two periods. Vast fields of red pixels were observed near the core built-up area, indicating vegetation loss occurred during the urban expansion in T1 (Figure 5b). Two large blocks of vegetation reduction occurred in the eastern and southern parts of the city due to the construction of two new subcities (Xiasa and Jiangnan). Furthermore, a more severe decrease in the greenness in T2 could be confirmed by the widespread yellow color around the existing built-up areas in the main city and in the two subcities. In Xiasa, a suburban college town, road networks and communities constructed in T1 were depicted in red (Figure 5e). The remaining vegetation along the road networks was replaced by new construction in T2, indicated by yellow. Meanwhile, some of the roads and communities represented by magenta suggest vegetation recovery in the form of road and community greening after 2002. In most WLSS regions, the vegetation coverage remained consistently high, as indicated in white. However, a much more complex color composition of blue, green, purple and magenta was observed in XNWP and reflects a complex but overall subtle increase in wetland vegetation coverage over the past two decades.

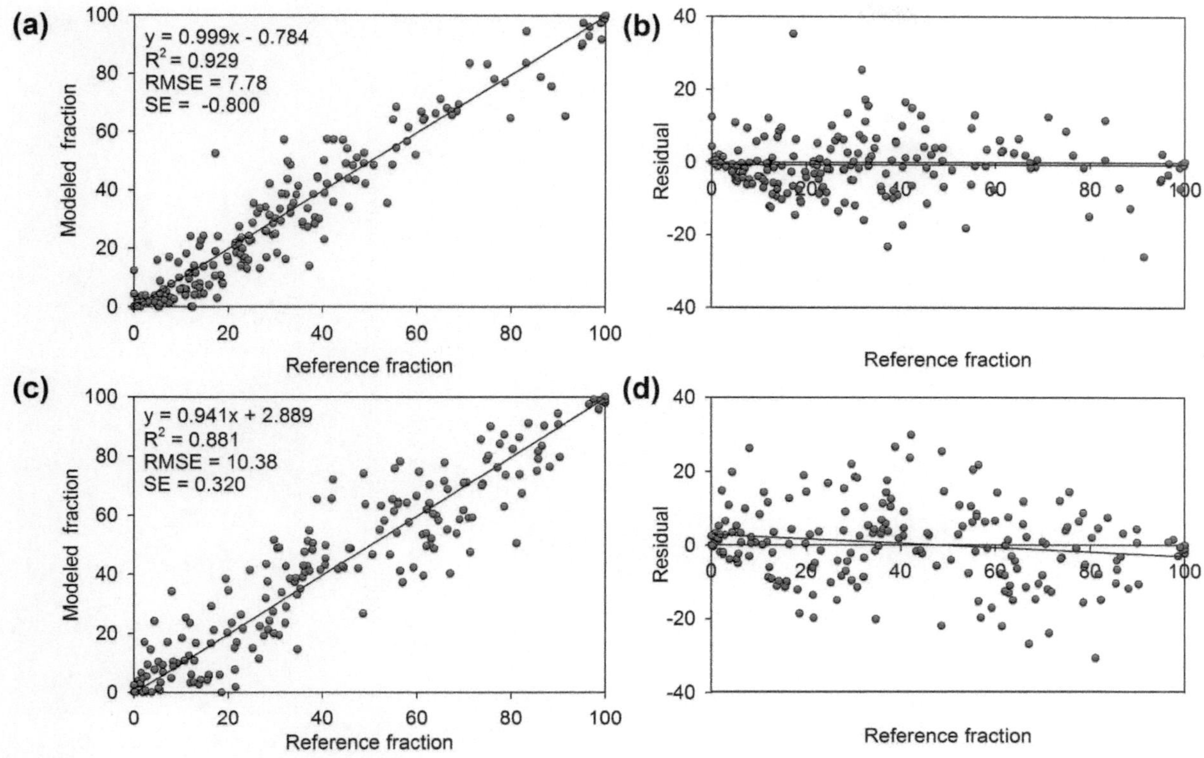

Figure 4. Comparisons between the reference and modeled fractions. (**a**) Vegetation fraction scatter plot for 2010, (**b**) vegetation fraction residuals for 2010, (**c**) vegetation fraction scatter plot for 2002, and (**d**) vegetation fraction residuals for 2002. Perfect agreement, represented by the 1:1 line, is displayed in the scatter plots. The best-fit line is displayed in the residual plots to indicate the general trends of overestimation and underestimation.

A clear spatial disparity in vegetation coverage change intensity was found (Figure 6) by analyzing the relationships between the change intensity and the distance to the urban center. In T1, the vegetation change pattern was dominated by a moderate increase within 2 km. At 2 km, the extreme decrease soared to over 20% and became the major change process. Two peaks appeared at approximately 4 km and 19 km, reflecting the vegetation loss due to the expansion of the core built-up area and construction of the new subcities. In T2, a greater intensity and extent of moderate increases was observed; this suggests that vegetation coverage more vigorously recovered in the core built-up area. However, a more severe extreme decrease occurred in the outer region of the core built-up area and subcities and peaked at approximately 11 km and 15 km. Therefore, the hotspot of the vegetation decline moved to the suburban area between the core built-up area and the subcities. The conversion, both negative and positive, was much more significant in the later decade. A pattern similar to the class of the extreme decreases could be found in the class of moderate decreases, while the class of extreme increases remained consistently low and showed no distinct pattern related to distance in either period.

General pattern of urban greenness accompanied by urban expansion

Twenty concentric belts radiating from the city center to the fringe were analyzed to provide insights into the spatial patterns of urban greenness change within and around the city during the two periods. In 1990, the average vegetation fraction remained low in belts 1–5 but increased significantly in belt 6; this highlights the difference in vegetation cover between the core built-up area and

Table 2. Total percent vegetated area and changes from 1990 to 2010 for the study area and the two scenic spots.

	1990	2002	2010	1990–2002 (per year)	1990–2002 (per year)
TPVA	64.61%	54.51%	42.76%	−0.84%	−1.53%
TPVA in XNWP	52.39%	54.44%	58.92%	0.17%	0.58%
TPVA in WLSS	85.51%	84.70%	83.13%	−0.07%	−0.20%

TPVA is the total percent of vegetated area, XNWP is the Xixi National Wetland Park, and WLSS is the West Lake Scenic Spots.

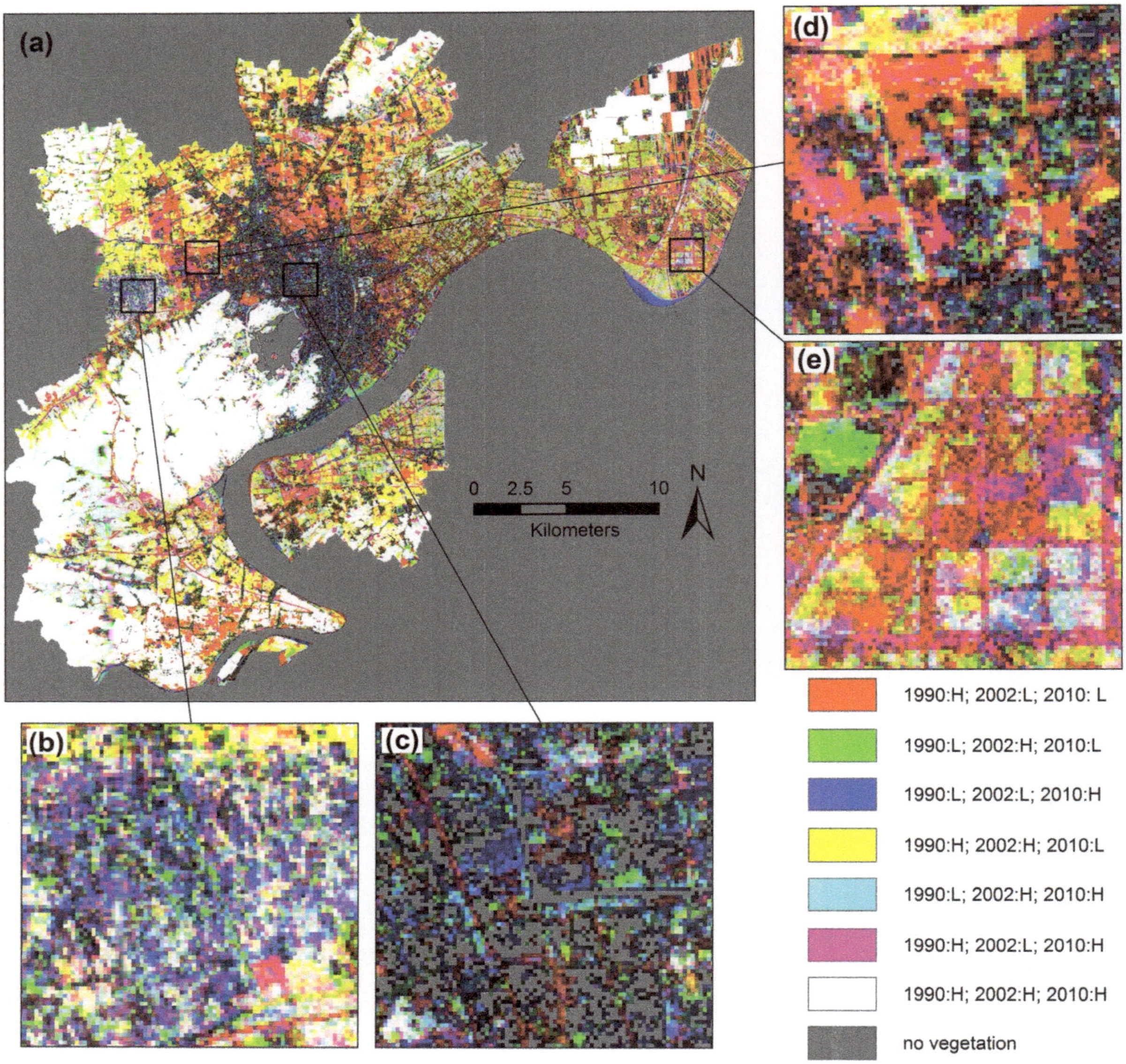

Figure 5. RGB color composite using vegetation fraction maps of 1990 (R), 2002 (G) and 2010 (B). The typical examples demonstrated are (a) the study area, (b) Xixi National Wetland Park, (c) the urban center, (d) residential communities and (e) the Xiasha suburban college town. Colors for the typical compositions of the vegetation fractions on the three dates are illustrated. H represents high vegetation fraction and L represents low vegetation fraction.

the suburbs. Because of urban expansion and sprawl, the boundary between the core built-up area and the suburbs was extrapolated to location near belt 9 in 2002. Meanwhile, a slight increase was detected for belts 1–5, indicating an improvement in the vegetation cover in the existing urban area. By 2010, a more obvious increase occurred in belts 1–9, along with a more severe decrease in belts 10–20. Additionally, the correlation between the vegetation abundance and the distance from the city center became weaker. These changes occurred because nearly all the regions within the twenty belts were highly urbanized, and only sparse native vegetation remained in belts 16–20. With the expansion and sprawl of built-up areas, urban greenness declined sharply in the urban fringe and the suburbs. A subsequent recovery of the vegetation cover occurred in the existing urban area, along with a decrease in the outer belts. This reflects the main process of conversion from natural vegetation to artificial urban greening because of urban expansion.

Urban greenness in the existing urban area contributes valuable ecosystem services which are related closely with the daily life of the local residents [55]. The lowest vegetation fraction was found in 1990 (Figure 7a: belts 1–5, 1990). Since then, the vegetation cover continued to increase in the older neighborhoods over the past two decades (belts 1–5 in T1 and belts 1–9 in T2), indicating sustained improvement in the urban environment can be a component of the urbanization. In 2010, the average vegetation fraction of the urban area was nearly double that in 1990. In addition, an obvious increase in the middle cover pixels occurred in belts 1–10, especially in belts 1–5 (Figure 7d), along with a decrease in the low cover pixels (Figure 7e). The result of change intensity analysis also showed that the changing pattern in urban center was dominated by moderate

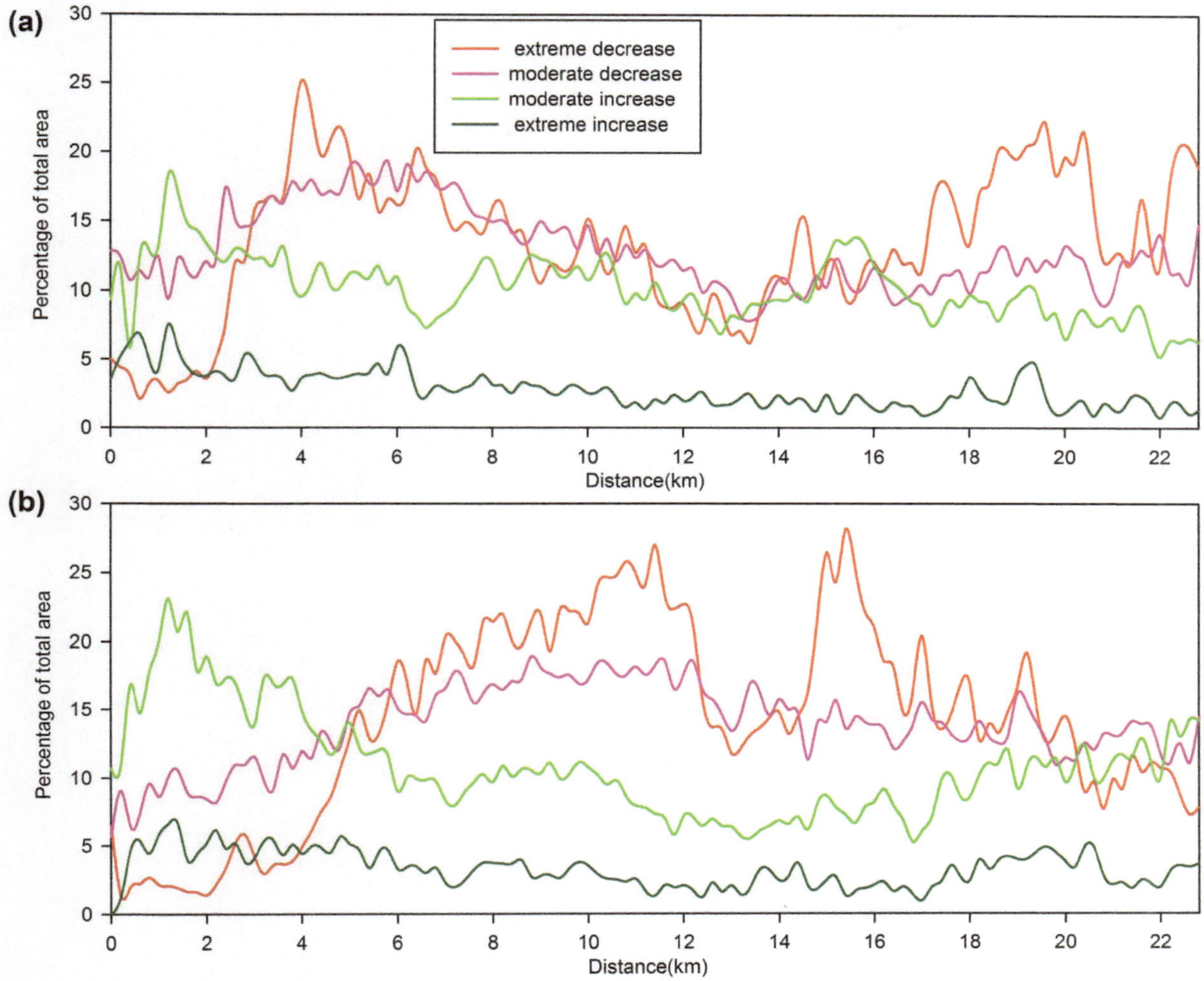

Figure 6. Spatial characteristics of the vegetation fraction change from the urban center over the two periods. (a) 1990–2002 and (b) 2002–2010. The vegetation change categories are shown by the different colors in the legend.

increase. These characteristics of vegetation fraction within a pixel further indicated those small green patches with size smaller than the pixel size of the Landsat data has mainly contributed to the increase of green space in existing urban area.

Discussion

MESMA and integration of the multiple analysis technique

The accuracy of the information acquired on vegetation cover has a direct influence on how well the trends and processes of greenness change are understood, especially for a diverse urban area. Urban greenness is generally highly fragmented in Chinese cities, resulting in a large number of small patches. Li et al. indicated that 86% of the patches of urban greenness were smaller than 900 m^2, and 37% were smaller than 100 m^2 in their study area [56]. Likewise, our study highlights the importance of the increasing area of small green patches, which may be difficult to detect by traditional classification techniques. An analysis based on directly classifying medium-resolution data would lead to an underestimation of the amount of urban greenness, which is crucial to urban areas with generally limited amounts of vegetation cover. This study demonstrated the advantage of MESMA in providing accurate and reliable subpixel information with physical meaning. However, this method still failed to acquire accurate information on the species and structure of urban greenness in multispectral images. Moreover, the method was unable to resolve nonlinear mixing effects, and it generally needed more supervision in the unmixing process. By integrating RGB-vegetation fraction model, the change intensity analysis and concentric analysis, detailed information on the direction, magnitude, location and pattern of the variation of urban greenness were able to be visually and quantitatively analyzed. These information have improved our understanding of how urban greenness evolved under rapid urbanization. The strength of MESMA is its ability to represent a wide variety of surface reflectance types as simple combinations of endmember abundances. Furthermore, because of the global availability and rich historical data archive of the Landsat data, analysis based on these techniques can be easily expanded to other cities around the world.

Greenness change in response to urbanization and greening policies

Despite the proven benefits of urban green spaces [3,4,5,6,7], widespread losses of these spaces have been reported in American

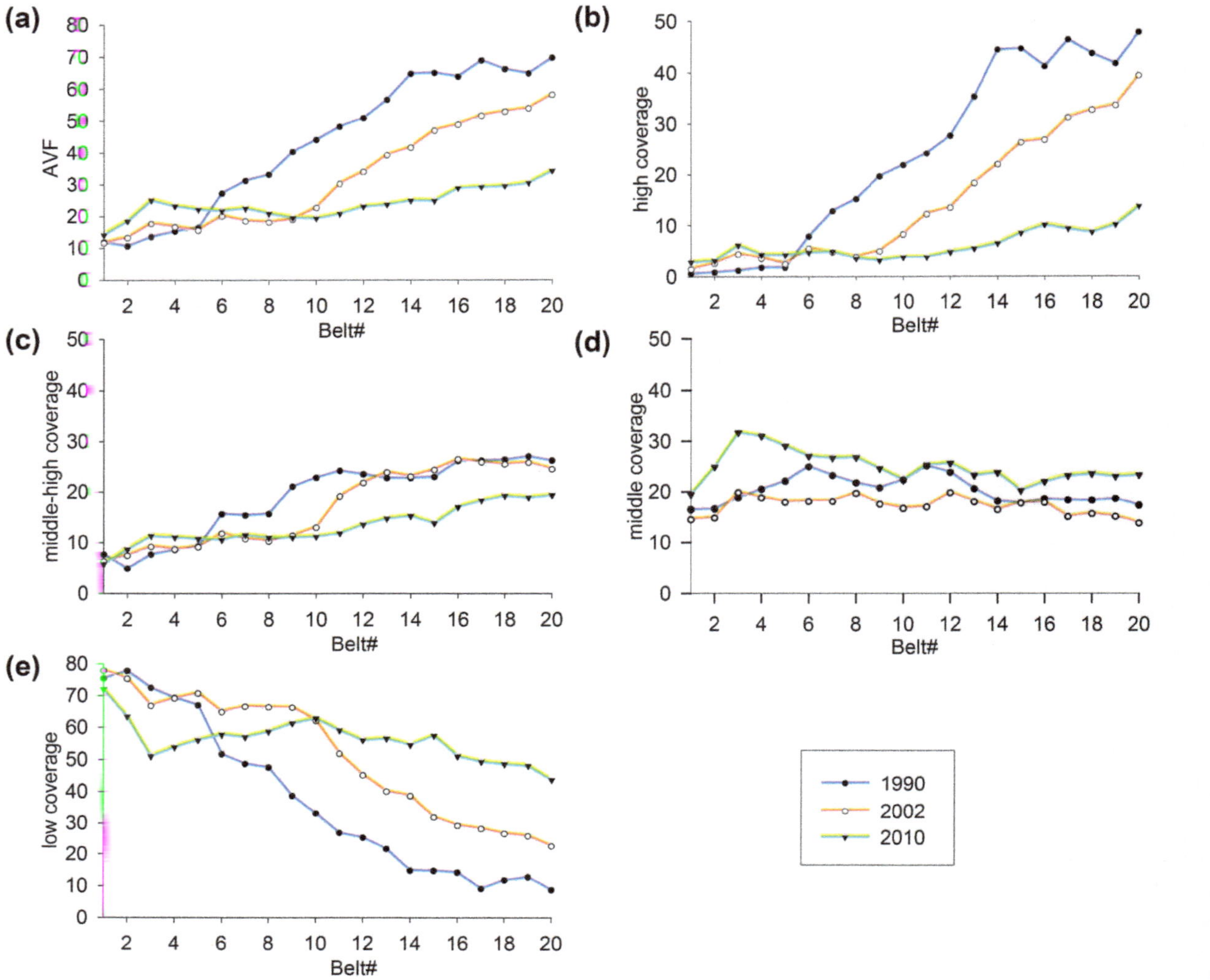

Figure 7. Concentric vegetation coverage analysis in each belt for each year. (a) Average vegetation fraction (AVF), (b) percentage of area of high coverage pixels, (c) percentage of area of middle-high coverage pixels, (d) percentage of area of middle coverage pixels, and (e) percentage of area of low coverage pixels.

and European cities [9,11]. In our study area, one-third of the vegetation cover that was present in 1990 was gradually lost over these two decades. The RGB-vegetation fraction model showed that most of the vegetation loss occurred in the urban fringe or in the new subcities. This trend agrees with the regional decline in vegetation cover caused by urbanization, as observed by Sun et al. [57]. In contrast to the overall decline throughout the study area, both the RGB-vegetation fraction model and concentric analysis demonstrated a consistent increase in urban greenness in the existing urban area. This confirmed the finding that Chinese cities have undergone a steady increase in urban greenness in the built-up areas [12]. We further demonstrated the contribution to the recovery of urban greenness by the increasing area of small green patches. The vegetation cover of the scenic areas remained relatively stable, even with a slight increase in XNWP, which reflects the effect of a scenic protection policy and an ecological renewal project. The changing pattern of greenness in the study area from 1990 to 2010 was a response to the combined effects of rapid urbanization and greening policies. Because of China's unique institutional character, cities have developed into modern metropolitan areas in a much different way than other regions. The government plays a key role in guiding urban planning. Greenness was very limited in the urban area in 1990, and most of the greenness was located in the West Lake Scenic Spots. This uneven distribution limited the access of urban residents to the ecological benefits of urban green space. The high degree of urbanization in the city center and the rapid growth of urban population exacerbated the demand for urban greenness but also placed limits on its development. The implementation of new urbanization strategies since 2000, such as "leapfrog development along the Qiantang River" and "crossing the Qiantang River and developing southward", triggered eastward sprawl and the construction of two subcities [54], resulting in the use of more space to reduce the population and building density and the availability of space for the creation and development of urban green space system. A green network system called "Two rings, Two axles, Six ecological zones" and a series of greening projects, e.g., the West Lake comprehensive protection project, the Grand Canal comprehensive protection project, and the Xixi wetland comprehensive protection project, have been conducted. These projects contributed to the accelerated recovery of urban greenness from 2002 to 2010 within the city. Meanwhile, innovative efforts such as retrofitting green space alongside formerly dilapidated canals, underneath and alongside main roads and railway lines were done to activate neglected spaces to restore lost green space [58]. In addition, the urban greening management regulation of Hangzhou set the minimum greening cover for new residential area to 30%. These factors explain the remarkable increase in the number of middle cover pixels.

Conclusions

Acquiring accurate and area-wide information remains a challenge in urban environments due to the extremely heterogeneous spatial patterns within cities. Subpixel methods can overcome this problem and provide valuable quantitative information. In this paper, we investigated the patterns of change in urban greenness in Hangzhou, China, over the past two decades using Landsat imagery from 1990, 2002, and 2010. Multiple endmember spectral mixture analysis (MESMA) was used to derive vegetation cover fractions (VCFs) for various segments of the urban area. Despite the complex spectral confusion of different land cover types in the urban area, MESMA could provide accurate and reliable subpixel information of the vegetation cover fraction. An RGB-vegetation fraction model, change intensity analysis and the concentric technique were integrated to characterize the detailed, spatial characteristics and overall pattern of the urban greenness change over time.

The study area has experienced a drastic loss of urban greenness. Nearly one-third of the vegetation coverage that was present in 1990 was gradually lost over the past two decades. Most of this loss is located in the urban fringe and suburban areas due to urban expansion and the construction of new subcities. No significant change occurred in scenic spots, except for a recovery of vegetation coverage in the Xixi National Wetland Park from 2002 to 2010. Meanwhile, a remarkable recovery in the vegetation coverage occurred in the existing urban area that was mainly attributed to the continuous increase in small green patches. These changing patterns were more obvious from 2002 to 2010 than from 1990 to 2002. The characteristics of the changing pattern of urban greenness over time and in different parts of the urban environment reveal the combined effects of rapid urbanization and greening policies.

This study demonstrated the unique role of medium-resolution satellite images in the longitudinal analysis of the evolution of urban greenness. Subpixel methods provided abundant meaningful information, and more analysis techniques could be developed based on such continuous information. Furthermore, regional comparisons of cities with different economic and natural conditions should be conducted using these techniques to investigate the relationships between urban greenness and urbanization.

Acknowledgments

We thank Shahtahmassebi Amirreza for his help during the study and the anonymous reviewers and the editor for their valuable comments, which helped improve our manuscript.

Author Contributions

Conceived and designed the experiments: MYG JSD KW. Performed the experiments: MYG XYZ. Analyzed the data: MYG XYZ. Contributed reagents/materials/analysis tools: MYG XYZ. Contributed to the writing of the manuscript: MYG XYZ. Designed/drew figures: MYG XYZ. Revised the manuscript: MYG JSD YH KW.

References

1. Grimm NB, Faeth SH, Golubiewski NE, Redman CL, Wu JG, et al. (2008) Global change and the ecology of cities. Science 319: 756–760.
2. Robinson SL, Lundholm JT (2012) Ecosystem services provided by urban spontaneous vegetation. Urban Ecosystems 15: 545–557.
3. Davies RG, Barbosa O, Fuller RA, Tratalos J, Burke N, et al. (2008) City-wide relationships between green spaces, urban land use and topography. Urban Ecosystems 11: 269–287.
4. Walsh CJ, Fletcher TD, Burns MJ (2012) Urban Stormwater Runoff: A New Class of Environmental Flow Problem. PLoS ONE 7: e45814.
5. Akbari H, Pomerantz M, Taha H (2001) Cool surfaces and shade trees to reduce energy use and improve air quality in urban areas. Solar Energy 70: 295–310.
6. Donovan GH, Butry DT, Michael YL, Prestemon JP, Liebhold AM, et al. (2013) The Relationship Between Trees and Human Health: Evidence from the Spread of the Emerald Ash Borer. American Journal of Preventive Medicine 44: 139–145.
7. Everson-Rose SA, Lewis TT (2004) Psychosocial factors and cardiovascular diseases. Annual Review of Public Health 26: 469–500.
8. Tan PY, Wang J, Sia A (2013) Perspectives on five decades of the urban greening of Singapore. Cities 32: 24–32.
9. Fuller RA, Gaston KJ (2009) The scaling of green space coverage in European cities. Biology Letters 5: 352–355.
10. Dallimer M, Tang ZY, Bibby PR, Brindley P, Gaston KJ, et al. (2011) Temporal changes in greenspace in a highly urbanized region. Biology Letters 7: 763–766.
11. Nowak DJ, Greenfield EJ (2012) Tree and impervious cover change in US cities. Urban Forestry & Urban Greening 11: 21–30.
12. Zhao JJ, Chen SB, Jiang B, Ren Y, Wang H, et al. (2013) Temporal trend of green space coverage in China and its relationship with urbanization over the last two decades. Science of the Total Environment 442: 455–465.
13. Yang J, Huang C, Zhang Z, Wang L (2013) The temporal trend of urban green coverage in major Chinese cities between 1990 and 2010. Urban Forestry & Urban Greening.
14. Zhou XL, Wang YC (2011) Spatial-temporal dynamics of urban green space in response to rapid urbanization and greening policies. Landscape and Urban Planning 100: 268–277.
15. Deng JS, Wang K, Hong Y, Qi JG (2009) Spatio-temporal dynamics and evolution of land use change and landscape pattern in response to rapid urbanization. Landscape and Urban Planning 92: 187–198.
16. Liu T, Yang XJ (2013) Mapping vegetation in an urban area with stratified classification and multiple endmember spectral mixture analysis. Remote Sensing of Environment 133: 251–264.
17. He C, Convertino M, Feng Z, Zhang S (2013) Using LiDAR Data to Measure the 3D Green Biomass of Beijing Urban Forest in China. PLoS ONE 8: e75920.
18. Powell RL, Roberts DA, Dennison PE, Hess LL (2007) Sub-pixel mapping of urban land cover using multiple endmember spectral mixture analysis: Manaus, Brazil. Remote Sensing of Environment 106: 253–267.
19. Small C (2001) Estimation of urban vegetation abundance by spectral mixture analysis. International Journal of Remote Sensing 22: 1305–1334.
20. Myint SW (2006) Urban vegetation mapping using sub-pixel analysis and expert system rules: A critical approach. International Journal of Remote Sensing 27: 2645–2665.
21. Foody GM, Boyd DS (1999) Detection of partial land cover change associated with the migration of inter-class transitional zones. International Journal of Remote Sensing 20: 2723–2740.
22. Herold M, Couclelis H, Clarke KC (2005) The role of spatial metrics in the analysis and modeling of urban land use change. Computers, Environment and Urban Systems 29: 369–399.
23. Patino JE, Duque JC (2013) A review of regional science applications of satellite remote sensing in urban settings. Computers, Environment and Urban Systems 37: 1–17.
24. Adams JB, Smith MO, Gillespie AR (1993) Imaging spectroscopy: Interpretation based on spectral mixture analysis. In: CM P, P E, ^editors. Remote geochemical analysis: elemental and mineralogical composition: New York: Cambridge Univ. pp. 145–166.
25. Roberts DA, Gardner M, Church R, Ustin S, Scheer G, et al. (1998) Mapping chaparral in the Santa Monica Mountains using multiple endmember spectral mixture models. Remote Sensing of Environment 65: 267–279.
26. Smith MO, Ustin SL, Adams JB, Gillespie AR (1990) Vegetation in deserts: I. A regional measure of abundance from multispectral images. Remote Sensing of Environment 31: 1–26.
27. Lu DS, Weng QH (2004) Spectral mixture analysis of the urban landscape in Indianapolis with landsat ETM plus imagery. Photogrammetric Engineering & Remote Sensing 70: 1053–1062.
28. Rashed T, Weeks JR, Roberts D, Rogan J, Powell R (2003) Measuring the physical composition of urban morphology using multiple endmember spectral

mixture models. Photogrammetric Engineering & Remote Sensing 69: 1011–1020.
29. Small C, Lu JWT (2006) Estimation and vicarious validation of urban vegetation abundance by spectral mixture analysis. Remote Sensing of Environment 100: 441–456.
30. Michishita R, Jiang Z, Xu B (2012) Monitoring two decades of urbanization in the Poyang Lake area, China through spectral unmixing. Remote Sensing of Environment 117: 3–18.
31. Powell RL, Roberts DA (2010) Characterizing urban land-cover change in Rondônia, Brazil: 1985 to 2000. Journal of Latin American Geography 9: 183–211.
32. Weng QH, Lu DS (2009) Landscape as a continuum: an examination of the urban landscape structures and dynamics of Indianapolis City, 1991–2000, by using satellite images. International Journal of Remote Sensing 30: 2547–2577.
33. Rashed T, Weeks JR, Stow D, Fugate D (2005) Measuring temporal compositions of urban morphology through spectral mixture analysis: toward a soft approach to change analysis in crowded cities. International Journal of Remote Sensing 26: 699–718.
34. Tang JM, Chen F, Schwartz SS (2012) Assessing spatiotemporal variations of greenness in the Baltimore-Washington corridor area. Landscape and Urban Planning 105: 296–306.
35. Hangzhou almanac 2011 (2012) Beijing: Fangzhi Press. 546 p.
36. USGS Global Visualization Viewer. Available: http://glovis.usgs.gov/. Accessed 2013 Oct 17.
37. Roberts DA, Batista G, Pereira J, Waller E, Nelson B (1999) Change Identification Using Multitemporal Spectral Mixture Analysis: Applications in Eastern Amazonia. In: Lunetta R, Elvidge C, ^editors. London: Taylor & Francis. pp. 318, 16.
38. Dennison PE, Halligan KQ, Roberts DA (2004) A comparison of error metrics and constraints for multiple endmember spectral mixture analysis and spectral angle mapper. Remote Sensing of Environment 93: 359–367.
39. Song CH (2005) Spectral mixture analysis for subpixel vegetation fractions in the urban environment: How to incorporate endmember variability? Remote Sensing of Environment 95: 248–263.
40. Roberts DA, Halligan K, Dennison PE (2007) VIPER Tools user manual (Version1.5) University of California at Santa Barbar a, 91P. Available: http://www.vipertools.org/. Accessed 2013 Aug 28.
41. Dennison PE, Roberts DA (2003) Endmember selection for multiple endmember spectral mixture analysis using endmember average RMSE. Remote Sensing of Environment 87: 123–135.
42. Tompkins S, Mustard JF, Pieters CM, Forsyth DW (1997) Optimization of endmembers for spectral mixture analysis. Remote Sensing of Environment 59: 472–489.
43. Okin GS, Roberts DA, Murray B, Okin WJ (2001) Practical limits on hyperspectral vegetation discrimination in arid and semiarid environments. Remote Sensing of Environment 77: 212–225.
44. Somers B, Asner GP, Tits L, Coppin P (2011) Endmember variability in Spectral Mixture Analysis: A review. Remote Sensing of Environment 115: 1603–1616.
45. Powell R (2011) Characterizing Urban Subpixel Composition Using Spectral Mixture Analysis. In: Yang X, edito. Urban Remote Sensing: Monitoring, Synthesis and Modeling in the Urban Environment. chichester,UK: Wiley-blackwell. pp. 111–128.
46. Boardman JW, Kruse FA, Green RO (1995) Mapping target signatures via partial unmixing of AVIRIS data. In AVIRIS Airborne Geoscience Workshop Proceedings. Pasadena.
47. Roberts DA, Dennison PE, Gardner ME, Hetzel Y, Ustin SL, et al. (2003) Evaluation of the potential of Hyperion for fire danger assessment by comparison to the Airborne Visible/Infrared Imaging Spectrometer. IEEE Transactions on Geoscience & Remote Sensing 41: 1297–1310.
48. Powell RL, Roberts DA (2008) Characterizing Variability of the Urban Physical Environment for a Suite of Cities in Rondonia, Brazil. Earth Interactions 12.
49. Dennison PE, Roberts DA (2003) The effects of vegetation phenology on endmember selection and species mapping in southern California chaparral. Remote Sensing of Environment 87: 295–309.
50. Michishita R, Gong P, Xu B (2012) Spectral mixture analysis for bi-sensor wetland mapping using Landsat TM and Terra MODIS data. International Journal of Remote Sensing 33: 3373–3401.
51. Lu DS, Moran E, Hetrick S (2011) Detection of impervious surface change with multitemporal Landsat images in an urban-rural frontier. ISPRS Journal of Photogrammetry and Remote Sensing 66: 298–306.
52. Wu CS, Yuan F (2007) Seasonal sensitivity analysis of impervious surface estimation with satellite imagery. Photogrammetric Engineering & Remote Sensing 73: 1393–1401.
53. Wu CS (2004) Normalized spectral mixture analysis for monitoring urban composition using ETM plus imagery. Remote Sensing of Environment 93: 480–492.
54. Wu K, Zhang H (2012) Land use dynamics, built-up land expansion patterns, and driving forces analysis of the fast-growing Hangzhou metropolitan area, eastern China (1978–2008). Applied Geography 34: 137–145.
55. Jim CY, Chen WY (2009) Ecosystem services and valuation of urban forests in China. Cities 26: 187–194.
56. Li XM, Zhou WQ, Ouyang ZY (2013) Relationship between land surface temperature and spatial pattern of greenspace: What are the effects of spatial resolution? Landscape and Urban Planning 114: 1–8.
57. Sun J, Wang X, Chen A, Ma Y, Cui M, et al. (2011) NDVI indicated characteristics of vegetation cover change in China's metropolises over the last three decades. Environmental monitoring and assessment 179: 1–14.
58. Wolch JR, Byrne J, Newell JP (2014) Urban green space, public health, and environmental justice: The challenge of making cities 'just green enough'. Landscape and Urban Planning.

24

Carbon Sequestration and Fertility after Centennial Time Scale Incorporation of Charcoal into Soil

Irene Criscuoli[1,7]*, Giorgio Alberti[2,3], Silvia Baronti[4], Filippo Favilli[5], Cristina Martinez[1,4], Costanza Calzolari[6], Emanuela Pusceddu[4], Cornelia Rumpel[7], Roberto Viola[8], Franco Miglietta[1,4]

1 FoxLab, Forest & Wood Science, Fondazione E.Mach, San Michele all'Adige (Trento), Italy, **2** MountFor Project Centre, European Forest Institute, San Michele all'Adige (Trento), Italy, **3** Department of Agricultural and Environmental Sciences, University of Udine, Udine, Italy, **4** Institute of Biometeorology, National Research Council, Firenze, Italy, **5** European Academy of Bolzano, Bolzano, Italy, **6** Institute of Research for Hydrogeological Protection, National Research Council, Sesto Fiorentino (FI), Italy, **7** Biogéochimie et écologie des milieux continentaux, CNRS-INRA-ENS-Paris 6, Thiverval-Grignon, France, **8** Centre for Research and Innovation, Fondazione E.Mach, San Michele all'Adige (Trento), Italy

Abstract

The addition of pyrogenic carbon (C) in the soil is considered a potential strategy to achieve direct C sequestration and potential reduction of non-CO_2 greenhouse gas emissions. In this paper, we investigated the long term effects of charcoal addition on C sequestration and soil physico-chemical properties by studying a series of abandoned charcoal hearths in the Eastern Alps of Italy established in the XIX century. This natural setting can be seen as an analogue of a deliberate experiment with replications. Carbon sequestration was assessed indirectly by comparing the amount of pyrogenic C present in the hearths (23.3±4.7 kg C m^{-2}) with the estimated amount of charcoal that was left on the soil after the carbonization (29.3±5.1 kg C m^{-2}). After taking into account uncertainty associated with parameters' estimation, we were able to conclude that 80±21% of the C originally added to the soil via charcoal can still be found there and that charcoal has an overall Mean Residence Time of 650±139 years, thus supporting the view that charcoal incorporation is an effective way to sequester atmospheric CO_2. We also observed an overall change in the physical properties (hydrophobicity and bulk density) of charcoal hearth soils and an accumulation of nutrients compared to the adjacent soil without charcoal. We caution, however, that our site-specific results should not be generalized without further study.

Editor: Ben Bond-Lamberty, DOE Pacific Northwest National Laboratory, United States of America

Funding: This study was financially supported by the European Commission through the EuroChar project (FP7-ENV-478 2010ID-265179) and AgroPyroGas funded by Regione Toscana, Italy. I.C. was supported by a Ph.D. grant from Fondazione Edmund Mach, Trento. The funders had no role in study design, data collection and analysis, decision to publish, or preparation of the manuscript.

Competing Interests: The authors have declared that no competing interests exist.

* E-mail: irene.criscuoli@fmach.it

Introduction

Thermo-chemical conversion of organic material under limited oxygen supply, within a certain range of temperatures (200–1200°C), transforms biomass into bio-oil and syngas, which may be used as an energy source, and produces a carbonaceous co-product (i.e. biomass-derived Pyrogenic-C or charcoal or biochar [1]) which has been proposed as a tool to mitigate climate change and improve soil fertility [2]. A recent study [3] quantified the theoretical carbon (C) sequestration potential of biochar following its incorporation in agricultural soils at a maximum rate of 50 Mg C ha^{-1} to a depth of 0.15 m as 1.8 Gt CO_2-$C_{equivalent}$ per year. This estimate corresponds to 12% of current global anthropogenic C emissions and includes: a) direct C sequestration, associated with the burial of recalcitrant organic C forms [4]; (b) potential reduction of N_2O and CH_4 emissions from soils associated with biochar application [5]; and (c) CO_2 emissions avoided due to fossil fuel substitution by the energy released by biomass during pyrolysis and gasification. Moreover, several studies have shown that the addition of biochar to both poor and fertile agricultural soils may have beneficial effects on plant yields, thus amplifying its environmental benefit. These effects are associated with improvements in soil physical [6] and chemical properties [7], microbiological activity [8], temperature increase due to changes in surface albedo [9], hormesis (i.e. favorable biological responses to low exposures; [10]), as well as combinations of several of these different drivers [8].

However, while short-term studies have confirmed the potential of biochar to increase C storage and to improve soil physico-chemical properties in the short-term, the long-term effects of incorporating large amounts of pyrogenic C into the soil remain rather elusive. The actual ability of biochar to act as a C sink into the soil remains controversial due to uncertainties related to its long term stability [11]. Thousand-year old charcoal residues identified in archeological sites and areas interested by wildfires have been considered a demonstration of its long term stability in soils [12] [13] even though some studies have outlined the fact that the amount of Pyrogenic Carbon measured in soils is much lower than what would have been expected according to other paleontological and archeological artifacts [14] or the frequency and intensity of fires [15]. Rapid transformations of charcoal in soils by abiotic and biotic oxidation can occur [16] and its stability varies according to the initial feedstock, the charring conditions, and the environmental characteristics of the burial site [17]. Very ancient charcoal deposits, such as Terra Preta de Indio in

Figure 1. Soil profile at the control site (Panel a) and the charcoal hearth (Panel b). The letters indicate different pedologic horizons. In the charcoal hearth the dark anthropogenic layer (Acoal; 0–10 cm) can be easily identified.

Amazonia [18] and Bronze Age human settlements of Terramare in the Po Valley in northern Italy [19], are still rich in C and are still fertile substrates [20] [21].

The present study aimed to explore the centennial time scale stability of pyrogenic C incorporated as charcoal in soil. To do this, we used Alpine areas where charcoal, produced in traditional charcoal piles, was added to the soil more than one century ago and was not mixed with other organic sources. Moreover, we were able to assess the effect on physio-chemical soil properties after char addition to soil.

Materials and Methods

Ethics Statement

Collection of soil samples was authorized by the Stelvio National Park and Trentino Forest Service.

Site Description and Soil Sampling

The study site is located in Val di Pejo (Trentino, Northern Italy; 46°20′16.18″ N, 10°36′07.02″ E) at an elevation ranging from 2120 to 2170 m a.s.l. The mean annual temperature at the site is 3.5°C and the mean annual precipitation is 903 mm [22]. The lowest precipitations are registered in January, while the highest are distributed between April and November.

Starting from the 16th century the area was subject to intensive wood resource exploitation for larch charcoal production which was subsequently used in the local iron industry. Production ceased in 1858 when a severe fire event destroyed the major iron foundry in the valley [13] [23]. Charcoal production was based on large forest clear-cuts, wood chopping and downhill transportation to artificial flat terraces (charcoal hearths) with an elliptical shape. Some of these hearths are still identifiable today as terraces where

Table 1. Parameters, coefficients and variables used to distinguish and quantify the different carbon pools in charcoal soil layer (means ± standard error, n = 3).

Parameter/coefficient/ variable	Definition	Unit	Value	Source(s)
$\delta^{13}C_{TOT}$	Isotopic signature of the bulk anthropogenic soil layer in hearths	‰	−24.72±0.14	Measured with IRMS
$\delta\ ^{13}C_{CHAR}$	Isotopic signature of charcoal fragments extracted from the anthropogenic soil layer in hearths	‰	−24.53±0.01	Measured with IRMS
$\delta\ ^{13}C_{SOM}$	Isotopic signature of soil organic matter of control soils	‰	−26.28±0.30	Measured with IRMS
$\delta\ ^{13}C\text{-}CO_{2HEARTHS}$	Isotopic signature of respired CO_2 from incubated hearth soils	‰	−25.22±1.13	Measured with CRDS and Keeling plot
$\delta\ ^{13}C\text{-}CO_{2CONTROL}$	Isotopic signature of respired CO_2 from incubated control soils	‰	−24.81±0.31	Measured with CRDS and Keeling plot
C_{TOT}	Carbon content of anthropogenic soil layer in hearths	kg C m^{-2}	26.2±5.3	Measured and calculated
C_{CHAR}	Pyrogenic carbon content of anthropogenic soil layer in hearths	kg C m^{-2}	23.3±4.7	Calculated with mass balance
C_{SOM}	Soil organic matter content of anthropogenic soil layer in hearths	kg C m^{-2}	2.9±0.6	Calculated with mass balance

Table 2. Total carbon and nutrient stocks in the control soils and charcoal hearths and the estimated amount added by carbonization calculated according to Eq. [2].

Element	control soils	charcoal hearths	p-value	Input by carbonization
C (kg m^{-2})	8.1±0.3	26.2±5.3	0.03	29±5
P (g m^{-2})	45.8±3.1	95±7	0.003	117±21
K (g m^{-2})	231±34	285±70	0.53	112±20
Ca (g m^{-2})	136±23	368±80	0.10	229±41
Mg (g m^{-2})	127±88	62±5	0.51	59±10
N (g m^{-2})	582±84	500±29	0.42	80±15

Mean ± standard error (n = 3). Results of the comparison between control and charcoal hearth (p-value) are also reported.

wood piles were prepared and subsequently covered by soil and tree branches [24]. The relatively large size and elliptical shape of the hearths suggest that more than one pile of wood was carbonized at a particular time as it was already well known that only circular piles could ensure a uniform and high quality charcoal production [25]. Wood carbonization required between four and ten days according to the dimension of the wood pile. After a two-day cooling down period, the charcoal was finally transported to the foundry.

Three hearths and three adjacent control areas with southerly aspects and unaffected by significant geo-morphological dynamics (erosion or sedimentation) or recent anthropogenic disturbances were selected. These are flat (2% slope) and have an average surface area (*s*) of 94 m^2. Soils within the control areas are shallow to moderately deep (35–70 cm), sandy-loam brown acid soils and with restricted areas of podzols (Lithic Dystrudept and Entic Haplorthod according to USDA, 2010 [26]) with an approximate 25% slope. Soils in the charcoal hearths show a truncated profile, with a shallow surface organic horizon approximately 2 cm deep covering a thicker (19.3±2.8 cm) black anthropogenic layer. This horizon contains a large amount of charcoal fragments and fine particles left after carbonization which have been subsequently incorporated and well mixed with the pre-existing soil in response to bioturbation [27] and freeze-thaw processes [28] (Figure 1).

Both control soils and charcoal hearths are, nowadays, covered by the same herbaceous vegetation dominated by *Nardus stricta L.* while the surrounding forest is dominated by *Larix decidua* L. and *Picea abies* L.

The exact date of charcoal production was assessed using a dendro-anthracological approach. This method relies on cross-dating tree ring widths in charcoal fragments with known tree chronologies and has been used previously by [29], who showed that the oldest and youngest tree rings identified in charcoal fragments at our study area were dated 1530 and 1858 respectively. This last date corresponds to the year in which a wildfire event down in the valley caused the destruction of industrial plants thus determining the interruption of the local iron industry and charcoal production in the area [13].

Soil Sampling and Chemical Analysis

The anthropogenic layer within the charcoal hearths was sampled using a manual soil corer at five different sampling points in each hearth. Similarly, the soil at approximately the same depth was sampled at five points in each control area. Soil samples were dried for 72 hours at 35°C and sieved to 2 mm. In the case of the charcoal hearths, the fraction of soil >2 mm was further separated into two subsamples, one including charcoal fragments and one including plant debris, roots and stones. All further analysis was completed on the five sampling points separately.

Soil pH was measured in a soil/water solution (1:2.5 ratio). Soil C and N contents were determined by dry combustion using a CHN elemental analyzer (© Perkin Elmer 2400 series II CHNS/O elemental analyzer). Total Ca, K, Mg, Na, P, and available Ca, K, Mg, P concentrations were determined for subsamples oven-dried at 105°C for 24 h according to the EPA method 3052 [30] and the filtered solutions were analyzed using an ICP-OES spectrophotometer (Varian Inc., Vista MPX). A further set of subsamples was used to assess NO_3^--N according to the method proposed by [31] and NH_4^+-N according to the method proposed by [32]. Hydrophobicity was measured following the method of [33]. Such term defines the affinity for soils to water controlling infiltration or wetting. It can be caused by coating of long-chained hydrophobic organic molecules on individual soil particles in response to the decay of organic matter but also to the diversity of soil micro-organisms. Increased hydrophobicity is normally observed after wildfires that leaves charcoal fragments at the soil surface.

C and N content, and total and available Ca, K, Mg, Na, P concentrations of >2 mm charcoal individual fragments were determined using the same methodology described above. Micrographs of those charcoal fragments were made using a Scanning Electron Microscope, XL 20 FEI SEM, with CRYO-GATAN ALTO 2100 technology on samples dried under vacuum, following standard procedures [34].

To enable a comparison between old and fresh charcoal, fragments of larch wood were carbonized in a muffle furnace at 400°, 500°, 600°, 860°C. The time needed to complete the carbonization corresponds to the time needed for the sample to stabilize its weight loss. C and N contents were determined on samples using the methodology described above.

Pyrogenic C Determination

The relative contribution of charcoal-C ($C_{CHAR} : C_{TOT}$) to total soil carbon (C_{TOT}) was estimated using a mass balance method [35] based on the $\delta^{13}C$ values of charcoal fragments excavated from the anthropogenic soil layer ($\delta\ {}^{13}C_{CHAR}$), the mean $\delta\ {}^{13}C$ of the entire layer ($\delta^{13}C_{TOT}$) and the $\delta\ {}^{13}C$ of the SOM contained in the adjacent control soils ($\delta^{13}C_{SOM}$) (Table 1):

$$\frac{C_{CHAR}}{C_{TOT}} = \frac{\delta^{13}C_{TOT} - \delta^{13}C_{SOM}}{\delta^{13}C_{CHAR} - \delta^{13}C_{SOM}} \quad (1)$$

Table 3. Total and available nutrient concentrations (mg kg^{-1} dry matter ± standard error; n = 3) measured in control soils, charcoal hearths, old and new charcoal fragments and larch wood.

	Control soils		Charcoal hearths		Old charcoal fragments	Fresh charcoal fragments	Larch wood
Element	**Total concentration**	**Available**	**Total concentration**	**Available**	**Total concentration**	**Total concentration**	**Total concentration**
Ca^{2+}	993±135	278±35	3300±185	1006±158	3438±275	5920±70	5334±53
K^{+}	1603±116	147±30	2463±69	279±120	885±66	2899±39	1415±77
Mg^{2+}	2739±73	80±12	2378±384	245±4	1215±76	1533±32	1158±12
Na^{+}	297±50	34±1	86±8	33±26	216±1	207±4	190±4
P	321±7	7±2	921±206	12±4	346±81	3005±53	2716±46
N-NO_3^{-}	1.96±0.74	–	2.34±0.89	–	–	–	–
N-NH_4^{+}	3.93±1.00	–	3.73±1.80	–	–	–	–

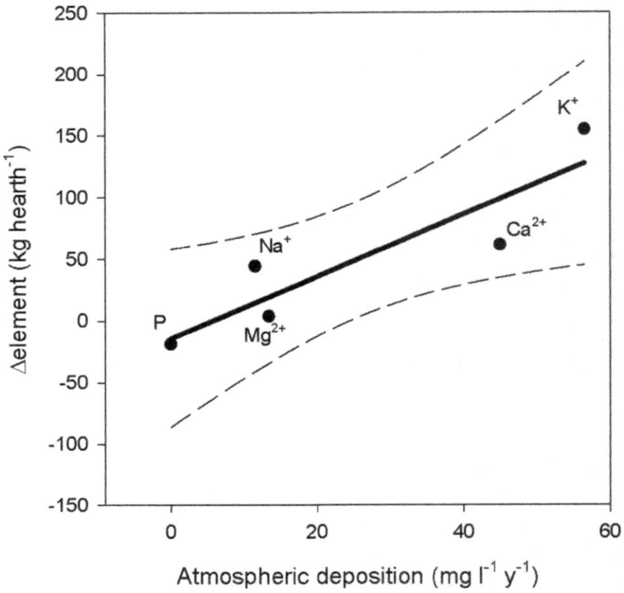

Figure 2. Correlation between average annual atmospheric deposition of P, K^+, Ca^{2+}, Mg^{2+}, Na^{2+} (mg l^{-1} y^{-1}) and the difference between the input of the same elements due to charcoal application in 1858 and the amount found today in hearth's soils (Δelement, kg $hearths^{-1}$) ($y = 2.50\times - 14.31$, $R^2 = 0.82$, $p = 0.035$). Dashed lines represent 95% confidence interval.

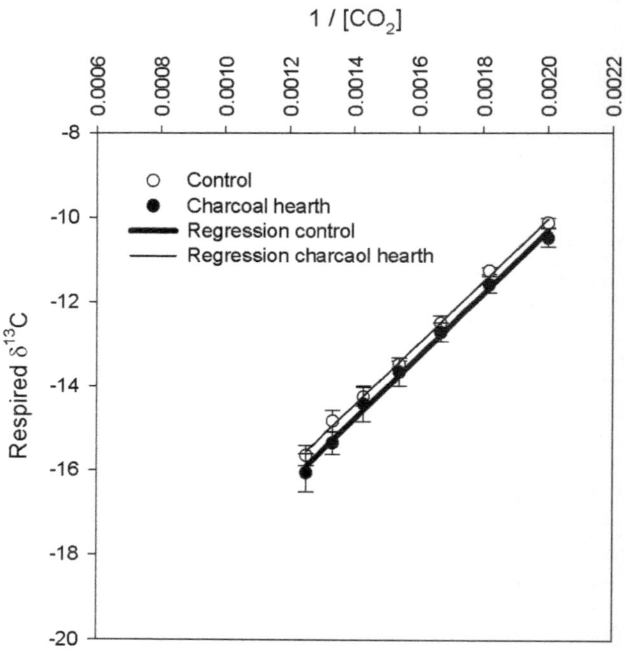

Figure 3. Keeling plots measured by CRDS showing the $\delta^{13}C$ of respired CO_2 fluxes versus the reciprocal of CO_2 concentration for control and charcoal hearth incubated soils ($\delta^{13}C_{CONTROL} = 7353*[CO_2]^{-1} - 24.8$, $R^2 = 0.99$; $\delta^{13}C_{CHARCOAL\ HEARTH} = 7467*[CO_2]^{-1} - 25.2$, $R^2 = 0.99$). Horizontal and vertical bars indicate standard deviations (n = 3).

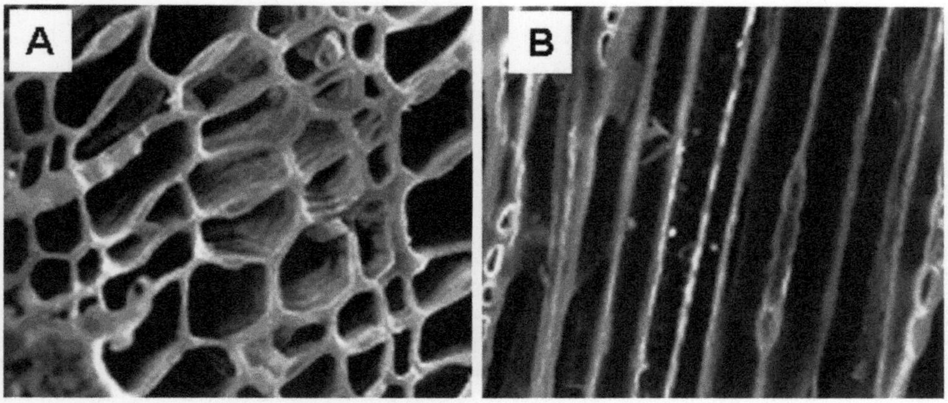

Figure 4. SEM micrographs showing the inner morphology of charcoal fragments and the absence of any microbes or plant debris. a) is a radial section b) a longitudinal section.

Stable C isotope ratio ($\delta^{13}C$) measurements were made on the fine fraction (<2 mm) of representative subsamples of control and hearth soils using an Isotope Ratio Mass Spectrometer (© Thermo Fischer Scientific, Delta V Plus) following total combustion in an elemental analyser (© EA Flash 1112 ThermoFinnigan).

The $\delta^{13}C$ signature of respired CO_2 from incubated charcoal hearth and control soils was measured using the Picarro G2131-*i* $\delta^{13}C$ High-precision Isotopic CO_2 Cavity Ring Down Spectrometer (CRDS) and Keeling plot method [36]. Representative subsamples (~250 cm^3; n = 3) were incubated in Erlenmeyer flasks at 40°C for 15 minutes. Air was continuously circulated from the flask to a pump and then back into the flask at a rate of 0.8–1.0 l min^{-1}. The CRDS was connected to the pump inlet tube and the air sub-sampled at 0.015 l min^{-1} for measurements of CO_2 concentration and $\delta^{13}C$. Sampling frequency was 0.5 Hz. To determine the $\delta^{13}C$ signature of the respired CO_2, the Keeling method was applied [36]. The intercept of the linear regression with the y-axis represents the isotopic signature of the source of the flux. Regression coefficients were calculated on averaged data at each 50 ppm interval of CO_2 concentration, starting from 450 to 800 ppm to establish a steady mixing within the flask. Finally, mean and standard deviation values of $\delta^{13}C$ were computed for both charcoal hearth and control soils (n = 3). We assumed that any difference in the $\delta^{13}C$ of SOM in control and charcoal hearths would also be reflected by a difference in the $\delta^{13}C$ of the respired CO_2.

Net C Sequestration

Ancillary information that is required to estimate the net C sequestration in hearths' soils was gathered from different sources:

- the amount of wood that was harvested and used for the carbonization process (W_s) was estimated on the basis of forest stand volume at three sub-alpine larch forests of varying ages. One forest was located on the slopes above the hearths of Val di Pejo, one in the nearby Val di Rabbi and one in Val Comasine. All forests had the same altitude (2000 m) and the latter two sites are known to be old-growth forests where 600 year old larch trees can still be found. Stand volume (m^3 ha^{-1}) was estimated using Light Detection And Ranging (LiDAR) measurements processed according to [37] and converted into biomass using an average wood basal density of 860 kg m^{-3}. For each forest, maximum wood stock was calculated using a fixed number of pixels;
- forest surface area that supplied each hearth during charcoal production (h), was determined by geo-referencing the charcoal hearths and drawing a 26 ha polygon on a digital orthophoto (Terraitaly™; © Compagnia Generale Ripreseaeree S.p.A. – Parma). h was calculated according to the following two criteria: i) the wood collection area should be located above each charcoal hearth up to the tree line; ii) the lateral boundaries of the area corresponded to the mean distance between two charcoal hearths (~100 m). Finally, the total area was divided by the total number of hearths identified in the area (n = 7).
- carbonization efficiency, i.e. the ratio between produced charcoal and used wood (q), was assumed to be equal to 20% according to [38];
- the fraction of charcoal that was left on the soil surface at the end of each carbonization cycle after char was removed by charcoal makers (w), was experimentally estimated using a modern analogue. A charcoal hearth that is currently in use

Table 4. Description of larch forests considered as analogues of the larch forest harvested for charcoal production in Val di Pejo (mean ± standard error).

Forest site	Coordinates	Elevation (m a.s.l.)	Forest age	Number of plots	Wood stock (W, t ha^{-1})
Val di Pejo	46°20′16.18″ N, 10°36′07.02″ E	2152	150	51	288±41
Val di Rabbi	46°26′39.45″ N, 10°45′59.56″ E	1864	500	20	213±27
Val Comasine	46°20′02.00″ N, 10°39′58.78″ E	2119	650	40	258±25

Data were determined using LiDAR measurements [37].

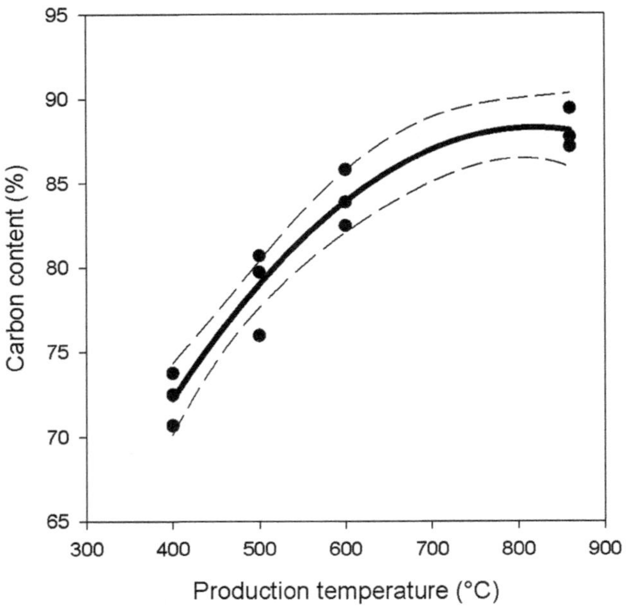

Figure 5. Carbon content of charcoal produced from larch wood at different temperatures. Wood samples were collected in close proximity to the hearths. Charcoal was produced in a muffle furnace at 400°, 500° 600° and 860°C. Dashed lines represent 95% confidence interval. ($Y = 26.9 + 0.15\ X - 9.2\ 10^{-5}\ X^2$; $r^2 = 0.94$; $p < 0.0001$).

was identified at short distance from the Val di Pejo, where expert charcoal makers repeat traditional charcoal production mainly for didactical purposes. They also recorded, year by year, the exact amount of wood used and of charcoal produced. The amount of pyrogenic C left on hearth soil was quantified using the loss-on-ignition (LOI) method [39] using four replicates randomly selected within the hearth area. w was determined as the ratio between the sum of C contained in the charcoal that was made over the last ten years and the corresponding amount of C that was found in the soil. The uncertainty was estimated as the standard error of the mean;

- C content of freshly produced larch wood charcoal (r_0) was also assessed experimentally. Known amounts of larch wood taken from wood disc collected in the proximity of Val di Peio charcoal hearths were pyrolized at different temperatures in a muffle furnace. The C concentration was measured on each charcoal sample using a CHN elemental analyzer. Measured C content data were fitted to production temperature using a second order polynomial relationship and C content at a reference temperature of 450°C was assumed to be an analog of the charcoal originally produced [40].

Uncertainty and Sensitivity Analysis

All data in the text and in the tables are reported as mean ± standard error (n = 3) if not differently indicated.

Gaussian error propagation technique (GEP) was used in error analysis to analytically determine the error or uncertainty produced by multiple and interacting measurements or variables. For this, the uncertainty associated to each measurement was calculated as standard error (se) of the mean (se = standard deviation/root square of the number of samples) and the classical error propagation theory and equations were used [41]. Error sensitivity analysis was also made by constructing an error budget [42], thus enabling further understanding of the error structure, i.e. the relative contribution of the errors associated with each parameter to the overall error estimate. Such sensitivity indices provide in this way a measure of the percentage rate of change in an output variable produced per unit percentage change in its input variable, an information that may be used critically to identify where error reduction of estimates may lead to lower uncertainty.

Table 5. Parameters, coefficients and variables used to estimate charcoal stability in soil (mean ± standard error; n = 3).

Parameter/ coefficient/variable	Definition	Unit	Value	Source(s)
q	Carbonization efficiency	–	0.20	[38]
w	Fraction of charcoal left on the ground at the end of carbonization	–	0.02±0.017	Measured with LOI
r_0	Carbon content of larch wood charcoal produced at 450°C	$g\ g^{-1}$	0.76±0.040	Measured with CHN and extrapolation
r_{155}	Carbon content of charcoal fragments found in the hearth's soil	$g\ g^{-1}$	0.60±0.032	Measured with CHN
W_s	Larch wood stock of the forest in Val di Pejo	$t\ ha^{-1}$	288±41	Measured (LiDAR) and calculated to convert from $m^3\ ha^{-1}$ to $t\ ha^{-1}$
h	Forest area for wood collection for charcoal production per hearth	ha $hearth^{-1}$	3.7	Measured on ortophotos
s	Surface area of charcoal hearths	m^2	94	Measured
C_{IN}	Pyrogenic carbon left on the ground at the time of the carbonization	$Kg\ C\ m^{-2}$	29.3±5.1	Calculated using Eq. [2]
-	Fraction of pyrogenic carbon lost since the time of the carbonization calculated with Eq. [2]	–	0.20±0.28	Calculated
-	Fraction of pyrogenic carbon lost since the time of the carbonization calculated with r_0 and r_{155}	–	0.21±0.04	Calculated
k_{CHAR}	Annual pyrogenic carbon decay rate in hearths	$years^{-1}$	0.0015±0.0003	Calculated
MRT	Mean residence time of charcoal in the soil	years	650±139	Calculated

Table 6. Uncertainty and sensitivity analysis results for the setimation of equation 3 parameters.

Variable	Variable uncertainty (standard error)	Relative contribution to the overall uncertainty in ΔC_{CHAR} in Eq.3
W_s	41	16%
w	0.002	11%
r_0	0.004	1%
Soil organic C content	0.03	18%
Soil bulk density	71	14%
Anthropogenic soil layer depth	0.04	26%
$\delta^{13}C_{TOT}$	0.14	11%
$\delta\ ^{13}C_{CHAR}$	0.31	3%
$\delta\ ^{13}C_{SOM}$	0.01	1%

Results and Discussion

Soil Bulk Density and Hydrophobicity

The anthropogenic layer of the charcoal hearth soils has a lower bulk density than control soils (0.60 ± 0.08 vs. 0.87 ± 0.12 Mg m^{-3}; $p<0.01$). Soil bulk density is an important indicator of physical soil quality, being linked to air capacity, resistance to root growth and capacity of storing and transmitting water [43]. Such a decrease in bulk density is associated with a 97.3% ±1.6 decrease in hydrophobicity. This is in line with the water infiltration data from charcoal production sites in Ghana [44], but in contrast with short term observations following biochar applications to soil [45] [46] that showed small but consistent increases in soil hydrophobicity which is largely controlled by the surface chemistry of fresh biochar particles [47]. We speculate that a prolonged residence time of charcoal caused substantial leaching or degradation of hydrophobic compounds [48], a shift in soil texture ([49], [44]), and a microbially-driven creation of functional groups [50]. Decreased hydrophobicity is known to increase water availability for plants and is also important for nutrient cycling, as it favors water infiltration into the soil and reduces runoff, thus preventing lateral nutients losses.

Nutrients and Carbon

Nutrient content (total and plant available fractions) is higher in the hearths than in the control soils (Table 2; Table 3). In particular, the total P-stock is 107% larger in hearths than in the control (95 ± 6 vs. 46 ± 3 g m^{-2}, $p=0.003$), while the plant available P is 24% higher (1.2 ± 0.04 g m^{-2} vs. 0.9 ± 0.3 g m^{-2}). The higher P content is not surprising as charcoal contains at least 20% of P originally contained in the wood (Table 2). If we assume that charcoal made in the middle of the XIX century had a P content comparable to that of modern charcoal (3.0 ± 0.05 g P kg^{-1}; Table 3), we may estimate that carbonization events led to the addition of 117 ± 21 g P m^{-2} (Table 2). In the absence of grass mowing, a virtually closed P-cycle can be hypothesized for these soils as P-leaching does not usually occur if the overall concentrations are low so that P can be considered "virtually immobile" [51] [52]. Nevertheless, large herbivores are known to be net P-exporters in alpine grasslands as they may preferentially graze in P-enriched areas and then release P as dung elsewhere [53]. This export largely depends on the grazing pressure, but it is unlikely to exceed the maximum value of 0.07 g P m^{-2} y^{-1} that was assessed experimentally in the Swiss Alps [53]. When scaled to the time that charcoal was added to the soil the amount of P exported would not have exceeded 11.5 g m^{-2}. Atmospheric deposition may have also contributed to the P balance as a result of long-range desert dust transport and as a consequence of atmospheric pollution, including biomass combustion [54]. Although the latter is known to be variable in time and space, this input is not large in the Eastern Alps, with less than 0.01 g P m^{-2} y^{-1} [55]. When measured today, the total amount of P contained in the anthropogenic layer of the hearth's soil is only 19% less than what was initially added during the carbonization events (95 ± 6 vs. 117 ± 21 g m^{-2}; Table 2). This highlights the fact that charcoal production was indeed a long-lasting source of P in an otherwise P-limited environment. P-fertilization persisted on a centennial time-scale and it is also interesting to observe that both organic and inorganic (extractable) fractions of P that were added to the soil are now mostly contained in the non-pyrogenic fraction of the SOM, as the P contained in old charcoal fragments is only 12% of that of modern laboratory-produced larch charcoal (Table 3).

Other nutrients, such as potassium (K) and calcium (Ca) are also more abundant in the charcoal hearth soils even though there are no significant differences with control soils due to the large spatial variability (Table 2). Furthermore K and Ca are 155% and 61% higher than what was added to the soil with charcoal (Table 2). This excess may be attributed to an higher retention of atmospheric K and Ca depositions which have been previously reported to be relevant in the Alpine region [56] and can be related to the higher cation exchange capacity (CEC) of charcoal [57]. A strong correlation between current atmospheric deposition rates and element excesses found in the hearth soil seems to confirm such hypotheses (Figure 2).

Total N content is not significantly different between hearth and control soils ($p=0.42$; Table 2) and no significant difference was found in the concentrations of mineral N (NO_3^-: $p=0.31$; NH_4^+: $p=0.92$; Table 3).

Total C content of the anthropogenic layer at the charcoal hearths is three times higher than that of the adjacent control soils ($p=0.03$; Table 1). The amount of C that is now contained in the anthropogenic soil layer (C_{TOT}, kg C m^{-2}) is the sum of pyrogenic (C_{CHAR}, kg C m^{-2}) and non-pyrogenic components (C_{SOM}, kg C m^{-2}) as carbonates are absent due to the low pH (4.2 ± 0.3 and 4.6 ± 0.3 in charcoal hearths and control areas, respectively). Charcoal fragments larger than 2 mm represent approximately 4.1 ± 1.7% (by weight) of the entire mass of the anthropogenic layer within the deeper soil horizons where no charcoal debris can be identified (Figure 1).

Any meaningful evaluation of net C sequestration achieved in charcoal hearths firstly requires a precise separation of C fractions contained in the charcoal (C_{CHAR}) and in the rest of the soil, followed by a accurate estimation of the C input at the time of the carbonization event.

Assessing the exact ratio between C_{CHAR} and C_{TOT} using sable C isotopes is problematic because:

- Equation 1 assumes a net difference in the isotopic signature of old charcoal and modern SOM. This is supported by the observation that wood formed before 1900 is on average ~1‰ less negative than that formed after 1950 [58] due to the recent rapid rise in δ^{13}C-depleted atmospheric CO_2 concentrations as a result of fossil fuel burning [59]. In addition, it has been reported that branch or stem wood of C_3 plants is generally enriched by 1–3‰ compared to leaves [60]. Despite the fact that limited ^{13}C-depletion may occur during wood carbonization at temperatures above 300°C [61], it could be expected that the charcoal fragments found in the hearth's soils are significantly enriched in the heavier C isotope compared to the more recent SOM pools that are mainly derived from the decomposition of litter formed more recently;
- Equation 1 assumes that $\delta^{13}C_{SOM}$ in the hearth's soils is equal to $\delta^{13}C_{SOM}$ in the control soils. Such equivalence cannot be assumed *a priori*, but CRDS-based δ^{13}C flux measurements demonstrated that the signature of respired CO_2 was not significantly different between control and charcoal hearth incubated soil ($p = 0.75$; Figure 3 and Table 1). Similarly, it may be assumed that the δ^{13}C of the less recalcitrant (non-pyrogenic) SOM fractions is the same in both hearth and control soils. Such an equivalence is further confirmed by the linearity observed by plotting the reciprocal of the C content of charcoal fragments, hearth and control soils versus their respective stable C isotopic ratios (slope = −11.3; intercept = −24.3; $r^2 = 0.98$; n = 9; $p < 0.0001$). The fact that the data points fall within close proximity to a straight line indicates that the two C pools (pyrogenic and non-pyrogenic C) are distinguished in the hearth's soils;
- The δ^{13}C of the charcoal fragments ($\delta^{13}C_{CHAR}$ in Equation 1) could be affected by the presence of organic debris or microorganisms in charcoal pores. The analysis of charcoal fragments with a Scanning Electron Microscope showed that, despite being exposed in soil for over 150 years, no organic or inorganic debris were present in the inner portion of the fragments ([34]; Figure 4). A recent study [62] showed that, in spite of the large empty space in charcoal pores, these were only sparsely populated by microorganisms three years after application to soil. This was attributed to the adsorption of inorganic and organic compounds which may cause blockage of the charcoal pores and thus prevent microorganisms penetrating the inner portion of fragments. Our data support this observation, showing that charcoal, also in the long term, does not provide a habitat for microbes and, accordingly, the δ^{13}C of the charcoal fragments is unaffected.

Based on the above assumptions and considerations, the fraction of pyrogenic C (C_{CHAR}) contained in the overall soil C (C_{TOT}) was finally estimated using Equation 1 to be equal to 89±9%, corresponding to an absolute amount of 23.3±4.7 kg of pyrogenic C m^{-2} (Table 1).

Carbon Sequestration

Once the amount of pyrogenic C is known, a reliable estimation of the C sequestration achieved in the hearth soil over centennial timescale requires a proper estimation of the initial C input (C_{IN}, kg C m^{-2}). This can be obtained according to equation:

$$C_{IN} = \frac{W_s \times h \times q \times w \times r_0}{s} * 1000 \qquad (2)$$

Solving Equation 2 is problematic mainly because of the uncertainty associated with the determination of parameters W_s, w and r_0:

- the total amount of wood biomass (W_s) that was used for carbonization cannot be directly estimated but requires the use of a proxy. The assumption can be made that the forest standing biomass one and half centuries ago was comparable to what is currently found in our study area (288±41 t ha^{-1}; Table 4). This assumption is partly confirmed by the fact that LIDAR-based standing volumes of two additional old-growth forest sites (>200 years) are comparable to the wood stock of the study area. The fact that tree volume is independent of stand age is not surprising and is confirmed by the usually reported plateau of forest wood stocks over centennial times scales [63] in alpine larch forests [64];
- the fraction of charcoal left on the soil after carbonization can only be estimated indirectly as there a no reliable sources reporting such a value. We assumed that an experimental assessment using modern charcoal hearth is the most reliable proxy of ancient charcoal hearths. The value obtained in this way ($w = 0.02 \pm 0.002$) was not far from judgment of two expert charcoal makers who unanimously said that no more than 2% charcoal fragments are normally left over the soil at the end of each carbonization cycle;
- the C content of new charcoal (r_0) is known to vary substantially with production temperature ($r^2 = 0.94$, $p < 0.0001$; Figure 5). Here again there are not historical data on temperature during charcoal production but FAO reports that the average temperature in traditional carbonization wood piles is around 450°C [40]. Using this value, r_0 was estimated to be equal to 0.76±0.004.

Based on the above assumptions and considerations, C_{IN} could be calculated to be equal to 29.3±5.1 kg C m^{-2} (Table 5) and the overall pyrogenic C lost was then quantified as:

$$\Delta C_{CHAR} = \frac{(C_{IN} - C_t)}{C_{IN}} \qquad (3)$$

where C_t is the actual pyrogenic C in the soil and t is time since the last char production at the charcoal hearth (153 years). Thus, the fraction of pyrogenic C lost since the time of carbonization was equal to 0.20±0.28 of C_{IN} (Table 5). This value is given by the sum of the direct charcoal degradation by biotic and abiotic factors and the lateral transport due to surface runoff and erosion occurring during and after char production, before a new soil layer covered the anthropogenic layer [65]. The presence of small amounts of minute charcoal fragments predominantly down-slope of the hearths seems to confirm the occurrence of such lateral flows.

Recent studies have used changes of the C-content of individual charcoal fragments buried in the soil as a proxy of the fraction of carbon lost from charcoal over long time scales [66]. Such method is certainly questionable, as it cannot distinguish between actual C-losses due to oxidation and C-dilution effects associated to the deposition of inorganic salts and minerals on those fragments.

Nevertheless, it is noteworthy, here, to highlight that the C content of individual charcoal fragments found in the charcoal hearth's soils ($r_{155}=0.60\pm0.03$ gC g^{-1}) is consistently lower than the C content of modern larch charcoal made in the laboratory ($r_0=0.76\pm0.04$ gC g^{-1}). The fact that the relative difference between r_0 and r_{155} (0.21) almost exactly matches our estimate of the fraction of C lost on centennial time scale (Equations 2 and 3) is likely coincidental but suggests that more detailed investigations on the oxidation of ancient charcoal fragments ([67]) and on mineral deposition on old charcoal fragments are needed.

The decay rate (k, years^{-1}) can be finally calculated as:

$$k=\frac{\ln\left(\frac{1}{1-\Delta C_{CHAR}}\right)}{t} \quad (4)$$

and is equal to 0.0015 ± 0.0003. Such a value is less than half of the value estimated in a recent meta-analysis of charcoal decomposition [68] and corresponds to a Mean Residence Time (MRT) of 650±139 years (Table 5).

Error sensitivity analysis performed on the results of Equation 2 and 3 illustrates the relative contribution of errors to the overall uncertainty of the result (Table 6). The data show that the largest contribution to the overall uncertainty is in the error associated to the anthropogenic soil layer depth. Such large error suggests that variance among the three replicated hearths may not be simply random but reflects, instead, some systematic effects possibly associated to differences in the amount of wood that was carbonized in each hearth. This is indeed a critical aspect, possibly requiring a re-analysis of the simplifying assumption of an even distribution of carbonization intensity (amount of wood carbonized) among the replicated hearths. Such re-analysis would in fact enable a substantial reduction of the uncertainty, while not affecting the means. Other important sources of error are related to the estimation of the C-content of the anthropogenic layers (C_{tot}) and in the estimated amount of wood that was used for the carbonization process (W_s). Error reduction would have been possible, in the first case, by increasing the number of replicates and more unlikely by increasing samplings in each replicate. Nevertheless, a relative large uncertainty would remain associated to the second case, as W_s was inferred on the basis of large simplification and that by no means new information could be retrieved to finally reduce its error.

Conclusions

This study presents two key messages. First, it provides novel insights into the long-term decomposition of pyrogenic C in the soil, by demonstrating that charcoal addition to soil is indeed a way to obtain substantial C-sequestration. Despite some inevitable uncertainty, we have shown that 23.3 ± 4.7 kg C m^{-2} of pyrogenic C are still present in the soil after an addition of 29.3 ± 5.1 kg C m^{-2} that was made in the middle of the XIX century. Carbon sequestration was estimated as 80±21% of the original added C. Secondly, our investigation provides substantial evidence that the avaliability of macro- and some micro-nutrients is higher in charcoal hearth soils over centennial timescales. This supports the common observation that the addition of various forms of pyrogenic C (biochar) increases soil fertility and plant yield, even in the long-term. The persistence of enhanced nutrient avaliability over centennial timescales is likely associated with mechanisms favoring their accumulation and improving soil water relations. Overall, this strongly supports the idea that the addition of biochar to soil is indeed a feasible, effective, and sustainable strategy to both sequester atmospheric C and to enhance crop yields.

However, we call for some caution on excessive generalization of our results: the hearths within the Val di Pejo are located in a mountainous area, at high elevations, and are exposed to peculiar climatic conditions, that are certainly different to the vast majority of areas where croplands are concentrated. Even if soil freeze-thaw cycles associated with large seasonal temperature fluctuations are likely to favour decomposition and C oxidation, it is not certain if and how the C decomposition rates that we observed are greater or smaller than in other climates. Additional caution is required since that the charcoal added to the studied hearths came from a single feedstock (larch wood) and was produced in traditional charcoal production systems. The fate and the effects of other feedstocks and of other production processes such as slow and fast pyrolysis, pyrogasification, and hydrothermal conversion may indeed create totally different biochar types, possibly behaving in different ways in the soil.

Charcoal hearths in Val di Pejo are certainly a unique resource for investigating the long term effect of pyrogenic C addition to the soil. The number of replications which are available, the long residence time of the charcoal in the soil, the accuracy of charcoal dating, the reliability of ancillary information on the sequence of events that were associated to charcoal production and its sudden cessation, contribute to the scientific value of these sites. A number of questions, not considered in the present study, could be addressed in the near future, such as, for instance, the effect of long-term charcoal burial on (i) its physio-chemical properties; (ii) non-CO_2 greenhouse gas fluxes (CH_4 and N_2O); (iii) plant productivity; and (iv) shifts and changes in soil biodiversity.

Acknowledgments

The authors would like to thank Federica Camin and Luca Ziller, Matteo Girardi, Michele Dalponte and Damiano Gianelle (E. Mach Foundation, San Michele a/Adige) Luisa Andrenelli, Adriano Baglio, Fabio Zuliani (Stelvio National Park) for their support in field and laboratory work, as well as Christa Backmeroff, Fabio Angeli, Dino Buffoni, Jean Christophe Lata, Marie Alexis and Katell Quenea for useful discussions.

Author Contributions

Conceived and designed the experiments: IC GA CR RV FM. Performed the experiments: IC GA SB FF CM CC EP FM. Analyzed the data: IC GA CC FM. Wrote the paper: IC GA CM FM.

References

1. Lehmann J, Gaunt J, Rondon M (2006) Bio-char Sequestration in Terrestrial Ecosystems – A Review. Mitigation and Adaptation Strategies for Global Change 11: 395–419. Available: http://link.springer.com/10.1007/s11027-005-9006-5. Accessed 23 May 2013.
2. Laird DA (2008) The Charcoal Vision: A Win–Win–Win Scenario for Simultaneously Producing Bioenergy, Permanently Sequestering Carbon, while Improving Soil and Water Quality. doi:10.2134/agronj2007.0161.
3. Woolf D, Amonette JE, Street-Perrott FA, Lehmann J, Joseph S (2010) Sustainable biochar to mitigate global climate change. Nature communications 1: 56. Available: http://www.pubmedcentral.nih.gov/articlerender.fcgi?artid=2964457&tool=pmcentrez&rendertype=abstract. Accessed 29 May 2013.
4. Joseph SD, Downie A, Munroe P, Crosky A, Lehmann J (2007) Biochar for Carbon Sequestration, Reduction of Greenhouse Gas Emissions and Enhancement of Soil Fertility; A Review of the Materials Science: 130–133.
5. Yanai Y, Toyota K, Okazaki M (2007) Effects of charcoal addition on N 2 O emissions from soil resulting from rewetting air-dried soil in short-term laboratory experiments. Soil Science and Plant Nutrition 53: 181–188. Available: http://www.tandfonline.com/doi/abs/10.1111/j.1747-0765.2007.00123.x. Accessed 28 May 2013.

6. Peng X, Ye LL, Wang CH, Zhou H, Sun B (2011) Temperature- and duration-dependent rice straw-derived biochar: Characteristics and its effects on soil properties of an Ultisol in southern China. Soil and Tillage Research 112: 159–166. doi:10.1016/j.still.2011.01.002.
7. Oguntunde PG, Fosu M, Ajayi AE, Van de Giesen N (2004) Effects of charcoal production on maize yield, chemical properties and texture of soil. Biology and Fertility of Soils 39: 295–299. Available: http://link.springer.com/10.1007/s00374-003-0707-1. Accessed 29 May 2013.
8. Lehmann J, Rillig MC, Thies J, Masiello C, Hockaday WC, et al. (2011) Biochar effects on soil biota – A review. Soil Biology and Biochemistry 43: 1812–1836. Available: http://linkinghub.elsevier.com/retrieve/pii/S0038071711001805. Accessed 25 May 2013.
9. Genesio L, Miglietta F, Lugato E, Baronti S, Pieri M, et al. (2012) Surface albedo following biochar application in durum wheat. Environmental Research Letters 7: 014025. Available: http://stacks.iop.org/1748-9326/7/i = 1/a = 014025?key = crossref.4c8f29d72864c065a47107c0fadd5c71. Accessed 29 May 2013.
10. Graber ER, Meller Harel Y, Kolton M, Cytryn E, Silber A, et al. (2010) Biochar impact on development and productivity of pepper and tomato grown in fertigated soilless media. Plant and Soil 337: 481–496. Available: http://link.springer.com/10.1007/s11104-010-0544-6. Accessed 29 May 2013.
11. Gurwick NP, Moore L a, Kelly C, Elias P (2013) A systematic review of biochar research, with a focus on its stability in situ and its promise as a climate mitigation strategy. PloS one 8: e75932. Available: http://www.pubmedcentral.nih.gov/articlerender.fcgi?artid = 3786913&tool = pmcentrez&rendertype = abstract. Accessed 17 December 2013.
12. Schmidt MWI, Skjemstad JO, Jäger C (2002) Carbon isotope geochemistry and nanomorphology of soil black carbon: Black chernozemic soils in central Europe originate from ancient biomass burning. Global Biogeochemical Cycles 16. doi:10.1029/2002GB001939.
13. Favilli F, Cherubini P, Collenberg M, Egli M, Sartori G, et al. (2010) Charcoal fragments of Alpine soils as an indicator of landscape evolution during the Holocene in Val di Sole (Trentino, Italy). The Holocene 20: 67–79. doi:10.1177/0959683609348850.
14. Kaal J, Martínez Cortizas A, Eckmeier E, Costa Casais M, Santos Estévez M, et al. (2008) Holocene fire history of black colluvial soils revealed by pyrolysis-GC/MS: a case study from Campo Lameiro (NW Spain). Journal of Archaeological Science 35: 2133–2143. doi:10.1016/j.jas.2008.01.013.
15. Masiello Ca (2004) New directions in black carbon organic geochemistry. Marine Chemistry 92: 201–213. Available: http://linkinghub.elsevier.com/retrieve/pii/S0304420304002051. Accessed 27 May 2013.
16. Zimmermann M, Bird MI, Wurster C, Saiz G, Goodrick I, et al. (2012) Rapid degradation of pyrogenic carbon. Global Change Biology 18: 3306–3316. Available: http://doi.wiley.com/10.1111/j.1365-2486.2012.02796.x. Accessed 29 May 2013.
17. Knicker H (2011) Pyrogenic organic matter in soil: Its origin and occurrence, its chemistry and survival in soil environments. Quaternary International 243: 251–263. doi:10.1016/j.quaint.2011.02.037.
18. Smith N (1980) Anthrosols and human carrying capacity in Amazonia. Annals of the Association of American Geographers 70: 553–566.
19. Cremaschi M, Pizzi C, Valsecchi V (2006) Water management and land use in the terramare and a possible climatic co-factor in their abandonment: The case study of the terramara of Poviglio Santa Rosa (northern Italy). Quaternary International 151: 87–98. doi:10.1016/j.quaint.2006.01.020.
20. Glaser B, Haumaier L, Guggenberger G, Zech W (2001) The "Terra Preta" phenomenon: a model for sustainable agriculture in the humid tropics. Naturwissenschaften 88: 37–41. Available: http://link.springer.com/10.1007/s001140000193. Accessed 24 May 2013.
21. Mercuri AM, Accorsi CA, Mazzanti MB, Bosi G, Cardarelli A, et al. (2006) Economy and environment of Bronze Age settlements – Terramaras – on the Po Plain (Northern Italy): first results from the archaeobotanical research at the Terramara di Montale. Vegetation History and Archaeobotany 16: 43–60. doi:10.1007/s00334-006-0034-1.
22. Di Piazza A, Eccel E (2012) Analisi di serie giornaliere di temperatura e precipitazione in Trentino nel periodo 1958–2010. Provincia Autonoma di Trento e Fondazione E. Mach.
23. Sonna R (n.d.) Le miniere di Comasine. Studi e Ricerche: 1–10.
24. Rizzi G (2010) Sentiero Val di Sole, le sfumature del verde. Trento: Tipografia Editrice Temi s.a.s.
25. Schenkel Y, Bertaux P, Vanwijnbserghe S, Carre J (1998) An Evaluation of the Mound Kiln Carbonization Technique. Biomass and Bioenergy 14: 505–516.
26. Smith RT, Atkinson K (1975) Techniques in pedology. London: Paul Elek (Scientific Books) Ltd.
27. Eckmeier E, Gerlach R, Skjemstad JO, Ehrmann O, Schmidt MWI, et al. (2007) Minor changes in soil organic carbon and charcoal concentrations detected in a temperate deciduous forest a year after an experimental slash-and-burn: 377–383.
28. Carcaillet C (2001) Are Holocene wood-charcoal fragments stratified in alpine and subalpine soils? Evidence from the Alps based on AMS 14C dates. The Holocene 11: 231–242. doi:10.1191/095968301674071040.
29. Backmeroff C (2013) Precisely dating iron-ore mining and its effects on an alpine valley: Summary of a dendrochronological investigation of charcoal hearths and relict woodland stands. In: Silvertant J ed., editor. Mining and Cultural Landscape –8th International Symposium on Archaeological Mining History. pp. 218–251.
30. USEPA (1996) Method 3052, Microwave assisted aced digestion of siliceous and organically based matrices: 1–20.
31. Vendrell P, Zupancic J (1990) Determination of soil nitrate by transnitration of salicylic acid. Communications in Soil Science and Plant Analysis 21.
32. Willis RB, Schwab GJ, Gentry CE (1993) Elimination of interferences in the colorimetric analysis of ammonium in water and soil extracts. Communications in Soil Science and Plant Analysis 24.
33. Letey J (1969) Measurement of contact angle, water drop penetration time, and critical surface tension. In: University of California, editor. Water repellent soils. Proceedings of the Symposium on Water Repellent Soils. Riverside. pp. 43–47.
34. Pusceddu E, Criscuoli I, Miglietta F (2013) Morphological investigation and physical characterization of ancient fragments of pyrogenic carbon. Journal of Physics: Conference Series 470: 012003. Available: http://stacks.iop.org/1742-6596/470/i = 1/a = 012003?key = crossref.442c46e432fd91c4a0ddb5660ebc00b8. Accessed 28 December 2013.
35. Del Galdo I, Six J, Peressotti A, Cotrufo F (2003) Assessing the impact of land-use change on soil C sequestration in agricultural soils by means of organic matter fractionation and stable C isotopes. Global change biology: 1204–1213.
36. Keeling CD (2011) The Concentration and Isotopic Abundances of Carbon Dioxide in the Atmosphere. Tellus A. Available: http://journals.sfu.ca/coaction/index.php/tellusa/article/view/9366.
37. Tonolli S, Dalponte M, Vescovo L, Rodeghiero M, Bruzzone L, et al. (2010) Mapping and modeling forest tree volume using forest inventory and airborne laser scanning. European Journal of Forest Research 130: 569–577. Available: http://link.springer.com/10.1007/s10342-010-0445-5. Accessed 29 May 2013.
38. Mantovani P (2006) Ricordi di un carbonèr. Da Bondone a Bissina. Memorie autobiografiche. Antolini, editor Antolini.
39. Ball DF (1964) Loss-on-ignition as an estimate. 15.
40. FAO (1987) Simple technologies for charcoal making. Rome: Mechanical Wood Products Branch Forest Industries Division FAO Forestry Department.
41. Lehrter JC, Cebrian J (2010) Uncertainty propagation in an ecosystem nutrient budget. Ecological applications: a publication of the Ecological Society of America 20: 508–524. Available: http://www.ncbi.nlm.nih.gov/pubmed/20405803.
42. Lo E (2005) Gaussian error propagation applied to ecological data: post-ice-storm-downed woody biomass. Ecological Monographs 75: 451–466. Available: http://www.esajournals.org/doi/abs/10.1890/05-0030. Accessed 28 December 2013.
43. Reynolds WD, Drury CF, Tan CS, Fox C a., Yang XM (2009) Use of indicators and pore volume-function characteristics to quantify soil physical quality. Geoderma 152: 252–263. doi:10.1016/j.geoderma.2009.06.009.
44. Oguntunde PG, Abiodun BJ, Ajayi AE, Van de Giesen N (2008) Effects of charcoal production on soil physical properties in Ghana. Journal of Plant Nutrition and Soil Science 171: 591–596. Available: http://doi.wiley.com/10.1002/jpln.200625185. Accessed 29 May 2013.
45. Scott DF, Van Wyk DB (1990) The effects of wildfire on soil wettability and hydrological behaviour of an afforested catchment. Journal of Htdrology 121: 239–256.
46. Lane PNJ, Croke JC, Dignan P (2004) Runoff generation from logged and burnt convergent hillslopes: rainfall simulation and modelling. Hydrological Processes 18: 879–892. doi:10.1002/hyp.1316.
47. Kinney TJ, Masiello C a., Dugan B, Hockaday WC, Dean MR, et al. (2012) Hydrologic properties of biochars produced at different temperatures. Biomass and Bioenergy 41: 34–43. Available: http://linkinghub.elsevier.com/retrieve/pii/S0961953412000438. Accessed 27 May 2013.
48. Briggs CM, Breiner J, Graham R (2005) Contributions of Pinus Ponderosa charcoal to soil chemical and physical properties. The ASA-CSSA-SSSA International Annual Meetings (November 6–10, 2005). Salt Lake City. pp. 1–13.
49. Glaser B, Lehmann J, Zech W (2002) Ameliorating physical and chemical properties of highly weathered soils in the tropics with charcoal - a review. Biology and Fertility of Soils 35: 219–230. Available: http://link.springer.com/10.1007/s00374-002-0466-4. Accessed 23 May 2013.
50. Lehmann J, Liang B, Solomon D, Lerotic M, Luizão F, et al. (2005) Near-edge X-ray absorption fine structure (NEXAFS) spectroscopy for mapping nano-scale distribution of organic carbon forms in soil: Application to black carbon particles. Global Biogeochemical Cycles 19: n/a–n/a. Available: http://doi.wiley.com/10.1029/2004GB002435. Accessed 22 May 2013.
51. Hesketh N, Brookes PC (2000) Development of an Indicator for Risk of Phosphorus Leaching. Journal of Environmental Quality 29: 105–110.
52. Flueck WT (2009) Biotic Translocation of Phosphorus: The Role of Deer in Protected Areas. Sustainability 1: 104–119. doi:10.3390/su1020104.
53. Schütz M, Risch AC, Achermann G, Thiel-Egenter C, Page-Dumroese DS, et al. (2006) Phosphorus Translocation by Red Deer on a Subalpine Grassland in the Central European Alps. Ecosystems 9: 624–633. doi:10.1007/s10021-006-0091-4.
54. Bergametti G, Remoudaki E, Losno R, Steiner E, Chatenet B (1992) Source, transport and deposition of atmospheric phosphorus over the northwestern Mediterranean. Journal of Atmospheric Chemistry 14: 501–513.
55. Mahowald N, Jickells TD, Baker AR, Artaxo P, Benitez-Nelson CR, et al. (2008) Global distribution of atmospheric phosphorus sources, concentrations and

deposition rates, and anthropogenic impacts. Global Biogeochemical Cycles 22. doi:10.1029/2008GB003240.
56. Tait D, Thaler B (2000) Atmospheric deposition and lake chemistry trends at a high mountain site in the eastern Alps. 59: 61–71.
57. Basso AS, Miguez FE, Laird D a., Horton R, Westgate M (2013) Assessing potential of biochar for increasing water-holding capacity of sandy soils. GCB Bioenergy 5: 132–143. doi:10.1111/gcbb.12026.
58. Saurer M, Cherubini P, Reynolds-Henne CE, Treydte KS, Anderson WT, et al. (2008) An investigation of the common signal in tree ring stable isotope chronologies at temperate sites. Journal of Geophysical Research 113: G04035. Available: http://doi.wiley.com/10.1029/2008JG000689. Accessed 29 May 2013.
59. Friedli H, Lötscher H, Oeschger H, Stauffer U, Siegenthaler B (1986) Ice core record of the 13C/12C ratio of atmospheric CO2 in the past two centuries. Nature 324: 237–238.
60. Badeck F-W, Tcherkez G, Nogués S, Piel C, Ghashghaie J (2005) Post-photosynthetic fractionation of stable carbon isotopes between plant organs–a widespread phenomenon. Rapid communications in mass spectrometry: RCM 19: 1381–1391. doi:10.1002/rcm.1912.
61. Bird MI, Ascough PL (2012) Isotopes in pyrogenic carbon: A review. Organic Geochemistry 42: 1529–1539. Available: http://linkinghub.elsevier.com/retrieve/pii/S0146638010002305. Accessed 22 May 2013.
62. Quilliam RS, Glanville HC, Wade SC, Jones DL (2013) Life in the "charosphere" – Does biochar in agricultural soil provide a significant habitat for microorganisms? Soil Biology and Biochemistry 65: 287–293. Available: http://linkinghub.elsevier.com/retrieve/pii/S0038071713002149. Accessed 17 October 2013.
63. Mencuccini M, Martínez-Vilalta J, Vanderklein D, Hamid H a, Korakaki E, et al. (2005) Size-mediated ageing reduces vigour in trees. Ecology letters 8: 1183–1190. doi:10.1111/j.1461-0248.2005.00819.x.
64. Poda A (1963) Tavola alsometrica locale del larice cresciuto in fustaia coetanea del bosco diel comune di Cesana Torinese (Torino). Piano di assestamento, decennio 1963–1972. In: Castellani CE, editor. Tavole Stereometriche Ed Alsometriche Costruite Per Boschi Italiani. Trento: ISAFA. p. 780.
65. Jaffé R, Ding Y, Niggemann J, Vähätalo A V, Stubbins A, et al. (2013) Global charcoal mobilization from soils via dissolution and riverine transport to the oceans. Science (New York, NY) 340: 345–347. Available: http://www.ncbi.nlm.nih.gov/pubmed/23599492. Accessed 27 May 2013.
66. Cheng C-H, Lehmann J, Engelhard MH (2008) Natural oxidation of black carbon in soils: Changes in molecular form and surface charge along a climosequence. Geochimica et Cosmochimica Acta 72: 1598–1610. Available: http://linkinghub.elsevier.com/retrieve/pii/S0016703708000306. Accessed 27 May 2013.
67. Naisse C, Alexis M, Plante A, Wiedner K, Glaser B, et al. (2013) Can biochar and hydrochar stability be assessed with chemical methods? Organic Geochemistry 60: 40–44. Available: http://linkinghub.elsevier.com/retrieve/pii/S0146638013000946. Accessed 28 December 2013.
68. Singh N, Abiven S, Torn MS, Schmidt MWI (2012) Fire-derived organic carbon in soil turns over on a centennial scale. Biogeosciences 9: 2847–2857. Available: http://www.biogeosciences.net/9/2847/2012/. Accessed 27 May 2013.

PERMISSIONS

All chapters in this book were first published in PLOS ONE, by The Public Library of Science; hereby published with permission under the Creative Commons Attribution License or equivalent. Every chapter published in this book has been scrutinized by our experts. Their significance has been extensively debated. The topics covered herein carry significant findings which will fuel the growth of the discipline. They may even be implemented as practical applications or may be referred to as a beginning point for another development.

The contributors of this book come from diverse backgrounds, making this book a truly international effort. This book will bring forth new frontiers with its revolutionizing research information and detailed analysis of the nascent developments around the world.

We would like to thank all the contributing authors for lending their expertise to make the book truly unique. They have played a crucial role in the development of this book. Without their invaluable contributions this book wouldn't have been possible. They have made vital efforts to compile up to date information on the varied aspects of this subject to make this book a valuable addition to the collection of many professionals and students.

This book was conceptualized with the vision of imparting up-to-date information and advanced data in this field. To ensure the same, a matchless editorial board was set up. Every individual on the board went through rigorous rounds of assessment to prove their worth. After which they invested a large part of their time researching and compiling the most relevant data for our readers.

The editorial board has been involved in producing this book since its inception. They have spent rigorous hours researching and exploring the diverse topics which have resulted in the successful publishing of this book. They have passed on their knowledge of decades through this book. To expedite this challenging task, the publisher supported the team at every step. A small team of assistant editors was also appointed to further simplify the editing procedure and attain best results for the readers.

Apart from the editorial board, the designing team has also invested a significant amount of their time in understanding the subject and creating the most relevant covers. They scrutinized every image to scout for the most suitable representation of the subject and create an appropriate cover for the book.

The publishing team has been an ardent support to the editorial, designing and production team. Their endless efforts to recruit the best for this project, has resulted in the accomplishment of this book. They are a veteran in the field of academics and their pool of knowledge is as vast as their experience in printing. Their expertise and guidance has proved useful at every step. Their uncompromising quality standards have made this book an exceptional effort. Their encouragement from time to time has been an inspiration for everyone.

The publisher and the editorial board hope that this book will prove to be a valuable piece of knowledge for researchers, students, practitioners and scholars across the globe.

LIST OF CONTRIBUTORS

Jinman Wang and Zhongke Bai
College of Land Science and Technology, China University of Geosciences, Beijing, People's Republic of China
Key Laboratory of Land Consolidation and Rehabilitation, Ministry of Land and Resources, Beijing, People's Republic of China

Peiling Yang
College of Hydraulic and Civil Engineering, China Agricultural University, Beijing, People's Republic of China

Ali Sadeghi and Barbara Mazzolai
Center for Micro-BioRobotics, Istituto Italiano di Tecnologia, Pontedera, Italy

Alice Tonazzini and Liyana Popova
Center for Micro-BioRobotics, Istituto Italiano di Tecnologia, Pontedera, Italy
The BioRobotics Institute, Scuola Superiore Sant'Anna, Pontedera, Italy

Liqun Zhu, Naijuan Hu and Zhengwen Zhang
College of Agriculture, Nanjing Agricultural University, Nanjing, China

Minfang Yang and Xinhua Zhan
College of Resources and Environmental Science, Nanjing Agricultural University, Nanjing, China

Vidya de Gannes and Gaius Eudoxie
Dept. Food Production, Faculty of Food and Agriculture, University of the West Indies, St. Augustine Campus, Republic of Trinidad and Tobago

William J. Hickey
O.N. Allen Laboratoryfor Soil Microbiology, Dept. Soil Science, University of Wisconsin-Madison, Madison, Wisconsin, United States of America

Yan Li, Gary W. Lawrence, Shien Lu and Clarissa Balbalian
Department of Biochemistry, Molecular Biology, Entomology and Plant Pathology, Mississippi State University, Mississippi State, Mississippi, United States of America

Vincent P. Klink
Department of Biological Sciences, Mississippi State University, Mississippi State, Mississippi, United States of America

Nu-Fang Fang
State Key Laboratory of Soil Erosion and Dryland Farming on the Loess Plateau, Northwest A & F University, Yangling, PR China
Institute of Soil and Water Conservation of Chinese Academy of Sciences and Ministry of Water Resources, Yangling, PR China

Zhi-Hua Shi
State Key Laboratory of Soil Erosion and Dryland Farming on the Loess Plateau, Northwest A & F University, Yangling, PR China
Institute of Soil and Water Conservation of Chinese Academy of Sciences and Ministry of Water Resources, Yangling, PR China
College of Resources and Environment, Huazhong Agricultural University, Wuhan, China

Ben-Jiang Yue and Ling Wang
Institute of Soil and Water Conservation of Chinese Academy of Sciences and Ministry of Water Resources, Yangling, PR China

Gu Feng, Zhenling Cui, Fusuo Zhang, Xinping Chen and Yan Deng, and Keru Chen
Center for Resources, Environment and Food Security, China Agricultural University, Beijing, China

Wan Teng and Yiping Tong
State Key Laboratory for Plant Cell and Chromosome Engineering, Institute of Genetics and Developmental Biology, Chinese Academy of Sciences, Beijing, China

Ai Zhan
State Key Laboratory of Soil Erosion and Dryland Farming on the Loess Plateau, Institute of Soil and Water Conservation, Chinese Academy of Sciences and Ministry of Water Resource, Yangling, China

Wen-Kun Huang, Jiang-Kuan Cui, Gang-Feng Wang, Ling-An Kong, Huan Peng and De-Liang Peng
State Key Laboratory for Biology of Plant Diseases and Insect Pests, Institute of Plant Protection, Chinese Academy of Agricultural Sciences, Beijing, P. R. China

Jian-Hua Sun
Tianjin Key Laboratory of Animals and Plants Resistance, College of Life Sciences, Tianjin Normal University, Tianjin, P. R. China

Shu-Long Chen
Institute of Plant Protection, Hebei Academy of Agricultural and Forestry, Baoding, Hebei Province, P. R. China

Tahir Hussain Awan and Bhagirath Singh Chauhan
Weed Science, Crop and Environmental Sciences Division, International Rice Research Institute (IRRI), Los Baños, Laguna, Philippines

Pompe C. Sta. Cruz
Crop Science Cluster, College of Agriculture, University of Philippines Los Baños, Laguna, Philippines

Xiangyang Sun, Suyan Li, Tao Zhang, Wei Zhang and Penghui Zhai
College of Forestry, Beijing Forestry University, Beijing, China

Linlin Wang
Forestry Research Institute of Hulun Buir, Hulun Buir, Inner Mongolia, China

Shenzhong Tian, Tangyuan Ning, Hongxiang Zhao, Bingwen Wang, Na Li and Zengjia Li
State Key Laboratory of Crop Biology, Shandong Key Laboratory of Crop Biology, National Engineering Laboratory for Efficient Utilization of Soil and Fertilizer Resources, Shandong Agricultural University, Taian, Shandong, China

Yu Wang
Shandong Rice Research Institute, Jinan, Shandong, China

Shuyun Chi
College of Mechanical and Electronic Engineering, Shandong Agricultural University, Taian, Shandong, China

Jifu Li, Jianwei Lu, Xiaokun Li, Tao Ren, Rihuan Cong and Li Zhou
College of Resources and Environment, Huazhong Agricultural University, Wuhan, China
Key Laboratory of Arable Land Conservation (Middle and Lower Reaches of Yangtse River), Ministry of Agriculture, Wuhan, China

Gangsheng Wang, Melanie A. Mayes and Lianhong Gu
Climate Change Science Institute, Oak Ridge National Laboratory, Oak Ridge, Tennessee, United States of America
Environmental Sciences Division, Oak Ridge National Laboratory, Oak Ridge, Tennessee, United States of America

Christopher W. Schadt
Climate Change Science Institute, Oak Ridge National Laboratory, Oak Ridge, Tennessee, United States of America
Biosciences Division, Oak Ridge National Laboratory, Oak Ridge, Tennessee, United States of America

Baoliang Chi
Dryland Agriculture Research Centre, Shanxi Academy of Agricultural Sciences, Taiyuan, Shanxi, China

Oene Oenema
Environmental Sciences, Wageningen University, Wageningen, The Netherlands

Wei Qin
Dryland Agriculture Research Centre, Shanxi Academy of Agricultural Sciences, Taiyuan, Shanxi, China
Environmental Sciences, Wageningen University, Wageningen, The Netherlands

Chongfeng Bu
Institute of Soil and Water Conservation, Northwest A & F University, Yangling, Shaanxi, China
Institute of Soil and Water Conservation, Chinese Academy of Sciences and Ministry of Water Resources, Yangling, Shaanxi, China

Shufang Wu
Institute of Soil and Water Conservation, Northwest A & F University, Yangling, Shaanxi, China
College of Water Resources and Architectural Engineering, Northwest A & F University, Yangling, Shaanxi, China

Yongsheng Yang
Institute of Soil and Water Conservation, Chinese Academy of Sciences and Ministry of Water Resources, Yangling, Shaanxi, China

Mingguo Zheng
Institute of Geographic Sciences and Natural Resources Research, Chinese Academy of Sciences, Beijing, China

Larry Giles and Gregory P. Asner
Department of Global Ecology, Carnegie Institution for Science, Stanford, California, United States of America

Christopher B. Field
Department of Global Ecology, Carnegie Institution for Science, Stanford, California, United States of America
Department of Biology, Stanford University, Stanford, California, United States of America

Broadbent
Department of Global Ecology, Carnegie Institution for Science, Stanford, California, United States of America
Department of Biology, Stanford University, Stanford, California, United States of America
Sustainability Science Program, Kennedy School of Government, Harvard University, Cambridge, Massachusetts, United States of America

lica M. Almeyda Zambrano
Department of Global Ecology, Carnegie Institution for Science, Stanford, California, United States of America
Sustainability Science Program, Kennedy School of Government, Harvard University, Cambridge, Massachusetts, United States of America
Department of Anthropology, Stanford University, Stanford, California, United States of America

Ange Eben N Rodolfo Dirzo
Department of Biology, Stanford University, Stanford, California, United States of America

Marlene Soriano
Instituto Boliviano de Investigacio´n Forestal, Santa Cruz de la Sierra, Bolivia

Harrison Ramos de Souza
Centro de Investigación y Preservación de la Amazonia, Universidad Amazónica de Pando, Cobija, Bolivia

Marielos Peña-Claros
Instituto Boliviano de Investigacio´n Forestal, Santa Cruz de la Sierra, Bolivia
Forest Ecology and Forest Management Group, Wageningen University, Wageningen, the Netherlands

Rachel I. Adams
Plant and Microbial Biology, University of California, Berkeley, California, United States of America

Sergio Saia, Gaetano Amato, Alfonso Salvatore Frenda, Dario Giambalvo and Paolo Ruisi
Dipartimento di Scienze Agrarie e Forestali, Universitádegli Studi di Palermo, Palermo, Italy

Antje Lauer, Joe Daryl Hugo Baal, Kayla Casimiro and Natasha Shroff
Department of Biology, California State University, Bakersfield, California, United States of America

Jorge Talamantes and Luis Jaime Medina
Department of Physics & Engineering, California State University, Bakersfield, California, United States of America

Laura Rosío Castañón Olivares
Laboratorio de Micología Médica, Facultad de Medicina, Universidad Nacional Autónoma de Meéxico, Mexico City, Mexico

Kirt W. Emery
County of Kern Public Health Services Department, Bakersfield, California, United States of America

W. David Walter
U.S. Geological Survey, Pennsylvania Cooperative Fish and Wildlife Research Unit, Pennsylvania State University, University Park, Pennsylvania, United States of America

Rick Smith and Mike Vanderklok
Animal Industry Division, Michigan Department of Agriculture and Rural Development, Lansing, Michigan, United States of America

Kurt C. VerCauteren
United States Department of Agriculture, Animal and Plant Health Inspection Services, Wildlife Services, National Wildlife Research Center, Fort Collins, Colorado, United States of America

Anja Madelen ödman and Pål Axel Olsson
Biodiversity, Department of Biology, Lund University, Lund, Sweden

Grzegorz żurek and Kamil Prokopiuk
Department of Grasses, Legumes and Energy Plants, Plant Breeding and Acclimatization Institute IHAR-PIB, Radzików, Błonie, Poland

Krystyna Rybka
Department of Plant Physiology and Biochemistry, Plant Breeding and Acclimatization Institute IHAR-PIB, Radzików, Błonie, Poland

Marta Pogrzeba and Jacek Krzyżak
Department of Environmental Biotechnology, Phytoremediation Team, Institute for Ecology of Industrial Areas, Katowice, Poland

David L. Jones, Paul Simfukwe and Paul W. Hill
School of the Environment, Natural Resources & Geography, Bangor University, Bangor, Gwynedd, United Kingdom

Robert T. E. Mills
Ecole Polytechnique Fédérale de Lausanne (EPFL), School of Architecture, Civil and Environmental Engineering (ENAC), Laboratory of Ecological Systems (ECOS), Lausanne, Switzerland

Bridget A. Emmett
Centre for Ecology and Hydrology, Environment Centre Wales, Bangor, Gwynedd, United Kingdom

Muye Gan, Xinyu Zheng and Ke Wang
College of Environmental and Resource Sciences, Zhejiang University, Hangzhou, Zhejiang Province, China

Jinsong Deng
College of Environmental and Resource Sciences, Zhejiang University, Hangzhou, Zhejiang Province, China
School of Civil Engineering and Environmental Sciences and School of Meteorology, University of Oklahoma, Norman, Oklahoma, United States of America

Yang Hong
School of Civil Engineering and Environmental Sciences and School of Meteorology, University of Oklahoma, Norman, Oklahoma, United States of America
State Key Laboratory of Hydroscience and Engineering, Department of Hydraulic Engineering, Tsinghua University, Beijing, China

Cristina Martinez and Franco Miglietta
FoxLab, Forest & Wood Science, Fondazione E.Mach, San Michele all'Adige (Trento), Italy
Institute of Biometeorology, National Research Council, Firenze, Italy

Irene Criscuoli
FoxLab, Forest & Wood Science, Fondazione E.Mach, San Michele all'Adige (Trento), Italy
Biogéochimie et écologie des milieux continentaux, CNRS-INRA-ENS-Paris 6, Thiverval-Grignon, France

Giorgio Alberti
MountFor Project Centre, European Forest Institute, San Michele all'Adige (Trento), Italy
Department of Agricultural and Environmental Sciences, University of Udine, Udine, Italy

Silvia Baronti and Emanuela Pusceddu
Institute of Biometeorology, National Research Council, Firenze, Italy

Filippo Favilli
European Academy of Bolzano, Bolzano, Italy

Costanza Calzolari
Institute of Research for Hydrogeological Protection, National Research Council, Sesto Fiorentino (FI), Italy

Cornelia Rumpel
Biogéochimie et écologie des milieux continentaux, CNRS-INRA-ENS-Paris 6, Thiverval-Grignon, France

Roberto Viola
Centre for Research and Innovation, Fondazione E.Mach, San Michele all'Adige (Trento), Italy

Index